FORESTS AND INSECTS

JOIN US ON THE INTERNET VIA WWW, GOPHER, FTP OR EMAIL:

WWW: http://www.thomson.com
GOPHER: gopher.thomson.com
FTP: ftp.thomson.com
EMAIL: findit@kiosk.thomson.com

A service of I(T)P®

FORESTS AND INSECTS

Edited by

Allan D. Watt

Institute of Terrestrial Ecology, Edinburgh Research Station, UK

Nigel E. Stork

Director of the Cooperative Research Centre for Tropical Rainforest Ecology and Management, James Cook University, Cairns, Australia

and

Mark D. Hunter

Institute of Ecology, University of Georgia, Athens, GA, USA

Published in association with the Royal Entomological Society

London · Weinheim · New York · Tokyo · Melbourne · Madras

Published by Chapman & Hall, 2–6 Boundary Row, London SE1 8HN, UK

Chapman & Hall, 2–6 Boundary Row, London SE1 8HN, UK

Chapman & Hall GmbH, Pappelallee 3, 69469 Weinheim, Germany

Chapman & Hall USA, 115 Fifth Avenue, New York, NY 10003, USA

Chapman & Hall Japan, ITP-Japan, Kyowa Building, 3F, 2-2-1 Hirakawacho, Chiyoda-ku, Tokyo 102, Japan

Chapman & Hall Australia, 102 Dodds Street, South Melbourne, Victoria 3205, Australia

Chapman & Hall India, R. Seshadri, 32 Second Main Road, CIT East, Madras 600 035, India

First edition 1997

© 1997 The Royal Entomological Society

Typeset in 10/12pt Palatino by Saxon Graphics Ltd, Derby

Printed in Great Britain at the University Press, Cambridge

ISBN   0 412 79110 2

Apart from any fair dealing for the purposes of research or private study, or criticism or review, as permitted under the UK Copyright Designs and Patents Act, 1988, this publication may not be reproduced, stored, or transmitted, in any form or by any means, without the prior permission in writing of the publishers, or in the case of reprographic reproduction only in accordance with the terms of the licences issued by the Copyright Licensing Agency in the UK, or in accordance with the terms of licences issued by the appropriate Reproduction Rights Organization outside the UK. Enquiries concerning reproduction outside the terms stated here should be sent to the publishers at the London address printed on this page.

The publisher makes no representation, express or implied, with regard to the accuracy of the information contained in this book and cannot accept any legal responsibility or liability for any errors or omissions that may be made.

A catalogue record for this book is available from the British Library

Library of Congress Catalog Card Number: 97-66332

∞ Printed on permanent acid-free text paper, manufactured in accordance with ANSI/NISO Z39.48-1992 (Permanence of Paper).

CONTENTS

List of contributors	vii
Preface	xi
Introduction	xiii

PART ONE   COLONIZATION OF TREES BY INSECTS — 1

1 Adaptations of phytophagous insects to life on trees, with particular reference to aphids — 3
  A.F.G. Dixon

2 Host specificity in forest insects — 15
  Simon M. Fraser

PART TWO   TEMPORAL AND SPATIAL POPULATION ECOLOGY — 35

3 Population dynamics of forest insects: are they governed by single or multiple factors? — 37
  T. Royama

4 The impact of parasitoids and predators on forest insect populations — 49
  Neil A.C. Kidd and Mark A. Jervis

5 Herbivore-induced responses in trees: internal vs. external explanations — 69
  Erkki Haukioja and Tuija Honkanen

6 Incorporating variation in plant chemistry into a spatially explicit ecology of phytophagous insects — 81
  Mark D. Hunter

7 Forest structure and the spatial pattern of parasitoid attack — 97
  Jens Roland, Phil Taylor and Barry Cooke

PART THREE   INSECTS IN FOREST ECOSYSTEMS — 107

8 Termites as mediators of carbon fluxes in tropical forest: budgets for carbon dioxide and methane emissions — 109
  David E. Bignell, Paul Eggleton, Lina Nunes and Katherine L. Thomas

9 Herbivory in forests: from centimetres to megametres — 135
  Margaret D. Lowman

PART FOUR   FOREST PESTS — 151

10 Comparative analysis of patterns of invasion and spread of related lymantriids — 153
    Pedro Barbosa and Paul W. Schaefer

11 Threats to forestry by insect pests in Europe — 177
    Keith R. Day and Simon R. Leather

12 Forest pests in the tropics: current status and future threats — 207
    Martin R. Speight

13 The impacts of climate change and pollution on forest pests — 229
    Maureen Docherty, David T. Salt and Jarmo K. Holopainen

PART FIVE   INSECT DIVERSITY — 249

14 Patterns of use of large moth caterpillars (Lepidoptera: Saturniidae and Sphingidae) by ichneumonid parasitoids (Hymenoptera) in Costa Rican dry forest — 251
    Daniel H. Janzen and Ian D. Gauld

15 Impact of forest loss and regeneration on insect abundance and diversity — 273
    Allan D. Watt, Nigel E. Stork, Paul Eggleton, Diane Srivastava, Barry Bolton, Torben B. Larsen, Martin J.D. Brendell and David E. Bignell

16 Beetle abundance and diversity in a boreal mixed-wood forest — 287
    John R. Spence, David W. Langor, H.E. James Hammond and Gregory R. Pohl

17 An overview of invertebrate responses to forest fragmentation — 303
    Raphael K. Didham

18 Impact of forest and woodland structure on insect abundance and diversity — 321
    Peter Dennis

19 *Ficus*: a resource for arthropods in the tropics, with particular reference to New Guinea — 341
    Yves Basset, Vojtech Novotny and George Weiblen

PART SIX   INSECT CONSERVATION — 363

20 Arthropods of coastal old-growth Sitka spruce forests: conservation of biodiversity with special reference to the Staphylinidae — 365
    Neville N. Winchester

21 Conservation corridors and rain forest insects — 381
    Christopher J. Hill

22 Insect conservation — 395
    Thomas E. Lovejoy, Nathaniel Erwin and Sarah Boren

Index — 401

CONTRIBUTORS

Pedro Barbosa
Department of Entomology
University of Maryland
College Park
Maryland 20742
USA

Yves Basset
c/o TROPENBOS
12E Garnett Street
Campbelville
Georgetown
Guyana

David E. Bignell
School of Biological Sciences
Queen Mary & Westfield College
University of London
Mile End Road
London E1 4NS
UK

Barry Bolton
Entomology Department
Natural History Museum
Cromwell Road
London SW7 5BD
UK

Sarah Boren
Smithsonian Institution
The Castle
The Mall
Washington DC20560
USA

Martin J.D. Brendell
Entomology Department
Natural History Museum
Cromwell Road
London SW7 5BD
UK

Barry Cooke
Department of Biological Sciences
University of Alberta
Edmonton
Alberta T6G 2E9
Canada

Keith R. Day
School of Environmental Studies
University of Ulster
Coleraine BT52 1SA
Northern Ireland

Peter Dennis
Ecology and Animal Science Group
Macaulay Land Use Research Institute
Craigiebuckler
Aberdeen AB15 8QH
UK

Raphael K. Didham
Biodiversity Division
Natural History Museum
Cromwell Road
London SW7 5BD

Current address:
Department of Biology
University of Delaware
Newark, DE 19716
USA

A.F.G. Dixon
School of Biological Sciences
University of East Anglia
Norwich
Norfolk NR4 7TJ
UK

Maureen Docherty
Division of Biological Sciences
Institute of Environmental and Biological Sciences
Lancaster University
Lancaster LA1 4YQ
UK

Current address:
School of Environmental Studies
University of Ulster
Cromore Road
Coleraine
County Londonderry
N. Ireland BT52 1SA

Paul Eggleton
Entomology Department
Natural History Museum
Cromwell Road
London SW7 5BD
UK

Nathaniel Erwin
National Museum of National History
Smithsonian Institution
The Castle
The Mall
Washington DC 20560
USA

Simon M. Fraser
NERC Centre for Population Biology
Imperial College at Silwood Park
Ascot
Berkshire SL5 7PY
UK

Current address:
Santa Fe Institute
1399 Hyde Park Road
Santa Fe NM 87501
USA

Ian D. Gauld
Biodiversity Representative
Natural History Museum
Cromwell Road
London SW7 5BD
UK

H.E. James Hammond
Department of Biological Sciences
CW-405A Biological Sciences Building
University of Alberta
Edmonton
Alberta T6G 2E9
Canada

Erkki Haukioja
Laboratory of Ecological Zoology
Department of Biology
University of Turku
FIN-20500 Turku
Finland

Christopher J. Hill
Cooperative Research Center for Tropical Rainforest Ecology and Management
Department of Zoology
James Cook University
Townsville
Queensland 4811
Australia

Jarmo K. Holopainen
Department of Environmental Sciences
University of Kuopio
PO Box 1627
SF-70211 Kuopio
Finland

Tuija Honkanen
Laboratory of Ecological Zoology
Department of Biology
University of Turku
FIN-20500 Turku
Finland

Mark D. Hunter
Institute of Ecology
University of Georgia
Athens, GA 30602-2202
USA

Daniel H. Janzen
Department of Biology
University of Pennsylvania
Philadelphia, PA 19104
USA

Mark A. Jervis
School of Pure and Applied Biology
University of Wales
Cardiff
UK

Neil A.C. Kidd
School of Pure and Applied Biology
University of Wales
Cardiff
UK

David W. Langor
Department of Biological Sciences
CW-405A Biological Sciences Building
University of Alberta
Edmonton
Alberta T6G 2E9
Canada

Torben B. Larsen
Entomology Department
Natural History Museum
Cromwell Road
London SW7 5BD
UK

Simon R. Leather
Department of Biology
Imperial College at Silwood Park
Ascot
Berkshire SL5 7PY
UK

Thomas E. Lovejoy
Smithsonian Institution
The Castle
The Mall
Washington DC 20560
USA

Margaret D. Lowman
Department of Research and Conservation
The Marie Selby Botanical Gardens
811 South Palm Avenue
Sarasota, FL 34236-7726
USA

Vojtech Novotny
Institute of Entomology CAS
University of S. Bohemia
Branisovska 31
37005 Ceske Budejovice
Czech Republic

Lina Nunes
Nucleo de Madeiras
Laboratorio Nacional de Engenharia Civil
Av. do Brasil 101
1799 Lisboa Codex
Portugal

Gregory R. Pohl
Northern Forestry Centre
Canadian Forest Service
5320-122nd Street
Edmonton
Alberta T6H 3S5
Canada

Jens Roland
Department of Biological Sciences
University of Alberta
Edmonton
Alberta T6G 2E9
Canada

T. Royama
Canadian Forest Service
Atlantic Forestry Centre
Natural Resources Canada
PO Box 4000
Fredericton
New Brunswick E3B 5P7
Canada

David T. Salt
Division of Biological Sciences
Institute of Environmental and Biological
   Sciences
Lancaster University
Lancaster LA1 4YQ
UK

Paul W. Schaefer
Agricultural Research Service
US Department of Agriculture
Beneficial Insects Introduction Research
   Laboratory
501 South Chapel Street
Newark, Delaware 19713
USA

Martin R. Speight
Department of Zoology
University of Oxford
South Parks Road
Oxford OX1 3PS
UK

John R. Spence
Department of Biological Sciences
CW-405A Biological Sciences Building
University of Alberta
Edmonton
Alberta T6G 2E9
Canada

Diane Srivastava
NERC Centre for Population Biology
Imperial College at Silwood Park
Ascot
Berkshire SL5 7PY
UK

Nigel E. Stork
Natural History Museum
Cromwell Road
London SW7 5BD
UK

Current address:
Cooperative Research Centre for Tropical Rain
   Forest Ecology and Management
James Cook University
PO Box 6811
Cairns QLD 4870
Australia

Phil Taylor
Department of Biology
Acadia University
Wolfville
Nova Scotia B0P 1X0
Canada

Katherine L. Thomas
School of Pure and Applied Biology
University of Wales Cardiff
PO Box 915
Cardiff CF1 3TL
UK

Allan D. Watt
Institute of Terrestrial Ecology
Edinburgh Research Station
Bush Estate
Penicuik
Midlothian EH26 0QB
UK

George Weiblen
Harvard University Herbaria
22 Divinity Avenue
Cambridge, MA 02138
USA

Neville N. Winchester
Department of Biology
University of Victoria
PO Box 1700
Victoria
British Columbia V8W 2Y2
Canada

PREFACE

The idea for this book emerged in 1993 when two of the editors (AW and NS) were in Cameroon working together on forest pests and the diversity of insects in natural forests and forest plantations. Our work had two aims. First, to determine whether the damage caused by pests of Terminalia ivorensis, a native West African timber tree species, was affected by establishing plantations in different ways, and second, to measure the impact of forest clearance and the conversion of forest to different types of plantation on insect diversity. Despite the contrasting aims of our research, there was much in common between them, particularly the potential link between pest outbreaks and insect diversity. This experience and our experience elsewhere (particularly in Scotland where AW and MH worked together on forest pests in the late 1980s) made us feel that it was time to bring entomologists working on different aspects of forest entomology together. We particularly wanted to produce a book with contributions from the traditional 'forest entomologists' concerned with managing forest pests, and entomologists who were concerned with the diversity and conservation of insects in temperate and tropical forests. A few recent books deal with these different aspects of forest entomology, such as Population Dynamics of Forest Insects (1990, edited by Watt, Leather, Hunter and Kidd, published by Intercept, Andover) and Canopy Arthropods (1997, edited by Stork, Adis and Didham, published by Chapman & Hall, London), but this is the first volume to bring the different elements of forest entomology together. The book contains chapters on the colonization of trees by insects, the population dynamics of insects in forests, the various roles insects perform in forest ecosystems, the management of pests in temperate and tropical forests, the diversity of insects in temperate and tropical forests, and the conservation of forest insects.

Most of the chapters in this book were presented as papers during the 18th Symposium of the Royal Entomological Society in London, September 1995. We thank the staff of the Royal Entomological Society, particularly Greg Bentley, Richard Lane, the President of the Society, and Emma Lindsay and other colleagues in the Institute of Terrestrial Ecology and the Natural History Museum who helped us with the organization and running of the Symposium. All of the chapters were subject to external review and we thank the many anonymous reviewers for their work. Finally, we thank Emma Lindsay and Amy Watt for their assistance, and Ward Cooper, Paul Gill and Martin Tribe of Chapman & Hall for their support and guidance in preparing this book.

INTRODUCTION

The principal aim of this book, and of the 18th Symposium of the Royal Entomological Society at which most of the chapters were presented, was to cover the full breadth of 'forest entomology'. In particular, we wanted to bring together two groups of people: those working on forest pests and those working on insect diversity and conservation. In the first group are traditional forest entomologists, concerned mainly with the impact and management of forest pests. Those in the second group are concerned with the diversity of insects in forests, and the conservation of this diversity. In this book, we have also added chapters by entomologists concerned with the various roles of insects in forest ecosystems, such as termites as producers of methane – a topic with global significance.

Although the ultimate aims of the various entomologists carrying out the research described in this book are very different, their research programmes have much in common. All entomologists working in forests face technical problems such as canopy access and identification of insect material. In addition, there are major themes in the underlying science which unite 'forest entomologists' of all varieties. For example, the topics of fragmentation and forest disturbance turn up in chapters on pests and on insect conservation. More generally, this book demonstrates that both those concerned with the management of pests and those concerned with the conservation of insect communities in forests depend on an understanding of population and community ecology. Not surprisingly, therefore, this book also covers recent advances in insect ecology and conservation.

The first two chapters (Part One) deal with contrasting aspects of the colonization of trees by insects. In Chapter 1, Tony Dixon considers the adaptations of aphids to life on trees, and in Chapter 2, Simon Fraser discusses recent trends in the exploitation of non-native conifer trees by native insects, a phenomenon which has led to the appearance of many new forest pests in Europe and elsewhere in the last 20 years (see also Chapters 10, 11 and 12).

The study of forest pests has been instrumental in the development of insect population ecology (Part Two). For example, insect population dynamics began, during the early 1900s, with Howard and Fiske's studies of the action of natural enemies introduced against exotic forest pests. Tom Royama (Chapter 3) describes how, since then, studies on forest pests such as the spruce budworm, *Choristoneura fumiferana*, have led to a better understanding of the processes involved in population outbreaks of forest insects. Much of the research on forest insects has been directed at either top-down (parasitism and predation) or bottom-up (plant resource constraints) processes. For example, using evidence from classical biological control introductions, key-factor analysis and models, Kidd and Jervis (Chapter 4) show that the role of natural enemies in the population dynamics of forest insects may often have been underestimated. Much of the impetus for research on bottom-up processes comes from Erkki Haukioja's research on delayed induced resistance in mountain birch. In Chapter 5, Haukioja and Hünkanen discuss the range of impacts that damage can have on trees and, therefore, on the population dynamics of insects that feed on those trees.

In recent years, there has been an increasing emphasis on spatial variation as an ecological factor. In Chapter 6, Mark Hunter demonstrates that spatial variation in plant chemistry can explain much of the spatial variation in the abundance of insects on oak and apple. A spatial perspective is also needed to understand the impact of natural enemies. For example, in Chapter 7 Jens Roland,

Phil Taylor and Barry Cooke show how the spatial structure (specifically, the degree of fragmentation) of forests can affect the impact of parasitism and disease on the population dynamics of the forest tent caterpillar.

The two chapters on insects in forest ecosystems (Part Three) demonstrate the scale of ecological processes involving insects in forests. David Bignell and co-workers (Chapter 8) discuss the mineralization of carbon (as CO_2 and methane) by termites in tropical forests. Termites are the most abundant insects found in tropical forests but the scale of their abundance and diversity in African forests has only recently emerged from research by Paul Eggleton and David Bignell (see also Chapter 15). Similarly, the scale of the impact of herbivorous insects in tropical forests has only become apparent as methods of access to the upper canopies of forests have been developed. In Chapter 9, Meg Lowman describes these developments and the findings of her research in forests in many parts of the world. She describes, for example, grazing levels of over 300% in Australian eucalyptus forest.

The chapter on herbivory in natural forests leads neatly into the section on forest pests (Part Four). This section focuses on the development of pest problems in different parts of the world. In the first chapter of the section (Chapter 10), and following on from Simon Fraser's earlier chapter on the colonization of new hosts by insect herbivores, Pedro Barbosa and Paul Schaefer consider the invasion of new areas by insect herbivores. Why some species spread more rapidly than others, and why some species become pests and others do not, are questions that have occupied entomologists of all kinds for many years – see, for example, Elton's 1958 book, *The Ecology of Invasions* (Methuen, London). Barbosa and Schaefer tackle these questions by focusing on the different temporal and spatial dynamics of four closely related lymantriid moths in North America. In Chapter 11, Keith Day and Simon Leather take a different geographical perspective by considering the threats posed by insect pests to forestry in Europe. Some 500 pests have been recorded in Europe, with problems developing as a result of indigenous insect species acquiring new hosts (see also Chapter 2), and other species extending their range. However, Chapter 11 also emphasizes how pest problems arise as a consequence of the move from natural forest to intensively managed forest. The plea by Day and Leather that pest problems could be avoided if there was a better understanding of how insect pests are affected by factors such as drought, soil type and silvicultural factors is strongly echoed by Martin Speight (Chapter 12). His chapter takes a very different geographical focus from the preceding two, but Speight shows that the problems posed by pests in the tropics can be as serious as in temperate forests, and have similar root causes. The final chapter in this section (Chapter 13) focuses on the future. Maureen Docherty and co-authors consider how the threats posed by forest pests may be affected by climate change and atmospheric pollution.

The remaining chapters in the book consider insect diversity in forests (Part Five) and the conservation of this diversity (Part Six), focusing on many different parts of the world. In Chapter 14, Dan Janzen and Ian Gauld demonstrate the scale of the research effort required to understand the diverse insect communities of tropical forests. They present the current level of understanding gained from their long-term study of the Lepidoptera and their ichneumonid parasites in the Santa Rosa National Park, Costa Rica, an area globally notable in the conservation of tropical forest flora and fauna.

The efforts of Janzen and others are needed because insect diversity, and biodiversity in general, in both temperate and tropical forests is threatened by forest clearance. Allan Watt and colleagues illustrate this point in Chapter 15 by describing the impact of deforestation in Cameroon on ant, butterfly, beetle and termite diversity. Deforestation (complete forest clearance) may be the most serious threat to forest biodiversity, but it is not the only one; the way that forests are managed can also have an impact on the insect communities within them. In Chapter 16, John Spence and his colleagues describe their research on the litter-dwelling and saproxylic beetle assemblages of the *Populus–Picea* forests of central Alberta, Canada. They show that many saproxylic beetle species essential in the early stages of breakdown of coarse woody material are restricted to old-growth forest, and that, therefore, the planned short rotation management of these forests may have serious consequences for the beetle fauna and the roles they play in the forest ecosystem.

Forest biodiversity is also threatened by fragmentation. In Chapter 17, Raphael Didham reviews research on the impact of fragmentation, concentrating on the processes of isolation and the creation of new forest edges. Using data on leaf-litter invertebrate communities collected in Central Amazonia, Didham shows that isolation is a much more important determinant of changes in species composition than are edge effects. The fragmentation theme is continued by Peter Dennis (Chapter 18), but with a very different geographical focus – 'farm' woodlands in the UK. Finally in this section, Chapter 19 by Yves Basset and others also has a different geographical focus: the Madang area of Papua New Guinea. More significantly, this chapter takes a very different perspective from other chapters in the book; it assesses the validity of the concept of 'keystone' tree species in maintaining biodiversity, in this case 15 species of fig (*Ficus* spp.).

All the chapters mentioned above touch on insect conservation. Didham, for example, discusses habitat size (area of forest fragment) requirements for insects; Watt and co-authors describe the relative merits of different types of reforestation for insect species richness; and Bassett *et al.* compare fig species as keystone species for insects, frugivorous mammals and birds. The last section of the book (Part Six) deals with insect conservation in more depth. In Chapter 20, Neville Winchester, focusing on the arthropod fauna from a Sitka spruce forest in the Carmanah Valley on Vancouver Island, Canada, shows that effective conservation programmes cannot be constructed without the description of species assemblages and the documentation of their habitat preferences. This chapter, like that of Janzen and Gauld, underlines the immensity of the task of describing, let alone understanding, the relationships within arthropod communities in forests. In Chapter 21, Chris Hill returns to the topic of forest fragmentation and specifically to the value of 'conservation corridors' as a strategy for ameliorating the negative impacts of forest fragmentation. Hill describes research on the value of corridors in Australian tropical rain forest for ants, butterflies and dung beetles.

The last chapter of the book (Chapter 22) is an overview of the threats to insect diversity. Tom Lovejoy and Nathaniel Erwin, show, firstly, that we need to understand the response of trees to climate change before we can fully understand the responses of insects to climate change. They then focus on the threat of species extinctions posed by deforestation and other forms of habitat loss, particularly the extinction of the many insects that perform important functions in ecosystems, such as decomposers and pollinators. They describe, for example, how the loss of insect pollinator species can seriously threaten crops such as Brazil nut plantations. This chapter also discusses important practical issues such as the measurement of insect diversity, indicator species or groups, hot spots of species richness and centres of endemism and priority setting. The chapter ends with a plea for an ecosystem management approach to the conservation of insects and other taxa.

We thank the staff of the Royal Entomological Society, particularly Greg Bentley, Richard Lane and Emma Lindsay, and the many referees who helped us with the organization and running of the Symposium and the editing of this book.

Allan D. Watt
Nigel E. Stork
Mark D. Hunter

PART ONE
COLONIZATION OF TREES BY INSECTS

ADAPTATIONS OF PHYTOPHAGOUS INSECTS TO LIFE ON TREES, WITH PARTICULAR REFERENCE TO APHIDS

A.F.G. Dixon

1.1 INTRODUCTION

The insect–plant relationship considered here is that between phytophagous insects and trees, and is restricted to insects that live on rather than in trees. In considering this topic two questions arise: in what features do trees differ from other plants, and which of these features have been important in shaping the adaptations of phytophagous insects? The relevant features are not only structural but also ecological. Adaptation is used here in a descriptive sense to indicate the result of a selective process.

As the different groups of phytophagous insects vary greatly in their length of life, rates of increase and ways of feeding, the adaptations they show to living on trees are likely to differ. Previous attempts to review such adaptations for particular taxa have concluded that it is premature (cf. Wagner and Raffa, 1993) or have tended to be mainly descriptive (e.g. Mattson, 1977; Niemalä *et al.*, 1981; Gaston and Reavey, 1989; Denno, 1994a,b; Lindstrom *et al.*, 1994; Hunter, 1995). Functional explanations depend on a good biological understanding, which is lacking for most groups of insects. One exception is the aphids, which are very well studied; therefore this chapter will be mainly concerned with aphid adaptations to living on trees.

The dictionary definition of a tree is a perennial plant with a single woody self-supporting stem or trunk usually unbranched for some distance above the ground. This refers to three features that could be important for aphids: trees are long lived; they have a large stem and branches; and they have a complicated architecture, with the leaves positioned well off the ground. Botanical text books tend to emphasize the woodiness and architectural complexity of trees (e.g. Ingrouille, 1992). For aphids, the woody nature of the stem in itself is not important; however, associated with woody stems and branches there is usually a thick bark, which could be important. Other features, which are likely to be important but which are not usually included in botanical definitions are abundance and the seasonally complementary growth patterns of trees and herbaceous plants.

The thesis of this chapter is that certain aspects of aphid biology – colour, size, macroptery and host alternation – have been shaped by the structure and ecology of trees.

1.2 ADAPTATIONS TO THE STRUCTURE OF TREES

1.2.1 SIZE

Aphids feed on phloem sap. They obtain the sap by inserting into the phloem elements of plants their mandibular and maxillary stylets, which together form a hollow needle-like feeding tube. The trunks and branches of trees are exploited as a source of

---

Forests and Insects. Edited by A.D. Watt, N.E. Stork and M.D. Hunter. Published in 1997 by Chapman & Hall, London. ISBN 0 412 79110 2.

food by aphids (Dixon et al, 1995). In order to feed on these structures aphids need to be able to reach the phloem elements deep beneath the bark. For this they need long stylets. When not in the feeding position the stylets are supported in a groove in the anterior surface of the labium or proboscis. The proboscis consists of four segments and projects from the head of the aphid and runs posteriad between the bases of the legs. For feeding, the stylets are exposed by telescoping the distal segment within the basal segments of the proboscis. *Stomaphis quercus* (L.) feeds on the trunks of mature oak trees and adults have a proboscis that is nearly twice the length of their body, which itself is nearly a centimetre long. The telescoping of the proboscis, necessary to expose the stylets, results in the invaginated proboscis running internally, the full length of the body. Thus to function a long proboscis needs to be associated with a long body. The proboscis and body length of 205 species of aphids, ranging in size from the smallest to the largest, are closely correlated (Figure 1.1; Dixon *et al.*, 1995). The minimum size at birth is determined by the length of the proboscis. Accepting this constraint on birth size, then, an optimal energy allocation model and empirical data indicate that the optimal adult size for maximizing r_m is approximately 15 times the birth size (Dixon and Kindlmann, 1994). Thus, although it is not known why the multiplication factor is 15, it is clear that the birth size of a species is a major determinant of adult size. Those species feeding on the trunk and branches of trees have to be large at birth in order to feed; as a consequence the adults are also large. Thus in aphids their mode of feeding makes it physically impossible for small aphids to feed on tissues very deep within a plant.

1.2.2 COLOUR

Aphids appear to be cryptically coloured, with those living on foliage a greenish colour and those on the trunk and branches of trees brownish. This impression is based on the species of several large genera like *Acyrthosiphon* and *Macrosiphum* (which feed predominantly on foliage) being mainly green in colour. There are notable exceptions, but these are usually aposematic species like *Aphis*

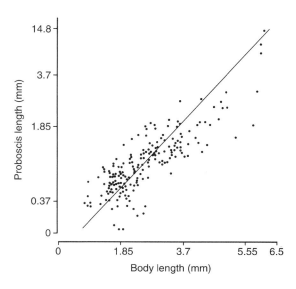

Figure 1.1 Lengths of the proboscees of 205 species of aphid in relation to lengths of their bodies ($r = 0.83$).

cytisorum and *A. nerii*, which sequester secondary plant substances as a defence against visual predators like birds (Dixon, 1985). It is also possible that dark colours absorb ultraviolet radiation, and so reduce tissue damage.

More rigorous evidence for colour offering some protection from visually hunting predators comes from the body colour of aphids that live on different parts of trees, e.g. birch and oak. The species living on the leaves are greenish and those on the trunk and branches brownish. However, relatively few species co-occur on each species of tree, and they often belong to different genera and subfamilies, which tends to confound the trend in colour. In some species, like *Periphyllus*, the first generation individuals, which hatch from overwintering eggs before the buds burst in spring, feed and reach maturity on the twigs and are a brownish colour; whereas individuals of the second generation, which are born on the unfurling and growing leaves and mature while feeding on them, are green.

Most of the Cinarini belong to the genus *Cinara*, all of which live on conifers. Blackman and Eastop (1994) list 200 species in this genus and cite the body colour in life of 135 species. Their colour, i.e. whether greenish or brownish, is associated with where they feed on their host plant. Those that live on the needles and young twigs of their host plants are more likely to be greenish (20%) compared with those that live on the trunk and branches (3%) ($\chi^2 = 11.3$; $P < 0.01$). If this analysis is extended to include all the species of the Cinarini for which the body colour is recorded ($n = 160$), then the percentages are 41% and 3%, respectively ($\chi^2 = 34.2$; $P < 0.01$). That is, although not all aphids that feed on foliage are green, nevertheless a greater proportion of them are 'green' than are the aphids that feed on the trunks and branches of conifers, which are predominantly a brownish colour.

It is therefore likely that visual predators, like birds, have been important in determining one of the adaptations shown by aphids to life on plants: their colour. A brownish colouration is a specific adaptation to living on trees as this colour is characteristic of the structures (trunks and branches) that distinguish trees from other plants.

1.3 ADAPTATION TO FOREST/WOODLAND ECOLOGY

All aphids are polyphenic in that each species and even clones consist of both winged and unwinged morphs. This alary dimorphism varies in incidence between species with the more fully winged species mainly associated with trees. This has been accounted for in terms of the architectural complexity (Waloff, 1983) and habitat fragmentation and abundance of trees (Dixon and Kindlmann, 1990; Dixon et al., 1993).

Waloff (1983) and Denno (1994a,b) present a case based on Heteroptera and Homoptera Auchenorrhyncha, in support of the hypothesis that host plant architectural complexity has been a major factor in the evolution of alary dimorphism in insects; i.e. tree-dwelling insects are more likely to be fully winged than those living on herbaceous plants. They argue that wings are needed for efficient exploitation of the markedly three-dimensional habitat associated with trees, as they enable displaced insects to relocate themselves. The possession of wings is thought to be more advantageous than any increased fecundity that might result from brachyptery or aptery.

More aphids of the family Drepanosiphidae are fully winged than in other families but even in this family the number of species in the UK that are fully winged only makes up 65% of those that live on trees (Dixon, 1984). In the four largest aphid families – Aphididae, Drepanosiphidae, Lachnidae and Pemphigidae – which make up 90% of all aphids, only 7% of the species living on shrubs and trees are fully winged. However, a significantly greater proportion of those living on shrubs and trees than on herbaceous plants are fully winged (Table 1.1). If the advantages of being fully winged when living on trees is strong then it is relevant to ask why the proportion of macropterous species living on trees is so low (7%).

1.3.1 MACROPTERY

In winged aphids the optimal partitioning of resources between soma, fuel and gonads depends on the timing and frequency of migration (Dixon et al., 1993). This is an optimization problem concerned with the risk associated with migration, the resources necessary to fuel the flight, the optimum time for migration, and whether an aphid should autolyse its wing muscles. The following is an illustration of the possible mechanisms.

The number of individuals in a population at a particular site at time t is $N = N(t)$, and the risk connected with migration is expressed as the probability p that an average individual will survive the migration. The risk is likely to be inversely proportional to the degree of habitat fragmentation. The reproductive success of an individual at a particular site is defined by rx f(N), where x is the number of its offspring and r the potential rate of increase of the offspring under optimal conditions, i.e. in high-quality sites and in the absence of density dependence. The function f describes all the negative influences that either the environment or the number of individuals present (density dependence) can have on the potential rate of growth of the population. The function f may also depend on site and time, as there may be differences between sites both in space and in time, t, as the quality of

Table 1.1 The number and proportion of species in the Aphididae, Drepanosiphidae, Lachnidae and Pemphigidae that live on trees and shrubs, and herbaceous plants, which have only fully winged or both winged and unwinged morphs ($\chi^2 = 27.5$, $P < 0.001$)

Host plants	Parthenogenetic females			
	Fully winged		Winged and unwinged morphs	
	No.	(Proportion)	No.	(Proportion)
Trees and shrubs	20	(0.07)	272	(0.93)
Herbaceous plants	1	(0.002)	437	(0.998)

the site undergoes, say, seasonal changes. However, this is neglected here.

The migratory strategy is expressed in terms of the proportion (q) of a clone that stays and reproduces at the original site, and the proportion $(1 - q)$ that attempts to locate and reproduce at a new site. The fitness (F) of the clone under such assumptions can be expressed as:

$$F = qrf(N) + (1 - q)prf(M) \qquad (1.1)$$

where M denotes the number of individuals of this species already present at the new site. The necessary condition for fitness to be maximized is $dF/dq = 0$, which gives:

$$f(N) = pf(M) \qquad (1.2)$$

Before plant quality deteriorates it is assumed that N increases with time. If only a small proportion of sites are occupied then a migratory individual is likely to arrive at an empty site ($M = 0$) and if f is a decreasing function of N, which is the usual case, then equation (1.2) defines the critical density, N_{crit}, for migration:

$$f(N_{crit}) = pf(0) \qquad (1.3)$$

For illustration one can assume a logistic growth in equation (1.3) and look for the dependence of the interval between flights on habitat fragmentation, represented by $1/p$ in Figure 1.2. From this figure it is evident that when $1/p$ is low, insects should fly often, and when $1/p$ is large and increasing the time to migration should increase dramatically. Thus species living in non-fragmented habitats (like forests dominated by one species of tree) where p is high, i.e. the cost of flight is low, should always be winged but invest relatively little in fuel and retain the ability to fly throughout adult life (i.e. retain their wing muscles), even though this results in a reduced fecundity and lower population growth rate. That is, as the cost of host (patch) transfer is low it is frequently more advantageous to locate another host (patch) than to stay. However, in those species that occupy very fragmented habitats (most herbaceous plants), i.e. the cost of flight is high, flight is only advantageous when the patch becomes very overcrowded. The optimum interval between flights in this situation may exceed the length of an individual's life. In such species there are advantages in a high investment in fuel but, on colonizing an empty habitat, to go in for wing muscle autolysis, use all the energy for reproduction, and produce unwinged offspring, i.e. switch from a migratory to a reproductive mode.

As predicted, nearly all the aphids that are fully winged live on abundant forest trees. In addition, most of the aphids that live on herbaceous plants,

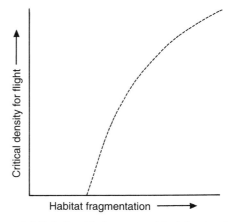

Figure 1.2 Relationship between critical density for flight and habitat fragmentation.

which are often spatially unpredictable resources, tend to go in for wing muscle autolysis after locating a suitable host plant, whereas those that live on trees that are spatially predictable and abundant resources retain their wing muscles. Associated with wing muscle autolysis is a dramatic and often very rapid increase in the size of the gonads. As wing muscles compete with the gonads for resources (Newton and Dixon, 1990), there are advantages in wing muscle autolysis and in investing all the resources in reproduction. Wing muscle autolysis has the additional advantage of preventing the accidental displacement of an aphid from its host plant – the 'diffusive losses' of Roff (1986).

The abundance per unit of resource of aphids living on predictable and abundant plants (e.g. forest trees) is also likely to differ from that of aphids living on herbaceous plants. The equilibrium population density of a species is seen as the outcome of the interaction between its rate of increase and the strength of the density-dependent factor acting upon it. Natural enemies do not appear to regulate the abundance of deciduous tree-dwelling aphids (Dixon, 1990; Dixon et al., 1996). Moreover, there is no evidence that the efficiency of the natural enemies of the various aphid species differ; therefore, differences between aphid species in their rate of population increase, r_m, or more particularly realized rate of population increase, R, appear to be the most likely cause of differences in abundance between species. The theoretical grounds for this is presented in Dixon and Kindlmann (1990).

Analysis of the empirical data for three species of tree-dwelling aphids has given the following difference equation for the between-year dynamics:

$$\log X_{(t+1)} = \log X_t + \log R - D\log X_t \quad (1.4)$$

where X_t and $X_{(t+1)}$ are the peak numbers in spring of year t and $(t+1)$, and D is the density-dependent factor. After antiloging, equation (1.4) gives:

$$X_{(t+1)} = RX_t^{(1+D)} \quad (1.5)$$

the equilibrium density (X^*) of which is:

$$X^* = R^{(1/D)} \quad (1.6)$$

What factors are likely to affect the degree to which r_m is realized? Dixon et al. (1987) argued that one such factor is the probability of finding a host plant $P(c)$, where c is the proportion of the ground covered by its host plant assuming aphids disperse regularly. Thus, the realized rate of increase, R, which includes losses incurred in dispersal is:

$$R = r_m P(c) \quad (1.7)$$

Given that the probability of finding a host plant $P(c)$ after T trials is (Dixon et al., 1987):

$$P(c) = 1 - (1 - c)^T \quad (1.8)$$

then the equilibrium density is given by:

$$X^* = \{r_m[1 - (1 - c)^T]\}^{(1/D)} \quad (1.9)$$

This indicates that, all other things being equal, the proportional cover of the host plant through its effect on realized r_m can markedly affect the abundance of an aphid (Figure 1.3a).

Empirical support for this comes from qualitative estimates of the abundance of the commonest of the indigenous deciduous tree-dwelling aphids of Britain, which all belong to the Drepanosiphidae, and of their host trees (Dixon and Kindlmann, 1990). The relation between the ranked abundance of an aphid and that of its host plant is given in Figure 3b. This tends to support the idea that plant abundance is a major factor determining aphid abundance ($r_s = 0.52$, $P < 0.05$).

Aphids are abundant when living on abundant trees and the ability of these species to avoid the more frequently experienced intraspecific competition by flying has possibly been another factor counterbalancing the cost of being able to fly and of flying. Aphids living in such habitats are frequently fully winged, although living in a permanent habitat. A good example is that of the tree-dwelling Drepanosiphidae, which can produce a rare apterous morph but nevertheless mainly produce only winged virginoparae. However, in the case of the uncommon *Drepanosiphum dixoni* H.R.L., which only lives on the leaves of *Acer campestre* L. that are in deep shade (i.e. in a very fragmented habitat), the virginoparae are mostly brachypterous, with the aphid switching to the production of the less fecund fully winged individuals only when it becomes abundant (Dixon, 1972a,b).

8 *Adaptations of phytophagous insects to life on trees*

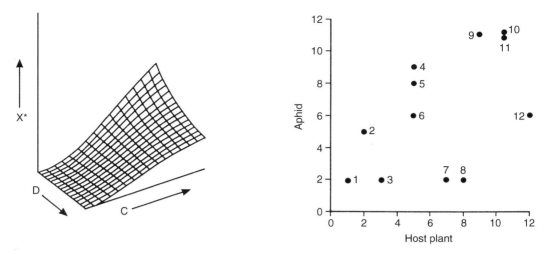

Figure 1.3 Relationship between host plant abundance and aphid abundance. (a) Theory: a three-dimensional surface showing the dependence of the equilibrium density (X^*) on both the strength of the density dependent factor(s) (D) and plant cover (C). (b) Empirical data: the rank abundance of 12 species of Drepanosiphidae in relation to the rank abundance of their host trees. (1) Oak aphid, *Tuberculoides annulatus* (Hart.); (2) beech aphid, *Phyllaphis fagi* (L.); (3) birch aphid, *Euceraphis punctipennis* (Zett.); (4) alder aphid, *Pterocallis alni* (de Geer); (5) sweet chestnut aphid, *Myzocallis castanicola* (Baker); (6) hornbeam aphid, *M. carpini* (Koch); (7) sycamore aphid, *Drepanosiphum platinoidis* (Schrank); (8) lime aphid, *Eucallipterus tiliae* (L.); (9) field maple aphid, *D. aceris* Koch; (10) turkey oak aphid, *M. boerneri* Stroyan; (11) holm oak aphid, *M. schreiberi* H.R.L. and Stroyan; (12) walnut aphid, *Chromaphis juglandicola* (Kaltenbach). (After Dixon and Kindlmann, 1990.)

When the host plant is uncommon and its distribution very fragmented, the losses incurred in migrating are likely to be very high and the aphid only infrequently abundant. Although the host plants here are most likely to be herbaceous, and temporary habitats, nevertheless it is in such habitats that aptery is likely to be favoured. On colonizing such vacant habitats, there are advantages for the colonist in wing muscle autolysis and in producing apterae in order to exploit the resource, with later generations switching to alate production when the plant eventually becomes overcrowded.

1.3.2 HOST ALTERNATION

One in 10 species of aphid show host alternation (Eastop, 1986), which involves regular seasonal movement between trees (primary) and herbaceous (secondary) host plants. Aphids are small insects and host alternation is costly, as few individuals will survive to reach the alternative host. In addition, although host alternation is common and has evolved independently on several occasions in the Aphidoidea, it is rare in the other groups of insects. This indicates a selection pressure that is peculiar to aphids (Moran, 1988). Aphids do best when their host plants are either actively growing or senescent, and badly when they are mature (Dixon, 1985). The timing of these developmental stages varies greatly between plants. Many trees start growing and are a source of food for aphids early in the year, long before herbaceous plants become available. However, as noted by Mordvilko (1908), trees generally are poor hosts in summer when their foliage is mature, but improve again in autumn when nutrients are recovered from the leaves prior to leaf fall. In contrast, many herbaceous plants tend to grow actively during summer, when trees are mature, and senesce before the trees shed their leaves. This complementary phenology could have supplied the potential for host alternation to evolve.

To account for host alternation Mordvilko (1928) proposed two hypotheses. The first is that specialization of the morph that hatches from an egg (fundatrix) to living on the primary host prevents the transfer of the whole cycle to the secondary host plant. That is, host alternation is not favoured by natural selection but maintained by constraint. This fundatrix specialization hypothesis was subsequently championed by Shaposhnikov (1955), Szelegiewicz (1978), Akimoto (1983) and Moran (1988, 1992). The second hypothesis is that of optimal host use, which views the seasonal host transfer as a means of maintaining high individual and therefore population growth rates throughout the year by utilizing the complementary growth patterns of woody and herbaceous plants.

The two hypotheses have been extensively reviewed by Moran (1988, 1992) and Mackenzie and Dixon (1991). Here further consideration will be given to the role host plant phenology and habitat fragmentation, along with the high rate of increase of aphids, might have played in the evolution of host alternation.

(a) Optimal host use hypothesis

The aphids that hatch from the overwintering eggs (fundatrices) of eight species of host-alternating Aphididae have been successfully reared on their secondary host plants. This has mainly been achieved by transferring the aphid to the secondary host plant at egg hatch (Dixon and Kundu, 1994). In *Cavariella aegopodii*, however, the eggs were laid and hatched and the emerging fundatrices reared to maturity on the secondary host plant. Although the fundatrices do not survive as well on the secondary host as they do on the primary host at budburst, they do survive and grow as well or better than those reared on the mature foliage of the primary host (Kundu and Dixon, 1995).

That is, contrary to the prediction of the fundatrix specialization hypothesis, there does not appear to be a physiological barrier to some host-alternating Aphididae transferring their whole life cycle over to the secondary host plant, i.e. becoming autoecious. Therefore, if host-alternating aphids are not constrained to this way of life, what are its advantages and under what conditions is it likely to have evolved?

Host alternation involves at least two host transfers each year in most species. Thus host-alternating clones need to be able to compensate for the large losses incurred in host transfer, otherwise there will be a strong selection pressure to simplify the life cycle and for the aphid to remain (become autoecious) on either the primary or the secondary host.

The population growth rates (Figure 1.4) and fecundities (Figure 1.5) recorded in the field for each generation of two host-alternating aphids reveal that the potential amplification in the number of individuals in each clone on the primary host in spring could compensate for losses incurred in the spring host transfer. Only as few as 1 in 100 are thought to make each transfer between hosts (Taylor, 1977). Thus these clones are better represented at the beginning of the season on the secondary host than they would have been if they had overwintered there. In two of these species the further amplification in numbers achieved on the primary host in autumn is probably not greater than it would have been if they had stayed on the secondary host (Figure 1.5). Thus the amplification in numbers in each clone achieved in spring on the primary host has to offset the cost of not just of one host transfer but both host transfers. By virtue of the very high fecundities achieved on the primary host in the spring, these aphids can potentially compensate for the huge losses incurred in making the two transfers.

The remigrants of most host-alternating Aphididae that have been studied do not feed on the primary host (Dixon and Kundu, 1994). This has been attributed to the short period remaining for ovulation and maturing additional embryos before leaf fall (Walters *et al.*, 1984). Those of the willow-carrot aphid, *Cavariella aegopodii*, are exceptional as they achieve a high fecundity, feed and mature half their embryos on the primary host (Figure 1.4). This difference in the reproductive behaviour of the remigrants is possibly associated with the pattern of leaf fall of their tree hosts. Willow (*Salix* spp.), the primary host of *C. aegopodii*, retains its leaves much later than do the primary hosts of other host-alternating aphids, e.g. *Malus*, *Prunus* and *Pyrus*, and as a consequence the aphid can further amplify its numbers on its pri-

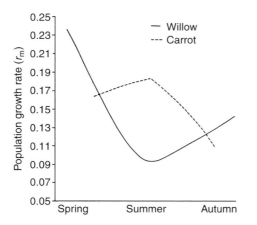

Figure 1.4 Population growth rate (r_m) achieved by the willow carrot aphid, *Cavariella aegopodii* (Scop.), when reared on willow and carrot in spring, summer and autumn. (After Kundu and Dixon, 1995.)

mary host. That is, the relative phenology of the two host plants will determine the advantage of host alternation.

As *C. aegopodii* can complete its life cycle on either its primary or its secondary host it is possible to estimate the advantage of host alternation over non-host alternation, assuming that all individuals of clones with the later strategy make one host change in the third generation between willow

trees. Figure 1.6 gives the relative success of the two strategies in terms of the number of eggs produced and indicates that if 1% are successful at each host transfer then host alternation is advantageous over non-host alternation even if there are three host transfers. If the success in locating a host is as little as 1 in 1000, host alternation involving two host transfers is still advantageous. In the field we can expect two or three host transfers, which, provided that the success of moving from one host plant to another is in the order of at least 1%, would result in host alternation being more advantageous than autoecy (Kundu and Dixon, 1995).

(b) Opportunity for host alternation

Given that trees in summer are generally conceded to be inferior to herbaceous plants as hosts for aphids (Dixon and Kundu, 1994) and that host alternation is adaptive (Kundu and Dixon, 1995), it is surprising that host alternation is not more common. A possible answer may be found in the association of host alternation with the dominant status of aphids' host plants. In British woodland, significantly fewer (6%) of the aphids associated with dominant species of trees (e.g. *Quercus* and *Fagus*) host-alternate compared with the aphids (58%) associated with subdominant tree species. Thus,

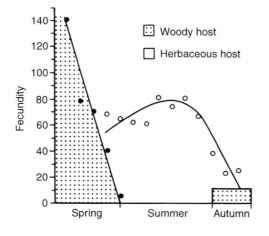
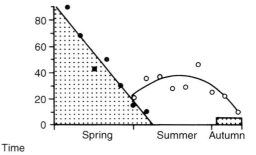

Figure 1.5 Fecundity of each of the generations that (a) *Hyperomyzus lactucae* and (b) *Melanaphis pyraria* achieve on their primary (solid circle) and secondary (open circle) hosts in spring and summer, respectively, and the fecundity achieved by the autumnal migrants of these species on their primary hosts. (After Karczewska, 1976, 1979.)

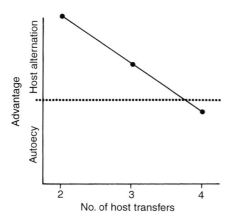

Figure 1.6 Relative advantage of host alternation over non-host alternation in *Cavariella aegopodii* in terms of eggs produced at end of a season, assuming different degrees of success in locating a new host and different numbers of host transfers for the host alternating form. (After Kundu and Dixon, 1994a.)

there appears to have been less opportunity for the evolution of host alternation in climax woodland situations dominated in summer by a few tree species. The closed canopy characteristic of such woodland prevents the development of herbaceous plants during summer when the 'quality' of the trees for aphids is low. In contrast subdominant trees like *Populus*, *Salix* and *Ulmus* live in more open habitats such as river banks, where there is often a rich herbaceous flora in summer. Thus opportunity in terms of an abundant woody and an abundant herbaceous plant with complementary growth patterns growing close together, i.e. a suitable plant community structure, could have been important in the evolution of host alternation (Dixon, 1990).

1.4 DISCUSSION

Aphids appear to maximize their fecundity by minimizing the size of their offspring, subject to the constraint that at birth they are large enough to be able to feed. Accepting this constraint, an optimal allocation of energy model for aphids and empirical data indicate that the optimal adult size for maximizing r_m is approximately 15 times the birth size (Dixon and Kindlmann, 1994). Aphids feed by tapping into phloem elements, which in trunks and branches of trees lie beneath a thick layer of bark. In order to reach this phloem aphids need long stylets, which are exposed by telescoping the segments of the proboscis back within their body (Dixon, 1985). Thus aphids feeding on these structures need to be large at birth. Therefore, for functional reasons, the largest species of aphids feed on trees, in particular the trunks and branches. As in other species, these large species are also cryptically coloured, but in this case brown rather than green.

In the various groups of Lepidoptera there is also a tendency for the larger species to live on trees but the trend is not always significant (Mattson, 1977; Niemalä *et al.*, 1981; Gaston and Reavey, 1989; Lindstrom *et al.*, 1994). The functional explanation offered for this trend is that when feeding on trees Lepidoptera are less likely to be constrained in their growth by the availability of food and can be large. However, there are notable exceptions with very large species developing on herbaceous plants.

Host location in aphids is risky as they are weak fliers and as a consequence have little control over their flight direction. Their short adult life makes host transfer even more hazardous. The best estimates of the risk involved in dispersal is that 99 out of every 100 fail to locate a host (Taylor, 1977; Ward *et al.*, 1997). Thus, habitat abundance/fragmentation, rather than permanency, is likely to have been a major factor shaping the migratory strategies of aphids, with species in fragmented habitats showing alary dimorphism and those in unfragmented habitats being fully winged. That is, aphids living on trees can show either migratory trait. If the host tree is dominant and abundant over a wide area, the cost of moving between host plants is low and the aphid is likely to be abundant and frequently experience intraspecific competition. Under these circumstances the retention of wings in all asexual generations appears to be advantageous.

Arboreal species of other insects, plant-hoppers, Heteroptera and Psocoptera also tend to be fully winged (Southwood and Leston, 1959; New, 1974; Waloff, 1983; Denno, 1994a,b). The fully winged state is seen as the most effective means of foraging and relocating resources following displacement in architecturally complex plants like trees (Waloff, 1983; Denno, 1994a,b). However, as many tree-dwelling aphids are of a similar size and mobility to these insects, but show alary dimorphism, it appears unlikely that it is architectural complexity on its own that determines macroptery in these insects but rather some form of ecological interconnectedness, as in aphids.

In marked contrast the few Lepidoptera (1%) that show reduced wings (apterous) are forest dwellers (Roff, 1990, 1994; Sattler, 1991; Hunter, 1995). Aptery is particularly associated with spring feeding in Geometridae and Lymantridae. The advantages of immobility and/or high fecundity in this situation is not clear.

Host alternation in insects is unique to aphids and has been attributed by Moran (1988), among others, to the specialization of the form that hatches from the egg to live on a woody plant, which is thought to be of a poorer quality than a herbaceous plant. However, as the losses in making two host transfers are likely to be very large a more likely explanation for host alternation being unique to aphids is that they can compensate for the great losses by virtue of their very high rates of population increase. Successful individuals can very rapidly increase the representation of a clone and compensate for the losses incurred in host transfer. This then raises the question: why is host alternation mainly between a woody primary host and an herbaceous secondary host? The fact that these plants show complementary growth patterns make such host transfers advantageous but does not account for why the woody host is the primary host. This may be due to many trees becoming vegetatively active long before herbaceous plants, which makes trees the hosts on which to start and finish a cycle. In addition, in all the cases in which the aphid performance has been monitored on both primary and secondary hosts, the best performance is recorded on the primary host (Dixon and Kundu, 1994). The more globular form of the fundatrix of host-alternating aphids might indicate that this is a general phenomenon. That is, contrary to a widely held view (e.g. Scriber, 1979; Scriber and Feeny, 1979; Moran, 1988) trees are not generally poorer quality hosts than herbaceous plants. In fact the quantum change in morphology of the fundatrix morph in those species that are now autoecious on former secondary host plants – 'perhaps the strongest support for the fundatrix-specialization

Table 1.2 Factors that shape the adaptations shown by aphids to living on trees

Aphid features	Tree features			
	(a) Structure			
	Trunk	Branches	Twigs	Leaves
Way of feeding: stylets enclosed in a proboscis	**Large** ....>			Small
Feeding site	**Brownish** ....>			Greenish
	(b) Ecology			
	Unfragmented habitat		Complementary phenology in a fragmented habitat	
Small size and short adult life	**Macroptery**		–	
High rate of population increase	–		**Host alternation**	

hypothesis of host alternation' (Moran, 1988) – is possibly more due to the quality of herbaceous plants being poorer than that of many trees at budburst.

In conclusion (Table 1.2), aphids have adapted to living on the trunks and branches of trees most strikingly in terms of their large size and brownish colour. Adaptations to forest and woodland ecology appear to have mainly affected the incidence of macroptery and host alternation. Continuous forest dominated by a single species appears to favour a higher incidence of species that are fully winged; and fragmented woodland, in which woody and herbaceous plants show complementary growth patterns, seems to favour a high incidence of aphid species showing host alternation. The existence of functional explanations for these traits in aphids supports the contention that the traits have a selective advantage. Functional explanations, however, depend on a good understanding of biology.

ACKNOWLEDGEMENTS

I am indebted to Bill Mattson and Pekka Niemalä for discussing this topic with me and introducing me to the ways of life of Lepidoptera and sawflies.

REFERENCES

Akimoto, S. (1983) A revision of the genus *Erisoma* and its allied genera in Japan (Homoptera, Aphidoidea). *Insecta Matsumarana* **27**, 37–106.

Blackman, R.L. and Eastop, V.F. (1994) *Aphids on the World's Trees*, C.A.B. International, Wallingford.

Denno, R.F. (1994a) Life history variation in planthoppers. In *Planthoppers: their Ecology and Management* (eds R.F. Denno and T.J. Perfect), pp. 163–215, Chapman & Hall, London.

Denno, R.F. (1994b) The evolution of dispersal polymorphisms in insects: the influence of habitats, host plants and mates. In *Proceedings of Memorial and International Symposium on Dispersal Polymorphism of Insects, Its Adaptation and Evolution, Okayama, Japan*, pp. 75–86.

Dixon, A.F.G. (1972a) Crowding and nutrition in the induction of macropterous alatae in *Drepanosiphum dixoni*. *J. Insect Physiol.* **18**, 459–464

Dixon, A.F.G. (1972b) Fecundity of brachypterous and macropterous alatae in *Drepanosiphum dixoni* (Callaphididae, Aphididae) *Ent. exp. et appl.* **15**, 335–340.

Dixon, A.F.G. (1984) Plant architectural complexity and alary polymorphism in tree-dwelling aphids. *Ecol. Ent.* **9**, 117–118.

Dixon, A.F.G. (1985) *Aphid Ecology*, Blackie, Glasgow.

Dixon, A.F.G. (1990) Ecological interactions of aphids and their host plants. In *Aphid–Plant Genotype Interactions* (eds R.K. Campbell and R.D. Eikenbary), pp. 7–19, Elsevier, Amsterdam.

Dixon, A.F.G. and Kindlmann, P. (1990) Role of plant abundance in determining the abundance of herbivorous insects. *Oecologia* **83**, 281–283.

Dixon, A.F.G. and Kindlman, P. (1994) Optimum body size in aphids. *Ecol. Ent.* **19**, 121–126.

Dixon, A.F.G. and Kundu, R. (1994) Ecology of host alternation. *Eur. J. Entomol.* **91**, 63–70.

Dixon, A.F.G., Kindlmann, P., Leps, J. and Holman, J. (1987) Why there are so few species of aphids, especially in the tropics? *Amer. Nat.* **129**, 580–582.

Dixon, A.F.G., Horth, S. and Kindlmann, P. (1993) Migration in insects: cost and strategies. *J. Anim. Ecol.* **62**, 182–190.

Dixon, A.F.G., Kindlmann, P. and Jarosik, V. (1995) Body size distribution in aphids: relative surface area of specific plant structures. *Ecol. Ent.* **20**, 111–117.

Dixon, A.F.G., Kindlmann, P. and Sequeira, R. (1996) Population regulation in aphids. In *Frontiers of Population Ecology* (eds R.B. Floyd, A.W. Sheppard and P.J. De Barr), pp. 103–114, CSIRO, Collingwood.

Eastop, V.F. (1986) Aphid plant associations. In *Coevolution and Systematics* (eds A.R. Stone and D.L. Hawksworth), pp. 35–54, Clarenden Press, Oxford.

Gaston, K.J. and Reavey, D. (1989) Patterns in the life histories and feeding strategies of British macrolepidoptera. *Biol. J. Linn. Soc.* **37**, 367–381.

Hunter, A.F. (1995) The ecology and evolution of reduced wings in forest macrolepidoptera. *Evol. Ecol.* **9**, 275–287.

Ingrouille, M. (1992) *Diversity and Evolution of Land Plants*, Chapman & Hall, London, 340 pp.

Karczewska, M. (1976) Studies on the development of the aphid *Longiunguis pyrarius* Pass. (Homoptera, Aphididae) *Polsk. Pismo Entomol.* **46**, 319–341.

Karczewska, M. (1979) *Bionomia i ekologia Hyperomyzus lactucae (L.) (Hom., Aphididae) na tle innych gatunk˘w mszyc wystepujacych na Ribes nigrum L.*, Roczniki Akademii Rolniczej w Poznaniu Rozprawy Naukowe **95**, 72 pp.

Kundu, R. and Dixon, A.F.G. (1995) Evolution of complex life cycles in aphids. *J. Anim. Ecol.* **64**, 245–255.

Lindstrom, J., Kaila, L. and Niemelä, P. (1994) Polyphagy and adult body size in geometrid moths. *Oecologia* **98**, 130–132.

Mackenzie, A. and Dixon, A.F.G. (1991) An ecological perspective of host alternation in aphids (Homoptera: Aphidinea: Aphididae) *Entom. Gen.* **16**, 265–284.

Mattson, W. (1977) Size and abundance of forest Lepidoptera in relation to host plant resources. *Coll. Int. C.N.R.S.* **265**, 429–441.

Moran, N.A. (1988) The evolution of host plant alternation in aphids: the evidence for specialization as a dead end. *Amer. Nat.* **132**, 681–706.

Moran, N.A. (1992) The evolution of aphid life cycles. *Ann. Rev. Ent.* **37**, 321–348.

Mordvilko, A. (1908) Beiträge zur Biologie der Pflanzenläuse, *Aphididae* Passerini. *Biol. Zentr.* **28**, 631–638.

Mordvilko, A.K. (1928) The evolution of cycles and the origin of heteroecy (migration) in plant-lice. *Ann. Mag. Nat. Hist.* **2**, 570–582.

New, T.R. (1974) *Psocoptera. Royal Entomological Society Handbooks for the Indentification of British Insects*, Part 7, 102 pp.

Newton, C. and Dixon, A.F.G. (1990) Embryonic growth rate and birth weight of the offspring of apterous and alate aphids: a cost of dispersal. *Ent. exp. et appl.* **55**, 223–229.

Niemalä, P., Hanhimaki, S. and Mannila, R. (1981) The relationship of adult size in noctuid moths (Lepidoptera, Noctuidae) to breadth of diet and growth form of host plants. *Ann. Ent. Fenn.* **47**, 17–20.

Roff, D.A. (1986) The evolution of wing polymorphism and its impact on life cycle adaptation in insects. In *The Evolution of Insect Life Cycles* (eds F. Taylor and P. Karban), pp 209–221, Springer, Berlin.

Roff, D.A. (1990) The evolution of flightlessness in insects. *Ecol. Monogr.* **60**, 389–421.

Roff, D.A. (1994) The evolution of flightlessness: is history important? *Evol. Ecol.* **8**, 639–657.

Sattler, K. (1991) A review of wing reduction in Lepidoptera. *Bull. Br. Mus. Nat. Hist. (Entomol.)* **60**, 243–285.

Scriber, J.M. (1979) Effects of leaf-water supplementation upon postingestive nutritional indices of forb-, shrub-, vine-, and tree-feeding Lepidoptera. *Ent. exp. et appl.* **25**, 240–252.

Scriber, J.M. and Feeny, P.P. (1979) Growth of herbivorous caterpillars in relation to feeding specialisation and to the growth form of their good plants. *Ecology* **60**, 829–850.

Shaposhnikov, G.K. (1955) The question of the migration of aphids from one host to another. *Trud. Zool. Inst. Acad. Nauk SSSR* **21**, 241–246.

Southwood, T.R.E. and Leston, D. (1959) *Land and Water Bugs of the British Isles*, F. Warne, London.

Szelegiewicz, H. (1978) Róznodomnsc (heteroecja) u mszyc jej pochodzenie i ewolnja. *Zes. Prob. Post. Nauk Roln.* **208**, 19–31.

Taylor, L.R. (1977) Migration and the spatial dynamics of an aphid, *Myzus persicae*. *J. Anim. Ecol.* **46**, 411–423.

Wagner, M.R. and Raffa, K.F. (1993) *Sawfly Life History Adaptations to Woody Plants*, Academic Press, San Diego, 581 pp.

Waloff, N. (1983) Absence of wing polymorphism in the arboreal, phytophagous species of some taxa of temperate Hemiptera: an hypothesis. *Ecol. Ent.* **8**, 228–232.

Walters, K.F.A., Dixon, A.F.G. and Eagles, G. (1984) Non-feeding by adult gynoparae of *Rhopalosiphum padi* and its bearing on the limiting resource in the production of sexual females in host alternating aphids. *Ent. Exp. et Appl.* **36**, 9–12.

Ward, S.A., Leather, S.R. and Pickup, J. (1997) Mortality during dispersal, and the conditions for host specificity in parasites: How many aphids find hosts? *J. Anim. Ecol.* In press.

HOST SPECIFICITY IN FOREST INSECTS 2

Simon M. Fraser

2.1 INTRODUCTION

The evolution of the diet breadths of herbivorous insects is a topic that has generated much research over the past few decades – work which is notable for its diversity of explanation and paucity of consensus. Since Ehrlich and Raven's (1964) classic work on chemical coevolution, plant chemistry has remained the factor considered to be the most important in the evolution of insect diet breadths (references in Strong, 1988). Other factors, such as natural enemies (Fox and Eisenbach, 1992; Price *et al.*, 1986), host abundance (Wiklund, 1982) and host attachment (Kennedy, 1986) have also been emphasized (Bernays and Graham, 1988), but no single factor emerges as the overriding determinant of host range. Arguments for a pluralistic approach to the problem of the evolution of host range are supported by distributions in the frequency of diet breadths (Jaenike, 1990), though there are problems with the quantification of host range, which are discussed below.

This chapter concentrates on factors affecting host use in forest insects; it argues that, for many tree-feeding species, the relative influences of many of the proposed determinants of host range will be different from those for the herb-feeding species with which most work in the area has been done. Using as 'indicator species' a group of angiosperm-feeding lepidopterans, which may be in the process of colonizing novel conifer hosts, the chapter examines the roles played by various life history, ecological and distributional characteristics in host range expansion. The generality of these results for the evolution of diet breadth is then discussed. (Throughout this chapter the terms 'host range expansion' and 'host shift' are used synonymously, following Lawton and Strong (1981) and Strong *et al.* (1984). Under this usage, a host shift does not necessarily entail the cessation of use of any existing host plant species.)

2.1.1 THE IMPORTANCE OF UNDERSTANDING DIET BREADTHS

Several areas of ecology would benefit from a better understanding of the factors that influence insect diet breadth, and that scope is increased when host range is used as a surrogate for niche breadth. One such area of application, that of host-associated speciation, is discussed here; the interrelated topics of herbivore species richness on host plants and the colonization of exotic species by insect herbivores are discussed in a later section.

Host-associated differentiation and speciation

Changes in host plant use (**host shifts**) can lead to the subdivision of populations, the formation of host races and ultimately new species; thus host-mediated changes in population structure have received much attention. Theoretically, mutations in genes concerned only with host selection behaviour could result in different genotypes utilizing different host plants, and genetic divergence between two populations (Bush, 1969). For example, sibling species of tephritid flies in the genus *Rhagoletis* were initially suggested to have formed as a result

---

Forests and Insects. Edited by A.D. Watt, N.E. Stork and M.D. Hunter. Published in 1997 by Chapman & Hall, London. ISBN 0 412 79110 2.

of reductions in gene flow between populations utilizing different hosts, because of changes in genes concerned with host selection (Bush, 1969).

However, such divergence is more likely to occur if other contributory factors are present. First, changes in the timing of the flight period due to differences in host plant phenology are often important. For example, more recent work on *Rhagoletis* shows that some of the genetic differentiation between populations feeding on apple and hawthorn can be attributed to differences in hatching time on the two hosts (Courtney, 1988; Feder and Bush, 1989; Feder *et al.*, 1993). Within a species, phenologically mediated differences between host-associated populations exist in parthenogenetic genotypes of the geometrid moth *Alsophila pometaria*, with the different genotypes showing allochronic separation matching the timing of budburst in their main hosts (Mitter *et al.*, 1979; Schneider, 1980).

Secondly, use of the plant as a mating location causes assortative mating (Bush and Diehl, 1982) which accelerates genetic divergence between populations. This is common, for example, in many dipteran gallers (Seitz and Komma, 1984; Stille, 1985; Waring *et al.*, 1990). Finally, host-associated differentiation is also promoted by low dispersal between host plants, which is not unusual in many small herbivorous insects – for example, in the microlepidopteran *Yponomeuta* (Menken, 1981) and in *Rhagoletis* (Feder *et al.*, 1993).

As a result of the prevalence of these factors that cause either temporal or spatial isolation of subpopulations, there is disagreement over whether most such speciations can be labelled as truly sympatric (Futuyma and Mayer, 1980; Bush and Diehl, 1982; Slatkin, 1987). Irrespective of semantics, understanding what determines the suite of plant species that are accepted as hosts by a herbivore has important implications for understanding short-term changes in host use, as well as the evolutionary dynamics of the insect–host-plant association.

2.1.2 PATTERNS IN THE DISTRIBUTION OF DIET BREADTH

We can learn something about the processes underlying diet breadth evolution by looking at patterns in the distribution of diet breadths in extant species. Jaenike (1990) presents data on the frequency distribution of diet breadths across species within a number of insect groups. These data show a truncated log-normal distribution, with a few species that are generalists and many specialists. Log-normal distributions are widespread in ecology (e.g. Preston, 1962; May, 1975) and one way in which they are formed is through the multiplicative contributions of many independent factors. This alone strongly suggests that we should not be seeking a single explanation for the evolution of diet breadth; what is required is a pluralistic approach that describes which factors are important, and how those factors interact (Barbosa, 1988; Thompson, 1988a).

(a) Phylogenetic analyses of diet breadth evolution

Discussions on the evolution of insect diet breadth would be meaningless if it was clear that host range does not change over evolutionary lineages. However, this is not the case. One of the few attempts to look at specialization in host use in a phylogenetic context (Thompson, 1994) suggests that specialists have evolved from generalists and vice versa in several lineages of insect herbivores, though the paucity and low resolution of the available phylogenetic data prohibit detailed analyses. Evidence from extant species also suggests that diet breadth can change within related taxa, for there is considerable variation in diet breadth within families and genera. For examples, the genus *Eupithecia* (Lepidoptera: Geometridae) in Britain contains species which range all the way from polyphagy (feeding on three or more plant families) to monophagy on a single genus.

(b) Lability in host use

There are frequent reports of new hosts being included in the diet of herbivorous insects (e.g. Shapiro and Masuda, 1980; Tabashnik, 1983; Thomas *et al.*, 1987; Berenbaum and Zangerl, 1991; Bowers *et al.*, 1992; DuMerle *et al.*, 1992), and the rapid colonization of introduced species by native herbivores (Strong, 1974b; Strong *et al.*,

1977, 1984; Auerbach and Simberloff, 1988; Welch and Greatorex-Davies, 1993; Fraser and Lawton, 1994). There is also considerable evidence of local adaptation to different hosts (Fox and Morrow, 1981; Hsiao, 1982; Scriber, 1983; Hare and Kennedy, 1986; Radtkey and Singer, 1995) in some species. All these suggest that host use in insects can be highly labile.

For species in which females select hosts for oviposition, the inclusion of a novel host plant in the diet of a herbivore presumably requires both behavioural and physiological adaptations (Bush, 1975). Behaviourally, ovipositing females and feeding larvae (Rausher, 1982) must recognize the plant as an acceptable host, and feeding larvae need to be physiologically capable of digesting host tissue and dealing with any allelochemicals present. Oviposition preference is assumed to be the more evolutionarily labile (e.g. Singer *et al.*, 1993) because it is controlled by a small number of genes (Thompson *et al.*, 1990). This, together with the observation that larvae can often develop successfully on species which females do not recognize as suitable hosts (e.g. Wiklund, 1975; Smiley, 1978; Karowe, 1990), suggests that host range expansion could occur with just a change in oviposition preference.

The relationship between the relative preference of an ovipositing female for different hosts, and subsequent larval performance on those hosts, is of obvious evolutionary importance here, and this 'preference–performance correlation' has been advocated as the crux of the problem in the evolution of the host associations of phytophagous insects (Thompson, 1988b). However, this scenario of females actively choosing host plants for their offspring does not apply to all insect herbivores. Some butterfly species scatter their eggs over grassland while on the wing (e.g. *Melanargia galathea* L.), while others oviposit on tree trunks close to herbaceous foodplants (*Argynnis paphia* L.) (Mansell and Newman, 1968). Some species of forest Lepidoptera show characteristics of the 'winter moth syndrome' (Schneider, 1980) – for example, *Lymantria dispar* L., *Alsophila pometaria* Harris and *Operophtera brumata* L. These have females that are flightless and may have vestigial wings, overwintering in the egg stage, and spring-feeding larvae that can disperse by ballooning (aerial dispersal on silk threads) (Barbosa *et al.*, 1989). In such species, many of which are important forest pests, host selection may be exercised by the differential dispersal of ballooning larvae. Care must be taken not to assume that all herbivorous insects fall into the category of choosy females selecting host plants for their relatively immobile larvae.

(c) Taxonomic and chemical restrictions on host shifts

The abundance of records of host shifts discussed above gives a misleading picture of their nature. The majority of hosts shifts that occur, in species whose diet breadths are narrow, are to plants which are closely related or chemically similar to their normal hosts. Numerous examples exist of families of herbivorous insects that are restricted to the members of a single plant family (examples in Futuyma, 1983), though speciation in plant hosts and their herbivores is not always closely linked – cladograms of hosts and herbivores are rarely congruent (Farrell *et al.*, 1992; Thompson, 1994). In addition, analyses of the herbivore faunas of different species reveal that host taxonomic affinity accounts for much of their faunal similarity (Holloway and Hebert, 1979; Neuvonen and Niemelä, 1983).

There are instances known of hosts shifts to species that are unrelated to previous hosts, but that share some chemical similarity. In Britain the white butterflies, *Pieris brassicae* and *P. rapae*, now feed on introduced nasturtiums (*Tropaeolum majus*) which contain mustard oils, as do their usual cruciferous hosts (Strong *et al.*, 1984). Butterflies in the *Papilio machaon* complex in North America, whose main hosts are umbellifers containing linear furanocoumarins, also feed on unrelated plants containing these chemicals (Berenbaum, 1981).

2.1.3 CORRELATES OF DIET BREADTH

Diet breadth does not change independently from other life history and ecological characteristics, of course – diet breadth is just one of a suite of insect characteristics that are interdependent. For example, there are significant interactions between diet breadth and overwintering stage and the timing of egg hatch (Gaston and Reavey, 1989), body size (Wasserman and Mitter, 1978; Niemelä et al., 1981) and the architectural type of the host plant (Futuyma, 1976). This last observation is of particular relevance to the question of host range evolution in forest insects, so deserves some discussion.

(a) Diet breadth and host plant growth form

One would expect herbivore species that feed on plants that are rare, unpredictable in time or hard to find, to be polyphagous, so that their chances of finding suitable hosts for oviposition and feeding are improved, while those whose hosts form a large, apparent and stable resource should be monophagous. These expectations are largely borne out in comparisons of herbivores on annual and perennial plants (Lawton and Strong, 1981), though comparisons with tree feeders often give the opposite result (Claridge and Wilson, 1976; Futuyma, 1976), or show no effect of plant architecture (Basset and Burckhardt, 1992). This is surely because the original hypothesis as it stands is too naïve, in failing to take into account evolutionary responses of the host plants to insect attack.

The wider diets of some tree-feeding species could be explained by the convergence of the chemical defences of trees towards constitutive systems (Feeny, 1976; Rhoades and Cates, 1976), though this is likely to be an over-simplification, and the role of tannins as constitutive defence chemicals is not clear cut (Bernays, 1981). But if true, this would enable tree feeders that have evolved tolerance of such constitutive defence systems to utilize a greater number of species. Unpredictability in the timing of budburst is another factor favouring polyphagy in spring feeders, whether this unpredictability results from selection pressure from herbivores (Tuomi et al., 1994) or from climatic factors (Campbell, 1989). For example, it is well known that a failure in synchrony between budburst and egg hatch in species such as the winter moth (*Operophtera brumata*) can lead to high mortality in neonate larvae (Varley and Gradwell, 1968) and that alternative host plants are used in years when synchrony is poor (Wint, 1983). Both these factors – host plant chemistry and the unpredictability of budburst – will select for tree feeders that are more generalist.

(b) Host abundance and distribution

Arguments relating to plant size and apparency, discussed in the previous section, will also apply to the effects of relative host abundance on insect diet, with the expectation that species utilizing less abundant hosts, which are unpredictable in space and time, will be more polyphagous (Cates, 1981; Lawton and Strong, 1981). Also on evolutionary time scales, the diets of insect species will be affected by their degree of exposure to potential host plants, a phenomenon which is considered in more detail in section 2.2.4. Bernays and Graham (1988) argue that host abundance is more a prerequisite for specialization than a cause of it, though on the ecological level consideration of the time constraints of ovipositing females suggests otherwise (Thompson, 1988a). Also at the ecological level, relative host abundance can certainly have a significant effect on host use in butterfly populations (Wiklund, 1982), with hosts that are hard to find being excluded from the diet.

(c) The quantification of insect diet breadth

A problem in comparative studies of insect diet breadth evolution is how to quantify the diet breadth of a species. The diets of individual species are known to show considerable variation over geographical scales (Scriber, 1983), and significant differences in host preference may even be apparent between populations which are quite close to each other (Singer et al., 1992). Diet breadth thus makes sense only in terms of host use by local populations (Fox and Morrow, 1981), in which diet should be considered to result from interactions between insect populations and their local host plants (Singer and Parmesan, 1993). Because of this, the diet breadth of the species as a whole combines that of many local populations, so that the species' host list may be considerably longer than that of a geographically restricted subset of the populations of that species. However, those data that we have on host use are usually at the species level, with host lists in the literature based on an accumulation of feeding observations over a large geographical range and a period of many years. In the analyses that follow, diet breadth is quantified using such species-level data.

A further problem in the quantification of diet breadth concerns the mapping between our classifications of plant species into taxa, and the groupings of species as insects perceive them. Comparative studies in the literature often use species counts but, as pointed out by Janzen (1979), insects take little heed of latin binomials – plants that we classify into separate species may appear chemically and morphologically indistinguishable to an insect, or, conversely, insects may perceive far more heterogeneity within plant populations than we can. In separating species into taxa it is unlikely that we have used those plant characteristics that are important for foraging insects (Barbosa, 1988). It is tempting to suggest that the solution to this problem is to judge similarity using bioassays with number of 'typical' species, but this is to assume that such species, and those of interest, use the same mechanisms for host finding and acceptance. Until we know more about the mechanisms that an insect uses to select hosts from non-hosts, counts of hosts taxa must suffice in providing an estimate of how host-specific a particular species is.

In summary, patterns in the distribution of diet breadths suggest that diet changes in response to a number of factors and their significant interactions, and those mechanisms which are most important for forest insects, such as host plant taxonomy and abundance, have been discussed. The next section uses data on a group of Lepidoptera that may be undergoing host range expansion to exotic conifers in the UK, to detect patterns in their ecological and life history characteristics which reveal why such shifts are taking place.

2.2 HOST RANGE EXPANSION IN FOREST INSECTS

2.2.1 INTRODUCTION

When a plant species is introduced into a new area, it recruits herbivores from the local fauna in an asymptotic fashion (Strong, 1974b; Strong et al., 1977), with those species that colonize first being polyphagous, external feeders (Turnipseed and Kogan, 1976), and host specialists only colonizing longer-established species (Andow and Imura, 1994). This process of herbivore colonization has rarely been studied in detail, despite its potential for revealing information about those factors that enable these sometimes dramatic changes in host use to occur. In addition, the process can be approached from both sides, by looking at the rate of herbivore accumulation on different exotic plant species (so saying something about the characteristics of the host plant that are important for host shifting), and at the insects species themselves; comparative analyses with non-shifting species can reveal the important characteristics of the insects. Both these approaches are taken here, with data on lepidopterans in the UK feeding on non-native conifer species, summarizing the results presented in Fraser and Lawton (1994) and extending the analyses to take into account the geographical range of the moth species, and their degree of encounter with conifers

2.2.2 METHODS

I analysed data on the life history and foodplant characteristics of over 500 species of British micro- and macrolepidoptera (data in Emmet, 1991; Bradley et al., 1973; Bradley et al., 1979; Skinner, 1984; plant data from Clapham et al., 1981), about 50 of which are undergoing host range expansion to introduced conifers (Hatcher and Winter, 1990; Hatcher, 1991). Only UK records were used in classifying the usual hosts of the moth species, to maintain consistency in the quality of host plant records, because host range expansions are assumed to be locally occurring phenomena, and insects are known to exploit different species of plants in different parts of their range (Fox and Morrow, 1981).

A **host range expansion** is defined as two or more records of a species, normally associated with angiosperms, feeding and being able to carry out larval development on conifers in the wild in Great Britain (Hatcher and Winter, 1990). These are denoted here as **shift species**. It is known that some such species use introduced conifers as primary hosts sustaining large populations, e.g. *Operophtera brumata* L. (Stoakley, 1985), *Alcis repandata* (L.) and *Odontopera bidentata* (Clerck) (Hatcher, 1991), and *Syndemis musculana* Hübn. (Styles, 1960); many other species may exploit conifers only rarely at the present time.

These shift species were compared in analyses with a set of 400 non-shifting species which were selected at random from the angiosperm-feeding moth fauna, denoted here as **control species**. In the absence of phylogenies for many of these groups, attempts were made to control for the effects of phylogeny by subsampling this control set in a way that matched the taxonomic distribution of the shift species (full details are given in Fraser and Lawton, 1994), to give a **taxonomically-constrained random sample** (TCRS).

Finally, comparisons with the set of 78 moth species that feed on the native British conifers (Scots pine, *Pinus sylvestris* L.; juniper, *Juniperus communis* L.; yew, *Taxus baccata* L.) are in places useful, and this set of species is termed the **primary conifer feeders**.

2.2.3 ANALYSES

(a) Diet breadth

As discussed above, the expectation is that polyphagous species are more likely to shift hosts, and this was tested by comparing the proportions of species in the control and shift sets that were **polyphages** (feeding on plants from three or more families), **oligophages** (hosts from two plant families), **family monophages** (hosts from a single plant family) and **genus monophages** (hosts from a single genus) (after Holloway and Hebert, 1979). The new conifer hosts were, of course, excluded from the data used to classify diets of the shifting species.

The results are presented in Figure 2.1, and show that there is a highly significant effect of diet breadth on shifting (G-test, 3 d.f., $G = 26.15$, $P < 0.001$), with a far greater than expected proportion of shifting species being polyphagous. Also note in this graph the abundance of polyphagous species in the control sample, in relation to the log-normal distributions of diet breadth discussed above, which is a result of the taxonomic sampling constraints – the control sample here is biased towards subfamilies with many polyphagous members, because it is from those subfamilies that the shift species come.

(b) Egg hatch and overwintering

Observations of one particular shifting species, the winter moth (*Operophtera brumata* L.) on its novel host, Sitka spruce (*Picea sitchensis*), suggest that the life stage in which the insect overwinters and the associated timing of egg hatch could be important factors in conifer colonization (Fraser, 1995); winter moth larvae feed exclusively on the new growth of spruce, and the synchrony of egg hatch with budburst is important for larval survival, though perhaps not quite so critically as on oak (Watt and McFarlane, 1991). Old conifer foliage is extremely tough, and may contain high levels of aromatic compounds, so may only be available to many angiosperm-adapted herbivores when young and soft. Thus we expect species (like the winter moth) that overwinter as eggs or young larvae and start feeding in spring to be abundant among the shifting species.

Again, these expectations are borne out by the data (Figure 2.2) which show that significantly more shifting species overwinter in the egg stage than the control species ($P < 0.001$ under a Monte Carlo test on the contingency table, with 5000 trials). Here, a

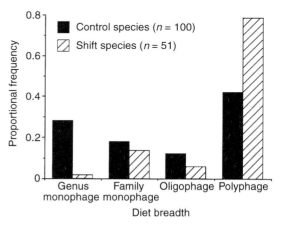

Figure 2.1 Proportions of species in the shift and control categories in different diet breadth classes (see text). Control species shown here are taxonomically constrained subsample of the larger control set.

comparison with the primary conifer feeders is instructive, to test whether being a spring feeder is beneficial for conifer feeding *per se*, or for being able to colonize conifers. The pattern of overwintering in these species follows more closely that of the control species, so it seems that spring feeding is an important factor for conifer colonization, and not conifer feeding itself. The possibility that these results could also stem from a confounding between diet breadth and overwintering stage was tested, and was unfounded – there was no significant interaction between diet breadth and overwintering stage ($P = 0.22$ under a Monte Carlo test on the contingency table, with 5000 trials). This is in contrast to the results of Gaston and Reavey (1989) who found significance, though no clear patterns, for this interaction with a larger data set of 600 species.

These results of overwintering stage are also reflected in the timing of egg hatch, with a significantly higher proportion of shifting species hatching in April than expected (Figure 2.3), even from the taxonomically constrained control sample (*G*-test, 5 d.f., $G = 13.57$, $P = 0.019$).

(c) Usual host plants

It comes as no surprise that the majority of the shifting species come from woody trees or shrubs, simply because expanding the host range from one tree species to another is more likely than from a herb to a tree (Figure 2.4; *G*-test, 3 d.f., $G = 18.01$, $P < 0.001$). What is more interesting is the distribution among plant families of the host plants of the control and shift species, which shows that an abundance of shifting species regularly feed on members of the Ericaceae (Figure 2.5), and to a lesser extent a number of other families.

The predominance of certain plant families, such as the Ericaceae and Myricaceae, in the diets of shifting species could occur for a number of reasons. First, it is consistent with the encounter-frequency hypothesis (Southwood, 1961; Gould, 1979; Strong *et al.*, 1984). For example, the Ericaceae includes genera such as *Calluna, Erica* and *Vaccinium* which are heath and moorland plants commonly encountered near recently planted conifers in the UK. Secondly, members of plant families with many shifting species may have physical, chemical and phenological characteristics similar to those of conifers – for example, adaptation to feeding on *Calluna,* an evergreen woody shrub with tough waxy leaves, may predispose a moth species towards conifer-feeding. More detailed studies of plant morphological and biochemical characteristics and their ecological associations are needed to clarify the more important factors.

(d) Geographical distributions

The previous section analysed the characteristics of species undergoing host shifts that are large in magnitude, as judged by the taxonomic distinctness of original and novel host plants, and thus reveal the most important factors enabling such shifts to take place. It suggested that ecological association (as well as diet breadth, egg hatch date and the architecture of the usual host plants) was an important contributor. Here, those analyses are expanded to include the geographical distribution of the moth species, and their encounter on a large scale with potential conifer hosts. This enables a comparison of the relative importance of diet breadth and geographical range.

It is commonly observed that there is a positive correlation between geographical range and local abundance (references in Gaston, 1990) which holds for many different plant and animal groups,

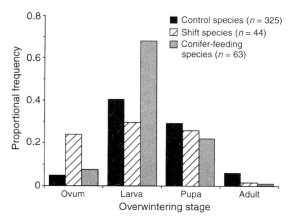

Figure 2.2 Comparison of frequency of species that overwinter in different stages, among the three experimental groups. Note that the conifer feeders are included here (see text).

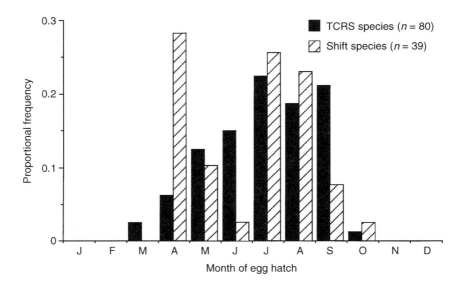

Figure 2.3 Comparison of month of egg hatch of taxonomically constrained random species and shifting species. Only those species with one generation a year are included.

and at different spatial scales (Bock, 1987). This correlation will tend to strengthen the patterns we seek here, for if species with greater geographical ranges, in which conifer encounter is highest, also maintain greater local abundances, that will increase the magnitude of their encounter rate with potential hosts. Brown (1984) discusses reasons for the range–abundance correlation in terms of niche breadth, for species able to exploit a wider range of resources will be both widespread and locally abundant, a hypothesis that can also be used to explain apparent deviations from the usual positive relationship (Gaston and Lawton, 1990). It also suggests that there should be a correlation between niche breadth and geographical range, a relationship that has been little studied (Gaston, 1990) but may be examined here by using diet breadth as a measure of niche width. As predicted, this range size–diet breadth relationship (Figure 2.6) shows a clear increase in diet breadth in more wide-ranging species. The significance of this interaction is tested in the analyses below.

Macrolepidoptera distributions

Geographical distribution data were obtained from the Biological Records Centre, Monks Wood, UK (Harding and Sheail, 1992). These consisted of presence/absence data for each 10 km square in the UK, based on records collected from 1960 to the present. Data were obtained for all the macrolepidopteran control and shift species in the present

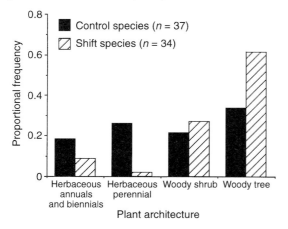

Figure 2.4 Frequency of members of taxonomically constrained sample of control species and the shifting species, classified according to architecture of their normal host plants.

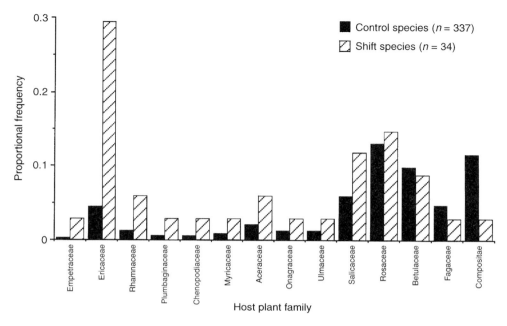

Figure 2.5 Distribution of shift and control species among plant families of their usual host plants. Note that oliphagous and polyphagous species may occur in more than one category here.

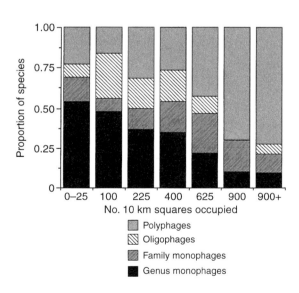

Figure 2.6 Changes in diet breadth over species with different range sizes. Note non-linear division of range categories, chosen to give approximately equal numbers of species in each category.

study, but not the microlepidopterans, for which distribution data at the 10 km square level are unavailable. In addition, only species which are resident in the UK were used in the analyses, excluding migrants and those of questionable status. A sample distribution, that of the winter moth, is shown in Figure 2.7a.

Conifer distributions in the UK

The conifer species of interest for this analysis are non-native species which are used largely for forestry purposes. Data on their distributions were obtained from the Forestry Commission's subcompartment database (Locke, 1987), a data set maintained for the management of FC-owned plantations, from which the hectarages of the taxa listed in Table 2.1 were calculated for each 10 km square in the UK. This database contains information only on extant crops on land managed by the Forestry Commission and omits non-FC plantations, though these can make up a substantial proportion of the total conifer area, from about 40–45% in Scotland

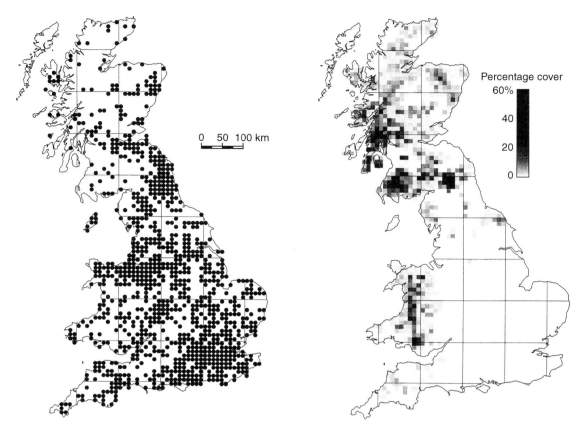

Figure 2.7 Sample distributions, as used in analyses of geographical range and conifer overlap. (a) Distribution of winter moth, *Operophtera brumata*, in terms of presence/absence in 10 km squares in UK. (b) Distribution of one of its novel host plants, Sitka spruce (*Picea sitchensis*) plotted as percentage cover in 10 km squares in UK.

and Wales to 70% in England (Locke, 1987; non-FC proportions after this survey was taken are likely to be higher). Data on non-FC planting is difficult to obtain, so for the purposes of this study the Forestry Commission data are assumed to be an accurate reflection of the relative abundances of these conifer species, both geographically and between taxa. A sample distribution, that of *Picea sitchensis*, is shown in Figure 2.7b.

Analysis

An analysis is performed to test the hypothesis that the magnitude of the overlap with potential conifer hosts (i.e. the exposure frequency) is important in shifting. The measure of conifer overlap used was the total area of all the conifer species in Table 2.1 in the squares in which the moth species occurs (**total conifer overlap**). Its explanatory power was compared with a simple count of occupied 10 km squares (a measure of **range size**) for each moth species. In addition, a comparison was made with the mean conifer area in the squares in which a moth species occured, namely total conifer overlap divided by range size, which was termed **mean conifer overlap**.

Shift and control species were compared using a logistic regression analysis, executed using the generalized linear modelling package GLIM (version 3.77, Royal Statistical Society, 1985), using shifting

Table 2.1 Conifer species for which Forestry Commission distribution data were available, giving number of 10 km squares in which each type occurs, and total area of each in the UK.

Species		Squares	Total ha
Corsican pine	*Pinus nigra* var. *maritima* (Aiton) Melville	548	36009
Lodgepole pine	*Pinus contorta* Douglas	841	102150
Norway spruce	*Picea abies* (L.) Karst	957	44751
Sitka spruce	*Picea sitchensis* (Bong.) Carr.	980	83805
Douglas fir	*Pseudotsuga menziesii* (Mirb.) Franco	763	23051
Western hemlock	*Tsuga heterophylla* (Raf.) Sarg.	657	6084
Firs	*Abies* spp.	657	4122
Larches	*Larix* spp.	1095	64582
Other pines	*Pinus* spp.	219	793
Other spruce	*Picea* spp.	225	495
Mixed conifers[a]		506	2875
Misc. conifers.[b]		516	3183

[a] A mixture of species not dominated by any one conifer type
[b] Other conifer species not specifically listed.

as a binary response variable with the logit link (Crawley, 1993). Since the continuous explanatory variables are so highly correlated ($n = 177$; correlation coefficient $r = 0.911$), they were fitted in turn to a model containing the other explanatory variables. The amount of variation explained by each was then compared, based on the change in deviance resulting from the removal of this effect from the full model. This procedure was repeated comparing the shift species with those in the taxonomically constrained random samples (TCRS). It should be borne in mind that the shift/non-shift binary data are low in information, and that the changes in scaled deviance, on which significance tests are based, are not approximated well by the χ^2 distribution on which they are tested with small sample sizes (Crawley, 1993). As a result, care should be taken in the interpretation of such tests, especially where significance is marginal.

Results

The results of the logistic regression analysis are presented in Table 2.2. In all cases (shift vs control or TCRS species) the measures of range size and conifer overlap enter the model significantly, and there is a significant interaction between these and diet breadth. The month of egg hatch also enters the model significantly, as was true for the simpler analysis in section 2.2.3(b), though it did not significantly interact with any of the other explanatory variables. Overwintering stage and the number of habitats occupied were also tested, but their main effects and interactions were not significant in these analyses.

The significant effect of range size shows that shifting species on average occupy a greater number of 10 km squares in the UK than non-shifting species. Similarly, the total area of conifers encountered by a species has a highly significant effect on host shifting (Table 2.2).

The main effect of diet, which is significant for the comparison using all species, becomes non-significant as sample sizes fall in the shift–TCRS comparison using range size. The interactions of range size and conifer overlap with diet are consistently significant, and suggest that the effect of range or conifer overlap on shifting differs in the different diet categories. However, the small sample sizes of shifting species with narrower diet breadths encourage caution in interpretation.

The analysis using total conifer overlap explains a greater amount of the total variation in shifting (57%, or 76% in the shift–TCRS comparison) than range size. The effect of mean conifer overlap (i.e. the mean coverage of conifers in squares in which a moth species occurs), which is not shown, was consistently non-significant.

Table 2.2 Analysis of deviance tables for the logistic regression analysis, using 'shifting' as a binary response variable with the logit link in GLIM.

Main effect	d.f.	Shift vs. control ($n = 145$)		Shift vs. TCRS ($n = 71$)	
		Deviance	P	Deviance	P
Range size		48%[a]		60%[a]	
Egg hatch month[b]	5	26.60	<0.001	21.98	<0.001
Diet	3	10.85	0.025	6.29	n.s.
Squares	1	9.00	0.005	10.71	0.005
Diet × squares	3	9.24	0.05	8.30	0.05
Total conifer overlap		57%[a]		76%[a]	
Egg hatch month[b]	5	32.38	<0.001	27.71	<0.001
Diet	3	11.07	0.025	11.18	0.025
Total conifer overlap	1	20.77	<0.001	21.49	<0.001
Diet × Tot. con.	3	9.24	0.05	12.26	0.01

[a] Percentage of the variation explained in this analysis.
[b] Egg hatch categories are March–April, May, June, July, August, September–October.
Results of four analyses are presented, comparing the shift species with all the control species (left), and the taxonomically-constrained random sample (right), using range size (top), and total conifer overlap (lower) as explanatory variables. Figures given are the change in deviance resulting from the removal of the specified effect from the full model, which are compared with χ^2 tables to give the p-values shown. The percentage of the variation explained by the full model is also shown. Only univoltine species are included. See text for details of the modelling procedure.

Discussion

It is clear that moth species undergoing host range expansion to conifers in the UK tend to be more widespread, polyphagous species. The mean range size of shifting and non-shifting species is significantly different, and diet breadth is also important. Although diet breadth is correlated with geographical range (Figure 2.6), it still explains a significant amount of variation after range has been taken into account (Table 2.2). The significant diet–range interactions (Table 2.2) suggest that the probability of shifting with greater geographical range increases more rapidly in species that are more polyphagous.

The amount of variation explained by measures of range size and conifer overlap meets expectation, based on an exposure-frequency hypothesis of host shifting – the total overlap of a moth species with potential conifer hosts is a better predictor of host shifting than simple insect range size (Table 2.2). The mean conifer coverage within occupied squares was not significant; clearly what is important is not the proportion of a species' range that overlaps conifers, but the magnitude of that overlap. This is entirely consistent with the exposure-frequency hypothesis, which predicts that the amount of exposure is the important factor in host colonization.

The month in which larvae hatch from the egg (in the univoltine species that were tested here) remains an important factor in the propensity to expand the host range to conifers. A likely reason for this, as discussed above, is the importance of the availability of newly flushed conifer foliage for spring-feeding insects: it may be the only foliage on which larvae can successfully feed. This adds a temporal component to the exposure-frequency hypothesis, namely that species in which the synchrony between larval feeding and the availability of digestible foliage on a potential host is maximal will be more likely to colonize that host.

2.2.4 SPECIES ACCUMULATION ON EXOTIC CONIFERS

(a) Species–area introduction

The species–area relationship has been widely used in the study of herbivore abundance on different host plants, and host plant area (i.e. geographical range) is a consistently good predictor of the number of herbivore species supported – for example, in oak leaf-miners (Opler, 1974), cynipid gall wasps (Cornell and Washburn, 1979), pests on sugar cane (Strong *et al.*, 1977) and insects feeding on the British Rosaceae (Leather, 1986). Regressions of the number of insect species against area occupied by host plants typically explain 30–60% of the variation in species richness (Strong *et al.*, 1984, p. 50), though some studies (e.g. of leafhoppers: Claridge and Wilson, 1981) have found only weak relationships.

Strong *et al.* (1984) suggest three related reasons why such patterns are observed, and a consideration of their predictions is necessary to determine which is the most applicable to the present study. Considering plant species as 'islands' in a 'sea' of other, unpalatable species allows an analogy with the **theory of island biogeography** (Preston, 1960, 1962; MacArthur and Wilson, 1967) to explain the species richness of herbivores feeding on plants (Janzen, 1968, 1973). This theory considers the equilibrium number of species on an island (or a host plant species) to be a dynamic balance between rates of immigration (colonization) and extinction, which are set by island size and isolation.

The **habitat diversity hypothesis** (Williams, 1943, so called by Connor and McCoy, 1979) states that widespread plant species are likely to occur in a greater number of habitats, and thus to encounter a greater range of potential colonizers. This is equivalent to a widening of the species pool, so may be expected to alter the equilibrium species richness of herbivores, as well as their rate of colonization. For example, the greater slopes of species–area regressions for cynipid gall wasps in California compared with the Atlantic coast have been attributed to the former's greater habitat heterogeneity in terms of topographical variety and altitudinal range (Cornell and Washburn, 1979).

Strong *et al.* (1984) also suggest that **encounter frequency** is important, stating that widespread plant species will suffer greater exposure to potential insect colonizers, and thus accrue over time a greater herbivore load. Encounter frequency is more a determinant of the rate of species accumulation on a host plant, with the asymptote set by the size of the regional species pool from which colonizers are drawn (since species accumulation has been shown to be asymptotic: Strong, 1974a,b; Strong *et al.*, 1977), rather than of equilibrium species richness. It is thus more appropriate for systems in which the hosts are thought to be far from their asymptotic species richness, which is almost certainly the case in the present study.

(b) Analysis

In order to detect the effects of host area on the number of herbivore colonists of exotic conifers, I investigated the relationship between the area of conifers and the number of shift species recorded from each. Native conifer feeders were excluded, because records of their feeding on exotic conifers were considered to be less comprehensive than those for the shifting species. Data on the abundance and distribution of the exotic conifers in the UK is given by Locke (1987), who presents data on the standing crop of conifers of different age classes in Britain, subdivided at the county level. These data allow the partitioning of the total area of each conifer species that was standing when the survey was taken (1979–1982) into areas in a number of age classes, namely trees which were from one to eight decades old. Then the regressions of the current number of shifting species with the area contributions of each age class, and the total area, may reveal something about the time scale over which these colonizations are taking place.

(c) Results and discussion

The results of the analyses, presented in Table 2.3, show that there are highly significant regressions of the number of shift species against tree area for the last two planting decades, and also with the total standing crop; one such relationship is shown

28 Host specificity in forest insects

Table 2.3 Summary of results from species–area regression analyses, of number of shift species on each conifer host against area of that conifer species in each of age classes given.

Years of planting	n	Slope	r^2	P
1951-60	7	0.51	0.488	n.s.
1961-70	7	0.81	0.831	p < 0.005
1971-80	7	0.60	0.894	p < 0.002
Total area*	7	0.88	0.727	p < 0.05

*Represents standing crop of various conifer species at time of data collection (1982)

in Figure 2.8. Regressions with the area of trees planted pre-1960 are not significant.

As explained above, the exposure frequency hypothesis can be used to make predictions about colonization rates, and those agree with the data presented here – widespread conifers are accumulating species more rapidly. In addition, the strongest relationships are obtained with areas planted over the past few decades (because of increasing rates of planting, these represent a high proportion of the total area), which is what would be expected if colonization continues at a rate set by current area. The stronger relationships for the periods 1961–1970 and 1971–1980 also suggest that tree age may be an important factor for colonization in some Lepidoptera species. This is supported by the observation that early outbreaks of

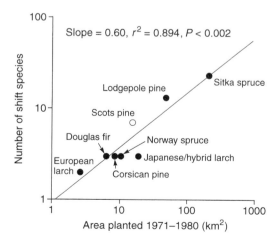

Figure 2.8 Species–area relationship for host-shifting Lepidoptera on exotic conifers (●) and Scots pine (○) in UK, using area of conifers in 1971–1980 age class.

the winter moth (*Operophtera brumata*) on a novel host, Sitka spruce (*Picea sitchensis*), were mostly associated with spruce plantations that were less than 12 years old (Stoakley, 1985), and outbreaks continue to be concentrated in young plantations (Fraser, 1995).

Problems associated with the statistics of the species area relationship (Preston, 1962; Loehle, 1990) are particularly acute here. First, sample sizes are unavoidably small (regressions are based on just 7 points). Second, there is likely to be some error in the estimates of conifer area, due to clear-felling and replanting, and errors and gaps in sampling, which violates one of the assumptions of regression analysis. Third, the lepidopteran fauna of conifer plantations in the UK is not well known and is likely to have been undersampled, because such plantations do not tend to attract amateur entomologists and because newly colonizing species will initially occur locally and at low density. For these reasons, this relationship has not been exhaustively analysed.

2.3 CONCLUSIONS

The majority of the host-shifting moth species in these analyses are polyphagous, tree-feeding species which, until now, have received little attention in terms of the factors that affect their host use. Standard models for the evolution of diet breadth, which concentrate on the relationship between oviposition preference and larval performance, and trade-offs between performance on different host plants, do not suit these types of species well. The analyses presented here suggest that for such species a different suite of factors may be important in determining patterns of host use.

One of the overriding results that emerge is that encounter frequency is very important for host shifting. Shifting species tend to feed on angiosperms which are associated with conifer afforestation in upland Britain, and they occur in habitats in which conifer planting is widespread (Fraser and Lawton, 1994). In addition, they tend to be not only wide-ranging species but also those wide-ranging species whose geographical overlap with conifers is highest (Table 2.2). As discussed above, local abundance and geographical range are highly correlated, so these shifting species are also likely to be those that are locally common, further increasing encounter frequencies.

Observations of colonization rates on different conifer species are the flip-side of this coin. Those conifers, like Sitka spruce, that are more widespread have accumulated more shifting species (Figure 2.8). More specifically, a number of host shifts may be attributed to the presence of a superabundance of a novel host plant. For example, Sitka spruce, the most abundant exotic conifer in the UK, covers 60% of the land surface in some 10 km squares in the UK (Figure 2.7), and it was proposed that this is the major factor in its recruitment of the winter moth as a new pest (Fraser, 1995). There are other observations of herbivores colonizing abundant crops that are not related to their normal hosts (e.g. Shapiro and Masuda, 1980).

The importance of egg hatch timing (section 2.2.3(b)) suggests that there is a temporal component to encounter frequency – there must be suitable foliage available to neonate larvae on novel hosts before they can start to use them. This phenological synchrony is an important criterion in alternative host use (DuMerle *et al*., 1992). The requirement for close synchrony with the host also opens up the possibility for that synchrony to break down, because of climatic or other effects, forcing synchrony-sensitive species to utilize alternative host plants in some years. This will tend to select for increased diet breadth in such species.

A surprisingly high proportion of the shifting species used in this analysis (six of 50 shifting species, compared with five of the 100 taxonomically matched control species) show a set of similar life history characteristics, which as a whole have been dubbed the 'winter moth syndrome' (Schneider, 1980). These characteristics include univoltinism, polyphagy, female flightlessness, overwintering as eggs, and dispersal of first instar larvae by ballooning. It appears that species such as the winter moth, gypsy moth (*Lymantria dispar* L.) and fall cankerworm (*Alsophila pometaria* Harris) have all converged on this type of strategy; and its independent evolution in species in three subfamilies of the Geometridae, the Lymantriidae and Psychidae, all of which inhabit stable forest habitats, suggests that ecological factors are important over and above the effects of phylogeny (Barbosa *et al*., 1989). Similarly, analyses of outbreaking and non-outbreaking species of North American Lepidoptera (Hunter, 1991, 1995) found that many outbreaking species share one or more of the 'winter moth syndrome' characteristics, showing an abundance of spring feeders, higher mean fecundities and greater diet breadths compared with non-outbreak species.

In these species, females are not choosy about their oviposition site and it is the dispersing larvae that exercise a rather inefficient form of host selection, by differential dispersal. Hence, determinants of host use are likely to be very different from those in the species on which models of evolution of host range have been based. The 'winter moth syndrome' species form only a small proportion of the lepidopteran fauna in this study, so it is useful to consider the generality of these results for the larger question of the evolution of host range. First, there are other species which, because of particular types of oviposition behaviour or other means of host choice (examples are given in section 2.1.2(b)) do not conform to the behaviour assumed by many models of diet breadth evolution. Second, there is a bias among species traditionally utilized in studies of host use towards those with narrower diet breadths, simply because these provide greater resolution between alternative hosts. This may have influenced our views of which mechanisms are important, and the current study goes some way towards redressing the balance in demonstrating that, in less constrained species, exposure to potential hosts in space and time is an important determinant of host use.

ACKNOWLEDGEMENTS

I thank John Lawton, Allan Watt and Kevin Gaston for comments on earlier versions of the manuscript. This chapter benefited greatly from suggestions by Mike Singer and an anonymous referee. Data for some of the analyses were provided by the Institute of Terrestrial Ecology and the Forestry Commission, and the work described here was carried out while the author was supported by a Natural Environment Research Council CASE Studentship, with the Institute of Terrestrial Ecology and the Forestry Commission.

REFERENCES

Andow, D.A. and Imura, O. (1994) Specialization of phytophagous arthropod communities on introduced plants. *Ecology* **75**, 296–300.

Auerbach, M. and Simberloff, D. (1988) Rapid leafminer colonization of introduced trees and shifts in sources of herbivore mortality. *Oikos* **52**, 41–50.

Barbosa, P. (1988) Some thoughts on 'The evolution of host range'. *Ecology* **69**, 912–915.

Barbosa, P., Krischik, V. and Lance, D. (1989) Life-history traits of forest-inhabiting flightlesss Lepidoptera. *Am. Midl. Nat.* **122**, 262–274.

Basset, Y. and Burckhardt, D. (1992) Abundance, species richness, host utilization and host specificity of insect folivores from a woodland site, with particular reference to host architecture. *Rev. Suisse De Zool.* **99**, 771–791.

Berenbaum, M. (1981) Effects of linear furanocoumarins on an adapted specialist insect (*Papilio polyxenes*). *Ecol. Entomol.* **6**, 345–351.

Berenbaum, M.R. and Zangerl, A.R. (1991) Acquisition of a native hostplant by an introduced oligophagous herbivore. *Oikos* **62**, 153–159.

Bernays, E.A. (1981) Plant tannins and insect herbivores: an appraisal. *Ecol. Entomol.* **6**, 353–360.

Bernays, E.A. and Graham, M. (1988) On the evolution of host specificity in phytophagous arthropods. *Ecology* **69**, 886–892.

Bock, C.E. (1987) Distribution–abundance relationships of some Arizona landbirds: a matter of scale? *Ecology* **68**, 124–129.

Bowers, M.D., Stamp, N.E. and Collinge, S.K. (1992) Early stage of host range expansion by a specialist herbivore, *Euphydryas phaeton* (Nymphalidae). *Ecology* **73**, 526–536.

Bradley, J.D., Tremewan, W.G. and Smith, A. (1973) *British Tortricoid Moths: Cochylidae and Tortricidae: Tortricinae*, The Ray Society, London.

Bradley, J.D., Tremewan, W.G. and Smith, A. (1979) *British Tortricoid Moths: Olethreutidae*, The Ray Society, London.

Brown, J.H. (1984) On the relationship between abundance and distribution of species. *Am. Nat.* **124**, 255–279.

Bush, G.L. (1969) Sympatric host race formation and speciation in frugivorous flies of the genus *Rhagoletis* (Diptera: Tephritidae). *Evolution* **23**, 237–251.

Bush, G.L. (1975) Sympatric speciation in phytophagous parasitic insects. In *Evolutionary Strategies of Parasitic Insects and Mites* (ed. P.W. Price), pp. 187–206, Plenum Press, New York.

Bush, G.L. and Diehl, S.R. (1982) Host shifts, genetic models of sympatric speciation and the origin of parasitic insect species. In *Proceedings of the Fifth International Symposium on Insect–Plant Relationships* (eds J.H. Visser and A.K. Minks), pp. 297–305, Pudoc, Wageningen, The Netherlands.

Campbell, I.M. (1989) Does climate affect host-plant quality? Annual variation in the quality of balsam fir as food for spruce budworm. *Oecologia* **81**, 335-340.

Cates, R.G. (1981) Host plant predictability and the feeding patterns of monophagous, oliphagous, and polyphagous insect herbivores. *Oecologia* **48**, 319–326.

Clapham, A.R., Tutin, T.G. and Warburg, E.F. (1981) *Excursion Flora of the British Isles*, Cambridge University Press, Cambridge.

Claridge, M.F. and Wilson, M.R. (1976) Diversity and distribution patterns of some mesophyll-feeding leafhoppers of temperate woodland canopy. *Ecol. Entomol.* **1**, 231–250.

Claridge, M.F. and Wilson, M.R. (1981) Host plant associations, diversity and species–area relationships of mesophyll-feeding leafhoppers of trees and shrubs in Britain. *Ecol. Entomol.* **6**, 217–238.

Connor, E.F. and McCoy, D. (1979) The statistics and biology of the species–area relationship. *Am. Nat.* **113**, 791–833.

Cornell, H.V. and Washburn, J.O. (1979) Evolution of the richness–area correlation for cynipid gall wasps on oak trees: a comparison of two geographic areas. *Evolution* **33**, 257–274.

Courtney, S.D. (1988) Heritable divergence of *Rhagoletis pomonella* host races by seasonal asynchrony. *Nature* **336**, 66–67.

Crawley, M.J. (1993) *GLIM for Ecologists*, Blackwell Scientific Publications, Oxford.

DuMerle, P., Brunet, S. and Cornic, J.F. (1992) Polyphagous potentialities of *Choristoneura murinana* (Hb.) (Lep., Tortricidae): A 'monophagous' folivore extending its host range. *J. Appl. Entomol.* **113**, 18–40.

Emmet, A.M. (1991) Life history and habits of the British Lepidoptera. In *The Moths and Butterflies of Great Britain and Ireland*, Vol. 7, Part 2 (eds A.M. Emmet and J. Heath), pp. 61–303, Harley Books, Colchester.

Ehrlich, P.R. and Raven, P.H. (1964) Butterflies and plants: a study in coevolution. *Evolution* **18**, 586–608.

Farrell, B.D., Mitter, C. and Futuyma, D.J. (1992) Diversification at the insect-plant interface. *BioScience* **42**, 34–42.

Feder, J.L. and Bush, G.L. (1989) A field test of differential host-plant usage between two sibling species of *Rhagoletis pomonella* fruit flies (Diptera: Tephritidae) and its consequences for sympatric models of speciation. *Evolution* **43**, 1813–1819.

Feder, J.L., Hunt, T.A. and Bush, L. (1993) The effects of climate, host-plant phenology and host fidelity on the genetics of apple and hawthorn infesting races of *Rhagoletis pomonella*. *Entomol. Exp. Appl.* **69**, 117–135.

Feeny, P.P. (1976) Plant apparency and chemical defence. *Rec. Adv. Phytochem.* **10**, 1–40.

Fox, L.R. and Eisenbach, J. (1992) Contrary choices: possible exploitation of enemy-free space by herbivorous insects in cultivated vs. wild crucifers. *Oecologia* **89**, 574–579.

Fox, L.R. and Morrow, P.A. (1981) Specialization: species property or local phenomenon? *Science* **211**, 887–893.

Fraser, S.M. (1995) The colonization of Sitka spruce by winter moth in Scotland. PhD Thesis, University of London.

Fraser, S.M. and Lawton, J.H. (1994) Host range expansion by British moths onto introduced conifers *Ecol. Entomol.* **19**, 127–137.

Futuyma, D.J. (1976) Food plant specialization and environmental predictability in Lepidoptera. *Am. Nat.* **110**, 285–292.

Futuyma, D.J. (1983) Selective factors in the evolution of host choice by phytophagous insects. In *Herbivorous Insects: Host Seeking Behaviour and Mechanisms* (ed. S. Ahmad), pp. 227–244, Academic Press, New York.

Futuyma, D.J. and Mayer, G.C. (1980) Non-allopatric speciation in animals. *Syst. Zool.* **29**, 254–271.

Gaston, K.J. (1990) Patterns in the geographical ranges of species. *Biol. Rev.* **65**, 105–129.

Gaston, K.J. and Lawton, J.H. (1990) Effects of scale and habitat on the relationship between regional distribution and local abundance. *Oikos* **58**, 329–335.

Gaston, K.J. and Reavey, D. (1989) Patterns in the life histories and feeding strategies of British macrolepidoptera. *Biol. J. Linn. Soc.* **37**, 367–381.

Gould, F. (1979) Rapid host range evolution in a population of the phytophagous mite *Tetranychus urticae* Koch. *Evolution* **33**, 791–802.

Harding, P.T. and Sheail, J. (1992) The Biological Records Centre: a pioneer in data gathering and retrieval. In *Biological Recording of Changes in British Wildlife* (ed. P.T. Harding), pp. 5–19, HMSO, London.

Hare, J.D. and Kennedy, G.G. (1986) Genetic variation in plant–insect associations: survival of *Leptinotarsa decemlineata* populations on *Solanum carolinense*. *Evolution* **40**, 1031–1043.

Hatcher, P.E. (1991) The conifer-feeding macro-Lepidoptera fauna of an English woodland as determined by larval sampling. *The Entomologist* **110**, 11–23.

Hatcher, P.E. and Winter, T.G. (1990) An annotated checklist of British conifer-feeding macro-Lepidoptera and their foodplants. *Entomol. Gaz.* **41**, 177–196.

Holloway, J.D. and Hebert, P.D.N. (1979) Ecological and taxonomic trends in macrolepidopteran host plant selection. *Biol. J. Linn. Soc.* **11**, 229–251.

Hsiao, T. (1982) Geographic variation and host plant adaptation of the Colorado potato beetle. In *Proceedings of the Fifth International Symposium on Insect–Plant Relationships* (eds J.H. Visser and A.K. Minks), pp. 315–324, Pudoc, Wageningen, The Netherlands.

Hunter, A.F. (1991) Traits that distinguish outbreaking and nonoutbreaking Macrolepidoptera feeding on northern hardwood trees. *Oikos* **60**, 275–282.

Hunter, A.F. (1995) The ecology and evolution of reduced wings in forest macrolepidoptera. *Evol. Ecol.* **9**, 275–283.

Jaenike, J. (1990) Host specialization in phytophagous insects. *Annu. Rev. Ecol. Syst.* **21**, 243–274.

Janzen, D.H. (1968) Host plants as islands in evolutionary and contemporary time. *Am. Nat.* **102**, 592–595.

Janzen, D.H. (1973) Host plants as islands. II. Competition in evolutionary and contemporary time. *Am. Nat.* **107**, 786–790.

Janzen, D.H. (1979) New horizons in the biology of plant defences. In *Herbivores: their Interaction with Secondary Plant Metabolites* (eds J. Rosenthal and D.H. Janzen), pp. 331–350, Academic Press, New York.

Karowe, D.N. (1990) Predicting host range evolution: colonization of *Coronilla varia* by *Colias philodice* (Lepidoptera: Pieridae). *Evolution* **44**, 1637–1647.

Kennedy, C.E.J. (1986) Attachment may be a basis for specialization in oak aphids. *Ecol. Entomol.* **11**, 291–300.

Lawton, J.H. and Strong, D.R. (1981) Community patterns and competition in folivorous insects. *Am. Nat.* **118**, 317–338.

Leather, S.R. (1986) Insect species richness of the British Rosaceae: the importance of host range, plant architecture, age of establishment, taxonomic isolation and species–area relationships. *J. Anim. Ecol.* **55**, 841–860.

Locke, G.M. (1987) *Census of Woodlands and Trees*, Report, HMSO, London.

Loehle, C. (1990) Proper statistical treatment of species–area data. *Oikos* **57**, 143–145.

MacArthur, R.H. and Wilson, E.O. (1967) *The Theory of Island Biogeography*, Princeton University Press, Princeton, New Jersey.

Mansell, E. and Newman, H. (1968) *The Complete British Butterflies in Colour*, Ebury Press, London.

May, R.M. (1975) Patterns of species abundance and diversity. In *Ecology and Evolution of Communities* (eds M.L. Cody and J.M. Diamond), pp. 81–120, Harvard University Press, Cambridge, Massachussets.

Menken, S.B. (1981) Host races and sympatric speciation in small ermine moths, Yponomeutidae. *Entomol. Exp. Appl.* **30**, 280–292.

Mitter, C., Futuyma, D.J., Schneider, J.C. and Hare, D.J. (1979) Genetic variation and host-plant relations in a parthenogenetic moth. *Evolution* **33**, 777–790.

Neuvonen, S. and Niemelä, P. (1983) Species richness and faunal similarity of arboreal insect herbivores. *Oikos* **40**, 452–459.

Niemelä, P., Hanhimäki, S. and Mannila, R. (1981) The relationship of adult size in noctuid moths (Lepidoptera: Noctuidae) to breadth of diet and growth form of host plants. *Ann. Entomol. Fenn.* **47**, 17-20.

Opler, P.A. (1974) Oaks as evolutionary islands for leaf-mining insects. *Am. Sci.* **62**, 67–73.

Preston, F.W. (1960) Time and space and the variation of species. *Ecology* **41**, 611–627.

Preston, F.W. (1962) The canonical distribution of commoness and rarity. *Ecology* **43**, 185–215, 410–432.

Price, P.W., Westoby, M., Rice, B. *et al.* (1986) Parasite mediation in ecological interactions. *Annu. Rev. Ecol. Sys.* **17**, 487–505.

Radtkey, R.R. and Singer, M.C. (1995) Repeated reversals of host preference evolution in a specialist insect herbivore. *Evolution* **49**, 351–359.

Rausher, M.D. (1982) Population differentiation in *Euphydryas editha* butterflies: larval adaptation to different hosts. *Evolution* **36**, 581–590.

Rhoades, D.F. and Cates, R.G. (1976) Towards a general theory of plant antiherbivore chemistry. *Rec. Adv. Phytochem.* **10**, 168–213.

Schneider, J.C. (1980) The role of parthenogenesis and female aptery in microgeographic, ecological adaptation in the fall cankerworm, *Alsophila pometaria* Harris (Lepidoptera: Geometridae). *Ecology* **61**, 1082–1090.

Scriber, J.M. (1983) Evolution of feeding specialization, physiological efficiency, and host races in selected Papilionidae and Saturniidae. In *Variable Plants and Herbivores in Natural and Managed Ecosystems* (eds R.F. Denno and M.S. McClure), pp. 373–412, Academic Press, New York.

Seitz, A. and Komma, M. (1984) Genetic polymorphism and its ecological background in Tephritid populations (Diptera: Tephritidae). In *Population Biology and Evolution* (eds K. Wöhrmann and V. Loeschcke), pp. 143–158, Springer-Verlag, Berlin.

Shapiro, A.M. and Masuda, K.K. (1980) The opportunistic origin of a new citrus pest. *Calif. Agric.* **34**. 4–5.

Singer, M.C. and Parmesan, C. (1993) Sources of variation in patterns of plant–insect association. *Nature* **361**, 251–253.

Singer, M.C., Ng, D., Vasco, D. and Thomas, C.D. (1992) Rapidly evolving associations among oviposition preferences fail to constrain evolution of insect diet. *Am. Nat.* **139**, 9–20.

Singer, M.C., Thomas, C.D. and Parmesan, C. (1993) Rapid human-induced evolution of insect–host associations. *Nature* **366**, 681–683.

Skinner, B. (1984) *Moths of the British Isles*, Viking, Harmondsworth.

Slatkin, M. (1987) Gene flow and the geographic structure of natural populations. *Science* **236**, 787–792.

Smiley, J. (1978) Plant chemistry and the evolution of host specificity: new evidence from Heliconius and Passiflora. *Science* **201**, 745–747.

Southwood, T.R.E. (1961) The evolution of the insect-host tree relationship – a new approach. *Proceedings of the XIth International Congress of Entomology, Vienna* **1**, 651–654.

Stille, B. (1985) Host plant specificity and allozyme variation in the parthenogenetic gall wasp *Diplolepis mayri* and its relatedness to *D. rosae* (Hymenoptera: Cynipidae). *Entomol. Gener.* **10**, 87–96.

Stoakley, J.T. (1985) Outbreaks of Winter Moth, *Operophtera brumata* L. (Lep., Geometridae) in young plantations of Sitka Spruce in Scotland. *Zeit. angew. Entomol.* **99**, 153–160.

Strong, D.R. (1974a) Nonasymptotic species richness models and the insects of British trees. *Proc. Natl. Acad. Sci.* **71**, 2766–2769.

Strong, D.R. (1974b) Rapid asymptotic species accumulation in phytophagous insect communities: the pests of Cacao. *Science* **185**, 1064–1066.

Strong, D.R. (ed.) (1988) Species feature: Insect host range. *Ecology* **69**, 885–915.

Strong, D.R., Lawton, J.H. and Southwood, T.R.E. (1984) *Insects on Plants*, Blackwell, Oxford.

Strong, D.R., McCoy, E.D. and Rey, J.R. (1977) Time and the number of herbivore species: the pests of sugarcane. *Ecology* **58**, 167–175.

Styles, J.H. (1960) *Syndemis musculana* Hübner (Lep., Tortricidae) in conifer plantations and forest nurseries in the British Isles. *Entomol. Gaz.* **11**, 144–148.

Tabashnik, B.E. (1983) Host range evolution: the shift from native legume hosts to alfalfa by the butterfly, *Colias philodice eriphyle*. *Evolution* **37**, 150–162.

Thomas, C.D., Ng, D., Singer, M.C. *et al.* (1987) Incorporation of a European weed into the diet of a North American herbivore. *Evolution* **41**, 892–901.

Thompson, J.N. (1988a) Coevolution and alternative hypotheses on insect/plant interactions. *Ecology* **69**, 893–895.

Thompson, J.N. (1988b) Evolutionary ecology of the relationship between oviposition preference and the performance of offspring in phytophagous insects. *Entomol. Exp. Appl.* **47**, 3–14.

Thompson, J.N. (1994) *The Coevolutionary Process*, University of Chicago Press, Chicago.

Thompson, J.N., Wehling, W. and Podolsky, R. (1990) Evolutionary genetics of host use in swallowtail butterflies. *Nature* **344**, 148–150.

Tuomi, J., Nilsson, P. and Astrom, M. (1994) Plant compensatory responses: bud dormancy as an adaptation to herbivory. *Ecology* **75**, 1429–1436.

Turnipseed, S.G. and Kogan, M. (1976) Soybean entomology. *Annu. Rev. Entomol.* **21**, 247–282.

Varley, G.C. and Gradwell, G.R. (1968) Population models for the winter moth. In *Insect Abundance. Symposium of the Royal Entomological Society of London* (ed. T.R.E Southwood), pp. 132–142, Symposium of the Royal Entomological Society of London No. 4, Blackwell Scientific Publications, Oxford.

Waring, G.L., Abrahamson, W.G. and Howard, D.J. (1990) Genetic differentiation among host-associated populations of the gallmaker *Eurosta solidaginis* (Diptera: Tephritidae). *Evolution* **44**, 1648–1655.

Wasserman, S.S. and Mitter, C. (1978) The relationship of body size to breadth of diet in some Lepidoptera. *Ecol. Entomol.* **3**, 155–160.

Watt, A.D. and McFarlane, A.M. (1991) Winter moth on Sitka spruce: synchrony of egg hatch and budburst, and its effect on larval survival. *Ecol. Entomol.* **16**, 387–390.

Welch, R.C. and Greatorex-Davies, J.N. (1993) Colonization of two Nothofagus species by Lepidoptera in southern Britain. *Forestry* **66**, 181–203.

Wiklund, C. (1975) The evolutionary relationship between adult oviposition preferences and larval host plant range in *Papilio machaon*. *Oecologia* **18**, 185–197.

Wiklund, C. (1982) Generalist vs. specialist utilization of host plants among butterflies. In *Insect Plant Relationships* (ed. R. van Emden), pp. 181–191, Blackwell Scientific Publications, Oxford.

Williams, C.B. (1943) Area and the number of species. *Nature* **152**, 264–267.

Wint, W. (1983) The role of alternate host-plant species in the life of a polyphagous moth, *Operophtera brumata* (Leopidoptera: Geometridae). *J. Anim. Ecol.* **52**, 439–450.

PART TWO
TEMPORAL AND SPATIAL POPULATION ECOLOGY

POPULATION DYNAMICS OF FOREST INSECTS: ARE THEY GOVERNED BY SINGLE OR MULTIPLE FACTORS?

T. Royama

3.1 INTRODUCTION

Are the population dynamics of forest insects governed by single or multiple factors? One way to answer this question might be to review recent literature. However, there have been several comprehensive review articles in recent years, individually or collectively published, on similar subjects (Barbosa and Schultz, 1987; Berryman, 1988; Myers, 1988; Price *et al.*, 1990; Watt *et al.*, 1990; Price, 1994). Adding another review would not significantly improve our understanding of the mechanisms underlying the population dynamics of forest insects.

There are many theories, conjectures and speculations to explain, for instance, what governs population cycles, e.g. the classic host–parasite interactions (Utida, 1957), the influence of weather (Wellington *et al.*, 1950; Thomson *et al.*, 1984), host–disease interactions (Anderson and May, 1980), host–virus interactions (Myers, 1993), defoliation–food quality relationships (Baltensweiler *et al.*, 1977; Baltensweiler and Fischlin, 1988; Haukioja and Neuvonen, 1987; Price *et al.*, 1990), and a 'maternal effect' on forest lepidoptera (Ginzburg and Taneyhill, 1994).

However, only a few sets of quantitative data, based on systematic, long-term studies, provide information for a reliable reconstruction of forest insect population dynamics. In many cases, the mechanism of population dynamics is not inferred from data. Rather, the data are used to support or to disprove preconceived hypotheses. These hypotheses are often tested by correlating the pattern of population dynamics with the ecological factors, or life history traits, that have been suggested in the hypotheses rather than by looking deeply into causal relationships.

An example is to seek correlation between insect outbreaks and weather records (Wellington *et al.*, 1950; Greenbank, 1963; Thomson *et al.*, 1984; Swetnam and Lynch, 1993). The particular danger here is that a correlation between two series of concurrent events is often spurious, especially when the length of observations is limited, as is the case with many sets of ecological data (Royama, 1992). Sometimes, population fluctuation is compared with weather fluctuation after the weather data have been 'smoothed' or 'filtered' by moving-average transformation. The transformation, however, creates an artificial cyclic trend in the weather data, resulting in an artificial correlation with the population cycles (Royama, 1978, 1984, 1992; Martinat, 1987). Also, there are problems in evaluating the role of natural enemies using key-factor analysis (Hassell, 1985; Price, 1987; Royama, 1996). Without looking into the causal relationship, no conclusion can be drawn (Price, 1987).

Forests and Insects. Edited by A.D. Watt, N.E. Stork and M.D. Hunter. Published in 1997 by Chapman & Hall, London. ISBN 0 412 79110 2.

A more fundamental problem in these idea-generating and testing-by-correlation approaches is that a natural system tends to be viewed through a narrow window of hypotheses. The gathering of information is likely to be fragmented and a comprehensive view is unlikely to emerge.

In forest insect populations with distinct generations, intergeneration dynamics are completely determined by the balance between the gain (by oviposition and immigration) and loss (by mortality and emigration) of members of the population. If we know the factors that determine these processes through life table or survivorship studies, there would be little room for speculation. Life table and survivorship studies, if technical difficulties are overcome, can provide the essential information. This paper will discuss certain aspects of analysis and inference of the factors involved using life table information.

3.2 LIFE TABLE AND SURVIVORSHIP STUDIES

Thomas Park (1949) remarked that 'a life table keeps the books on death'. Unfortunately, ecological life tables tend to have adopted this actuarial format which is concerned primarily with death. For example, in the well-known method of key-factor analysis (Varley and Gradwell, 1960, 1963, 1970), the main emphasis is in the analysis of 'generation survival' (Dempster, 1975; Southwood, 1978). Often, no attention is paid to the way a new cohort is recruited. Thus, the most crucial issue of population dynamics – the rate of change in the population – is not adequately analysed.

Some authors have been sceptical about the life table approach (Haukioja and Neuvonen, 1987; Myers, 1993). It is true that there are problems in constructing life tables of natural populations (Price, 1987; Bellows *et al.*, 1992), and in identifying and evaluating certain types of factors (Royama, 1981; Hassell, 1985), or in a particular method of life table analysis, such as key-factor analysis (Royama, 1996). Nonetheless, life table information, or more precisely, keeping the books on gain and loss, no matter how technically difficult to gather and interpret the information, is the basis of the inference and comprehension of the population process.

Insect life table studies were pioneered by Morris and Miller (1954) on the spruce budworm, *Choristoneura fumiferana*, by Varley and Gradwell (1963) on the winter moth, *Operophtera brumata*, and by Klomp and Gruys (1965) on the pine looper, *Bupalus piniaria*. Many similar studies have followed (Dempster, 1975, 1983; Podoler and Rogers, 1975; Bellows *et al.*, 1992). The following sections look at two of these classic examples, the spruce budworm and the pine looper, to see how many factors have to be taken into account to comprehend their dynamics.

3.3 SPRUCE BUDWORM OUTBREAK PROCESSES

3.3.1 BASIC COMPONENTS OF THE OUTBREAK CYCLE

Figure 3.1 is a schema of population change from one generation to the next. Starting with eggs at the beginning of generation t, say $X(t)$ after log transformation, moth density is determined through generation survival, the log of which is $H_g(t)$. Then, $X(t+1)$, log eggs of the new generation, is produced through the log recruitment rate, $H_r(t)$. The log rate of change in egg density from generation t to $t+1$, say $R(t)$, is equal to the difference $X(t+1) - X(t)$, which is, in turn, equal to the sum $H_g(t) + H_r(t)$. We see that at least two sets of factors are involved in determining intergeneration changes in population: one determining H_g and the other, H_r. One might ask, then: is one set more important than the other set? There is no simple answer.

Figure 3.2 shows the annual changes in $H_g(t)$, $H_r(t)$, $R(t)$ and $X(t)$ in spruce budworm observed at one study plot in northwestern New Brunswick, Canada, from 1945 to 1959, known as the Green River Study (Morris, 1963), reanalysed by Royama (1984, 1992). (Note that the above variables are logarithms of the original numbers; also, the time variable, t, may be dropped unless necessary.) The rise and fall of egg density (X) are translated into the rate of change (R) that declined from, on average, positive (indicating a population increase) to,

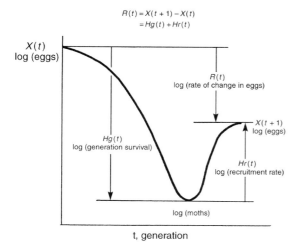

Figure 3.1 Schema of population changes from eggs in one generation to those in following generation. Downward arrows indicate negative value; upward arrows, positive value. For convenience, $R(t) < 0$, since $X(t+1) < X(t)$.

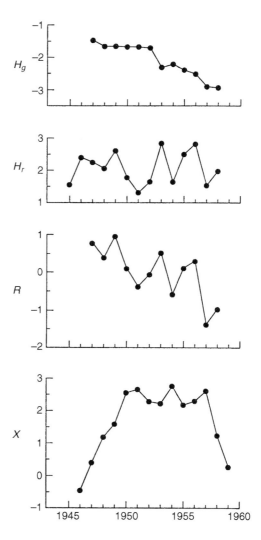

Figure 3.2 Annual changes in generation survival (H_g), recruitment rate (H_r), rate of change in egg density (R) and egg density (X) of spruce budworm population observed in northwestern New Brunswick, Canada (the Green River Study); all in common logarithms. (After Royama, 1996.)

on average, negative (indicating a population decline), although fluctuating erratically about the downward trend. Recalling that $R = H_g + H_r$ as in Figure 3.1, the downward trend in R is attributed to the same trend in H_g, and the erratic fluctuation about the trend is due to the fluctuation in H_r. We see that H_g and H_r played qualitatively different roles in determining population changes: H_g, generation survival, determined the basic pattern of the budworm outbreak cycle, while H_r, recruitment rate, could only influence the amplitude of the cycle. For instance, a large recruitment during the increasing phase could bring the population to an even higher level and vice versa.

Thus, if one is interested in how a budworm population rises and falls cyclically, the set of factors that determine generation survival is a major concern. On the other hand, to understand why some local populations increase more quickly to a higher level and more severely damage a given forest stand, the variation in recruitment is also a major concern. Evidently, to determine which set of factors is major in the present case involves subjectivity. It will now be shown that several different sets of factors are involved in determining the annual changes in recruitment rate and in generation survival.

3.3.2 COMPONENTS OF RECRUITMENT RATE AND GENERATION SURVIVAL

Recruitment rate in budworm (defined as the ratio: all eggs laid in the habitat / all moths locally emerged) is determined by two major factors: average oviposition rate and net moth dispersal (a balance between emigration and immigration). In the present example, average fecundity varied from 100 to 200 per female, depending on the severity of defoliation, but the recruitment rate fluctuated well beyond the range of variation in fecundity. Balance between the immigration and emigration of egg-carrying moths exerted a determining influence on the fluctuation in recruitment rate (Royama, 1992).

Two types of mortality factors govern generation survival: loss of larvae that mainly occurs during their dispersal and a complex of natural enemies throughout their life cycle. A substantial proportion of larvae is lost during their dispersal. This occurs twice in their life cycle: once immediately following hatching in search of wintering sites and again after spring emergence in search of feeding sites. Although the loss of the young larvae varies considerably from year to year, its pattern does not conform to the declining trend in generation survival (Royama, 1984). Therefore, it is not a direct cause of the outbreak cycle, though it might have a hidden role as a modifier (section 3.4). Thus, what primarily governs budworm outbreak processes is the action of a complex of natural enemies, the major type being insect parasitoids (Royama, 1992; Eveleigh, unpublished data).

While the action of natural enemies might be the principal factor underlying the cyclic pattern of budworm population changes, the pattern is unlike what we expect from the theory of simple predator-prey systems (Hassell, 1978; Taylor, 1984; Royama, 1992). We are left with many unanswered questions. Why is the length of a budworm cycle so long (on average, 30–40 years)? Why does an outbreak, once begun, tend to persist so long (8 or even 10 years)? Why is it so widespread, covering a substantial part of northeastern North America? And so on.

3.3.3 LENGTH OF A CYCLE

Although the complex of insect parasitoids is the principal factor that governs the budworm cycles, their control efficacy is rather low. This results in an unusually long cycle. A major reason for the low efficacy of the parasitoid complex is the existence of an amazingly rich fauna of hymenopteran hyperparasitoids. In three study plots in New Brunswick alone, we have so far recorded 26 species of hyperparasitoids which attacked 18 of the 33 primary parasitoid species (Eveleigh *et al.*, 1994). Furthermore, the dipteran primary parasitoids (tachinids) that pupate in the soil are also heavily attacked by mammalian and invertebrate predators (Eveleigh, unpublished data). As the rate of increase in these primary parasitoids is reduced by the action of their own natural enemies, it will take more generations for them to increase. This enables the budworm population to keep increasing, albeit slowly; it could even take 20 years or longer to reach an extremely high density (figure 9.2 in Royama, 1992).

3.3.4 PERSISTENCE OF AN EPIDEMIC POPULATION

Why does a high-density budworm population tend to persist for as long as 8 or even more than 10 years? If the cycle was generated by a simple predator–prey interaction process, the population would start declining once the parasitoid complex caught up with the host at the peak of the cycle rather than being sustained at a plateau (Figure 3.2, bottom graph). A likely reason is that poor feeding conditions due to heavy defoliation at high larval density promotes moth dispersal (Delisle and Hardy, 1994; McNeil *et al.*, 1995). This slows down a further increase, if not causing a reduction, in the local budworm population (Royama, 1992). Furthermore, the host trees – spruces (*Picea* spp.) and balsam fir (*Abies balsamea*) – are capable of producing new shoots for several years even though they are defoliated completely every year (Piene, 1989, 1991; Ostaff and MacLean, 1989). Thus, high budworm populations are often maintained for many years.

3.3.5 CAUSE OF POPULATION DECLINE

The Green River life table study (Morris, 1963) revealed that a population decline was caused mainly by low survival from third instar larva to pupa. Nevertheless, no single factor (e.g. a particular species of parasitoid, predator or pathogen) played a determining role (Royama, 1992). Our ongoing study (E.S. Eveleigh, chief investigator, unpublished results) has confirmed this tendency during the most recent outbreak period (1980–1988). There was an overall increase in the collective action of several parasitoids that attacked late instar larvae and pupae. Most of these late parasitoids were bivoltine and required alternate (not alternative) hosts in which to overwinter. This implies that a decline in the budworm population depends a great deal on an increase in the complex of the bivoltine parasitoids, which, in turn, depends on an increase in their alternate hosts.

Carolin (1980), quoted by Mason (1987), noticed that a group of lepidoptera associated with the western spruce budworm (*Choristoneura occidentalis*) increased following a budworm increase. A similar situation may apply to the eastern spruce budworm (*C. fumiferana*) system, allowing the bivoltine parasitoids to increase and, eventually, enable the collective action of all natural enemies to bring down the budworm population (Royama, 1992).

Synchronous fluctuations among several forest insects have been noticed by other authors (Blitz and Ross, 1958; Harris *et al.*, 1982; Roland and Embree, 1995). Also, the graphs of the outbreak history of several pine-feeding lepidoptera in Germany, compiled by Klimetzek (1990; his figures 2, 3 and 4) shows a good degree of synchrony.

This is no coincidence in light of the 'Moran effect' (so called after Moran, 1953; Royama, 1984, 1992) which synchronizes independently fluctuating local populations of a given species, or even different species populations. Widespread outbreak of an insect can be viewed as being a result of synchrony among independently fluctuating local populations over large areas.

3.3.6 CYCLE SYNCHRONY AND WIDESPREAD OUTBREAKS: THE MORAN EFFECT

Suppose that several regional populations cycle in a similar manner due to the same (or similar) intrinsic, density-dependent process (e.g. a host–parasitoid interaction) though independently of one another. The Moran theorem stipulates that the phase of these independent cycles will be synchronized under density-independent factors if such factors are correlated between the localities. The important point here is that the cause of population cycles is independent of the cause of their synchrony.

One such correlated factor in spruce budworm is the density-independent component of recruitment rate. A major part of the recruitment rate is influenced by moth dispersal which depends on the degree of defoliation. However, dispersal is also influenced by the prevailing weather conditions at the time of moth flight (Greenbank *et al.*, 1980). Consequently, the recruitment rates of populations over considerable areas tend to fluctuate in unison, i.e. correlated with each other, even if the mean rates differ among the local populations (Royama, 1984). In other words, while the interaction with the natural-enemy complex is the principal cause of budworm population cycles, correlated fluctuations in recruitment rate tend to synchronize local budworm cycles, resulting in widespread outbreaks.

Fischlin (1983) conjectured that even a small rate of moth exchange between local populations of the larch budmoth (*Zeiraphera diniana*) can result in synchrony. This might be so. However, the Moran theorem, which does not depend on exchanges of individuals between populations by migration, provides an explanation, not only for within-species synchrony, but also for between-species synchrony if the species involved are similarly influenced by a certain extrinsic, density-independent factor.

The Moran theorem has been somewhat misunderstood by some authors who seemed to think that the particular effects of density-independent influences, such as catastrophic weather influences (Berryman, 1981; Barbour, 1990) or a particular sequence of weather conditions (Keith, 1974), were causes of synchronization. Such assumptions are unnecessary. There just needs to be a density-independent factor that is correlated among local habitats (Royama, 1992).

3.3.7 SUMMARY OF INTERPRETATION OF SPRUCE BUDWORM OUTBREAK PROCESSES

1. The basic outbreak cycle is largely determined by survival from third instar larvae to pupae. The factors involved are the complex of natural enemies, in particular, the complex of insect parasitoids.
2. The erratic fluctuation in recruitment rate, largely determined by the balance between immigration and emigration of egg-carrying moths, is a major perturbation of the basic cycle, influencing cycle amplitude.
3. A long cycle length is attributed to the efficacy of the parasitoid complex being reduced by high hyperparasitism, and also to uncertain availability of non-budworm, alternate hosts for bivoltine parasitoids.
4. Defoliation-dependent moth emigration and ability of host trees to withstand heavy defoliation tend to maintain the budworm population at a high level for several years.
5. A budworm population declines from epidemic to endemic state when many elememts of the natural enemy complex, particularly bivoltine insect parasitoids, happen to increase more or less simultaneously.
6. Widespread outbreaks occur because populations over a large area tend to be synchronized due, most probably, to the Moran effect – the effect of an extrinsic density-independent factor whose action is correlated between populations. Correlation in recruitment rate among the populations is a likely mechanism because moth dispersal is under the influence of correlated meteorological conditions.

3.4 HIDDEN ROLE OF A FACTOR

Section 3.3.2 mentioned that mortality of young spruce budworm larvae is not a direct cause of population cycles but it might play a hidden role. This is because the factor that does cause the cyclic trend (e.g. an interaction with natural-enemy complex) may depend on the larval density prior to the enemies' attack. If so, a change in the mean loss of young larvae might influence the density-dependent action of natural enemies which could, in turn, change the pattern of population fluctuation.

This may explain the difference in population cycles between spruce budworm and a closely related, ecologically similar species, the jack pine budworm, *Choristoneura pinus*. The two species even share many common parasitoids and pathogens (Stairs, 1960; Thomson, 1959; Nealis, 1991, 1995). However, an average cycle length in *C. pinus* tends to be much shorter – about 10 years (Volney, 1988) – compared with a longer than 30-year cycle in *C. fumiferana*. Nealis and Lomic (1994) attribute the shorter cycle of *C. pinus* to one significant difference in ecology between the two species. Second instar *C. pinus* larvae, after their spring emergence, must feed on the pollen cones of their host trees to ensure good survival, whereas *C. fumiferana* larvae can feed on old foliage without an adverse effect. When the initial density of the *C. pinus* larvae is reduced because of a poor production of pollen cones, parasitism could increase because of a higher ratio of parasitoids to host larvae. This could, in turn, reduce the cycle length of *C. pinus*. As Nealis and Lomic (1994) argue, the production of pollen cones in jack pine depends itself on the level of defoliation by *C. pinus* in previous seasons and, thus, the insect–host plant interaction contributes to the observed pattern of population cycles.

3.5 LIFE TABLES OF THE PINE LOOPER

A 60-year record of pine looper population fluctuations at Letzlingen, Germany (Schwerdtfeger, 1935; Varley, 1949), exhibits a persistent cyclic pattern (Figure 3.3a). A 30-year record from northern Scotland (Barbour, 1985, 1988, 1990) exhibits a similar pattern. A 15-year record from the Netherlands (Klomp, 1966) (Figure 3.3b) also shows a sign of cyclic pattern, although it is rather too short to be definite. Nonetheless, its pattern resembles those in some sections of similar length

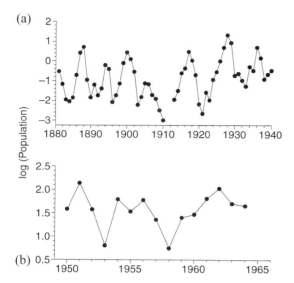

Klomp's (1966) study provides numerical life table data for the first 14 years. Figure 3.4, drawn using Klomp's data, consists of the intergeneration rate of change in egg density (R), partitioned into: H_1, survival from egg to first instar larva; H_2, larval survival to September; H_3, larval survival from September to October; H_4, larval survival from October to pupa in December; H_5, survival from pupa to emergence of moths in the following June; and H_r, the recruitment rate (the ratio: all eggs laid / all moths emerged).

Klomp's (1966) own representation, rearranged in Dempster (1975), consists of the conventional K-value (the negative value of the log generation survival, H_g, which is equal to the sum of H_1 to H_5) partitioned into five k-values (equal to the negative values of H_1 to H_5). It lacks what correspond to R and H_r in Figure 3.4. Incidentally, the straightforward H-values have been used rather than the conventional k-values as the negative values of corresponding Hs.

A glance at Figure 3.4 reveals no particular developmental stage exerting a determining influence on the temporal fluctuation in the rate of change in population, R. Thus, Dempster (1975) concluded that no single factor played a major role.

Figure 3.3 Annual changes in pine looper populations observed at (a) Letzlingen, Germany (after Varley, 1949) and (b) de Hoge, Veluwe, Netherlands (after Table 23 in Klomp, 1966).

in the Letzlingen and Scottish population cycles. A much less distinct pattern is exhibited by a population from Bavaria, Germany, and another from southwest Germany (Klimetzek, 1990).

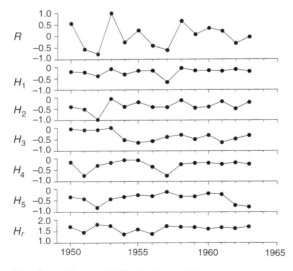

Figure 3.4 Pine looper life tables from Klomp (1966, his Table 23). R, annual rate of change in egg density; H_1, survival from egg to first instar larva; H_2, larval survival to September; H_3, larval survival from September to October; H_4, larval survival from October to pupa in December; H_5, survival from pupa to emergence of months; and H_r, recruitment rate; all in common logarithms.

However, if the rate of change of the population was partitioned into a number of stage-survival rates (Hs), the contribution of each H to R would become proportionately smaller because of the fact that R is the exact sum of all Hs. In other words, the more detailed is the life table, the less likely it is that a given H will play a key role. Therefore, a mere glance at Figure 3.4 does not give much insight.

Insight is gained, however, if we plot each H-value against R (Figure 3.5): most H-values that correspond to $R > 0$ (indicating a population increase) tend to be above average. An exception is H_3, though the tendency holds true for its values corresponding to the four highest R-values. This means that a comparatively large increase in the population was a result of above-average recruitment rate and survival at all stages. This, in turn, implies a simultaneous reduction in the adverse actions of all major factors that influence survival and recruitment throughout the looper's life cycle.

In contrast, among the H-values corresponding to $R < 0$ (indicating a population decrease), H_1, H_2 and H_4 (left-hand side of Figure 3.5) are mostly low for the lower range of $R < 0$, whereas H_3, H_5 and H_r (right-hand side) do not show that tendency. Thus, only the group on the left-hand side consistently contributed to a large decline in population whenever it happened. This means that a population decrease is not a result of an increase in adverse effects of all factors involved. Also, the major cause of population decline in one year is not necessarily the same in another year's decline. The above analysis confirms Dempster's (1975) earlier conclusion that no single factor played a dominant role in determining the dynamics of the Dutch population.

Klomp and Gruy (1965) argued that 'in the pine looper numbers are most probably regulated through mutual interference between larvae' and Klomp (1966) asserted: 'The results of the present study clearly support the theory of self-regulation more than that of the parasite–predator thesis.' In

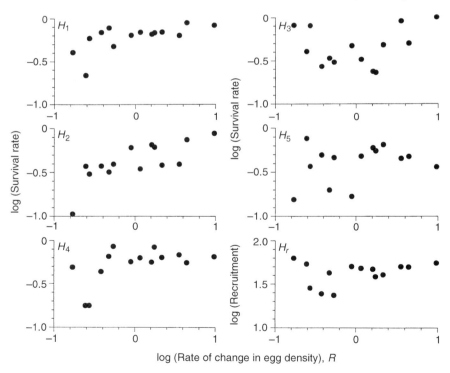

Figure 3.5 Plots of H_1–H_5 and H_r in Figure 3.4 against R.

contrast, Barbour (1988) argued that pupal parasitism was a major cause of the Scottish population cycle, and Dempster (1975) insisted on 'the combined effect of a large array of [unspecified] factors'.

Klomp's assertion is not convincing. A 'self-regulation' mechanism is essentially a first-order density-dependent process: in terms of the relationship in Figure 3.1, this is a process in which, given the density at generation $t - X(t)$, say – the density at the following generation, $X(t+1)$, is uniquely determined, save the influence of random variation. In contrast, for a population to be cyclic, there must be at least two solutions for $X(t+1)$, given $X(t)$: one greater (when the population is increasing), and the other less (when decreasing), than $X(t)$ (for details, see Royama, 1992). Thus, a self-regulating process, as a first-order density-dependent process, is unlikely to generate a series of population cycles as in Figure 3.4.

On the other hand, Klomp's life tables for the Dutch population are not consistent with Barbour's (1988) observation that pupal parasitism was a major cause of the Scottish population cycles. In fact, the rate of change in the Dutch population (R) is only poorly correlated with pupal survival (H_5) in Figure 3.5.

As far as the analysis in Figure 3.5 implies, it was mortality factors in other developmental stages, H_1, H_2 and H_4, that were more consistent contributors to the observed pattern of the Dutch population. The cyclical pattern, as exhibited by the Letzlingen and Scottish populations, is perhaps basically a result of the combined effects of many natural enemies at all stages of the host life cycle, as in the eastern spruce budworm. Unfortunately, the effects of many natural enemies were not adequately quantified in Klomp's study. The information available for the pine looper at the moment is still insufficient for drawing a conclusion, despite the fact that considerable time and resources have been invested in the study.

3.6 DISCUSSION

There seems to be no simple answer to the question: is it single or multiple factors that govern forest insect population dynamics? There are many aspects that are of interest in population dynamics, although a particular aspect may be controlled by one factor. But we must consider many aspects to understand, say, the cause(s) of the cyclical occurrence of outbreaks. As exemplified by the case of spruce budworm, these aspects could be length, amplitude, synchrony of a cycle, etc. These aspects are governed by different factors, and the same probably applies to other forest insects.

Since the early development of population ecology, there have been diverse theories concerning the dynamics of populations, e.g. climatic control theories, biotic regulation theories, and comprehensive theories (Solomon, 1949; Ito, 1978). Most of these theories were concerned, basically, with broad, qualitative aspects of population dynamics, especially the regulation of populations. Also, most of them relied on intuitive arguments rather than on rigorous analyses of quantitative data, and this resulted in continuing controversies. If we conduct long-term, comprehensive studies, there would be less room for speculation.

The most difficult problem we face, under the present unfavourable economic situation, is the logistics of conducting a long-term study to obtain comprehensive data. Without such data, speculations and conjectures will continue. We must somehow find a way to break this vicious circle.

ACKNOWLEDGEMENTS

I am grateful to Vince Nealis, Dan Quiring and Eldon Eveleigh, whose comments and advice were useful for improving the manuscript.

REFERENCES

Anderson, R.M. and May, R.M. (1980) The population dynamics of microparasites within invertebrate hosts. *Philosophical Transactions of the Royal Society London, Series B* **291**, 451–524.

Baltensweiler, W., Benz, G., Bovey, P. and Delucchi, V. (1977) Dynamics of larch budmoth populations. *Annual Review of Entomology* **22**, 79–100.

Baltensweiler, W. and Fischlin, A. (1988) The larch budmoth in the Alps. In *Dynamics of Forest Insect Populations* (ed. A.A. Berryman), pp. 331–351, Plenum Press, New York.

Barbosa, P. and Schultz, J.C. (eds) (1987) *Insect Outbreaks*, Academic Press, London.

Barbour, D.A. (1985) Patterns of population fluctuation in the pine looper moth *Bupalus piniaria* L. in Britain. In *Site Characteristics and Population Dynamics of Lepidopteran and Hymenopteran Forest Pests* (eds D. Boven and J.T. Stockley), Forestry Commission Research and Development Paper 135, pp. 8–20.

Barbour, D.A. (1988) The pine looper in Britain and Europe. In *Dynamics of Forest Insect Populations* (ed. A.A. Berryman), pp. 291–308, Plenum Press, New York.

Barbour, D.A. (1990) Synchronous fluctuations in spatially separated populations of cyclic forest insects. In *Population Dynamics of Forest Insects* (eds A.D. Watt, S.R. Leather, M.D. Hunter and N.A.C. Kidd), pp. 339–346, Intercept, Andover.

Bellows, T.S. Jr, van Driesche, R.G. and Elkinton, J.S. (1992) Life-table construction and analysis in evaluation of natural enemies. *Annual Review of Entomology* 37, 587–614.

Berryman, A.A. (1981) *Population Systems*, Plenum Press, New York.

Berryman, A.A. (ed.) (1988) *Dynamics of Forest Insect Populations*, Plenum Press, New York.

Berryman, A.A. (1995) Population cycles: a critique of the maternal and allometric hypotheses. *Journal of Animal Ecology* 64, 290–293.

Blitz, W.E. and Ross, D.A. (1958) Population trends of some common loopers (Geometridae) on douglas-fir, 1949–1956, in the Okanagan–Shuswap area. *Forest Biology Division, Ottawa, Bi-monthly Progress Report* 14, 2–3.

Carolin, V.M. (1980) Larval densities and trends of insect species associated with spruce budworms in buds and shoots in Oregon and Washington. *US Forest Service, Research Paper, PNW* 273, 1–18.

Delisle, J. and Hardy, M. (1994) Effects of larval food quality on *Choristoneura fumiferana* reproductive biology. *Annual Meeting of the Entomological Society of Canada, Winnipeg, 15–19 October, 1994.*

Dempster, J.P. (1975) *Animal Population Ecology*, Academic Press, London.

Dempster, J.P. (1983) The natural control of populations of butterflies and moths. *Biological Review* 58, 461–481.

Eveleigh, E., McCarthy, P., Pollock, S. *et al.* (1994) Hyperparasitism in populations of eastern spruce budworm, *Choristoneura fumiferana* (Clem.), in New Brunswick. In *Proceedings of the Combined Meeting of the Northeastern Forest Pest Council and the Northeastern Forest Insect Work Conference, Manchester, New Hampshire, March 21–23, 1994*, p. 23.

Fischlin, A. (1983) Modelling of alpine valleys, defoliated forests, and larch budmoth cycles: the role of moth migration. In *Mathematical Models of Renewable Resources* (ed. R. Lamberson), Vol. II, pp.102–104, Proceedings of the 2nd Pacific Coast Conference on Mathematical Modelling of Renewable Resources, University of Victoria, Victoria, British Columbia, Canada.

Ginsburg, L.R. and Taneyhill, D.E. (1994) Population cycles of forest lepidoptera: a maternal effect hypothesis. *Journal of Animal Ecology* 63, 79–92.

Greenbank, D.O. (1963) The development of the outbreak. In *The Dynamics of Epidemic Spruce Budworm Populations* (ed. R.F. Morris), *Memoirs of the Entomological Society of Canada*, 31, 19–23.

Greenbank, D.O., Schaefer, G.W. and Rainey, R.C. (1980) Spruce budworm (Lepidoptera: Tortricidae) moth flight and dispersal: new understanding from canopy observations, radar, and aircraft. *Memoirs of the Entomological Society of Canada*, **110**.

Harris, J.W.E., Dawson, A.F. and Brown, R.G. (1982) The western hemlock looper in British Columbia 1911–1980. *Canadian Forestry Service Information Report BC-X-***234**, 3–18.

Hassell, M.P. (1978) *The Dynamics of Arthropod Predator–Prey Systems*, Princeton University Press, Princeton, New Jersey.

Hassell, M.P. (1985) Insect natural enemies as regulating factors. *Journal of Animal Ecology* 54, 323–334.

Haukioja, E. and Neuvonen, S. (1987) Insect population dynamics and induction of plant resistance: the testing of hypotheses. In *Insect Outbreaks* (eds P. Barbosa and J.C. Schultz), pp. 411–432, Academic Press, London.

Ito, Y. (1978) *Comparative Ecology*, Cambridge University Press, Cambridge.

Keith, L.B. (1974) Some features of population dynamics of mammals. *Proceedings of the International Congress of Game Biologists* 11, 17–58.

Klimetzek, D. (1990) Population dynamics of pine-feeding insects: a historical study. In *Population Dynamics of Forest Insects* (eds A.D. Watt, S.R. Leather, M.D. Hunter and N.A.C. Kidd), pp. 3–10, Intercept, Andover.

Klomp, H. (1966) The dynamics of a field population of the pine looper, *Bupalus piniaria* L. (Lep., Geom.). *Advances in Ecological Research* 3, 207–305.

Klomp, H. and Gruys, P. (1965) The analysis of factors affecting reproduction and mortality in a natural population of the pine looper, *Bupalus piniarius* L. In *Proceedings of the 12th International Congress of Entomology, London, 1964*, pp. 369–372.

Martinat, P.J. (1987) The role of climatic variation and weather in forest insect outbreaks. In *Insect*

Outbreaks (eds P. Barbosa and J.C. Schultz), pp. 241–268, Academic Press, London.

Mason, R.P. (1987) Nonoutbreak species of forest lepidoptera. In *Insect Outbreaks* (eds P. Barbosa and J.C. Schultz), pp. 31–57, Academic Press, London.

McNeil, J.N., Cusson, M., Delisle, J. *et al.* (1995) Physiological integration of migration in Lepidoptera. In *Migration: Physical Factors and Physiological Mechanisms* (eds V.A. Drake and A.G. Gathehouse), pp. 279–302, Cambridge University Press, Cambridge.

Moran, P.A.P. (1953) The statistical analysis of the Canadian lynx cycle. II Synchronisation and meteorology. *Australian Journal of Zoology* **1**, 291–298.

Morris, R.F. (ed.) (1963) *The Dynamics of Epidemic Spruce Budworm Populations. Memoirs of the Entomological Society of Canada* **31**.

Morris, R.F. and Miller, C.A. (1954). The development of life tables for the spruce budworm. *Canadian Journal of Zoology* **32**, 283–301.

Myers, J.H. (1988) Can a general hypothesis explain population cycles of forest lepidoptera? *Advances in Ecological Research* **18**, 179–242.

Myers, J.H. (1993) Population outbreaks in forest lepidoptera. *American Entomologist* **81**, 240–251.

Nealis, V.G. (1991) Parasitism in sustained and collapsing populations of the jack pine budworm, *Choristoneura pinus* Free. (Lepidoptera: Tortricidae), in Ontario, 1985–1987. *Canadian Entomologist* **123**, 1065–1075.

Nealis, V.G. (1995). Population biology of the jack pine budworm. In *Proceedings of the Jack Pine Budworm Symposium, 1995, Winnipeg*. Canadian Forest Service Information Report, Nor-X-342, pp. 55–71.

Nealis, V.G. and Lomic, P.V. (1994) Host-plant influence on the population ecology of the jack pine budworm, *Choristoneura pinus* (Lepidoptera: Tortricidae). *Ecological Entomology* **19**, 367–373.

Ostaff, D.P. and MacLean, D.A. (1989) Spruce budworm population, defoliation, and changes in stand condition during uncontrolled spruce budworm outbreak on Cape Breton Island, Nova Scotia. *Canadian Journal of Forest Research* **19**, 1077–1086.

Park, T. (1949) Populations. In *Principles of Animal Ecology* (W.C. Allee, A.E. Emerson, O. Park, T. Park, and K.P. Schmidt), pp. 263–435, W.B. Saunders, London.

Piene, H. (1989) Spruce budworm defoliation and growth loss in young balsam fir: defoliation in spaced and unspaced stands and individual tree survival. *Canadian Journal of Forest Research* **19**, 1211–1217.

Piene, H. (1991) The sensitivity of young white spruce to spruce budworm defoliation. *Northern Journal of Applied Forestry* **8**, 168–171.

Podoler, H. and Rogers, D. (1975) A new method for the identification of key factors from life-table data. *Journal of Animal Ecology* **44**, 85–114.

Price, P.W. (1987) The role of natural enemies in insect populations. In *Insect Outbreaks* (eds P.Barbosa and J.C. Schultz), pp. 287–312, Academic Press, London.

Price, P.W. (1994) Phylogenetic constraints, adaptive syndromes, and emergent properties: from individual to population dynamics. *Researches on Population Ecology* **36**, 3–14.

Price, P.W., Cobb, N., Craig, T.P. *et al.* (1990). Insect herbivore population dynamics on trees and shrubs: new approaches relevant to latent and eruptive species and life table development. In *Insect–Plant Interactions*, Vol. II (ed. E.A. Bernays), pp. 1–38, CRC Press, Boca Raton, Ann Arbor, Boston, Mass.

Roland, J. and Embree, D.G. (1995) Biological control of the winter moth. *Annual Review of Entomology* **40**, 475–492.

Royama, T. (1978) Do weather factors influence the dynamics of spruce budworm populations? *Canadian Forestry Service Bi-monthly Research Notes* **34**, 9–10.

Royama, T. (1981) Evaluation of mortality factors in insect life table analysis. *Ecological Monographs* **51**, 495–505.

Royama, T. (1984) Population dynamics of the spruce budworm *Choristoneura fumiferana*. *Ecological Monographs* **54**, 429–462.

Royama, T. (1992) *Analytical Population Dynamics*, Chapman & Hall, London.

Royama, T. (1996) A fundamental problem in key factor analysis. *Ecology* **77**, 87–93.

Schwerdtfeger, F. (1935) Studien uber den Massenwechsel einiger Forstschadlinge. I. Das Klima der Schadgebiete von *Bupalus piniarius*, *Panolis flammea* und *Dendrolimus pini* in Deutschland. *Zeitschrift fur Forst- und Jagdwissenschafften* **67**, 15–38, 85–104, 449–82, 513–540.

Solomon, M.E. (1949) The natural control of animal populations. *Journal of Animal Ecology* **18**, 1–35.

Southwood, T.R.E. (1978) *Ecological Methods*, 2nd edn, Chapman & Hall, London.

Stairs, G.R. (1960) Infection of the jack pine budworm, *Choristoneura pinus* Freeman, with a nuclear polyhedrosis virus of the spruce budworm, *Choristoneura fumiferana* (Clemens), (Lepidoptera: Tortricidae). *Canadian Entomologist* **92**, 906–908.

Swetnam, T.W. and Lynch, A.M. (1993). Multicentury, regional-scale patterns of western spruce budworm outbreaks. *Ecological Monographs* **63**, 399–424.

Taylor, R.J. (1984) *Predation*, Chapman & Hall, London.

Thomson, A.J., Shepherd, R.F., Harris, J.W.E. and Silverside, R.H. (1984) Relating weather to outbreaks of western spruce budworm *Choristoneura occidentalis* (Lepidoptera: Tortricidae) in British Columbia. *Canadian Entomologist* **116**, 375–381.

Thomson, H.M. (1959) A microsporidian infection in the jack-pine budworm, *Choristoneura pinus* Free. *Canadian Journal of Zoology* **37**, 117–120.

Utida, S. (1957) Cyclic fluctuation of population density intrinsic to the host–parasite system. *Ecology* **38**, 442–449.

Varley, G.C. (1949) Population changes in German forest pests. *Journal of Animal Ecology* **18**, 117–122.

Varley, G.C. and Gradwell, G.R. (1960) Key factors in population studies. *Journal of Animal Ecology* **29**, 399–401.

Varley, G.C. and Gradwell, G.R. (1963) The interpretation of insect population changes. *Proceedings of the Ceylon Association for the Advancement of Science* **18**, 142–156.

Varley, G.C. and Gradwell, G.R. (1970) Recent advances in insect population dynamics. *Annual Review of Entomology* **15**, 1–24.

Volney, W.J.A. (1988) Analysis of historic jack pine budworm outbreaks in the Prairie provinces of Canada. *Canadian Journal of Forest Research* **18**, 1152–1158.

Watt, A.D., Leather, S.R., Hunter, M.D. and Kidd, N.A.C. (eds) (1990) *Population Dynamics of Forest Insects*, Intercept, Andover.

Wellington, W.G., Fettes, J.J., Turner, K.B. and Belyea, R.M. (1950) Physical and biological indicators of the development of outbreaks of the spruce budworm. *Canadian Journal of Research, D* **28**, 308–331.

THE IMPACT OF PARASITOIDS AND PREDATORS ON FOREST INSECT POPULATIONS

Neil A.C. Kidd and Mark A. Jervis

4.1 INTRODUCTION

The subject of forest insects remains one of the major potential sources of long-term data with which to explore the population dynamics of insects and role of parasitoids and predators in suppressing and regulating numbers – we use the term 'forest insects' loosely to include phytophagous insects of forest, plantation and woodland trees, though we occasionally broaden the definition even further where some data are difficult to separate. Forest insects are good candidates for examining the factors responsible for low population levels, given the apparent low frequency of pests amongst tree insect species. [We accept that insects may be accorded pest status without them being necessarily abundant (Southwood, 1977; Conway, 1976) but ignore this complication for the sake of argument.] Using the UK tree insect fauna (from data in Winter, 1983, and Bevan, 1983) as a not necessarily representative sample, we find that only around 10% (143 out of 1405) of forest insects constitute either minor or major pests. This raises two central questions:

1. How important are parasitoids and predators in suppressing and regulating forest insect populations? Are they less or more important than those in other ecosystems?

2. What is the relative importance of top-down (parasitism and predation) and bottom-up (plant resource constraints) regulation?

This chapter examines the role of parasitoids and predators in forest insect population dynamics, and highlights some of the problems and pitfalls that constrain progress. First, it briefly reviews those sources of evidence traditionally used to support the view that such natural enemies have a major role to play in the dynamics of forest insects: classical biological control introductions; predator–prey and parasitoid–host models; and field studies. Where possible, forest insect examples are emphasized. No attempt is made to include the wide literature on forest insect pathogens in this chapter, though this group of natural enemies can also have a major impact on forest insect population dynamics (e.g. Berryman *et al.*, 1990).

4.2 EVIDENCE FROM BIOLOGICAL CONTROL

Some of the best demonstrations of the impact of parasitoids and predators are provided by cases of 'classical' biological control, and forestry has provided its share of examples. Some biological control introductions against forest insects that have resulted in successful pest control are listed in Table 4.1.

Forests and Insects. Edited by A.D. Watt, N.E. Stork and M.D. Hunter. Published in 1997 by Chapman & Hall, London. ISBN 0 412 79110 2.

Table 4.1 Classical biological control introductions that have resulted in successful control of forest, woodland or plantation pests.

Pest	Agent	Crop	Country	Result	Ref
Homoptera					
Adelges piceae and *A.nordmannianae* (Adelgidae)	*Scymnus impexus* (Coccinellidae)	Silver fir	Sweden	S	7
Asterolecanium variolosum (Asterolecaniidae)	*Habrolepis dalmani* (Encyrtidae)	Oak	Tasmania Chile	S C	22 24
Cavariella aegopodii (Aphididae)	*Aphidius sp.* (Braconidae)	Willow	Australia	C	3
Cinara fresai (Lachnidae)	*Pauesia cinaravora* (Braconidae)	Pines	S.Africa	S	16
Gossyparia spuria (Eriococcidae)	*Coccophagus insidiator* (Aphelinidae)	Elm	USA	S	3
Myzocallis annulatus (Callaphididae)	*Aphelinus flavus* (Aphelinidae)	Oak	Tasmania	S	8
Pineus boerneri (Adelgidae)	*Leucopis obscura*	*Pinus radiata*	Chile	S	24
	L.tapiae (Chamaemyiidae)	Pines	N.Zealand	S	23
Pineus pini (Adelgidae)	*Leucopis obscura* (Chamaemyiidae)	Conifers	Hawaii	S	4
Lepidoptera					
Coleophora laricella (Coleophoridae)	*Bassus pumila* (Braconidae) *Chrysocharis laricinellae* (Eulophidae)	Larch	Canada USA	C& S&	9 17
Evagora starki (Gelechiidae)	*Dicladocerus sp.* (Coccinellidae)	Lodgepole pine	Canada	S	9
Lymantria dispar (Lymantriidae)	*Ooencyrtus kuwanae* (Encyrtidae)	Oak	Spain	P	7
Operophtera brumata (Geometridae)	*Agrypon flaveolatum* (Ichneumonidae) *Cyzenis albicans* (Tachinidae)	Deciduous trees	Canada	C&	2
Oxydia trychiata (Geometridae)	*Telenomus alsophilae* (Scelionidae)	Cypress	Colombia	C	1

Host (Family)	Parasitoid (Family)	Tree	Location	Code	Ref
Phyllonorycter messaniella (Gracillariidae)	*Pholetesor circumscriptus* (Braconidae) *Achrysocharoides splendens* (Eulophidae)	Oak	N.Zealand	C&	20
Stilpnotia salicis (Lymantriidae)	*Cotesia melanoscela* (Braconidae) *Meteorus versicolor* (Braconidae)	*Populus, Salix*	Canada USA USA	S S& S&	2 18 18
Coleoptera					
Dendroctonus micans (Scolytidae)	*Rhizophagus grandis* (Rhizophagidae)	Spruce	CIS	P	5
Gonipterus scutellatus (Curculionidae)	*Anaphes nitens* (Mymaridae)	*Eucalyptus*	Kenya S. Africa St.Helena Madagascar Mauritius Italy France N.Zealand	S C S C S S P S	6 6 6 6 6 14 14 12
Ips grandicollis (Scolytidae)	*Roptrocerus xylophagorum* (Pteromalidae)		Australia	P	11
Trachymela tincticollis (Chrysomelidae)	*Enoggera reticulata* (Pteromalidae)	*Eucalyptus*	S.Africa	C	21
Hymenoptera					
Heterarthrus nemoratus (Tenthredinidae)	*Chrysocharis laricinellae* (Eulophidae) *Phanomeris phyllotomae* (Braconidae)	Birch	USA	S	3
Gilpinia hercyniae (Diprionidae)	*Exenterus abruptorius* *E.amictorius* *E.confusus* *E.vellicatus* (Ichneumonidae)	Spruce	Canada	S&	9
Neodiprion sertifer (Diprionidae)	*Dahlbominus fuscipennis* (Eulophidae)	Pine	Canada	P&	9
Pristiphora geniculata (Tenthredinidae	*Olesicampe geniculata* (Ichneumonidae)	Mountain ash	Canada	S	15
Sirex noctilio (Siricidae)	*Ibalia leucospoides* (Ibaliidae) *Megarhyssa nortoni* (Ichneumonidae) *Rhyssa persuasoria* (Ichneumonidae)	Conifers	N.Zealand Tasmania	S& P&	13 11,19

Pest	Agent	Crop	Country	Result	Ref
Diptera					
Phytomyza ilicis (Agromyzidae)	*Epilampsis gemma* (Eulophidae) *Sphegigaster flavicornis* (Pteromalidae)	Holly	Canada	P	10

Abstracted from the BIOCAT database, in which the degree of control is categorized as follows (definitions from DeBach, 1971):
 C = complete control, i.e. the need for further control methods is virtually eliminated over large areas;
 S = substantial control, i.e. economic savings are less because either the crop is less important, or the control area is smaller, or occasional insecticide treatment is necessary;
 P = partial control i.e. the introduction has resulted in a reduction in the frequency of outbreaks and/or the frequency of pesticide applications.
 C&, S&, P&: multiple agent introductions.
References: (1) Bustillo and Drooz (1977); (2) Canada (1971); (3) Clausen (1978); (4) Culliney *et al.* (1988); (5) Evans (1987); (6) Greathead (1971); (7) Greathead (1976); (8) Luck (1981); (9) McGugan and Coppel (1962); (10) McLeod (1962); (11) Morgan (1989); (12) Nuttall (1989a); (13) Nuttall (1989b); (14) OPIE (1986); (15) Quednau (1990); (16) van Rensburg (1992); (17) Ryan (1990); (18) Sailer (1983); (19) Taylor (1978); (20) Thomas and Hill (1989); (21) Tribe (1992); (22) Wilson (1960); (23) Zondag and Nuttall (1989); (24) Zuniga (1985).

The outcomes of classical biological control introductions are, as in other 'crop' systems, currently classified by some practitioners according to whether they failed (no establishment of agent), or resulted in establishment of the agent without any control, or with (a) complete control (the need for other control methods virtually eliminated over large areas), (b) substantial control (economic savings somewhat less, because either the crop was less important, or the area controlled was smaller or else occasional insecticide treatment was necessary) or (c) partial control (the frequency of pest outbreaks reduced and/or the frequency of pesticide application) (DeBach, 1971) (Table 4.1). Ecologists, by contrast, have employed q-values (mean pest density following introduction of natural enemy divided by the mean pest density before introduction) to indicate the degree of pest suppression (Beddington *et al.*, 1978; Lawton and McNeill, 1979).

Neither the three control categories (a)–(c) listed above, nor q-values tell us much about the dynamics of the natural enemy–pest interactions. Sadly, with the exception of some programmes such as those against the winter moth in Canada (Embree and Roland, 1994) and the larch casebearer in the USA (Ryan, 1990), sufficiently detailed and precise quantitative data to provide meaningful interpretations of natural enemy impact are generally lacking. Even at the most fundamental experimental level – namely, the use of control plots – some studies are found wanting. If control plots are not used, depression of the pest population cannot necessarily be attributed to the established natural enemy: agent introduction and pest suppression may be merely coincidental (Gould *et al.*, 1992a).

The major constraints on obtaining adequate data from biological control introductions are funding and the rapidity with which some pest problems arise (Waage and Mills, 1992), preventing the more traditional (and desirable) 'before and after' studies. One way of partly compensating for such constraints is to examine within-season population changes and/or changes in pest age structure immediately following introduction (Gould *et al.*, 1992a,b, on ash whitefly). However, such an approach may be limited in what it can tell us about natural enemy impact in the long term, which may be significant, as in some cases several generations may elapse before a new stable state is achieved (Roland, 1994).

4.3 EVIDENCE FROM DEDUCTIVE MODELS

There is no doubt that deductive modelling has led to an enormous improvement in our understanding of the potential intricacies of parasitoid–host and predator–prey interactions. Some of the main features that have been examined so far, and their potential effects on stability and equilibrium levels, are reviewed by Hassell (1978), Waage and Hassell (1982), Hassell and Waage (1984) and Kidd and Jervis (1996). What should be noted at this stage is that a wide range of natural enemy attributes are predicted to have a stabilizing influence on host/prey population dynamics. In particular, non-random searching behaviour and spatial heterogeneity in parasitism have been given a great deal of attention.

Recently, insect ecologists concerned with the role that parasitoids and predators play in population dynamics have broadened their perspective, taking account of tri-trophic interactions between the host plant, the herbivore and its natural enemies (Price 1992; Schultz, 1992; Gutierrez et al., 1990, 1994; see also Lawton and McNeill, 1979). A useful deductive modelling framework, hitherto overlooked by most proponents of the tri-trophic approach, is the 'synoptic population model' of Southwood and Comins (1976), developed for the purpose of comparing different herbivore species and herbivore types with respect to their population growth characteristics and habitat stability (see also Berryman, 1978, for another such model designed to predict insect epidemics). Food plant constraints are not explicitly included in this model, but net population growth, in the absence of natural enemies, can be taken as a direct correlate.

In the synoptic model so-called r-strategists – those with rapid growth rates and high dispersal capability, adapted to unstable habitats – are unlikely to be strongly influenced by natural enemies, with respect to the theoretical population growth contour (Figure 4.1a). Such species have a poorly developed 'natural enemy ravine' – associated natural enemies have numerical responses that are too weak to suppress prey population growth, and host/prey numbers are instead constrained by availability of resources at the 'crash valley' side of the graph. At the other extreme are K-strategists – those with slow population growth rates and low dispersal capability, adapted to stable habitats and maintaining population densities at or close to the carrying capacity, and therefore by definition unaffected by natural enemies to any significant degree.

Between these extremes lie intermediate species characterized by a pronounced natural enemy ravine. The natural enemies of such species are likely to have numerical responses sufficient to suppress prey populations below the zero growth isocline, producing a region on the graph (Figure 4.1b) bounded by local stability at S and U. If, however, the natural enemy complexes are disturbed, or prey populations are translocated to areas where natural enemies are absent, intermediate species escape the natural enemy ravine (at unstable U) to occupy the 'epidemic ridge' of pest status. Only after reinstatement of natural enemies can the pest be returned to the stability of the natural enemy ravine. Southwood (1977) considered forest insect pests to lie in the intermediate part of the r–K continuum, with habitat stability perhaps pushing them towards the K-end. Certainly, the results of classical biological control introductions (section 4.1) fit quite well with the above interpretation of natural enemy impact.

The double or 'coincident' equilibrium view, implicit in the synoptic model, has also been invoked to explain the periodic outbreaks shown by many forest insect pests (e.g. Mason and Wickman, 1988; Mattson et al., 1988; Montgomery and Wallner, 1988). Although natural enemies may regulate at least some of these populations whilst they are in the low density endemic phase (e.g. birds for spruce budworm, Holling et al., 1976; Peterman et al., 1979) other factors (biotic or abiotic) may periodically conspire to raise population growth sufficiently to lift numbers into the epidemic phase, where serious damage can occur. The dynamics of the pine sawfly, Neodiprion sertifer, for example, have recently been interpreted in this way (Hanski, 1987; Larsson et al., 1993). With this species, low density populations appear to be regulated by pupal predators such as small mammals. Larsson et al. (1993) have suggested that the strong defensive reactions shown by the gregarious larvae may promote survival and lift the average population growth rate to levels close to the unstable 'escape point', such that even slight perturbations to egg or larval survival may be sufficient to initiate an outbreak. One factor which could be significant in this

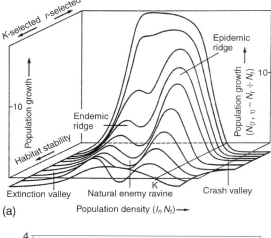

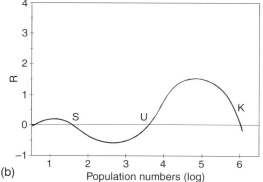

Figure 4.1 (a) Representation of synoptic population model of Southwood and Comins (1976) (see text for explanation). (b) Section through population growth contour of (a) to illustrate lower and upper stability boundaries (S and U, respectively) lying within the 'natural enemy ravine' and the high density resource equilibrium (K).

respect is the concentration of resin acids in the needles, which may vary considerably between trees and sites. High concentrations of these acids reduce larval survival and may reduce the risk of outbreak, by depressing population growth rates well below the 'escape point'. Supportive evidence for this interpretation comes from the fact that outbreaks are more frequent on poor soils (Larsson and Tenow, 1984), which in turn are more likely to be associated with low resin acid concentrations (Bjorkman *et al.*, 1991).

Although regulation by natural enemies (especially generalist predators) has been proposed by Southwood and Comins (1976) and others to account for the 'natural enemy ravine' and local stability at S (Figure 4.1b), other explanations are possible. Berryman (1978), for example, using similar topographical analogies to those of Southwood and Comins (1976), refers to the sub-zero growth isocline as 'lake equilibrium' and describes how some bark beetle populations can be regulated within this region by competition for suitable host trees. Once the populations attain a certain threshold density ($= U$) they can overcome host resistance and outbreak as more trees become susceptible as breeding sites. The important point to bear in mind is that 'coincident equilibria' at two distinct population densities may be derived in a number of ways.

4.4 EVIDENCE FROM FIELD STUDIES

Various techniques have been used to assess the impact of parasitoids and predators on insect populations in the field (review: Kidd and Jervis, 1996), but by far the most important has been the suite of methods associated with key factor analysis of life table data. Compared with those associated with arable and horticultural crops, tree insect populations make good subjects for key factor analysis, being relatively persistent in the same long-lived habitats over many generations. Of the 65 analyses examined by Price (1987) and Stiling (1988), 26 (35%) are concerned with tree-feeding insects. Among these, density-dependent mortality from natural enemies contributed to regulation in eight cases, whilst in 11 cases natural enemies (usually parasitoids) were key factors driving changes in abundance between generations. Given the classical delayed density-dependent action of many parasitoid species, their action as key factors is perhaps not too surprising. A more recent and more comprehensive analysis of life table data recorded 70 tree-dwelling species out of a total number of 124 species studied (56.4%) (Cornell and Hawkins, 1995). This analysis provided no evidence to implicate natural enemy mortality as a major factor affecting survivorship in late successional species (including those associated with forests and plantations) compared with early successional species (see also section 4.7). Unfortunately, no assessment of density-dependence was carried out, so it is not possible to determine the frequency of natural enemy regulation from the forest insects cases cited.

Despite the undoubted popularity of the key factor analysis technique, it has a number of recognized limitations and drawbacks (Kidd and Jervis, 1996). For example, the use of key factor analysis to determine the frequency of density-dependent regulation by natural enemies (Dempster, 1983; Dempster and Pollard, 1986) has led to concern about how sensitive or appropriate the technique is for detecting density-dependence. Strongly regulated populations may show little variation from equilibrium, making statistical detection difficult (Gould et al., 1990), while stochastic variation may obscure real density-dependent relationships altogether (Hassell, 1985, 1987). Moreover, density-dependence on a spatial, as opposed to a temporal, scale will not be detected, unless the sampling programme is specifically designed for this purpose (Hassell, 1987).

How much do natural enemies contribute to maintaining low stable populations and to suppressing outbreaks? With all three lines of evidence concerning the role of natural enemies – biological control, deductive models and field studies – there has been a tendency to place emphasis on the importance of regulation by natural enemies, but there are other hypotheses to account for low, stable populations. These are examined in the next two sections.

4.5 IS REGULATION BY NATURAL ENEMIES NECESSARY?

It should not be forgotten that non-regulating, natural enemy mortality (i.e. temporally density-independent) may also have an important role to play in insect population dynamics. For example, in interpreting the successful outcomes of classical biological control introductions, regulation by natural enemies (as in Figure 4.2a) does not necessarily have to be invoked to explain successful control;

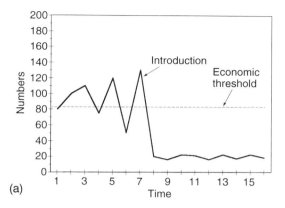
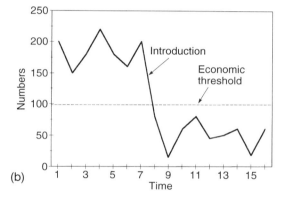
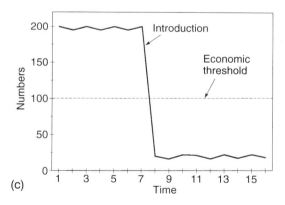

Figure 4.2 Three possible interpretations of results of classical biological control introductions: (a) conventional view; (b, c) two alternative interpretations (see text for further explanation).

mere suppression of numbers can have the same effect. There are three reasons for this:

56 *Impact of parasitoids and predators on forest insects*

1. Economic success may not necessarily rely on a stable equilibrium, as long as the dynamic profile of the pest after natural enemy introduction does not exceed the economic threshold (Figure 4.2b).
2. The pest may already be stabilized by density-dependent factors prior to natural enemy introduction, at a level above the economic threshold, and this stabilization may be maintained after introduction (Figure 4.2c). (Note that (1) and (2) probably apply to a small minority of cases.)
3. A density-independent mortality by natural enemies may suppress pest numbers sufficiently for them to be regulated by other density-dependent factors, a point made with the aid of a simple analytical model by Kidd and Lewis (1987) for an ecologically comparable vertebrate predator–prey example (Figure 4.3).

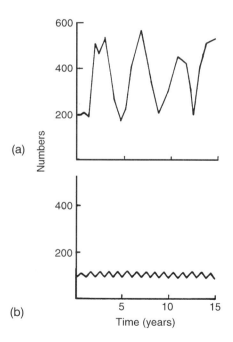

Figure 4.3 Population dynamics of theoretical vole population simulated by time-delay logistic model and with carrying capacity alternating between 200 and 600 individuals every 6 months. (a) No additional density-independent mortality applied; (b) a 60% mortality applied each time-step of 2 months. (From Kidd and Lewis, 1987.)

Furthermore, Roland (1994) suggested that the real contribution by introduced parasitoids to the control of the winter moth in British Columbia may be simply to provide sufficient mortality for other regulating factors to come into play (see the alternative explanation of Bonsall and Hassell, 1995, and the reply by Roland, 1995). Such non-regulating mortality may be usefully considered as an important 'co-factor' in regulation.

4.6 'BOTTOM-UP' AND 'TOP-DOWN' REGULATION

An alternative view of population regulation ('bottom-up' as opposed to 'top-down') has emphasized the importance of resource constraints in the population dynamics of herbivores, relegating natural enemies to a secondary or minor role. Failure to detect density-dependence in natural enemy-induced mortality has been used as evidence to support the bottom-up view (Dempster, 1983; Dempster and Pollard, 1986; and more recently, Harrison and Cappuccino, 1995), although, as we have seen, this failure may reflect in many cases the inadequacy of data or statistical analysis, rather than a true absence of density-dependence (Hassell, 1985). Even so, as explained above, regulation *per se* may not be necessary for natural enemies to have a significant impact in constraining insect numbers below the ceiling set by resources. It seems to us that the use of the term 'regulation' is somewhat misleading in the top-down/bottom-up debate. What is at issue is whether phytophagous insect populations are commonly or routinely constrained by plant resources alone, or whether the impact of natural enemies is sufficient to keep numbers well below the resource ceiling. Whilst the issue of bottom-up as opposed to top-down 'regulation' has been hotly debated in recent years (see also Wratten, 1992; Price, 1992; Hunter and Price, 1992), it is possible to go some way towards reconciling both viewpoints under the Southwood and Comins (1976) synoptic model framework within which we may expect natural enemy regulation to be lacking for sound ecological reasons (section 4.3). Put simply, many of the species in which natural enemy regulation has not been detected may have temporarily escaped the 'natural enemy ravine' or may be sufficiently r-selected, 'boom-and-bust' species to experience minimal constraints from natural enemies.

Several studies have been carried out in sufficient experimental detail to show a minimal impact by natural enemies, even at low endemic levels (review: Price, 1992). Price (1988, 1990), for example, showed that regulation in *Euura* sawflies, which form galls on willow shoots, is mediated through intraspecific competition for a limited number of suitable high quality shoots, making the population relatively rare. Parasitoids are restricted in their effectiveness by their inability to parasitize larvae in large galls, which are produced on high quality shoots. Price has proposed that the crucial difference separating such 'latent' species from their more abundant counterparts may be their ability to discriminate selectively between plant parts of varying quality for optimal larval survival; species with the potential to outbreak tend to lay eggs much more indiscriminately on foliage, sometimes even on different host plant species. This provides the foundation for high densities to develop given the appropriate conditions favouring high survival and/or fecundity.

The pattern of population dynamics shown by *Euura* spp. conforms, in our view, to the extreme K-selected syndrome of the Southwood and Comins (1976) synoptic model and appears to be common to many forest insects (Price, 1992). Thus again, we can see ways in which the 'bottom up'/'top-down' dichotomy can be reconciled within a single model of population dynamics. An examination of detailed published studies on 32 forest pest species reveals that 12 are reported by the authors to have population densities largely determined by resource constraints, usually host plant-related, with a secondary, minor or negligible role of natural enemies (Table 4.2). We refer to these as group A species. On the other hand, natural enemies are considered to play a major role (regulative, suppressive or destabilizing) in 20 cases (group B species). Of course, it should not be taken that our small sample is in any way representative of forest insects as a whole. Clearly, pest species predominate in such an analysis because they are the ones that receive most attention from researchers, and this is likely to bias the sample towards either one or the other group. Many of the group B species appear to occupy a 'ravine to ridge' contour, as explained in detail for *Neodiprion sertifer* (section 4.2). It is notable that this form of dynamics appears to be particularly common amongst the Lepidoptera examined and suggests that a more comprehensive survey may reveal some interesting patterns in relation to taxonomy or life history.

When considering those group A species for which natural enemies are thought to have a minor or negligible role to play, some caution needs to be exercised in accepting this interpretation. The reasons are two-fold:

1. The action of natural enemies in many of these studies may be poorly understood, even within the constraints of key factor analysis. For example, percentage parasitism is frequently used as an indicator of parasitoid impact, but estimates may give a very misleading impression of true mortality from this source, unless certain protocols are followed to take account, for example, of partial asynchrony between parasitoid and host (van Driesche, 1983; van Driesche *et al.*, 1991).

2. Mortality estimates, even if properly obtained, may give the wrong impression of the true effect of a natural enemy on prey or host population dynamics. On the one hand, even apparently low mortalities, if they are spatially or temporally density-dependent, may have a significant regulating impact (Hassell, 1985, 1987) – a self-evident point, the significance of which is often not appreciated. On the other hand, even low density-independent mortality can have a super-proportional impact on the depression of host population levels, depending on the type of action of other mortality factors.

To illustrate the latter point, we used a simple stage-structured simulation model of a generalized insect population, taking the following form:

$$E_{(t+1)} = F_t * R \qquad (4.1)$$
$$L_{(t+1)} = fE_{(t+1)} \qquad (4.2)$$
$$P_{(t+1)} = fL_{(t+1)} \qquad (4.3)$$

Table 4.2 Factors considered by authors to be important in determining observed forest insect population densities.

Species	Natural enemies	Resource constraints	Author
Diptera			
Lasiomma melania (larch conefly) (A)	Negligible	Competition for plant resources important	Roques, 1988
Dasyneura laricis (larch gall midge) (A)	Negligible	Plant resources important	Isaev *et al.*, 1988
Hemiptera			
Florinia externa *Nuculaspis tsugae* (hemlock scales) (B)	Parasitism important in Japan (not in US)	Plant resources important	McClure, 1988
Cryptococcus fagisuga (beech scale) (A)	Negligible	Plant resources important	Wainhouse and Gate, 1988
Adelges piceae (balsam woolly adelgid) (A)	Negligible	Plant resources important	Hain, 1988
Cinara pinea (large pine aphid) (A)	Secondary	Plant resources important	Kidd, 1988, 1990a,b
Drepanosiphon platanoides (sycamore aphid) (A)	Secondary	Plant resources important	Dixon, 1979
Eucallipterus tiliae (lime aphid) (A)	Secondary	Plant resources important	Dixon and Barlow, 1979
Cardiaspina albitextura (white lace lerp) (B)	Suppressive	Induced defences at high infestation levels	Morgan and Taylor, 1988
Lepidoptera			
Operophtera brumata (winter moth) (B)	Suppressive, regulative, destabilizing	–	Varley and Gradwell, 1960 Roland, 1994
Rhyacionia frustrana (pine tip moth) (B)	Suppressive, regulative	Occasionally at high density	Berisford, 1988
Oporinia autumnata (autumnal moth) (B)	Suppressive, destabilizing	Induced defences important	Haukioja *et al.*, 1988
Orgyia pseudotsugata (Douglas-fir tussock moth) (B)	Suppressive, regulative	Occasionally at high density	Mason and Wickman, 1988
Lymantria monacha (nun moth) (B)	Suppressive	At high density	Bejer, 1988
Coleophora laricella (larch casebearer) (B)	Suppressive	At high density	Long, 1988
Panolis flammea (pine beauty moth) (B)	Suppressive, regulative?	At high density	Watt and Leather, 1988

Hyblaea puera (teak defoliator) (A)	Minor	–	Nair, 1988
Bupalus pinarius (pine looper) (B)	Suppressive, regulative, destabilizing	At high density	Barbour, 1988
Choristoneura fumiferana (spruce budworm) (B)	Suppressive, regulative	At high density	Mattson *et al.*, 1988
Zeiraphera diniana (larch budmoth) (B)	Suppressive, secondary	At high density	Baltensweiler and Fischlin, 1988
Lymantria dispar (gypsy moth) (B)	Suppressive, regulative	At high density	Montgomery and Wallner, 1988
Hymenoptera			
Diprion pini (pine sawfly) (B)	Suppressive, regulative, destabilizing	At high density	Geri, 1988
Neodiprion sertifer (pine sawfly) (B)	Suppressive, regulative	At high density	Larsson *et al.*, 1993
Sirex noctilio (pine woodwasp) (B)	Suppressive, regulative?	At high density/absence of natural enemies	Madden, 1988
Coleoptera			
Monochamus alteratus (pine sawyer) (A)	Negligible?	Plant resources important	Kobayashi, 1988
Dendroctonus micans (great spruce bark beetle) (B)	Suppressive, regulative?	Plant resources important	Gregoire, 1988
Ips typographus (spruce bark beetle) (A)	Minor	Plant resources important	Christiansen and Bakke, 1988
Dendroctonus ponderosae (mountain pine beetle) (A)	Minor	Plant resources important	Raffa, 1988
Dendroctonus frontalis (southern pine beetle) (B)	Suppressive	Plant resources important	Flamm *et al.*, 1988
Scolytus ventralis (fir engraver) (B)	Suppressive, secondary	Plant resources important	Berryman and Ferrell, 1988
Trypodendrum lineatum (striped ambrosia beetle) (A)	Negligible	Plant resources important	Borden, 1988

The terms used and interpretations made from these studies are our own.
Negligible: effect considered too small to have a significant effect on dynamics.
Minor: small but significant effect on dynamics.
Secondary: effect may have a measurable and/or modifying effect on dynamics, but other factors more important.
Suppressive: a significant effect in reducing population levels.
Regulative: mortality shows measurable direct density-dependence.
Destabilizing: measurable density-disturbing effect or delayed density-dependence.
(A) = Type A species – population densities largely determined by resource constraints (usually plant-related), with a secondary, minor or negligible role of natural enemies.
(B) = Type B species – population densities largely determined by natural enemies.

$$A_{(t+1)} = fP_{(t+1)} \quad (4.4)$$
$$F_{(t+1)} = A_{(t+1)} * 0.5 \quad (4.5)$$

where E_t, L_t, P_t, A_t and F_t represented the numbers of eggs, larvae, pupae, adults and female adults, respectively. R was the average number of eggs produced per female. Functions governing survival from one stage to the next took two forms: density-independent (acting first and described by a proportional survival constant) and density-dependent. The equation defining density-dependent survival was of the general form:

$$S = C * \exp(-B * N) \quad (4.6)$$

where S is proportional survival, C is the density-dependence constant (maximum survival, where $C = 1$, is 100%), B is the coefficient defining the strength of density-dependence, and N is the number entering the stage in question.

With 30% density-independent mortality acting on the eggs and a weak density-dependence acting on the larvae and pupae (say, from intraspecific competition), severe fluctuations in numbers from generation to generation are produced. With an additional 10% density-independence acting on the eggs (inflicted by parasitoids, for example), the population fluctuations are quickly damped out and there is stability at a level 31% below the highest numbers achieved before parasitoid introduction. (Figure 4.4a). In a second simulation (Figure 4.4b), we alternately increased then decreased, by a factor of two, the strength of density-dependence acting on the larvae every seven generations. This was done in order to mimic the effects of changing quality or quantity of plant resources on the insect acting through increased or decreased competition. Adding 10% density-independent mortality to the eggs resulted in a stable equilibrium at the higher resource level, alternating with lower-density damping oscillations at the reduced resource level. Adding another 10% to the egg mortality (= 20%) to simulate density-independent parasitism created an unexpected reversal of the previous relationship between population levels and resource availability, i.e. a higher population at lower resource levels. The latter result echoes the finding of van Hamburg and Hassell (1984) that parasitoid-induced mortality acting early in the insect's life cycle can influence the impact of later density dependence, in some cases increasing the host population above the parasitoid-free level.

Although our model simulations are very simple, they serve to illustrate the point that low recorded mortality from natural enemies does not necessarily equate with low impact. We can cite some evidence from field populations to support

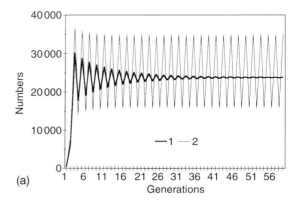

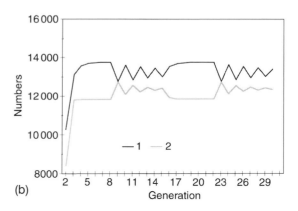

Figure 4.4 Dynamics of a simple simulation model with three immature stages, a 50:50 sex ratio and 50 eggs produced per female per generation; 30% density-dependence is applied to egg stage and a weak density-dependence to the larval (B = 0.0001, C = 1) and pupal (B = 0.00015, C = 1) stages. (a) Simulations given (1) with and (2) without an additional 10% density-independent mortality acting on the eggs. (b) Strength of density-dependence applied to larval stage is alternately strengthened and weakened by factor of 2 every seven generations; pupal density-dependence: B = 0.0005, C = 1; (1) and (2) as in (a).

this view. Ryan (1990), for example, indicates that the percentage reduction of larch casebearer, *Coleophora laricella*, after the introduction of two parasitoid species into Oregon was 98.5%, compared with a smaller maximum recorded 60% mortality from parasitism. For the winter moth in Vancouver, Roland (1994) has calculated that levels of parasitism as low as 12–14% may be sufficient to maintain moth numbers at low levels where regulation by predators can be achieved. We interpret this to mean that a slight drop in this level of parasitism (say, to less than 10%) may increase the risk of outbreak. Again, we are seeing here a high natural enemy impact coupled with low natural enemy-inflicted mortality. The unravelling of such effects in any consistent way is likely to require a level of investigation beyond that usually carried out, and may need to involve a combination of simulation modelling and natural enemy exclusion experiments (to isolate natural enemy effects from other effects) over a number of generations.

Even amongst those species for which natural enemies are considered to have some role (albeit secondary) to play in herbivore dynamics, the interactions between top-down and bottom-up effects may be more complex than simple theory would predict. For example, peak densities of the Scots pine lachnid, *Cinara pinea*, can vary considerably from year to year and tree to tree and these variations are determined by a combination of natural enemy and tree chemistry effects. An extensive simulation analysis has revealed the operating mechanisms (Kidd, 1988, 1990a,b), which briefly are as follows. Spring egg numbers are an important determinant of peak densities during the year, and these in turn are most affected by predation during the latter part of the previous year (winter mortality being fairly constant). Removing predation from the model can result in predicted populations the following year rising by as much as 10-fold, being constrained by seasonal changes in tree chemistry. However, these plant constraints may vary considerably from tree to tree and year to year and can impose a severe restriction on peak numbers attained, even in the absence of predation. The result is that populations will tend to track the between-year variations in plant quality peculiar to individual trees, with natural enemies having a significant effect in lowering peak densities only in those years where tree chemistry favours potentially high numbers.

In another study, not given in Table 4.2, Kato (1994) has shown that populations of the agromyzid fly *Chromatomyia suikazuirae*, which mines the leaves of the forest honeysuckle, *Lonicera gracilipes*, are regulated by bottom-up larval mortality and top-down pupal mortality acting alternately on successive generations.

In both of these examples it is clearly impossible to arrive at simplistic conclusions as to the relative importance of bottom-up and top-down effects. Given Schultz's (1992) view that in most systems bottom-up and top-down factors are probably acting in concert, it is perhaps more important to understand the different ways in which the two types of effect can interact, with a view to predicting the dynamic 'syndromes' associated with particular species. At present it is too early to make even tentative generalizations in this direction, but Schultz (1992) has made a bold beginning using simple graphical models to illustrate the main types of interaction that might be possible.

4.7 ARE NATURAL ENEMIES MORE IMPORTANT IN FORESTS THAN IN OTHER SYSTEMS?

This section again uses the synoptic model of Southwood and Comins (1976) to examine the importance of natural enemies in suppressing forest insect numbers, compared with those in other crop systems. There has already been mention of the detailed analysis of life tables carried out by Cornell and Hawkins (1995), in which they conclude that the impact of natural enemy mortality appears to be no different amongst late successional (*K*-selected) species (including forest insects) from that amongst early successional (*r*-selected) species. Here, this conclusion is re-examined using an independent data set of biological control introductions. These data are potentially very useful in that:

- a large body of information exists, which makes comparisons between systems possible;

- the frequency of successful pest control can perhaps be taken as a comparative measure of natural enemy impact.

The validity of the second point is supported by the analysis of Hawkins et al. (1993), who showed there to be a strong association between maximum percentage parasitism of the pest in the area of introduction, and the outcome of biological control (the exceptions to the natural enemy-induced mortality–impact relationship discussed earlier would account for at least some of the high degree of variability in this relationship). If natural enemies are less important in some systems than in others (possibly due to a greater frequency of bottom-up regulation), we might expect the frequency of biocontrol successes to be lower.

Following Southwood (1977), Greathead (1986) used the historical record of classical biological control introductions to assess the effectiveness of biological control among different crop systems, relating it to habitat stability. Using analyses by Beirne (1975), Hall and Ehler (1979) and Hall et al. (1980), Greathead categorized pests of forests as more K-selected (given a presumed greater habitat stability) and therefore less amenable to biological control compared with pests of orchards. Given this logic, forest insects should be positioned at the extreme K-end of the synoptic model response surface (Figure 4.1), whereas orchard insects should lie in the intermediate zone where top-down effects predominate. The data presented by Greathead supported this classification, in that success rate appeared higher for orchards and 'other perennials', though the data for establishment rates were equivocal.

We have carried out a more up-to-date and extensive analysis using the BIOCAT database developed by the International Institute of Biological Control (Greathead and Greathead, 1992), which includes all known biological control introductions (over 4000) worldwide. The BIOCAT database is increasingly being used as a resource for providing both guidelines for future biological control practice and ecological insight (Hawkins, 1994; Jervis et al., 1992, 1996a; Mills, 1994; Waage, 1990; Waage and Mills, 1992). Figure 4.5 shows that in terms of natural enemy establishment rate and success rate, forests, together with field and horticultural crop systems, compare unfavourably with both orchard and amenity systems. Also, in terms of the proportion of successes that have resulted in complete control of the pest, forests compare poorly with orchards, though the numbers involved are too few to reveal significance. Thus, Greathead's categorization of forest pests as more difficult to control than orchard pests would appear to be borne out by our analysis. If the frequency of successful pest control is taken as a comparative measure of natural enemy impact it does appear that natural enemies are less important in forest systems than in some others. It is important to stress at this point that we make no assumptions as to the regulating capacity of natural enemies from these data. As explained above (sections 4.2 and 4.4), natural enemy impact may be regulative and/or suppressive. In terms of the synoptic population model the natural enemies could act within the 'natural enemy ravine' as stabilizing factors or alternatively as density-independent cofactors. The latter could suppress pest numbers to a point (within the bounds of the 'ravine') where density-dependent factors can then operate.

However, biological control data could represent a biased picture of parasitoid–host population dynamics in forest systems:

- One possible criticism is that classical biological control is an artificial process. However, an analysis by Hawkins (1993) has shown that success rates correlate with parasitoid-induced mortality levels in biologically similar types of herbivore studied in their native ranges, suggesting that to generalize from biological control to 'natural control' is valid.
- It could be argued that the difference between forests and orchard/amenity systems in establishment rate and success rate may be largely explained by the smaller representation of Homoptera amongst forest pests targeted for biocontrol (N.J. Mills, personal communication, based on analysis of BIOCAT database). In classical biological control, both establishment rate and success rate are higher for Homoptera than for the endopterygote orders (Hall and Ehler, 1979; Hall et al., 1980), despite the sug-

Are natural enemies more important in forests? 63

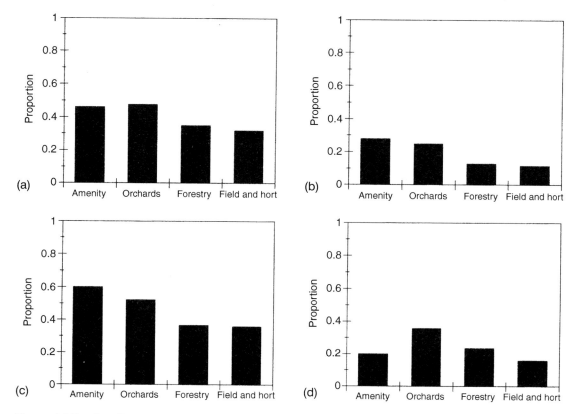

Figure 4.5 Results of BIOCAT database analysis showing how different crop systems compare with respect to establishment rate and success rate in classical biological control. (a) Establishments as a proportion of the 3210 introductions whose outcomes are known; (b) successes as a proportion of introductions; (c) successes as a proportion of establishments; (d) proportion of successes resulting in complete control. In (a)–(c), all groups are significantly different from each other ($P < 0.05$), except for amenity vs orchards, and field and horticulture vs forestry. In (d), none of the groups was significantly different.

gestion (Table 4.2) that forest Homoptera appear to be less top-down constrained than forest endopterygotes. (We caution against too much significance being attached to the latter observation, as there are insufficient data.)

- Forest pests targeted for biological control may not be representative of forest pests as a whole in their position on the r–K continuum. Most forest insects targeted are pests by virtue of the high population numbers they attain. This could be taken to indicate that they ought to be more, rather than less, amenable to biological control – especially given that in most cases they are maintained at low levels by natural enemies in their native ranges. However, there are some pests which Southwood (1977) categorizes as K-pests that become abundant through disturbance to the herbivore–plant interaction, i.e. high population numbers result from a reduction in food plant constraints. It would be useful to determine, from among pests that have been targeted for classical biological control, which species are K-inclined and which are r-inclined, and then to compare success rates among these.

Perhaps the Southwood and Comins (1976) synoptic model cannot explain differences in the outcome of classical biological control introductions among different crop systems. This certainly applies to establishment rate, which, as noted, is poor for forest systems compared with orchard and amenity systems. Why natural enemies introduced against forest pests should be poor not only in controlling pests but also in becoming established is not clear. One possible explanation is that forests, and particularly plantations, are relatively deficient in floral and other food sources for female natural enemies. The availability of such food sources is likely to influence the survival and reproductive capacity of parasitoids and predators (Powell, 1986; Jervis *et al.*, 1993, 1996b).

4.8 CONCLUSIONS

- Parasitoids and predators can have a significant impact in suppressing and/or regulating forest insect populations. This conclusion applies to both indigenous and introduced natural enemies.
- Parasitoids and predators may also be responsible for perturbations in forest pest populations. This is especially true of parasitoids which, by virtue of their often delayed density-dependent action, may drive population cycles.
- Even where their action is temporally density-independent, parasitoids and predators may have an important role as regulating 'co-factors'.
- The relative importance of natural enemies (top-down effects) and plant resources (bottom-up effects) on forest insect dynamics varies considerably between pest species, but can be reconciled within a single model of population dynamics. In many cases, however, the true impact of natural enemies and the complexity of their interactions with hosts and prey may have been underestimated.
- Comparing biological control establishment and success rates for different crop systems, forests rank poorly beside orchard and amenity systems. It could be argued that this reflects a lower importance of parasitoids and predators in the dynamics of forest insects, but again, other interpretations are possible.

ACKNOWLEDGEMENTS

We thank David Greathead for advice on using the BIOCAT database, Tim Winter and Simon Leather for help with data collection, Nick Mills and Brad Hawkins for useful discussion, and Nick Mills for his comments on the manuscript.

REFERENCES

Barbour, D.A. (1988) The pine looper in Britain and Europe. In *Dynamics of Forest Insect Populations* (ed. A. Berryman), pp. 291–308, Plenum Press, New York.

Baltensweiler, W. and Fischlin, A. (1988) The larch budmoth in the Alps. In *Dynamics of Forest Insect Populations* (ed. A. Berryman), pp. 331–351, Plenum Press, New York.

Beddington, J.R., Free, C.A. and Lawton, J.H. (1978) Modelling biological control: on the characteristics of successful natural enemies. *Nature* **273**, 513–519.

Beirne, B.P. (1975) Biological control attempts by introductions against pest insects in the field in Canada. *Canadian Entomologist* **107**, 225–236.

Bejer, B. (1988) The nun moth in European spruce forests. In *Dynamics of Forest Insect Populations* (ed. A. Berryman), pp. 211–231, Plenum Press, New York.

Berisford, C.W. (1988) The nantucket pine tip moth. In *Dynamics of Forest Insect Populations* (ed. A. Berryman), pp. 141–161, Plenum Press, New York.

Berryman, A. (1978) Towards a theory of insect epidemiology. *Researches on Population Ecology* **19**, 181–196.

Berryman, A. and Ferrell, G.T. (1988) The fir engraver beetle in western states. In *Dynamics of Forest Insect Populations* (ed. A. Berryman), pp. 555–577, Plenum Press, New York.

Berryman, A.A., Millstein, J.A. and Mason, R.R. (1990) Modelling douglas-fir tussock moth population dynamics: the case for simple theoretical models. In *Population Dynamics of Forest Insects* (eds A.D. Watt, S.R. Leather, M.D. Hunter and N.A.C. Kidd), pp. 369–380, Intercept, Andover.

Bevan, D. (1983) *Forest Insects*, Forestry Commission Handbook 1, HMSO, London.

Bjorkman, C., Larsson, S. and Gref, R. (1991) Effects of nitrogen fertilization on pine needle chemistry and sawfly performance. *Oecologia* **86**, 202–209.

Bonsall, M. and Hassell, M.P. (1995) Identifying density-dependent processes: a comment on the regulation of winter moth. *Journal of Animal Ecology* **64**, 781–784.

Borden, J.H. (1988) The striped ambrosia beetle. In *Dynamics of Forest Insect Populations* (ed. A. Berryman), pp. 579–596, Plenum Press, New York.

Bustillo, A.E. and Drooz, A.T. (1977) Cooperative establishment of a Virginia (USA) strain of *Telenomus alsophilae* on *Oxydia trychiata* in Colombia. *Journal of Economic Entomology* **70**, 767–770.

Canada (1971) *Biological Control Programmes Against Insects and Weeds in Canada 1959–1968*, Technical Communication No. 4, Commonwealth Agricultural Bureaux, Farnham Royal.

Christiansen, E. and Bakke, A. (1988) The spruce bark beetle of Eurasia. In *Dynamics of Forest Insect Populations* (ed. A. Berryman), pp. 479–503, Plenum Press, New York.

Clausen, C.P. (1978) *Introduced Parasites and Predators of Arthropod Pests and Weeds: A World Review*, Agriculture Handbook No. 480, US Department of Agriculture, Washington DC.

Conway, G. (1976) Man versus pests. In *Theoretical Ecology* (ed. R.M. May), pp. 257–281, Blackwell Scientific Publications, Oxford.

Cornell, H.V. and Hawkins, B.A. (1995) Survival patterns and mortality sources of herbivorous insects: some demographic trends. *American Naturalist* **145**, 563–593.

Culliney, T.W., Beardsley, J.W. and Drea, J. (1988) Population regulation of the eurasian pine adelgid (Homoptera: Adelgidae) in Hawaii. *Journal of Economic Entomology* **81**, 142–147.

DeBach, P. (1971) The use of imported natural enemies in insect pest management ecology. *Proceedings of the Tall Timbers Conference on Ecological Animal Control by Habitat Management* **3**, 211–233.

Dempster, J.P. (1983) The natural control of populations of butterflies and moths. *Biological Reviews* **58**, 461–481.

Dempster, J.P. and Pollard, E. (1986) Spatial heterogeneity, stochasticity and the detection of density dependence in animal populations. *Oikos* **46**, 413–416.

Dixon, A.F.G. (1979) Sycamore aphid numbers: the role of weather, host and aphid. In *Population Dynamics* (eds R.M. Anderson, B.D. Turner and L.R. Taylor), pp. 105–120, Blackwell Scientific Publications, Oxford.

Dixon, A.F.G. and Barlow, N.D. (1979) Population regulation in the lime aphid. *Journal of Animal Ecology* **67**, 225–237.

van Driesche, R.G. (1983) The meaning of 'per cent parasitism' in studies of insect parasitoids. *Environmental Entomology* **12**, 1611–1622.

van Driesche, R.G., Bellows, T.S., Elkinton, J.S. *et al.* (1991) The meaning of percentage parasitism revisited: solutions to the problem of accurately estimating total losses from parasitism in a host generation. *Environmental Entomology* **20**, 1–7.

Embree, D.G. and Roland, J. (1994) Biological control of the winter moth. *Annual Review of Entomology* **40**, 475–492.

Evans, H.F. (1987) Biological control of *Dendroctonus micans* in the USSR. *Entopath News* **89**, 5–7.

Flamm, R.O., Coulson, R.N. and Payne, T.L. (1988) The southern pine beetle. In *Dynamics of Forest Insect Populations* (ed. A. Berryman), pp. 531–553, Plenum Press, New York.

Geri, C. (1988) The pine sawfly in central France. In *Dynamics of Forest Insect Populations* (ed. A. Berryman), pp. 377–405, Plenum Press, New York.

Gould, J.R., Elkinton, J.S. and Wallner, W.E. (1990) Density dependent suppression of experimentally created gypsy moth, *Lymantria dispar* (Lepidoptera: Lymantriidae), populations by natural enemies. *Journal of Animal Ecology* **59** 213–233.

Gould, J.R., Bellows, T.S. and Paine, T.D. (1992a) population dynamics of *Siphoninus phillyreae* in California in the presence and absence of a parasitoid, *Encarsia partenopea*. *Ecological Entomology* **17**, 127–134.

Gould, J.R., Bellows, T.S. and Paine, T.D. (1992b) Evaluation of biological control of *Siphoninus phillyreae* (Haliday) by the parasitoid *Encarsia partenopea* (Walker), using life-table analysis. *Biological Control* **2**, 257–265.

Greathead, D.J. (1971) *A Review of Biological Control in the Ethiopian Region*, Technical Communication No. 5, Commonwealth Agricultural Bureaux, Farnham Royal.

Greathead, D.J. (1976) *A Review of Biological Control in Western and Southern Europe*, Technical Communication No. 7, Commonwealth Agricultural Bureaux, Farnham Royal.

Greathead, D. (1986) Parasitoids in classical biological control. In *Insect Parasitoids* (eds J. Waage and D. Greathead), pp. 289–318, Academic Press, London.

Greathead, D.J. and Greathead, A.H. (1992) Biological control of insect pests by parasitoids and predators: the BIOCAT database. *Biocontrol News and Information* **13**, 61N–68N.

Gregoire, J.-C.(1988) The greater European spruce beetle. In *Dynamics of Forest Insect Populations* (ed. A. Berryman), pp. 455–478, Plenum Press, New York.

Gutierrez, A.P., Hagen, K.S. and Ellis, C.K. (1990) Evaluating the impact of natural enemies: a multitrophic perspective. In *Critical Issues in Biological*

Control (eds M. Mackauer, L.E. Ehler and J. Roland), pp. 81–109, Intercept, Andover.

Gutierrez, A.P., Mills, N.J., Schreiber, S.J. and Ellis, C.K. (1994) A physiologically based tritrophic perspective on bottom-up–top-down regulation of populations. *Ecology* **75**, 2227–2242.

Hain, F.P. (1988) The balsam woolly adelgid in North America. In *Dynamics of Forest Insect Populations* (ed. A. Berryman), pp. 87–109, Plenum Press, New York.

Hall, R.W. and Ehler, L.E. (1979) Rate of establishment of natural enemies in classical biological control. *Entomological Society of America Bulletin* **25**, 280–282.

Hall, R.W., Ehler, L.E. and Bisabri-Ershadi, B. (1980) Rates of success in classical biological control of arthropods. *Entomological Society of America Bulletin* **26**, 111–114.

van Hamburg, H. and Hassell, M.P. (1984) Density dependence and the augmentative release of egg parasitoids against graminaceous stalkborers. *Ecological Entomology* **9**, 101–108.

Hanski, I. (1987) Pine sawfly population dynamics: patterns, processes, problems. *Oikos* **50**, 327–335.

Harrison, S. and Cappuccino, N. (1995) Using density-manipulation experiments to study population regulation. In *Population Dynamics* (eds N. Cappuccino and P.W. Price), pp. 131–147, Academic Press, San Diego.

Hassell, M.P. (1978) *The Dynamics of Predator–Prey Interactions*, Princeton University Press, Princeton.

Hassell, M.P. (1985) Insect natural enemies as regulating factors. *Journal of Animal Ecology* **54**, 323–334.

Hassell, M.P. (1987) Detecting regulation in patchily distributed animal populations. *Journal of Animal Ecology* **56**, 705–713.

Hassell, M.P. and Waage, J.K. (1984) Host–parasitoid population interactions. *Annual Review of Entomology* **29**, 89–114.

Haukioja, E., Neuvonen, S., Hanhimäki, S. and Niemalä, P. (1988) The autumnal moth in Fennoscandia. In *Dynamics of Forest Insect Populations* (ed. A. Berryman), pp. 163–178, Plenum Press, New York.

Hawkins, B.A. (1993) Refuges, host population dynamics and the genesis of parasitoid diversity. In *Hymenoptera and Biodiversity* (eds J. LaSalle and I.D. Gauld), pp. 235–256, CAB International, Wallingford.

Hawkins, B.A. (1994) *Pattern and Process in Host–parasitoid interactions*, Cambridge University Press, Cambridge.

Hawkins, B.A., Thomas, M.B. and Hochberg, M.E. (1993) Refuge theory and biological control. *Science* **262**, 1429–1432.

Holling, C.S., Jones, D.D. and Clark, W.C. (1976) Ecological policy design: a case study of forest and pest management. *International Institute of Applied Systems Analysis Conference* **1**, 139–158.

Howard, L.O. and Fiske. W.F. (1911) The importation into the United States of the parasites of the gipsy-moth and the brown-tail moth. *Bulletin of the Bureau of Entomology of the United States Department of Agriculture* **91**, 1–312.

Huffaker, C.B., Berryman, A.A. and Laing, J.E. (1984) Natural control of insect populations. In *Ecological Entomology* (eds C.B. Huffaker and R.L. Rabb), pp. 359–398, John Wiley and Sons, New York.

Hunter, M.D. and Price, P.W. (1992) Playing chutes and ladders: heterogeneity and the relative roles of bottom-up and top-down forces in natural communities. *Ecology* **73**, 724–732.

Isaev, A.S., Baranchikov, Y.N. and Malutina, V.S. (1988) The larch gall midge in seed orchards of South Siberia. In *Dynamics of Forest Insect Populations* (ed. A. Berryman), pp. 29–44, Plenum Press, New York.

Jervis, M.A., Kidd, N.A.C., McEwen, P. et al. (1992) Biological control strategies in olive pest management. In *Research Collaboration in European IPM Systems* (ed. P.T. Haskell), pp. 31–39, BCPC Monograph No. 52.

Jervis, M.A., Kidd, N.A.C., Fitton, M.G. et al. (1993) Flower-visiting by hymenopteran parasitoids. *Journal of Natural History* **27**, 67–105.

Jervis, M.A., Hawkins, B.A. and Kidd, N.A.C. (1996a) The usefulness of destructive host feeding parasitoids in classical biological control: theory and observation conflict. *Ecological Entomology* **21**, 41–46.

Jervis, M.A., Kidd, N.A.C. and Heimpel, G.E. (1996b) Parasitoid feeding ecology and biological control – a review. *Biocontrol News and Information* **17**, 11N–26N.

Kato, M. (1994) Alternation of bottom-up and top-down regulation in a natural population of an agromyzid leafminer, *Chromatomyia suikazurae*. *Oecologia* **97**, 9–16.

Kidd, N.A.C. (1988) The large pine aphid on Scots Pine in Britain. In *Dynamics of Forest Insect Populations* (ed. A. Berryman), pp. 317–327, Plenum Press, New York.

Kidd, N.A.C. (1990a) The population dynamics of the large pine aphid, *Cinara pini* (Mordv.). I. Simulation of laboratory populations. *Researches on Population Ecology* **32**, 189–208.

Kidd, N.A.C. (1990b) The population dynamics of the large pine aphid, *Cinara pini* (Mordv.). II Simulation of field populations. *Researches on Population Ecology* **32**, 209–226.

Kidd, N.A.C. and Jervis, M.A. (1996) Population dynamics. In *Insect Natural Enemies: Practical Approaches to their Study and Evaluation* (eds M.A. Jervis and N.A.C. Kidd), pp. 213–284, Chapman & Hall, London.

Kidd, N.A.C. and Lewis, G.B. (1987) Can vertebrate predators regulate their prey? *American Naturalist* **130**, 448–453.

Kobayashi, F. (1988) The Japanese pine sawyer. In *Dynamics of Forest Insect Populations* (ed. A. Berryman), pp. 431–454, Plenum Press, New York.

Larsson, S. and Tenow, O. (1984) Areal distribution of a *Neodiprion sertifer* (Hym., Diprionidae) outbreak on Scots pine as related to stand condition. *Holarctic Ecology* **7**, 81–90.

Larsson, S., Bjorkmann, C. and Kidd, N.A.C. (1993) Outbreaks in diprionid sawflies: why some species and not others? In *Sawfly Life History Adaptations to Woody Plants* (eds M. Wagner and K.F. Raffa), pp. 453–483, Academic Press, San Diego.

Lawton, J.H. and McNeill, S. (1979) Between the devil and the deep blue sea. In *Population Dynamics* (eds R.M. Anderson, B.D. Turner and L.R. Taylor), pp. 223–244, Blackwell Scientific Publications, Oxford.

Long, G.E. (1988) The larch casebearer in intermountain Northwest. In *Dynamics of Forest Insect Populations* (ed. A. Berryman), pp. 233–242, Plenum Press, New York.

Luck, R.F. (1981) Parasitic insects introduced as biological control agents for arthropod pests. In *CRC Handbook of Pest Management in Agriculture*, Vol. II (ed. D. Pimentel), pp. 125–284, CRC Press, Boca Raton.

Madden, J.L. (1988) Sirex in Australasia. In *Dynamics of Forest Insect Populations* (ed. A. Berryman), pp. 407–429, Plenum Press, New York.

Mason, R.R. and Wickman, B.E. (1988) The Douglas-fir tussock moth in the interior Pacific northwest. In *Dynamics of Forest Insect Populations* (ed. A. Berryman), pp. 1–28, Plenum Press, New York.

Mattson, W.J., Simmons, G.A. and Witter, J.A. (1988) The spruce budworm in eastern North America. In *Dynamics of Forest Insect Populations* (ed. A. Berryman), pp. 309–330, Plenum Press, New York.

McClure, M.S. (1988) The armoured scales of hemlock. In *Dynamics of Forest Insect Populations* (ed. A. Berryman), pp. 45–65, Plenum Press, New York.

McGugan, B.M. and Coppel, H.C. (1962) *A Review of the Biological Control Attempts Against Insects and Weeds in Canada: Part II Biological Control of Forest Insects, 1910–1958*, Technical Communication No. 2, Commonwealth Agricultural Bureaux, Farnham Royal.

McLeod, J.H. (1962) *A Review of Biological Control Attempts Against Insects and Weeds in Canada, Part I. Biological control of pests of of crops, fruit trees, ornamentals and weeds in Canada up to 1959*, Commonwealth Institute of Biological Control Technical Communication No. 2, 1–33.

Mills, N.J. (1994) Biological control: some emerging trends. In *Individuals, Populations and Pattern in Ecology* (eds S.R. Leather, A.D. Watt, N.J. Mills and K.F. Walters), pp. 213–222, Intercept, Andover.

Montgomery, M.E. and Wallner, W.E. (1988) The larch cone fly in the French alps. In *Dynamics of Forest Insect Populations* (ed. A. Berryman), pp. 353–375, Plenum Press, New York.

Morgan, F.D. (1989) Forty years of *Sirex noctilio* and *Ips grandicollis* in Australia. *New Zealand Journal of Forestry Science* **19**, 198–209.

Morgan, F.D. and Taylor, G.S. (1988) The white lace lerp on southeastern Australia. In *Dynamics of Forest Insect Populations* (ed. A. Berryman), pp. 129–140, Plenum Press, New York.

Nair, K.S.S. (1988) The teak defoliator in the Kerala, India. In *Dynamics of Forest Insect Populations* (ed. A. Berryman), pp. 267–289, Plenum Press, New York.

Nuttall, M.J. (1989a) *Gonipterus scutellatus* Gyllenhal, gum tree weevil (Coleoptera: Curculionidae). In *A Review of Biological Control of Invertebrate Pests and Weeds in New Zealand 1874–1987* (eds P.J. Cameron, R.L. Hill, J. Bain and W.P. Thomas), pp. 267–269, CAB International, Wallingford.

Nuttall, M.J. (1989b) *Sirex noctilio* F., sirex wood wasp (Hymenoptera: Siricidae). In *A Review of Biological Control of Invertebrate Pests and Weeds in New Zealand 1874–1987* (eds P.J. Cameron, R.L. Hill, J. Bain and W.P. Thomas), pp. 299–306, CAB International, Wallingford.

OPIE (Office pour l'Information Eco-entomologique) (1986) Lutte biologique l'aide d'insectes entomophages. *Cahiers de Liaison* **20**, 1–48.

Peterman, R.M., Clark, W.C. and Holling, C.S. (1979) The dynamics of resilience: shifting stability domains in fish and insect systems. In *Population Dynamics* (eds R.M. Anderson, B.D. Turner and L.R. Taylor), pp. 321–341, Blackwell, Oxford.

Powell, W. (1986) Enhancing parasitoid activity in crops. In *Insect Parasitoids* (eds J. Waage and D. Greathead), pp. 319–340, Academic Press, London.

Price, P.W. (1987) The role of natural enemies in insect populations. In *Insect Outbreaks* (eds P. Barbosa and J.C. Schultz), pp. 287–312, Academic Press, New York.

Price, P.W. (1988) Inversely density-dependent parasitism: the role of plant refuges for hosts. *Journal of Animal Ecology* **57**, 89–96.

Price, P.W. (1990) Evaluating the role of natural enemies in latent and eruptive species: new approaches to life table construction. In *Population Dynamics of Forest Insects* (eds A.D. Watt, S.R. Leather, M.D. Hunter and N.A.C. Kidd), pp. 221–232, Intercept, Andover.

Price, P.W. (1992) Plant resources as the mechanistic basis for insect herbivore population dynamics. In *Effects of Resource Distribution on Animal–plant Interactions* (eds M.D. Hunter, T. Ohgushi and P.W. Price), pp. 139–173, Academic Press, San Diego.

Quednau, F.W. (1990) Introduction in eastern Canada of *Olesicampe geniculata* Quednau and Lim, an important biological control agent of the mountain ash sawfly, *Pristiphora geniculata* (Hartig). *Canadian Entomologist* **122**, 921–934.

Raffa, K.F. (1988) The mountain pine beetle in western North America. In *Dynamics of Forest Insect Populations* (ed. A. Berryman), pp. 505–530, Plenum Press, New York.

van Rensburg, N.J.(1992) The black pine aphid: a success story. *Plant Protection News* **28**, 5–6.

Roland, J. (1994) After the decline: what maintains low winter moth density after successful biological control? *Journal of Animal Ecology* **63**, 392–398.

Roland, J. (1995) Response to Bonsall and Hassell 'Identifying density-dependent processes: a comment on the regulation of winter moth'. *Journal of Animal Ecology* **64**, 785–786.

Roques, A. (1988) The larch cone fly in the French alps. In *Dynamics of Forest Insect Populations* (ed. A. Berryman), pp. 1–28, Plenum Press, New York.

Ryan, R.B. (1990) Evaluation of biological control: introduced parasites of larch casebearer (Lepidoptera: Coleophoridae) in Oregon. *Environmental Entomology* **19**, 1873–1881.

Sailer, R.I. (1983) Beneficial foreign species of Hymenoptera known to be established in the 48 contiguous United States (unpublished typescript), 10 pp.

Schultz, J.C. (1992) Factoring natural enemies into plant tissue availability to herbivores. In *Effects of Resource Distribution on Animal–plant Interactions* (eds M.D. Hunter, T. Ohgushi and P.W. Price), pp. 175–197, Academic Press, San Diego.

Southwood, T.R.E. (1977) The relevance of population dynamic theory to pest status. In *The Origins of Pest, Parasite, Disease and Weed Problems* (eds J.M. Cherrett and G.R. Sagar), pp. 35–54, Blackwell Scientific Publications, Oxford.

Southwood, T.R.E. and Comins, H.N. (1976) A synoptic population model. *Journal of Animal Ecology* **45**, 949–965.

Stiling, P.D. (1988) Density-dependent processes and key factors in insect populations. *Journal of Animal Ecology* **57**, 581–594.

Taylor, K.L. (1978) Evaluation of the insect parasitoids of *Sirex noctilio* (Hymenoptera: Siricidae) in Tasmania. *Oecologia* **32**, 1–10.

Thomas, W.P. and Hill, R.L. (1989) *Phyllonorycter messaniella* (Zeller), oak leaf-miner (Lepidoptera: Gracillariidae). In *A Review of Biological Control of Invertebrate Pests and Weeds in New Zealand 1874–1987* (eds P.J. Cameron, R.L. Hill, J. Bain and W.P. Thomas), pp. 289–293, CAB International, Wallingford.

Tribe, G.D. (1992) Neutralisation of the eucalyptus tortoise beetle. *Plant Protection News* **29**, 5.

Varley, G.C. and Gradwell, G.R. (1960) Key factors in population studies. *Journal of Animal Ecology* **29**, 399–401.

Waage, J.K. (1990) Ecological theory and the selection of biological control agents. In *Critical Issues in Biological Control* (eds M. Mackauer, L.E. Ehler and J. Roland), pp. 135–157, Intercept, Andover.

Waage, J.K. and Hassell, M.P. (1982) Parasitoids as biological control agents – a fundamental approach. *Parasitology* **84**, 241–268.

Waage, J.K. and Mills, N.J. (1992) Biological control. In *Natural Enemies: The Population Biology of predators, Parasites and Diseases* (ed. M.J. Crawley), pp. 412–430, Blackwell, Oxford.

Wainhouse, D. and Gate, I.M. (1988) The beech scale. In *Dynamics of Forest Insect Populations* (ed. A. Berryman), pp. 67–85. Plenum Press, New York.

Watt, A.D. and Leather, S.R. (1988) The pine beauty moth in scottish lodgepole pine plantations. In *Dynamics of Forest Insect Populations* (ed. A. Berryman), pp. 243–266, Plenum Press, New York.

Wilson, F. (1960) *A Review of the Biological Control of Insects and Weeds in Australia and Australian New Guinea*, Technical Communication No. 1, Commonwealth Agricultural Bureaux, Farnham Royal.

Winter, T. (1983) *A Catalogue of Phytophagous Insects and Mites on Trees in Great Britain*, Forestry Commission Booklet 53, Forestry Commission, Edinburgh.

Wratten, S.D. (1992) Population regulation in insect herbivores – top-down or bottom-up? *New Zealand Journal of Ecology* **16**, 145–147.

Zondag, R. and Nuttall, M.J. (1989) *Pineus laevis* (Maskell), pine twig chermes or pine wooly aphid (Homoptera: Adelgidae). In *A Review of Biological Control of Invertebrate Pests and Weeds in New Zealand 1874–1987* (eds P.J. Cameron, R.L. Hill, J. Bain and W.P. Thomas), pp. 295–297, CAB International, Wallingford.

Zuniga, E. (1985) Ochenta anos de control biologico en Chile. *Agricultura Technica* **45** 175–182.

HERBIVORE-INDUCED RESPONSES IN TREES: INTERNAL VS. EXTERNAL EXPLANATIONS

Erkki Haukioja and Tuija Honkanen

5.1 INTRODUCTION

Herbivory regularly alters foliage quality either by leaving only the rejected plant tissue or because the plant's chemical, physical or ecological characteristics change after damage. Although herbivory may make plant quality better, or cause no changes at all (Roland and Myers, 1987; Rossiter *et al.*, 1988; Karban and Myers, 1989; Haukioja, 1990), the most fascinating alterations make plants more resistant to further herbivory, i.e. induce resistance or defence. Induced resistance may be true defence of plants, designed by natural selection for that purpose, or a by-product of plant recovery from the damage which fortuitously deters herbivores (e.g. Karban, 1993). True induced defences have been found particularly in forbs – for example, in tobacco (Baldwin, 1988; Baldwin and Callahan, 1993), tomato (Edwards *et al.*, 1985; Duffey and Felton, 1991; Jongsma *et al.*, 1994) and cotton (Karban and Carey, 1984; Karban and Niiho, 1995). Also certain induced responses of trees seem to be defensive in a strict sense – for example, in the foliage of alder (Baur *et al.*, 1991; Seldal *et al.*, 1994) and poplar (Parsons *et al.*, 1989; Davis *et al.*, 1991), and in conifer trunks (Raffa and Berryman 1987; Lewinsohn *et al.*, 1991a,b). Their defensive nature usually has been inferred from their specific functions, short triggering times and the spreading of the response from undamaged to intact parts of the plant. Such responses may be primarily targeted against pathogens but they also affect insects, even in species-specific ways (Nicholson and Hammerschmidt, 1992; Stout *et al.*, 1994).

For insect ecologists, very interesting herbivore-induced reactions in host plants are the delayed damage-triggered responses whose effects may be seen in insect performance in the growth season(s) following the damage. In many cases the effects of herbivory have been restricted to the damaged part of the plant only (e.g. Tuomi *et al.*, 1988a; Sprugel *et al.*, 1991). This makes them unlikely to protect the whole individual plant, the genet, although they may still be useful for the modules (Tuomi *et al.*, 1988a, 1990, 1991). Accordingly, it is not clear at all whether the delayed induced responses are evolutionarily defensive. This chapter deals with damage-induced responses in the quality of tree foliage with potential repercussions to multi-year fluctuations in herbivore populations. Because of our basically zoological motivation, we start with two case studies in which cyclical fluctuations of herbivore populations occur in host trees with demonstrated herbivore-induced responses. Our aims are to examine different types of induced responses and the hypotheses to explain them, and to show that a null hypothesis based on plant physiology may

Forests and Insects. Edited by A.D. Watt, N.E. Stork and M.D. Hunter. Published in 1997 by Chapman & Hall, London. ISBN 0 412 79110 2.

largely explain the observed variance in studies of delayed induced plant responses.

5.2 OUTBREAKS AND CYCLES OF FOREST INSECTS VS. DAMAGE-INDUCED RESPONSES OF HOST TREES

5.2.1 INSECTS

Population fluctuations in forest insects range from irregular outbreaks to cycles with peak densities emerging at predictable intervals. The former type of fluctuations characterize many sawflies and lepidopterans living on coniferous and deciduous tree species, and the outbreaks may last from one to a few seasons (Berryman, 1988, 1996). The most regular insect population cycles documented so far are those of the tortricid *Zeiraphera diniana*, the larch budmoth, defoliating European larch (*Larix decidua*) and of the geometrid *Epirrita* (= *Oporinia*) *autumnata*, the autumnal moth, on mountain birch (*Betula pubescens* ssp. *tortuosa*). Note that a demonstration of regular cyclicity demands observations for quite long periods. Documentation of the two cases is based on long historical data series, both covering more than a century (Tenow, 1972; Baltensweiler *et al.*, 1977; Baltensweiler 1993). Both cycles peak regularly at intervals of 8–9 years (*Zeiraphera*, Baltensweiler *et al.*, 1977) and 9–10 years (*Epirrita*, Haukioja *et al.*, 1988).

These two cases share a number of other common features. Hosts are deciduous trees, defoliated early in the season. Within the common range of the moth and the host, regular population cycles occur only within a restricted zone of the tree distribution, characterized by marginal growth conditions and strongly seasonal environments: in larch at altitudes of 1700–2000 m (Baltensweiler *et al.*, 1977) and in mountain birch in the northern part of the distribution (Haukioja *et al.*, 1988). In both cases, less regular peaks occur in much larger areas. Heights of peak densities vary tremendously, and regular cyclicity is a regional and not a tree or stand specific event: individual stands – particularly in the case of the mountain birch – are not defoliated during each successive cycle peak (Tenow and Bylund, 1989; Bylund, 1995).

5.2.2 TYPES OF HERBIVORE-INDUCED RESPONSE IN TREES: RELEVANCE FOR INSECT OUTBREAKS AND CYCLES

For multi-year fluctuations in herbivore densities only damage-induced responses in foliage that operate with density dependence and produce long time lags can be relevant (May, 1976). Induced types of resistance (**IR**) may, by definition, be density dependent but time lags of IR have to be longer than the generation time of the herbivore for IR to be germane for multi-annual cycles. For zoological reasons, induced resistance has been classified into two forms: delayed (**DIR**) and rapid induced resistance (**RIR**) (Haukioja, 1982). Their separation is based on the match between two time scales: the duration of the increased level of induced resistance in the tree, and the generation time of the herbivore, damage by which led to IR. Note that from the plant's point of view these two types of induced resistance may be identical, but only DIR is potentially relevant for multi-annual cycles and outbreaks.

Not all damage by herbivores makes future plant quality worse, i.e. induces resistance. Many attempts to demonstrate DIR have failed, particularly when conifers have been studied (Karban and Myers, 1989; Haukioja, 1991; Niemelä and Tuomi 1993). The strongest evidence for DIR comes from deciduous trees, specifically the mountain birch and European larch (Benz, 1974; Ruohomäki *et al.*, 1992). These are the two described systems with regular cycles. However, even if DIR does not operate, herbivore-induced responses by trees may still be relevant for multi-annual insect cycles and outbreaks. Paradoxically, this may be true even in cases in which changes diametrically opposed to induced resistance occur, i.e. an increase in foliage quality after herbivory. Such an increase in foliage quality after damage has been called resource manipulation (Craig *et al.*, 1986), induced amelioration (Haukioja *et al.*, 1990) or induced susceptibility (Karban and Niiho, 1995). We adopt the last term, and below use the abreviation **IS**.

Herbivore-induced susceptibility in foliage quality is particularly challenging for ecologists for two reasons. First, adaptive explanations for why trees respond to herbivory by becoming more palatable and, in a way, attracting more herbivory, have to assume strong constraints or trade-offs. Secondly, IS can contribute to cycles and outbreaks, irrespective of whether it operates with (**DIS**) or without (**RIS**) time lags, i.e. whether the damage causing insects themselves experience the RIS, or if it is the following generation of herbivores that benefits. In the latter case, DIS as a result of a specific and deliberate way of damaging the plant may be an adaptive trait for the insect if the benefits are restricted to kin (Tuomi et al., 1994).

5.3 IS DIR A RESULT OF CARBON/NUTRIENT IMBALANCE OR IS IT A DEFENCE?

Researchers (Tuomi et al., 1990; Bryant et al., 1991a, 1993) have often viewed the nature of DIR as a clearcut dichotomy: either DIR is a defence, i.e. it is caused by selection for resistance, or DIR is a passive by-product caused by damage-driven changes in the relative availability of foliar carbon vs. nutrients, particularly nitrogen. These alternatives are not exclusive, but operate at different levels of abstraction (Haukioja and Neuvonen, 1985) (Table 5.1). The following paragraphs review some experiments which have tested DIR and DIS in birches. We start from the original dichotomy between defence and carbon/nutrient imbalance, in spite of the logical error of such a contrast (Schultz, 1988).

Loss of foliar biomass always means loss of nutrients and carbon. On the nutrient-poor sites where *Epirrita autumnata* cycles occur, the basic assumption (Tuomi et al., 1984, 1990) has been that trees lose a larger proportion of nutrients than carbon in defoliation, and that nutrient deficiency is the reason for symptoms of DIR: shortage of mineral nutrients impedes tree growth but extra carbon can still accumulate. If nutrient shortage under conditions of excess carbon availability is the cause of DIR, adding nutrients to defoliated trees, or preventing them from getting carbon, should alleviate the symptoms of DIR. Indeed, in Alaskan paper birch (Bryant et al., 1993) and partially in oak (Hunter and Schultz, 1995), fertilization eliminated or decreased symptoms of DIR, in accordance with the prediction of the carbon/nutrient hypothesis.

In the case of mountain birch, the same proposition has been tested in two independent experiments. In the first, fertilization was not found to alleviate DIR (Haukioja and Neuvonen, 1985). This experiment was critized for not being long enough for fertilizers to affect the trees (Bryant et al., 1993). Another experiment, with both fertilizer and shading treatments applied to trees for three successive growth seasons, failed to reveal interactions between fertilizing/shading and DIR. DIR was highly significant, but was also significant in nitrogen-fertilized trees (Ruohomäki et al., 1996). Consequently, extra nutrients or shading did not eliminate DIR in the mountain birch, contrary to the prediction by the carbon/nutrient hypothesis.

The difference in response between Alaskan and Finnish birches may be real – different birch species and populations behaving in different ways, which would be analogous to the two oak species studied by Hunter and Schultz (1995). But it is possible that methodological details in the birch experiments have contributed to the discrepancy. If we want specifically to test whether DIR in

Table 5.1 The basic domain of hypotheses used to explain delayed induced resistance (DIR)

Basic domain	Active defence	Carbon/nutrient balance	Growth/ differentiation	Phytocentric	Sink/source
Physiological		x	x	x	(x)
Whole plant functioning				(x)	x
Ecological		(x)			
Evolutionary	x				

birch is a result of insect-caused nutrient deficiency, we should scale nutrient additions so that they exactly revert the disturbed balance between nutrients and carbon. That is difficult, and dosing fertilizers in a way that does not alter other tree functions may be impossible. In the mountain birch experiment by Ruohomäki *et al.* (1996), nutrient additions sufficiently replaced lost nutrients because shoot growth of the trees was significantly enhanced. In the experiment by Bryant *et al.* (1993), the idea was to provide trees with enough nutrients to increase tree growth so that almost all available carbon was bound to new biomass and less carbon would be left for defence – thus removing symptoms of DIR. With the huge amounts of nitrogen that Bryant *et al.* (1993) used, this very probably happened. But the design tested more whether leaves in vigorously growing shoots are good for *Rheumaptera* larvae than whether DIR is caused by a loss of nutrients in defoliation.

Some of the basic assumptions about DIR in mountain birch as a consequence of nutrient deficiency (Bryant *et al.*, 1983; Tuomi *et al.*, 1988b) may not be totally correct. Birches have been mentioned as an example of rapidly growing trees (Bryant *et al.*, 1991b) which lose more nitrogen than carbon in early season defoliations. But in the field, *Epirrita* outbreaks on mountain birch occur in mature to overmature stands (Bylund, 1995). Such trees are growing slowly, and many mature trees do not grow at all each year (unpublished data). More importantly, defoliation, in addition to removing nutrients in foliage, also removes carbon and the carbon-acquiring machinery of the tree. We are not aware of any study which has measured both losses in a tree's nutrient vs. carbon capitals and changes in the acquisition of new nitrogen and carbon. Therefore, it is not clear to what extent defoliated birches really suffer from nutrient deficiency and to what extent carbon deficiency occurs; hence the relevance of fertilization experiments in proving or falsifying the validity of the carbon/nutrient hypothesis remains unclear.

Tests of the defensive nature of DIR are, in practice, also difficult to conduct. In practice, we can try to demonstrate two items relevant for evolutionary defences: first, the possible heritability and variation in DIR and, second, fitness gains of trees with strong DIR when severe defoliation occurs. To our knowledge, the genetic basis of DIR is not known for any tree species. The above-mentioned three-year experiment by Ruohomäki *et al.* (1996) failed to find tree progeny/defoliation interactions in DIR although mother tree effects were significant, for example, in contents of foliage phenolics. For practical reasons fitness consequences of DIR in birch, or in other long-lived trees, have not been studied.

In summary, neither defensive explanations nor hypotheses based on availability of carbon vs. nutrients provide general explanations of DIR in birches. However, Table 5.2 shows that there are other potential explanations, and that explanations operate at different levels, ranging from physiological to evolutionary. Since all the upper level explanations have to be consistent with mechanisms at lower levels, we shall explore the physiological background of induced responses. We particularly emphasize the need to understand the interplay between plant parts and the integrated individual plant (Geiger, 1987; Haukioja, 1991). For this reason, we start by briefly reviewing some elements of whole plant physiology. We first consider what is relevant when a plant is defoliated, and what are the ramifications for herbivore-induced responses.

5.4 WHAT DOES DEFOLIATION DO FOR A TREE?

Defoliation removes part of the mineral capital of the tree, part of its carbon reserves and photosynthetic machinery that would have captured more energy and carbon and, indirectly, nutrients. But simultaneously herbivores may also damage tissues that are implements of the hormonal control system by which plants regulate intake and allocation of resources (Haukioja *et al.*, 1990). Two lines of approach have emphasized connections between induced resistance and plant physiology: the modularity approach (Tuomi *et al.*, 1988a) and the phytocentric view (Coleman 1986; Coleman and Jones, 1991). The former takes care of the repetitive, multicellular units of which plants are formed (White 1979). The latter identifies a number of

Table 5.2 Comparison among the resource-based (carbon/nutrient balance and growth/differentiation) hypotheses and the sink/source hypothesis

Hypothesis	Observation/ assumption	Selective factors	Consequences of selective factors	Auxiliary assumptions	DIR/DIS
Carbon/ nutrient balance	Trade-off in the allocation of phenylalanine	Resource availability	Inherent growth rate Evergreen/deciduous	Sink/source dynamics	By-product of carbon/nutrient losses
	Negative correlation between phenolics and nitrogen		Photosynthetic rate	Apical dominance	
Growth/ differentiation	Simultaneous growth/ differentiation not possible				
Sink/source	Apical dominance Sink/source dynamics	Factors favouring integration of the modular structure	Whole plant integration in the use of resources	Growth rate Age Resource availability	By-product of factors favouring integration of the modular structure

aspects of plant physiology (acquisition, partitioning and allocation of resources, separation of genetic vs. environmental factors, source–sink relations). In the following we try to refine the mechanistic consequences of plant physiological regulation to herbivore-induced responses. We call this approach the sink/source hypothesis (Honkanen and Haukioja 1994; Haukioja and Honkanen 1996).

We shall emphasize the effects of herbivory on the functioning of plants as coherent individuals, and the fact that this seems to happen via herbivore damage on tissues which regulate the physiological system integrating individual modules into plant individuals. We discuss whether such herbivore-mediated influences on the physiology of mountain birch and larch are so characteristic that such internal factors could explain the occurrence of delayed induced responses just in the two tree species. Since regular insect cycles occur only in a restricted range of forests formed by birch and larch, we study whether the harsh, marginal and strongly seasonal environments could interact with tree properties.

5.4.1 TREE STRUCTURE AND FUNCTION

Trees are constructed of modules. Originally, tree modularity was a predominantly morphological concept, indicating that trees are composed of repetitive, multicellular entities – modules (White, 1979; Hallé, 1986; Jerling, 1985; Vuorisalo and Tuomi, 1986). Modules are assumed to be semi-independent, particularly in their carbon metabolism (Watson and Casper, 1984), and they are connected to each other, and to the roots, by a common plumbing system that attaches modules in the same vascular duct closely to each other (Watson and Caspar, 1984; Jones et al., 1993). The functional role of modules has to be considered at two levels, that of the modules themselves and that of the individual plant, often the genet. The latter does not directly emerge from the former (Geiger, 1987; Haukioja, 1991). In spite of modularity, plant individuals are usually tightly integrated entities (Chapin, 1991; Chapin et al., 1993). Plant form and integration result from hormonal regulation among modules (Sachs, 1991, 1993; Sachs et al., 1993).

Although details of the hormonal control are incompletely known, emergence of a coherent plant individual demands that hormone production of modules and the reception of and responses to these hormones by other modules have to follow rules which lead to an optimal or close to optimal shape of the whole individual. Plant shape and form largely depend on apical dominance and lateral inhibition, and these are known to be under genetic control (Cline, 1991). Therefore, the regulative rules are common for the whole individual plant, the genet. In spite of the genetic basis, the rules cannot produce direct unconditional instructions but, instead, facultative ones, taking simultaneously into account relevant module-specific external cues (light and nutrient availability at a particular meristem) and internal cues (particularly proximity of other meristems). When modules and shoots at the most favourable sites gain dominance and form strong sinks, resources are allocated in the direction of these sinks. Simultaneously, modules in less favourable sites are inhibited or retarded; consequently, integration and form within a tree are automatically created (Haukioja, 1991; Sachs *et al.*, 1993). Because of the genetic component, both tree form and integration are able to evolve by natural selection.

The sink/source hypothesis recognizes the importance of nutrients and carbon, but via their effects on regulation of plant physiology. In other words, hormonal control produces physiological decisions to allocate more to growth when nitrogen availability is high. This happens via stronger sinks. However, for different types of plants the same nitrogen additions may produce different results: low shrubs never grow to be the size of a fir irrespective of the amount of nutrients. Thus it is the hormonal system which determines the use of nutrients, not amount of nutrients *per se* (Table 5.2).

5.4.2 SEASONAL SHIFTS IN BIRCH RESOURCE ALLOCATION

Understanding the damage-induced responses in foliage requires knowledge about the physiological roles of leaves at the time of damage (Haukioja *et al.*, 1990; Coleman *et al.* 1994; Haukioja and Honkanen, 1996). We emphasize which sinks the damaged leaves were provisioning. Since the sink–source hierarchies within trees vary during the course of the season, timing of the damage in relation to seasonal alterations in resource allocation patterns of the tree has to be recognized.

In mountain birch, shifts in resource allocation can occur based on the consequences of environmental variation, and of damage to individual leaves (Haukioja *et al.* 1990; Ruohomäki *et al.*, 1996). Birch leaf-out in spring is a simultaneous flushing of leaves in short shoots. Since birches may flush when the soil is still frozen, leaf expansion in spring has to start with resources sequestered in the previous year or years. In other words, the amount of resources possible to use for the spring flush has been determined by the end of the previous growing season. In the far north, the limiting environmental factor is temperature, and accordingly the temperature sum of the previous year will modify average leaf size: leaves are small after cold seasons and large after warm ones (Haukioja *et al.*, 1985; Hanhimäki *et al.*, 1995). The spring flush of short-shoot leaves shifts from sinks to sources at the age of a few days (Valanne and Valanne, 1984). The young source leaves first provision the dominant sink in the developing bud in the short shoot itself. As in birches in general (Kozlowski *et al.*, 1991), the bud contains preformed leaf meristems for the next year. Therefore, the number of short-shoot leaves was already fixed in the year before the flush. After the short-shoot buds have matured, short-shoot leaves feed developing long shoots. Growth of long shoots, and of their leaves, happens with resources drawn from outside the shoot. Accordingly, elimination of even several leaves from a long shoot does not retard its growth (Kozlowski and Clausen, 1966; Ruohomäki *et al.*, in press). Provisioning of lateral meristems, roots and reserves of the tree presumably takes place in mid to late summer (Kozlowski *et al.*, 1991).

In summary, short-shoot leaves shift from sinks to sources, then provision their local meristems and, after that, meristems in other branches, stem and roots. This has two implications. First, consequences of damage may be found in the leaf itself, but also in plant parts whose meristems were supported by the damaged leaves. Second, the degree of independence of shoot modules depends on where the strong sinks are located. If they are within the boundaries of a morphological module, the module is fairly independent in carbon economy. If the sinks are situated outside the module, module independence is reduced or totally lacking. Since the location of dominant sinks within a tree changes within a season, the degree of independence of modules is a seasonally variable trait. In early summer, shoot modules may be quite independent in their carbon economy, whereas in late season the whole tree is integrated and shoots have lost their functional independence (Sprugel et al., 1991).

Since resources among tree modules are transferred on the basis of the strength that the particular meristematic sinks are able to create (Foyer, 1987; Clifford, 1992), intra-tree allocation of resources is based on competition among sinks (Wareing, 1959). If the number of growing leaves is reduced, the remaining leaves get more resources from a store of fixed size. But aftermaths of damage to an individual leaf also depend on its position in the sink/source hierarchy of the shoot. Apical meristems control basal meristems, a manifestation of apical dominance (Thimann and Skoog, 1934). Therefore, loss of a given amount of leaf mass, and of tree resources, has variable consequences depending on whether dominant or subordinate meristems were removed. This indicates, first, that tissues in different positions in meristem hierarchies may respond differently to similar damage (Marquis, 1992; Honkanen and Haukioja, 1994). Second, damage to leaves before, during or after they switch from sinks to sources may differ (Coleman, 1986; Haukioja et al., 1990; Coleman et al., 1994). Indeed, small losses of basal buds or basal sink leaves in mountain birch have no measurable consequences (Haukioja et al., 1990; Senn and Haukioja, 1994), presumably since the access to resources by basal leaves was restricted by the apical leaves. Biomass losses of similar size to apical buds or apical sink leaves releases the basal leaves from apical dominance, and during leaf flush they take dramatically more resources than usual, becoming large and rich in nutrients (Haukioja et al., 1990; Senn and Haukioja, 1994; Danell, Haukioja, Huss-Danell, unpublished data).

5.4.3 A NULL HYPOTHESIS FOR DIR AND DIS IN MOUNTAIN BIRCH

The above scenario of seasonal shifts in the sink–source regulative system offers a null hypothesis for damage-induced responses in mountain birch (Haukioja and Honkanen, 1996). By a null hypothesis, we mean an explanation that directly emerges from plant physiology. Upper level explanations, like evolutionary ones, need not identify all the mechanisms but must not be in disagreement with actual mechanisms and constraints. Essentially, the null hypothesis indicates that, after damage, resource availability for a plant module varies for three reasons. First, storage and/or currently produced photosynthates go down. This indicates an absolute shortage of resources at the tree level. Second, if the damage weakens local meristems, these meristems may not create strong enough sinks to compete successfully over resources in the common pool of the tree. This indicates a relative shortage of resources for a meristem. Third, when dominant meristems take less resources from a fixed common pool, other meristems have access to more.

That the relative and not only the absolute deficiency of resources is crucial is shown by the very different consequences that temporally and spatially variable foliar damage can cause. If dormant apical buds of birch or recently flushed apical sink-phase leaves are damaged, the rest of the foliage improves for herbivores (Danell and Huss-Danell, 1985): the remaining basal leaves become larger and more nutritious than they are normally. In other words, they show symptoms of DIS. But if foliar damage takes place a little later, at the time when leaves already have a positive net photosynthesis, the effects are seen in the weakening of preformed leaf meristems. These produce the leaves for the next year, and as such leaves cannot successfully compete for resources in the common pool, they remain small and less nutritious. This equates with symptoms described as DIR. Since the change of mountain birch leaves from sinks to sources happens within a few days, induced tree responses may also shift from DIS to DIR just because of the timing of an experiment. If leaf damage is further postponed, to the time when strong sinks occur in long shoots and lateral meristems, foliar damage causes relatively mild, or no, local responses (Tuomi *et al.*, 1988a). Therefore, conditions experienced by the whole plant (e.g. alterations in nutrition) may be translated quite specifically, and in variable ways, to plant responses relevant for herbivores.

Accordingly, the hormonal control of resource allocation and seasonal shifts in provisioning different meristems may give rise to a full range of different herbivore-induced responses: DIR, DIS or no responses at all. As the physiological features of the mountain birch *per se* produce the response, they do not demand more specific explanations. However, since the poor foliage quality described as DIR is accompanied by both lower nitrogen and higher phenolic contents (Tuomi *et al.*, 1988a), herbivores have the potential to select for less nutritious and more toxic compounds.

5.4.4 FACTORS MODIFYING DIR/DIS

If damage-induced responses, DIR or DIS, or both, contribute to the regular insect cycles in mountain birch and larch, there should be internal traits in these tree species at the particular environments which promote regular between-year time lags to population dynamics of the insects. Because most insect herbivore species on birch or larch do not have cyclical fluctuations in their population densities, traits of the insects must also be relevant. Both *Epirrita autumnata* and *Zeiraphera diniana* consume foliage in spring before, during and after the source/sink shift in foliage. Deciduousness of the host plants in a highly seasonal environment means that, unlike evergreens, there is only a single leaf age class to feed primordial meristems for the following year's foliage. Therefore, both resource allocation and building preformed meristems extend their effects on the quality of foliage between years. In other words, the carry-over effects of foliar damage from year to year are more probable in short growing seasons in strongly seasonal environments than under conditions in which current resources may be rapidly used to build new growth.

Ruohomäki *et al.* (1992) analysed defoliation experiments on mountain birch for the efficacy of DIR, which was measured as a decrease in larval performance in trees defoliated the year before. Temperature of the previous season did not explain the observed variation among years. Since spring resources of the mountain birch depend on the temperature sum of the previous growing season, it seems likely that the absolute resource availability of birch, as such, was not important for their DIR. Interestingly, annual variation in the strength of DIR, i.e. the difference in larval performance on control vs. previously defoliated trees, was more attributed to an annual variation in larval performance on control than on previously defoliated trees. On previously defoliated trees, larvae produced light pupae each year, while on control trees they produced light or heavy pupae. This indicates that defoliation reduced leaf quality the next year to a relatively invariable low value and therefore the calculated strength of DIR depended more on how much better larvae grew on undefoliated control trees than on previously defoliated trees (Ruohomäki *et al.*, 1992). Therefore, factors producing annual variation into suitability of leaves were important.

Timing of source/sink shifts is known to be sensitive to selection pressures exerted by herbivores (Dyer *et al.*, 1991). However, the short growing seasons for mountain birch and larch constrain the degrees of freedom for timing of source/sink shifts. The only possible sequence is: foliage growth / preformation of leaf meristems for next year's foliage/accumulation of other resources. Furthermore, building the preformed meristems has to happen within a short time interval and therefore quite synchronously. This makes it possible for disturbances to meristem maturation to have potentially large effects.

5.5 DISCUSSION

5.5.1 METHODOLOGICAL ASPECTS

The null hypothesis emphasizes the importance of detailed botanical knowledge in studies trying to unravel causal reasons behind herbivore-induced responses in trees. Seemingly small variations in defoliation treatments may lead to very different results ranging from DIR to IS, or to no response at all. For instance, differences of a few days in timing of damage and location of leaves in apical vs.

basal portions of shoots may have profound effects on the outcome. Such variations in experimental procedures may even reverse the direction of plant responses but remain unrecognized by the researcher (Haukioja et al., 1990).

Herbivory on a single branch or ramet may serve as an example. Early season foliar damage leads to DIR or DIS in only that unit (Haukioja et al., 1990). This could result from their own resource stores in shoot or branch modules (Tuomi et al., 1990), or from a reduced sink strength of meristems supported by the damaged leaves (Honkanen and Haukioja, 1994). By using Scots pine, Honkanen and Haukioja (1994) demonstrated that branch-level defoliation had a stronger retarding effect on foliage in a branch than similar damage extended for all branches. Since larger foliage losses led to smaller consequences, it is hard to see that the outcome would have resulted from amounts or ratios of lost nutrients. But the result is consistent with the importance of whole-tree physiology: branch-level defoliation creates stronger competitive asymmetries within canopies than defoliation covering all the branches.

Consequences of certain types of damage may not be seen earlier than the year after the treatment was conducted, leading researchers into potentially erroneus conclusions. For instance, damage to growing long-shoot leaves in birch does not affect shoot growth (Kozlowski and Clausen, 1966; Haukioja et al., 1990; Ruohomäki et al., in press). The explanation for this ostensibly paradoxical observation is that the growth of long shoots takes place from resources coming from outside the shoot. Still, damage to long-shoot leaves is not inconsequential because long-shoot leaves feed their auxilliary buds; therefore, shoots from the auxilliary buds of damaged leaves develop less well. But the outcome does not become obvious until the year after the damage, when leaves and shoots from these buds develop (Haukioja et al., 1990; Ruohomäki et al., in press).

5.5.2 POLARIZED OPINIONS ABOUT THE SAME SET OF CORRELATIONS

Researchers studying induced responses of trees have arrived at surprisingly disparate explanations. This partially relates to the common habit of oversimplifying and defending the importance of those aspects that are emphasized. Secondly, the hypotheses may be so versatile and contain so many auxilliary assumptions that the boundary between predictions emerging from the hypotheses and those based on attached, generally true assumptions and on indisputable empirical correlations becomes blurred.

For herbivore-induced responses in trees, four hypotheses offer basically similar predictions and are therefore quite hard to separate from each other. All four agree that rapidly growing plant parts – presumably because of their low content of phenolics and high nitrogen content – provide good diets for herbivores, and this approaches the plant vigour hypothesis of Price (1991).

For DIR, the carbon/nutrient availability hypothesis (Bryant et al., 1993) and the growth/differentiation hypothesis (Herms and Mattson, 1992) both offer very general explanations and see the negative correlation between phenols and nutrients as a causal one, caused by shortage of nutrients which retards tree growth (Table 5.2). High phenol accumulation is the consequence of growth retardation. In addition, both hypotheses employ plant physiology via concepts like apical dominance. The phytocentric view (Jones and Coleman, 1991) resembles closely the growth/differentiation hypothesis, although it more fully recognizes the importance of plant physiology. Our sink/source hypothesis emphasizes the same physiological mechanisms, but particularly stresses ontogenetic and functional constraints, and the mechanistic role of hormonal regulation in the use of resources at both the module and whole plant level, iterating a basic tenet by Kozlowski (1969) that 'tree growth appears to be limited more by the rate of carbohydrate conversion to new tissues, rather than solely by the amount of available foods'.

REFERENCES

Baldwin, I.T. (1988) The alkaloid responses of wild tobacco to real and simulated herbivory. *Oecologia* **77**, 378–381.

Baldwin, I.T. and Callahan, P. (1993) Autotoxicity and chemical defense: nicotine accumulation and carbon gain in solanaceous plants. *Oecologia* **94**, 534–541.

Baltensweiler, W. (1993) Why the larch bud-moth cycle collapsed in the subalpine larch–cembran pine forests in the year 1990 for the first time since 1850. *Oecologia* **94**, 62–66.

Baltensweiler, W., Benz, G., Bovey, P. and DeLucchi, V. (1977) Dynamics of larch budmoth populations. *Annu. Rev. Ent.* **22**, 79–100.

Baur, R., Binder, S. and Benz, G. (1991) Nonglandular leaf trichomes as short-term inducible defense of the grey alder, *Alnus incana* (L), against the chrysomelid beetle, *Agelastica alni* L. *Oecologia* **87**, 219–226.

Benz, G. (1974) Negative Rückkoppelung durch Raum- und Nahrungskonkurrenz sowie zyklische Veränderung dr Nahrungsgrundlage als Regelprinzip in der Populationsdynamik des Grauen Lärchenwicklers, *Zeiraphera diniana* (Guenée) (Lep., Tortricidae). *Z. ang. Ent.* **76**, 196–228.

Berryman, A.A. (ed.) (1988) *Dynamics of Forest Insect Populations. Patterns, causes, implications*, Plenum Press, New York, 603 pp.

Berryman, A.A. (1996) What causes populatin cycles of forest Lepidoptera? *Trend. Ecol. Evolut.* **11**, 28–32.

Bryant, J.P., Chapin, F.S. III and Klein, D.R. (1983) Carbon/nutrient balance of boreal plants in relation to vertebrate herbivory. *Oikos* **40**, 357–368.

Bryant, J.P., Danell, K., Provenza, F.D. *et al.* (1991a) Effects of mammal browsing on the chemistry of deciduous woody plants. In *Phytochemical Induction by Herbivores* (eds D.W. Tallamy and M.J. Raupp), pp. 135–154, John Wiley, New York.

Bryant, J.P., Heitkoning, I., Kuropat, P. and Owen-Smith, N. (1991b) Effects of severe defoliation on the long-term resistance to insect attack and on leaf chemistry in six woody species of the southern African savanna. *Amer. Natur.* **137**, 50–63.

Bryant, J.P., Reichardt, P.B., Clausen, T.P. and Werner, R.A. (1993) Effects of mineral nutrition on delayed inducible resistance in Alaska paper birch. *Ecology* **74**, 2072–2084.

Bylund, H. (1995) Long-term interactions between the autumnal moth and mountain birch: the roles of resources, competitors, natural enemies, and weather. PhD Dissertation, Sveriges Lantbruksuniversitet, Uppsala.

Chapin, F.S. III (1991) Integrated responses of plants to stress. *BioScience*, **41**, 29–36.

Chapin, F.S. III, Autumn, K. and Pugnaire, F. (1993) Evolution of suites of traits in response to environmental stress. *Am. Nat* **142**, 78–92.

Clifford, P.E. (1992) Understanding the source–sink concept of phloem translocation. *J. Biol. Educ.* **26**, 112–116.

Cline, M.G. (1991) Apical dominance. *Bot. Rev.* **57**, 318–358.

Coleman, J.S. (1986) Leaf development and leaf stress: increased susceptibility associated with sink–source transition. *Tree Physiol.* **2**, 289–299.

Coleman, J.S. and Jones, C.G. (1991) A phytocentric perspective of phytochemical induction by herbivores. In *Phytochemical Induction by Herbivores* (eds D.W. Tallamy and M.J. Raupp), pp. 3–45, John Wiley & Sons, New York.

Coleman, J.S., McConnaughay, K.D.M. and Ackerly, D.D. (1994) Interpreting phenotypic variation in plants. *Trend. Ecol. Evolut.* **9**, 187–191.

Craig, T.P., Price, P.W. and Itami, J.K. (1986) Resource regulation by a stem-galling sawfly on the arroyo willow. *Ecology* **67**, 419–425.

Danell, K. and Huss-Danell, K. (1985) Feeding by insects and hares on birches earlier affected by moose browsing. *Oikos* **44**, 75–81.

Davis, J.M., Gordon, M.P. and Smit, B.A. (1991) Assimilate movement dictates remote sites of wound-induced gene expression in poplars. *Proc. Natl. Acad. Sci USA* **88**, 2393–2396.

Duffey, S.S. and Felton, G.W. (1991) Enzymatic antinutritive defenses of the tomato plant against insects. In *Naturally Occurring Pest Bioregulators* (ed. P.A. Hedin), pp. 166–197, American Chemical Society, Washington, DC.

Dyer, M.I., Acra, M.A., Wang, G.M. *et al.* (1991) Source–sink carbon relations in two *Panicum coloratum* ecotypes in response to herbivory. *Ecology* **72**, 1472–1483.

Edwards, P.J., Wratten, S.D. and Cox, H. (1985) Wound-induced changes in the acceptability of tomato to larvae of *Spodoptera littoralis*: a laboratory bioassay. *Ecol. Entomol.* **10**, 155–158.

Foyer, C.H. (1987) The basis for source–sink interaction in leaves. *Plant Phys. Biochem.* **25**, 649–657.

Geiger, D.R. (1987) Understanding interactions of source and sink regions of plants. *Plant. Physiol. Biochem.* **25**, 659–666.

Hallé, F. (1986) Modular growth in seed plants. *Philos. Trans. R. Soc. Lond. B.* **313**, 77–87.

Hanhimäki, S., Senn, J. and Haukioja, E. (1995) The convergence in growth of foliage-chewing insect species on individual mountain birch trees. *J. Anim. Ecol.* **64**, 543–552.

Haukioja, E. (1982) Inducible defences of white birch to a geometrid defoliator, *Epirrita autumnata*. In *Insect–plant Relationships* (eds J.H. Visser and A.K. Minks), pp. 199–203, PUDOC, Wageningen.

Haukioja, E. (1990) Induction of defenses in trees. *Annu. Rev. Ent.* **36**, 25–42.

Haukioja, E. (1991) The influence of grazing on the evolution, morphology and physiology of plants as modular organisms. *Philos. Trans. R. Soc. Lond. B.* **333**, 241–247.

Haukioja, E. and Honkanen, T. (1996) Why are tree responses to herbivory so variable?. In *Dynamics of Forest Herbivory: Quest for Pattern and Principle* (eds W.J. Mattson, P. Niemelä and M. Rousi), pp. 1–10, US For. Gen. Tech. Rept. NC-183, US Forest Service, Minneapolis.

Haukioja, E. and Neuvonen, S. (1985) Induced long-term resistance in birch foliage against defoliators: defensive or incidental? *Ecology* **66**, 1303–1308.

Haukioja, E., Niemelä, P. and Sirén, S. (1985) Foliage phenols and nitrogen in relation to growth, insect damage, and ability to recover after defoliation, in the mountain birch *Betula pubescens* ssp. *tortuosa*. *Oecologia* **65**, 214–222.

Haukioja, E., Neuvonen, S., Hanhimäki, S. and Niemelä, P. (1988) The autumnal moth in Fennoscandia. In *Dynamics of Forest Insect Populations. Patterns, Causes, Implications* (ed. A.A. Berryman), pp 163–178, Plenum Press, New York.

Haukioja, E., Ruohomäki, K., Senn, J. *et al.* (1990) Consequences of herbivory in the mountain birch (*Betula pubescens* ssp. *tortuosa*): importance of the functional organization of the tree. *Oecologia* **82**, 238–247.

Herms, D.A. and Mattson, W.J. (1992) The dilemma of plants: to grow or defend. *Quart. Rev. Biol.* **67**, 283–335.

Honkanen, T. and Haukioja, E. (1994) Why does a branch suffer more after branch-wide than after tree-wide defoliation? *Oikos* **71**, 441–450.

Hunter, M.D. and Schultz, J.C. (1995) Fertilization mitigates chemical induction and herbivore responses within damaged oak trees. *Ecology* **76**, 1226–1232.

Jerling, L. (1985) Are plants and animals alike? A note on evolutionary plant population ecology. *Oikos* **45**, 150–153.

Jones, C.G. and Coleman, J.S. (1991) Plant stress and insect herbivory: toward an integrated perspective. In *Response of Plants to Multiple Stresses* (eds H.A. Mooney, W.E. Winner and E.J. Pell), pp. 249–280, Academic Press, San Diego.

Jones, C.G., Hopper, R.F., Coleman, J.S. and Krischik, V.A. (1993) Control of systematically induced herbivore resistance by plant vascular architecture. *Oecologia* **93**, 452–456.

Jongsma, M.A., Bakker, P.L., Visser, B. and Stiekema, W.J. (1994) Trypsin inhibitor activity in mature tobacco and tomato plants is mainly induced locally in response to insect attack, wounding and virus infection. *Planta* **195**, 29–35.

Karban, R. (1993) Costs and benefits of induced resistance and plant density for a native shrub, *Gossypium thurberi*. *Ecology* **74**, 9–19.

Karban, R. and Carey, J.R. (1984) Induced resistance of cotton seedlings to mites. *Science* **225**, 53–54.

Karban, R. and Myers, J.H. (1989) Induced plant responses to herbivory. *Annu. Rev. Ent.* **20**, 331–348.

Karban, R. and Niiho, C. (1995) Induced rersistance and susceptibility to herbivory: plant memory and altered plant development. *Ecology* **76**, 1220–1225.

Kozlowski, T.T. (1969) Tree physiology and forest pests. *J. For.* **67**, 118–123.

Kozlowski, T.T. and Clausen, J.J. (1966) Shoot growth characteristics of heterophyllous woody plants. *Can. J. Bot.* **44**, 827–843.

Kozlowski, T.T., Kramer, P.J. and Pallardy, S.G. (1991) *The Physiological Ecology of Woody Plants*, Academic Press, San Diego, California.

Lewinsohn, E., Gijzen, M. and Croteau, R. (1991a) Defense mechanisms of conifers. Differences in constitutive and wound-induced monoterpene biosynthesis among species. *Plant Physiol.* **96**, 44–49.

Lewinsohn, E., Gijzen, M., Savage, T.J. and Croteau, R. (1991b) Defense mechanisms of conifers. Relationship of monoterpene cyclase activity to anatomical specialization and oleoresin monoterpene content. *Plant Physiol.* **96**, 38–43.

Marquis, R.J. (1992) A bite is a bite is a bite? Constraints on response to folivory in *Piper arieianum* (Piperaceae). *Ecology* **73**, 143–152.

May, R.M. (1976) Models for single populations. In *Theoretical Ecology. Principles and Applications* (ed. R.M. May), pp. 4–25, Blackwell, Oxford.

Nicholson, R.L. and Hammerschmidt, R. (1992) Phenolic compounds and their role in disease resistance. *Annu. Rev. Phytopathol.* **30**, 369–389.

Niemelä, P. and Tuomi, J. (1993) Sawflies and inducible resistance of woody plants. In *Sawfly Life History Adaptations to Woody Plants* (eds M.R. Wagner and K.F. Raffa), pp. 211–227, Academic Press, New York.

Parsons, T.J., Bradshaw, H.D. Jr and Gordon, M.P. (1989) Systemic accumulation of specific mRNAs in response to wounding in poplar trees. *Proc. Natl. Acad. Sci USA* **86**, 7895–7899.

Price, P.W. (1991) The plant vigor hypothesis and herbivore attack. *Oikos* **62**, 244–251.

Raffa, K.F. and Berryman, A.A. (1987) Interacting selective pressures in conifer-bark beetle systems: a basis for reciprocal adaptations? *Am. Nat.* **129**, 234–262.

Roland, J. and Myers, J.H. (1987) Improved insect performance from host-plant defoliation: winter moth on oak and apple. *Ecol. Entomol.* **12**, 409–414.

Rossiter, M., Schultz, J.C. and Baldwin, I.T. (1988) Relationships among defoliation, red oak phenolics, and gypsy moth growth and reproduction. *Ecology* **69**, 267–277.

Ruohomäki, K., Hanhimäki, S., Haukioja, E. *et al.* (1992) Variability in the efficacy of delayed inducible resistance in mountain birch. *Entomol. exp. appl.* **62**, 107–115.

Ruohomäki, K., Chapin, F.S. III, Haukioja, E. *et al.* (1996) Delayed inducible resistance in mountain birch in response to fertilization and shade. *Ecology* **77**, 2302–2311.

Ruohomäki, K., Haukioja, E., Repka, S. and Lehtilä, K. (in press) Value of leaves: effects of damage to individual leaves on growth and reproduction of mountain birch shoots. *Ecology* (in press).

Sachs, T. (1991). *Pattern Formation in Plant Tissues*, Cambridge University Press, Cambridge.

Sachs, T. (1993) The role of auxin in plant organization. *Acta Hortic.* **329**, 162–168.

Sachs, T., Novoplansky, A. and Cohen, D. (1993) Plants as competing populations of redundant organs. *Plant Cell Environ.* **16**, 765–770.

Schultz, J.C. (1988) Plant responses induced by herbivores. *Trend. Ecol. Evolut.* **3**, 45–49.

Seldal, T., Dybwad, E., Andersen, K. and Högstedt, G. (1994) Wound-induced proteinase inhibitors in grey alder (*Alnus incana*): a defence mechanism against attacking insects. *Oikos* **71**, 239–245.

Senn, J. and Haukioja, E. (1994) Reactions of the mountain birch to bud removal: effects of severity and timing, and implications for herbivores. *Funct. Ecol.* **8**, 494–581.

Sprugel, D.G., Hinckley, T.M. and Schaap, W. (1991) The theory and practice of branch autonomy. *Ann. Rev. Ecol. Syst.* **22**, 309–340.

Stout, M.J., Workman, J. and Duffey, S.S. (1994) Differential induction of tomato foliar proteins by arthropod herbivores. *J. Chem. Ecol.* **20**, 2575–2594.

Tenow, O. (1972) The outbreaks of *Oporinia autumnata* Bkh. and *Operophthera* spp. (Lep., Geometridae) in the Scandinavian mountain chain and northern Finland 1862–1968. *Zool. Bidr. Uppsala* **2**, 1–107.

Tenow, O. and Bylund, H. (1989) A survey or winter cold in the mountain birch/*Epirrita autumnata* system. *Mem. Soc. F. F. Fenn.* **65**, 67–72.

Thimann, K.V. and Skoog, F. (1934) On the inhibition of bud development and other functions of the growth substance in *Vicia faba*. *Proc. R. Soc. Lond. B.* **114**, 317–339.

Tuomi, J., Niemelä, P., Haukioja, E. *et al.* (1984). *Oecologia* **61**, 208–210.

Tuomi, J., Niemelä, P., Rousi, M. *et al.* (1988a) Induced accumulation of foliage phenols in mountain birch: branch response to defolication? *Amer. Natur.* **132**, 602–608.

Tuomi, J., Niemelä, P., Chapin, F.S. III *et al.* (1988b) Defensive responses of trees in relation to their carbon/nutrient balance. In *Mechanisms of Woody Plant Defenses against Insects. Search for Pattern* (eds W.J. Mattson, J. Levieux and C. Bernard-Dagan), pp. 57–72, Springer, New York.

Tuomi, J., Niemelä, P. and Sirén, S. (1990) The Panglossian paradigm and delayed inducible accumulation of foliar phenolics in mountain birch. *Oikos* **59**, 399–410.

Tuomi, J., Fagerström, T. and Niemelä, P. (1991) Carbon allocation, phenotypic plasticity, and induced defenses. In *Phytochemical Induction by Herbivores* (eds D.W. Tallamy and M.J. Raupp), pp. 85–104, John Wiley, New York.

Tuomi, J., Haukioja, E., Honkanen, T. and Augner, M. (1994) Potential benefits of herbivore behaviour inducing amelioration of food-plant quality. *Oikos* **70**, 161–166.

Valanne, N. and Valanne, T. (1984) The development of the photosynthetic apparatus during bud burst and leaf opening in two subspecies of *Betula pubescens* Ehr. *Rep. Kevo Subarct. Res. Stat.* **19**, 1–10.

Vuorisalo, T. and Tuomi, J. (1986) Unitary and modular organisms: criteria for ecological division. *Oikos* **47**, 382–391.

Wareing, P.F. (1959) Problems of juvenility and flowering in trees. *J. Linn. Soc. Lond. Bot.* **56**, 282–289.

Watson, M.A. and Casper, B.B. (1984) Morphological constraints on patterns of carbon distribution in plants. *Annu. Rev. Ecol. Syst.* **15**, 233–258.

White, J. (1979) The plant as a metapopulation. *Annu. Rev. Ecol. Syst.* **10**, 109–145.

INCORPORATING VARIATION IN PLANT CHEMISTRY INTO A SPATIALLY EXPLICIT ECOLOGY OF PHYTOPHAGOUS INSECTS

Mark D. Hunter

6.1 INTRODUCTION

In recent years, the study of ecological interactions has become increasingly spatially explicit. The re-emergence of metapopulation models (Lande, 1987; Gilpin and Hanski, 1991; Hanski *et al.*, 1994; Sjögren, 1994) that modify Levins' (1969) original model, the debate on the causes and consequences of spatial density dependence (Hassell *et al.*, 1987; Stewart-Oaten and Murdoch, 1990; Elkinton *et al.*, 1990) and the exploitation of source–sink models in conservation biology (Pullium, 1988; Watkinson and Sutherland, 1995) have brought spatially explicit population ecology into the mainstream of the ecological literature. In one sense, ecology has 'come of age' because investigators are now willing to embrace, rather than to ignore, spatial heterogeneity in the action of ecological factors or processes. In the minds of some authors at least, space is the 'final frontier' of ecological theory (Kareiva, 1994; Tilman, 1994; Holmes *et al.*, 1994; Molofsky, 1994; Goldwasser *et al.*, 1994).

There is a danger, however, that history will repeat itself as the young discipline of spatial ecology matures. What might be called 'classic' population ecology suffers from two historical weaknesses: a theoretical construct that under-represents the importance of variation in resource availability and quality (Price, 1992; Foster *et al.*, 1992; Hunter and Price, 1992) and a development led more by the conveniences of modelling than the rigour of experimentation and validation (Price and Hunter, 1995). There are clear signs that spatially explicit models may be following the same developmental pattern (Doak and Mills, 1994). Certainly, there is a lack of balance between the importance placed on mortality factors versus factors that influence resource availability in both theoretical and empirical studies. For example, much emphasis has been placed on spatial variation in mortality factors, such as the action of natural enemies. The continuing debate on the circumstances under which spatial density-dependent mortality will stabilize or destabilize an insect population is one example of this (Hassell *et al.*, 1987; Stewart-Oaten and Murdoch, 1990). The 'top-down' component of spatial models is well developed, and there is some considerable understanding of mechanism – the functional and numerical responses of particular natural enemies are often explicit components of population models and, in the best cases, are based on empirical data on the behaviour and reproduction of parasites and predators (Gould *et al.*, 1990).

The 'bottom-up' component of spatially explicit population ecology is not nearly as well developed, and is far less mechanistic. In most models,

---

Forests and Insects. Edited by A.D. Watt, N.E. Stork and M.D. Hunter. Published in 1997 by Chapman & Hall, London. ISBN 0 412 79110 2.

for example, the reproductive rate of a species is held constant across space (Gilpin and Hanski, 1991). Patch size and the distance between adjacent patches are often the only parameters linked to the potential population size of a particular prey species; patch quality is rarely considered. Although source–sink models have as their premise greater reproduction or dispersal from high quality 'source' patches (Pulliam, 1988), the lack of mechanism in such models leaves them open to a variety of interpretations (Watkinson and Sutherland, 1995). While both patch size and isolation are critical parameters in spatially explicit models (Hanski et al., 1994), they do not address the mechanisms that determine the distribution of organisms in space, nor their potential rates of increase and population equilibria. An understanding of patch quality is the only way to address the mechanisms by which resource availability influences the dynamics of species in a spatially complex environment (Thomas et al., 1996). If a realistic balance between resource availability and the action of natural enemies is wanted in theoretical and empirical models (Hunter and Price, 1992; Power, 1992) the mechanistic basis by which resource availability influences the distribution of organisms in space must be incorporated explicitly.

6.2 SPATIAL VARIATION IN RESOURCE QUALITY AND FOREST INSECTS

Like all organisms, forest insects are exposed to significant spatial and temporal variability in the ecological forces that influence their population dynamics and community structure. Recent texts on forest insects reveal a wealth of information on how variation in both the efficacy of natural enemies and the quality of host plants influence tree-feeding insects (Barbosa and Schultz, 1987; Barbosa and Wagner, 1989; Speight and Wainhouse, 1989; Watt et al., 1990). With all of this information available, it might be expected that a consensus would emerge regarding the relative importance of top-down and bottom-up forces for forest insect populations. Yet the debate continues to rage about the evolutionary (Bernays and Graham, 1988; Schultz, 1988; Thompson, 1988)

and ecological (Hunter and Price, 1992; Price, 1992; Schultz, 1992; Price and Hunter, 1995) significance of plant quality and natural enemies for the distributions and abundances of phytophagous insects. This debate is yet to have much impact on the development of the new, spatially explicit ecology because variation in bottom-up forces is essentially ignored in these models.

Many different characteristics of trees have been shown to influence the spatial distributions of forest insects. Foliar chemistry (Cates and Zou, 1990; Gershenzon, 1994), budburst phenology (Du Merle and Mazet, 1983; Hunter, 1992), leaf-fall phenology (Connor et al., 1994; Mopper and Simberloff, 1995), and tree genotype (Whitham, 1989; Preszler and Boecklen, 1994; Mopper et al., 1995) are just some of the factors that can generate spatial variation in insect densities. These are also the host-derived factors that can influence the only four parameters capable of causing population change: birth, death, immigration and emigration. As such, they provide the potential mechanistic basis for bottom-up effects in spatially explicit models. Variation in these host plant parameters can occur at multiple spatial scales, from the level of the leaf (Hunter, 1987) up to variation in tree quality among different landscapes (Schultz et al., 1990). We might also expect insect herbivores to respond to spatial variation in plant traits at the same range of spatial scales.

For the purposes of this chapter, however, discussion will be restricted to variation in one plant trait (plant chemistry, defensive and nutritional) at one spatial scale (differences among individual trees within a population), and the consequences for the distributions of insect herbivores among trees will be considered. It is widely believed that plant chemistry is important to the colonization, survival and reproduction of many insect herbivores (Rosenthal and Janzen, 1979; Denno and McClure, 1983; Tallamy and Raupp, 1991). The term 'plant chemistry' is used loosely here to include potential allelochemicals (tannins, terpenoids, phenolic glycosides, etc.), and foliar protein. Plant chemicals may have multiple functions for a particular plant species but, for present purposes, must have some influence on herbivore distributions; that influence can be either positive or negative (Rosenthal and Berenbaum, 1991). This chapter is an initial attempt to answer the following questions:

1. **Is spatial variation in plant chemistry of general importance to the many different herbivores that feed on a single host tree species?** If spatial variation in plant chemistry is only important to a few members of any community, then there may be no inclination to consider it to be a critical component of spatial models in general.
2. **Does significant reduction in genetic variability among individual trees dilute the importance of spatial variation in plant chemistry?** If variation in plant chemistry is shown to be an important determinant of insect distribution in systems with artificially low genetic variability, then it is likely to be important in most natural systems where genetic variability should be higher.
3. **How might spatial variation in plant quality be incorporated into a spatially explicit ecology of forest insects?** Both technical and theoretical difficulties must be overcome to include spatial variation in plant quality in spatially explicit models.

6.3 DOES PLANT QUALITY REALLY MATTER? A GRAPHICAL MODEL

There remains a misconception among some population ecologists that, for insect herbivore populations dominated by the action of natural enemies, variation in plant quality cannot influence the herbivore population. This misconception needs to be rectified before plant quality (and consequently spatial variation in plant quality) is recognized as a fundamental feature of forest insect population dynamics. A simple graphical model, based on Ricker (1954) curves of recruitment but simplified for the current discussion, is presented here to illustrate the importance of variation in plant quality for a herbivore population across a range of densities.

In Figure 6.1a, the rate of mortality of the insect herbivore is seen to vary with herbivore density, with predation dominating at relatively low herbivore densities and with intraspecific competition, and perhaps disease, dominating at higher densities. Note that predation is density dependent, up to some threshold, and that the natural enemy therefore has the potential to maintain the herbivore population below its carrying capacity. Simply adding the two curves together (Figure 6.1b) shows how the rate of mortality should vary across a range of herbivore densities.

If it is then considered how birth rate might change with density (Figure 6.1c), some decline in birth rate might be expected as density increases. Many studies have shown declines in pupal weight and fecundity of forest insects with increasing density (e.g. Rossiter *et al.*, 1988; Hunter and Willmer, 1989). In Figure 6.1c three host plants are represented, varying from high to low quality. Again, there is considerable empirical evidence that the fecundity of forest insects varies with host plant quality (Barbosa and Schultz, 1987; Watt *et al.*, 1990). In general, then, birth rate declines with increasing density and with decreases in host plant quality.

By combining the mortality and natality curves (Figure 6.1d) we can visualize the equilibrium population densities of the herbivore. Where the natality and mortality curves intersect, birth rate = death rate, and population equilibrium occurs. There are five potential equilibrium densities in Figure 6.1d, one of which (U) is unstable. K is the carrying capacity of the herbivore population on the high quality host plant. On the intermediate and low quality hosts, low birth rates never allow the herbivore population to 'escape' from the effects of predation (N1 and N2). Note, however, that even when the herbivore population is regulated below its carrying capacity by the natural enemy, any change in plant quality will still cause a shift in the equilibrium density of the herbivore population (N1 to N3). In other words, there is no such thing as a herbivore population immune to the effects of variation in plant quality, even if its dynamics are dominated by predation. The equilibrium density of the population depends on both natural enemies and host plant quality simultaneously – a balance of top-down and bottom-up forces (see Belovsky and Joern, 1995, for a similar theoretical argument

84 *Incorporation variation in plant chemistry*

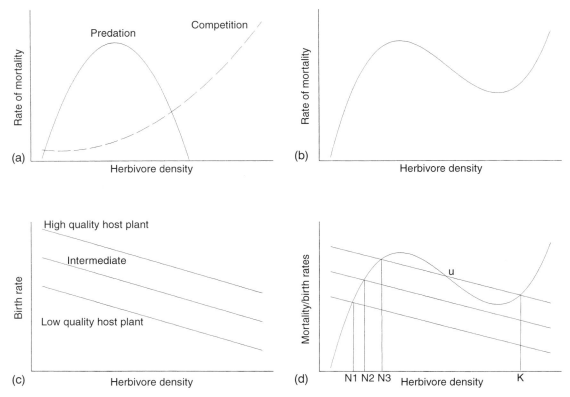

Figure 6.1 Model of combined roles of mortality factors and food quality in determining equilibrium density of insect herbivores. Note that, even when herbivore density is maintained below carrying capacity by a natural enemy, variation in food quality still causes variation in equilibrium density. (See text for details.)

with empirical support). In this model, plant quality directly affects herbivore reproduction. It should be noted that plant quality may also influence herbivore mortality, either in a density-dependent fashion (this would be included in the competition curve in Figure 6.1a), or else in a density-independent fashion. The latter would not influence equilibrium density, but could cause significant fluctuations around the equilibrium (e.g. Varley *et al.*, 1973).

The variation in plant quality represented by the three hosts in Figure 6.1d could be variation among plant species, among individual plants within one population, or temporal variation in host plant quality. In all cases, such variation in quality has an impact on the population ecology of the herbivore species. I would argue that spatial variation in plant quality is a near-ubiquitous feature of insect–plant systems, that it will cause spatial variation in natality and mortality rates, and that spatially explicit population models based solely on patch size rather than patch quality do not accurately reflect natural systems.

6.4 SPATIAL VARIATION IN FOLIAGE CHEMISTRY AND OAK HERBIVORES

If spatial variation in plant quality is to be incorporated into modern population ecology, we need to demonstrate that it is of general importance to insect herbivores. The vast majority of studies of plant chemistry-herbivore interactions have focused on the responses of one particular herbivore species to one or a few plant secondary compounds. While these studies have been critical to the development of the field of insect–plant interactions, they naturally contain some bias in the choice of insect and/or plant of study. An alternative approach is to consider the effects of variation in secondary chemistry on all the members of an insect herbivore community. The strength of this approach is that it assesses the generality of the importance of variation in plant quality for insect populations (Stiling and Rossi, 1995), irrespective of their economic, social or conservation status. One corresponding weakness, however, is that such studies are necessarily correlative rather than mechanistic – there are too many species involved for experiments on each one. Recognizing this limitation, the results of a study of the effects of spatial variation in foliar phenolic concentration in oak trees on a community of oak-feeding herbivores are presented below. Although the results are correlative in nature, they support the position that variation in plant chemistry affects most of the herbivore species in a community.

6.4.1 METHODS

During 1991, the insect fauna and foliar chemistry of 60 oak saplings, 2–4 m in height, were sampled in a west-facing mixed hardwood stand in Center County, Pennsylvania. The saplings varied considerably in morphology, and represented a mixture of 'pure' and hybrid individuals of red and black oak, *Quercus rubra* and *Q. velutina*, respectively. Hybridization between these species is a natural phenomenon in this region and reflects the variation in plant quality to which oak herbivores are typically exposed. While efforts to genotype these individual oaks continue, they have been pooled into one 'red oak group' for the purposes of the analyses presented here.

On June 3, 1991, 10–15 leaves were removed haphazardly from each sapling, pooled to a single sample per tree, and immediately flash-frozen in liquid nitrogen in the field. Samples were then lyophilized and ground to a fine powder for analysis. Five estimates of foliage quality were derived from these samples: gallotannins (as potassium iodate-reactive phenolics); proanthocyanidins (condensed tannin estimate); total phenolics (Folin–Denis method); foliar astringency (radial diffusion assay); and protein content. The methods for these analyses, or references to them, are given by Hunter and Schultz (1993).

Also on June 3, 1991, complete arthropod counts were made from the same individual trees. Each leaf on each tree was examined for arthropods, which were identified to species where possible. Many species, however, were represented by one or two individuals only, and all arthropods were categorized into one of the following six functional groups: gall-formers, leaf-miners, leaf-chewers, sap-suckers, ants, or other invertebrate predators (mostly spiders). Although grouping species by guild does not allow the series of single-species comparisons that would be ideal for such a study, it is a compromise that allows the inclusion of rare species and avoids multiple statistical tests on low density herbivores. After transformation to normalize the sampling data (log transformations for gallers and suckers, square-root transformations for the other guilds), the densities of the four herbivore guilds were standardized to density per 100 leaves and analysed separately by stepwise multiple regression procedures (SAS Institute, 1985). The five chemical measures and the two predator guilds were independent variables in each of the four models ($N = 60$ trees).

6.4.2 RESULTS

There was considerable variation in the concentrations of foliar phenolics among individual trees (Table 6.1), whereas the protein content of foliage varied much less. Proanthocyanidin concentrations and foliar astringency were particularly variable, with 13- and 23-fold variation among individual trees, respectively (Table 6.1).

The densities of gall-forming insects among trees were related to foliar astringency and proanthocyanidin concentrations only ($F = 4.74$; d.f. = 2,56; $P = 0.013$). Gall densities increased with increasing foliar astringency and increasing condensed tannin concentrations (Figure 6.2). No other variables entered into the stepwise model.

Table 6.1 Variation in the concentrations of foliar phenolic and protein from 60 red oak trees.

Variable	N	Mean	Min.	Max.	S.D.	Fold
Gallotannin	60	26.36	7.03	50.95	13.49	7.2 ×
Proanthocyanidin	59	7.39	2.75	36.24	5.41	13.2 ×
Astringency	60	1.88	0.17	3.88	0.74	22.8 ×
Total phenolics	60	9.96	4.65	18.01	3.48	3.9 ×
Protein	60	7.09	5.80	8.99	0.57	1.5 ×

S.D. = standard deviation.
Fold = fold variation from the lowest to the highest concentration.
Gallotannin and total phenolics expressed as % tannic acid equivalents.
Proanthocyanidin expressed as % quebracho tannin equivalents.
Astringency is cm^2 diffusion.
Protein is mg % BSA equivalents.

Leaf-miner densities showed the same apparent responses to foliar phenolics as did gall-formers, with increasing miner densities on trees with high foliar astringency and high proanthocyanidin concentrations (F = 10.49; d.f. = 2,56; P = 0.0002; Figure 6.3). Again, no other variables entered the stepwise model.

The densities of sap-sucking insects were unrelated to spatial variation in the phenolic concentrations of individual trees. Indeed, only non-ant

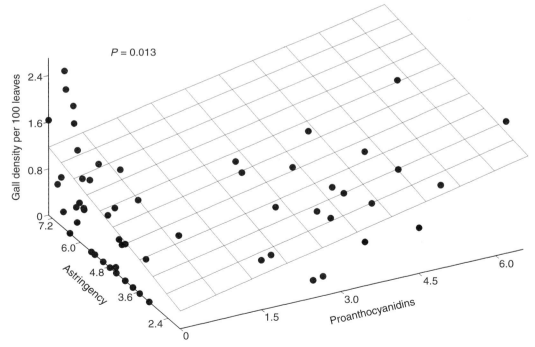

Figure 6.2 Relationship between gall density (log 10) and foliar phenolics from 60 oak trees in Pennsylvania. Proanthocyanidins are percentage quebracho tannin equivalents; astringency is cm^2 radial diffusion.

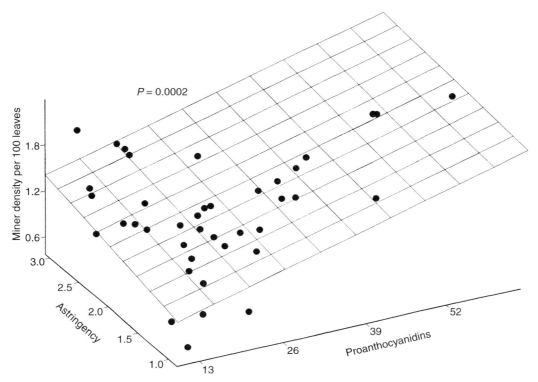

Figure 6.3 Relationship between leaf-miner density (square-root transformed) and foliar phenolics from 60 oak trees in Pennsylvania. Proanthocyanidins are percentage quebracho tannin equivalents; astringency is cm^2 radial diffusion.

predator densities were correlated with sap-sucker densities, and the relationship was positive ($F = 14.41$; d.f. = 1,58; $P = 0.0004$; Figure 6.4). We found no evidence that ants were tending sap-suckers in this system; there was no positive correlation between the densities of sap-suckers on individual trees and the densities of ants on those trees.

Leaf-chewer densities were related to both non-ant predator densities and foliar gallotannin concentrations ($F = 8.21$; d.f. = 2,57; $P = 0.0007$). The relationship with predators was positive and relationship with gallotannin was negative (Figures 6.5a and 6.5b, respectively). No other variables entered the stepwise model.

6.4.3 CONCLUSIONS – OAK STUDY

Several conclusions can be drawn from the results presented above. First, three out of four oak herbivore guilds exhibited spatial distributions among trees that were significantly correlated with concentrations of allelochemicals. This supports the view that tree chemistry is of general importance to

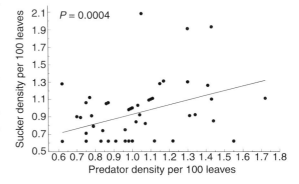

Figure 6.4 Relationship between sap-sucker density (log 10) and arthropod predator density (square-root transformed) from 60 oak trees in Pennsylvania.

88 *Incorporation variation in plant chemistry*

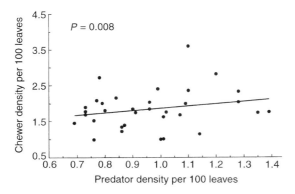

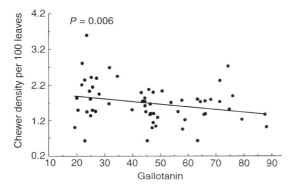

Figure 6.5 Relationship between leaf-chewer density (square-root transformed) and (a) arthropod predator density (square-root transformed) and (b) foliar gallotannin concentrations from 60 oak trees in Pennsylvania. Gallotannins are percentage tannic acid equivalents.

the spatial distribution of forest insects and not an idiosyncratic feature of species-by-species studies. Second, both guilds of enclosed feeders showed the same responses – positive correlations with foliar astringency and proanthocyanidins. Other studies have suggested that gallers and miners, living in humid environments within leaf tissue, may obtain some protection from their pathogens on high tannin foliage (Taper *et al.*, 1986; Taper and Case, 1987). Third, the densities of enclosed feeders were unrelated to the densities of arthropod predators on trees. In contrast, the exposed suckers and chewers both exhibited statistical relationships with arthropod predators. For both guilds, the relationships were positive, suggesting that arthropod predators track the abundances of exposed feeders in space, either by reproduction or by movement. Fourth, sucker densities were unrelated to oak chemistry in this study, suggesting that phloem-feeding insects may be able to avoid certain tissue-bound allelochemicals. Finally, leaf-chewer densities exhibited a negative correlation with gallotannin concentrations. Free-living leaf-chewers are unlikely to avoid tissue-bound metabolites, and may only rarely obtain an anti-pathogenic benefit from consuming high tannin foliage (for example, at the peaks of population cycles or during outbreaks; Rossiter, 1987; Foster *et al.*, 1992; Hunter and Schultz, 1993). The negative relationship between chewer density and hydrolysable tannins is consistent with published work that has shown negative effects of oak tannins on chewer growth, survival and reproduction (Schultz and Baldwin, 1982; Rossiter *et al.*, 1988).

What are the sources of individual variation in foliage secondary metabolites in oak? Like most traits expressed by organisms, resistance to herbivores probably depends upon tree genetics, the environment in which the tree is growing, and gene-by-environment interactions (Fritz and Simms, 1992; Weis and Campbell, 1992). The carbon–nutrient balance hypothesis (Chapin, 1980; Bryant *et al.*, 1983) makes some predictions about how oak chemistry should vary with environmental variation. Specifically, oaks growing under high light or low nutrient conditions should have higher foliar tannin concentrations than trees growing under low light or high nutrient conditions. Although there are some clear limitations to the carbon–nutrient balance hypothesis (Muzika, 1993; Muzika and Pregitzer, 1992), it is possible that some of the spatial variation in oak phenolic chemistry described above results from spatial variation in resource availability for the trees.

This was tested by manipulating the nutrient availability for *Q. rubra* and *Q. prinus* saplings growing in a common garden adjacent to the field site in Pennsylvania (above). Treatment trees received 140 g of Osmocote timed-release fertilizer in April, and their foliage chemistry was sampled on July 16 of the same year and compared with controls. Results were broadly consistent with the carbon–nutrient balance hypothesis: fertilized oaks had lower levels of gallotannins (both species) and lower levels of condensed tannins (*Q. prinus* only) than did control trees (Hunter and Schultz, 1995; Figure 6.6). It is probable that natural spatial variation in resource availability for oaks – for example, changes in soil quality along a gradient – will influence foliage chemistry and herbivore distributions among trees. Another environmental variable that can influence oak chemistry is defoliation itself. Oaks respond to insect herbivory by increasing the concentrations of tannins in their foliage (Rossiter *et al.*, 1988). In the fertilization experiment just described, fertilization 'turned off' phenolic induction in chestnut oak. In other words, spatial variation in nutrient availability is probably important for spatial variation in the strength of induction responses in oak and polyphenol concentrations in general.

6.5 SPATIAL VARIATION IN APPLE FOLIAR PHENOLICS AND EFFECTS ON HERBIVORE DENSITY

It could be argued that the results presented above are atypical because trees that hybridize readily, like oaks, may exhibit above average genetic variability, and therefore above average variation in foliar chemistry among trees. Commercial apple orchards provide a suitable comparison for the oak study because the trees have been selected for low genetic variability (we expect Golden Delicious apples to taste much the same each time that we buy them) and environmental variability is minimized within a 'stand' by common management practices. Moreover, commercial apple trees are grafted on common root stocks that should further reduce variation among individual trees. Although commercial apple orchards are clearly not forest systems, they provide a useful comparison with oaks: if there is still meaningful variation in foliage chemistry among apple trees in an orchard, and if that variation is of ecological significance to insect herbivores, then spatial variation in foliar chem-

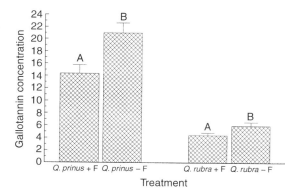

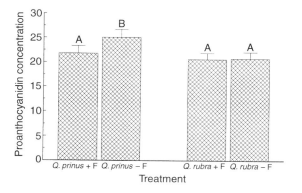

Figure 6.6 Comparisons of (a) gallotannin concentrations and (b) proanthocyanidin concentrations in foliage of *Quercus prinus* and *Q. rubra* saplings. Trees were either fertilized (+ F) or left as controls (– F). Proanthocyanidins are percentage quebracho tannin equivalents; gallotannins are percentage tannic acid equivalents. ($N = 10$ for each treatment.)

istry should be likely in most natural forest systems.

Two important phenolic compounds in the foliage of *Malus* species are the dihydrochalcone glycoside, phloridzin, and its aglycone, phloretin (Williams, 1964; Growchowska and Ciurzynska, 1979). While many different physiological and ecological roles have been suggested for phloridzin and phloretin (MacDonald and Bishop, 1952; Raa, 1968; Challice and Williams, 1970; Klinghauf, 1971; Hunter, 1975; Bassuk *et al.*, 1981), they are considered by some to be potent antiherbivore allelochemicals (Montgomery and Arn, 1974). During

1991, I measured variation in phloretin and phloridzin concentrations among apple trees from five different cultivars, and investigated the effects of the two compounds on a common Lepidopteran pest of Pennsylvania apple orchards, the tufted apple bud moth, *Platynota idaeusalis* (Tortricidae).

6.5.1 METHODS

On July 3, 1991, foliage samples were collected from 18 individual trees from each of five apple cultivars growing in an experimental orchard in Adams Co., Pennsylvania. The trees, planted in 1978 and arranged in mixed cultivar rows, were growing on Arendtsville gravelly loam soil. Cultivars (rootstocks) sampled were Delicious (M7A), Golden Delicious (M7A), Stayman (M7A), Rome Beauty (MM111) and Yorking (MM106). Three leaf-types were sampled from each of the 90 trees: fruiting spurs, non-fruiting spurs and terminal shoot leaves. Five spurs or shoots were collected at head height from five branches approximately evenly spaced around the tree, and pooled into one sample per leaf-type per tree (270 total samples). Collected leaves were immediately flash-frozen in liquid nitrogen, stored at –40°C, lyophilized and ground to a fine powder. Extracts were prepared and analysed according to Hunter and Hull (1993). Briefly, phloridzin and phloretin concentrations were estimated by reverse phase high-performance liquid chromatography against commercial standards.

Also on July 3, 1991, counts of *P. idaeusalis* larvae were made from each tree using a timed search method (Meagher and Hull, 1986). Counts were made from the base (10 minutes) and top (10 minutes) of each tree for a total of 20 minutes searching per tree, and larval distributions among the three leaf-types (above) were recorded. Finally, using a laboratory bioassay, the performance of *P. idaeusalis* larvae was compared on lima bean-based diets containing zero, 5% or 10% dry weight phloridzin (Hunter *et al.*, 1994).

6.5.2 RESULTS

As expected, there was less variation in foliar phenolic chemistry among apple trees than among the oak trees described above (Table 6.2a,b). Nonetheless, phloridzin concentrations varied by as much as five-fold among individuals within a cultivar (fruiting spurs on Yorking) and phloretin concentrations varied by as much as 15-fold among individuals (fruiting spurs on Stayman). In general, phloretin concentrations varied more among trees than did phloridzin concentrations (Table 6.2a,b). This level of variation among individuals within a cultivar is surprising given the reduced genetic variation, controlled growing environment and common root stocks used in the orchard.

Densities of *P. idaeusalis* larvae among trees increased with increasing concentrations of phloridzin in terminal shoot leaves (Figure 6.7). Larval densities were higher on shoot leaves during July than on the other two leaf-types (Hunter *et al.*, 1994). The positive correlation between foliar phloridzin concentrations and larval density may be explained by the results of the laboratory bioassays: both net and gross growth efficiencies of larvae increase on artificial diets containing increasing concentrations of phloridzin (Hunter *et al.*, 1994; Figure 6.8).

6.5.3 CONCLUSIONS – APPLE STUDY

Trees growing under very controlled conditions still exhibit significant spatial variation in their foliar chemistry, and this variation appears to influence the spatial distributions of insect herbivores among trees. Orchard systems are much simpler than nat-

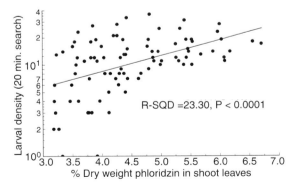

Figure 6.7 Relationship between density of *Platynota idaeusalis* on apple trees and concentration of phloridzin in terminal-shoot leaves on trees ($N = 90$ trees).

Table 6.2 Variation in the concentrations μg of compound per mg dry weight of foliage of (a) phloridzin and (b) phloretin in the foliage from five different apple cultivars

Cultivar	Leaf-type	N	Mean	Min.	Max.	SD	Fold
(a) PHLORIDZIN							
Red Delicious	Fruit	18	43.58	34.68	49.93	4.04	1.4 ×
	Shoot	18	43.45	35.40	50.17	4.03	1.4 ×
	Spur	18	43.56	32.55	63.62	6.75	2.0 ×
Golden Delicious	Fruit	18	40.40	29.42	50.75	6.04	1.7 ×
	Shoot	18	46.17	32.02	99.62	15.34	3.1 ×
	Spur	18	38.64	31.32	50.38	5.20	1.6 ×
Stayman	Fruit	18	44.51	36.62	54.03	5.26	1.5 ×
	Shoot	18	45.28	34.83	60.83	6.51	1.7 ×
	Spur	18	43.91	34.55	52.42	5.09	1.5 ×
Rome Beauty	Fruit	18	56.17	45.55	75.27	6.77	1.7 ×
	Shoot	17	57.33	48.55	67.05	4.73	1.4 ×
	Spur	18	55.79	45.32	67.77	5.23	1.5 ×
Yorking	Fruit	16	45.37	25.82	134.87	25.27	5.2 ×
	Shoot	18	36.24	31.80	41.48	3.51	1.3 ×
	Spur	16	37.38	30.08	48.73	4.93	1.6 ×

Cultivar	Leaf-type	N	Mean	Min.	Max.	SD	Fold
(b) PHLORETIN							
Red Delicious	Fruit	18	0.17	0.08	0.32	0.07	3.8 ×
	Shoot	18	0.15	0.08	0.35	0.07	4.2 ×
	Spur	18	0.15	0.08	0.23	0.05	2.8 ×
Golden Delicious	Fruit	18	0.06	0.05	0.22	0.04	4.3 ×
	Shoot	18	0.15	0.05	0.32	0.11	6.3 ×
	Spur	18	0.13	0.05	0.30	0.09	6.0 ×
Stayman	Fruit	18	0.19	0.05	0.77	0.18	15.3 ×
	Shoot	18	0.18	0.05	0.48	0.13	9.7 ×
	Spur	18	0.13	0.05	0.35	0.10	7.0 ×
Rome Beauty	Fruit	18	0.15	0.08	0.40	0.08	4.8 ×
	Shoot	17	0.11	0.08	0.20	0.04	2.4 ×
	Spur	18	0.12	0.08	0.20	0.04	2.4 ×
Yorking	Fruit	16	0.24	0.05	0.67	0.23	13.3 ×
	Shoot	18	0.14	0.05	0.35	0.10	7.0 ×
	Spur	16	0.13	0.05	0.40	0.12	8.0 ×

Leaf-type: Fruit = fruiting spurs
Shoot = terminal shoot leaves
Spur = non-fruiting spurs
SD = standard deviation
Fold = fold variation from the lowest to the highest concentration

ural forest systems, and it seems likely that the increased environmental heterogeneity and genetic variability in forests will result in greater variation in chemistry among individual trees. Again, the results support the hypothesis that the distributions of insect herbivores among trees are affected by the quality of foliage, and that this is likely to be a regular feature of insect–tree associations.

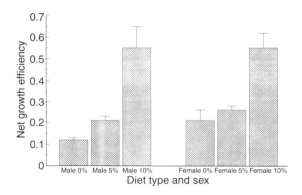

Figure 6.8 Net growth efficiency of male and female *Platynota idaeusalis* larvae raised on artificial diet containing 0, 5 or 10% phloridzin by dry weight.

6.6 GENERAL DISCUSSION

In recent years, many authors have become disillusioned with approaches to the population ecology of species that ignore the mechanistic basis of population change (see contributions in Cappuccino and Price, 1995). Mechanistic studies are important because only by understanding mechanism can we hope to predict, accurately and repeatedly, the changes that specific animal or plant populations are likely to make. Experimental approaches to understanding the processes that influence the birth, death and movement of species should provide the mechanistic basis for a new synthesis in population ecology (Price and Hunter, 1995). Population models must inevitably play a role in this new synthesis, but only if they represent accurately the biology of the species that they hope to represent. With the increasing emphasis on spatial modelling in ecology, understanding the mechanisms that determine the distributions of organisms in space has never been more important (Thomas *et al.*, 1996).

It has been argued here that the nutritional and defensive chemistry of plants is an important determinant of the distributions of forest insects in space. An integrated approach to balancing the roles of top-down and bottom-up forces in spatially explicit models of herbivore distribution should, of course, be based on broad estimates of plant quality (nutrition, phenology, etc.), not just defence. The graphical model presented in Figure 6.1 suggests why we should care about such variation in plant quality. It predicts, for example, that the equilibrium densities around which natural enemies will regulate their prey populations will depend on the quality of foliage for the herbivore in a given habitat patch. For relatively sessile organisms such as leaf-miners, gall-formers and scale insects, the 'patch' in question may be an individual tree (Edmunds and Alstad, 1978; Hanks and Denno, 1994; Strauss and Karban, 1994). For other insect species, the patch size may be larger, but the concept remains the same: spatial variation in food quality will influence the dynamics of the species. Large-scale variation in plant quality may even result in species with dramatically different dynamics in different habitats. Belovsky and Joern (1995) suggest, for example, that orthopteran populations are endemic in one area and epidemic in another because of spatial variation in food quality. Models that fail to incorporate this spatial variation in plant quality will also fail to represent accurately the dynamics of the species.

Other authors have described spatial variation in plant chemistry (Dolinger *et al.*, 1973; Lincoln and Langenheim, 1979; Rodman and Chew, 1980; Feeny and Rosenberry, 1982; Lincoln and Mooney, 1984) and its effects on insect herbivore populations. Svata Louda (University of Nebraska, Lincoln) pioneered studies of the effects of environmental variation on plant chemistry, and the consequences for herbivory (Louda and Rodman 1983a,b; Louda *et al.*, 1987). Louda has shown, for example, that gradients of water availability in prairie systems of only 25 m in length cause regular and linear effects on the methylglucosinolate concentrations of *Cleome serrulata* (Capparaceae). In turn, methylglucosinolate concentrations influence the levels of leaf damage and seed predation that plants suffer (Louda *et al.*, 1987). Likewise, experimentally induced increases in light availability result in decreases in specific glucosinolate compounds in *Cardamine cordifolia* and cause four-fold increases in subsequent herbivory (Louda and Rodman, 1983a). Similarly, Zangerl and Berenbaum (1990) have demonstrated that there is variation in furanocoumarin production both within and among populations of wild parsnip. Furanocoumarin concentration determines in large part the susceptibility of plants to defoliation by parsnip webworm (Zangerl and Berenbaum, 1993): attacked plants contain less than half of the concentrations of two furanocoumarins in unattacked plants. In forest systems, Cates and Zou (1990) have shown that variation in terpene production within and among populations of Douglas fir determine the susceptibility of trees to western spruce budworm. Individual trees with high foliar camphene, linalool and bornyl acetate concentrations are resistant to attack by this forest insect, and transfer experiments indicate that female pupal weights are much lower on resistant trees.

Simply put, spatial variation in plant chemistry appears to be a regular feature of plant–insect interactions, and there is solid evidence that plant chemistry is one mechanism determining the distribution and abundance of insect herbivores. What steps need to be taken to incorporate spatial variation in plant quality into our understanding of the population ecology of forest insects?

First, we need to measure population-level variation in plant quality. Although it is both expensive and time-consuming to collect, prepare and analyse foliage chemistry samples, modern analytical techniques such as HPLC have made such studies feasible. We need to characterize the chemistry of many individuals within a population of plants and, ideally, many different plant populations. Jack Schultz (Pennsylvania State University) has described this as the 'natural history' of plant allelochemistry and it is a prerequisite for any real understanding of links between plant quality and herbivore population dynamics (Schultz *et al.*, 1990). Second, we need experimental studies that describe how various plant compounds influence the four critical parameters of population change: birth, death, immigration and emigration. Such studies will provide the mechanistic basis for incorporating plant quality into population models (Price and Hunter, 1995).

Finally, we need population models that use the results of experimental studies as their foundation. Foster *et al.* (1992), for example, modified the classic Anderson and May (1981) epizootiology model of insect disease for the gypsy moth by: measuring natural variation in tree polyphenol concentrations; determining experimentally the relationships among foliar polyphenols, gypsy moth fecundity and gypsy moth mortality from virus: and adding parameters to the original model so that mortality rates and fecundity could vary according to the phenolic concentration of particular host species. The results of the model suggested that gypsy moth dynamics were intimately connected to the defensive chemistry of the host plants (Foster *et al.*, 1992). It is a relatively simple exercise to vary the reproductive rate of a species in a model according to a 'distribution of fecundity' generated from know spatial variation in plant quality. This is exactly the approach that will allow an accurate assessment of the role of plant resources for the population ecology of forest insects, and lead to a balanced view of top-down and bottom-up forces in population dynamics. With models of spatial dynamics still in their infancy, there will never be a better time for phytochemists and insect ecologists to influence the development of the new, spatially explicit ecology, and to avoid the top-down bias that permeates much of our population theory.

ACKNOWLEDGMENTS

I would like to thank Allan Watt, Kitti Reynolds, Rebecca Klaper, Rebecca Forkner, and three anonymous reviewers for helpful comments on an earlier draft of this paper. Field and laboratory assistance were provided by J.A. Schofield, C.M. Gourlie, J. Haller, L.M. Hamil, S. Heckman, C. LaCross, R.E. Kleeman, J.G. Adkins, K.M. Woller and P. Lasko. This research was supported by NSF grant BSR-89-18083 and USDA grant 91-34103-6398.

REFERENCES

Anderson, R.M. and May, R.M. (1980) Infectious diseases and population cycles of forest insects. *Science* **210**, 658–661.

Barbosa, P. and Schultz, J.C. (1987) *Insect Outbreaks*, Academic Press, New York.

Barbosa, P. and Wagner, M.R. (1989) *Introduction to Forest and Shade Tree Insects*, Academic Press, San Diego.

Bassuk, N.L., Hunter, L.D. and Howard, B.H. (1981) The apparent involvement of polyphenol oxidase and phloridzin in the production of apple rooting cofactors. *Journal of Horticultural Science* **56**, 313–322.

Belovsky, G.E. and Joern, A. (1995) Regulation of rangeland grasshoppers: differing dominant mechanisms in space and time. In *Population Dynamics: New Approaches and Synthesis* (eds N. Cappuccino and P.W. Price), pp. 359–386, Academic Press, San Diego.

Bernays, E. and Graham, M. (1988) On the evolution of host specificity in phytophagous arthopods. *Ecology* **69**, 886–892.

Bryant, J.P., Chapin, F.S. and Klein, D.R. (1983) Carbon/nutrient balance of boreal plants in relation to vertebrate herbivory. *Oikos* **40**, 357–368.

Cappuccino, N. and Price, P.W. (1995) *Population Dynamics: New Approaches and Synthesis*, Academic Press, San Diego.

Cates, R.G. and Zou, J. (1990) Douglas-fir (*Pseudotsuga menziesii*) population variation in terpene chemistry and its role in budworm (*Choristoneura occidentalis* Freeman) dynamics. In *Population Dynamics of Forest Insects* (eds A.D. Watt, S.R. Leather, M.D. Hunter and N.A.C. Kidd) pp. 169–182, Intercept, Andover.

Challice, J.S. and Williams, A.H. (1970) A comparative biochemical study of phenolase specificity in *Malus*, *Pyrus* and other plants. *Phytochemistry* **9**, 1261–1269.

Chapin, F.S. III (1980) Nutrient allocation and responses to defoliation in tundra plants. *Arctic and Alpine Research* **12**, 553–563.

Connor, E.F., Adams-Manson, R.H., Carr, T.G and Beck, M.W. (1994) The effects of host plant phenology on the demography and population dynamics of the leaf-mining moth, *Cameraria hamadryadella* (Lepioptera: Gracillariidae). *Ecological Entomology* **19**, 111–120.

Denno, R.F. and McClure, M.S. (1983) *Variable Plants and Herbivores in Natural and Managed Systems*, Academic Press, New York.

Doak, D.F. and Mills, L.S. (1994) A useful role for theory in conservation. *Ecology* **75**, 615–626.

Dolinger, P.M., Ehrlich, P.R., Fitch, W.L. and Breedlove, D.E. (1973) Alkaloid and predation patterns in Colorado lupine populations. *Oecologia* **13**, 191–204.

Du Merle, P. and Mazet, R. (1983) Stades phenologiques et infestation par *Tortrix viridana* L. (Lepidoptera: Tortricidae) des bourgeons du chene pubescent et du chene vert. *Acta Oecologica* **4**, 47–53.

Edmunds, G.F. and Alstad D.N. (1978) Coevolution in insect herbivores and conifers. *Science* **199**, 941–945.

Elkinton, J.S., Gould, J.R., Ferguson, C.S. *et al.* (1990) Experimental manipulation of gypsy moth density to assess impact of natural enemies. In *Population Dynamics of Forest Insects* (eds A.D. Watt, S.R. Leather, M.D. Hunter and N.A.C. Kidd), pp. 275–287, Intercept, Andover.

Feeny, P. and Rosenberry, L. (1982) Seasonal variation in the glucosinolate content of North American *Brassica nigra* and *Dentaria* species. *Biochemical Systematics and Ecology* **10**, 23–32.

Foster, M.A., Schultz, J.C. and Hunter, M.D. (1992) Modeling gyspy moth–virus–leaf chemistry interactions: implications of plant quality for pest and pathogen dynamics. *Journal of Animal Ecology* **61**, 509–520.

Fritz, R.S. and Simms, E.L. (1992) *Ecology and Evolution of Plant Resistance*, University of Chicago Press, Chicago.

Gershenzon, J. (1994) Metabolic costs of terpenoid accumulation in higher plants. *Journal of Chemical Ecology* **20**, 1281–1328.

Gilpin, M.E. and Hanski, I. (1991) *Metapopulation Dynamics: Empirical and Theoretical Investigations*, Academic Press, London.

Goldwasser, L., Cook, J. and Silverman, E.D. (1994) The effects of variability on metapopulation dynamics and rates of invasion. *Ecology* **75**, 40–47.

Gould, J.R., Elkinton, J.S. and Wallner, W.E. (1990) Density dependent suppression of experimentally created gypsy moth, *Lymantria dispar* (Lepidoptera: Lymantriidae) populations by natural enemies. *Journal of Animal Ecology* **59**, 213–233.

Growchowska, M.J. and Ciurzynska, W. (1979) Differences between fruit-bearing and non-bearing apple spurs in activity of an enzyme system decomposing phloridzin. *Biologica Planta* **21**, 201–205.

Hanks, L.M. and Denno, R.F. (1994) Local adaptation in the armored scale insect *Pseudaulacaspis pentagona* (Homoptera: Diaspididae). *Ecology* **75**, 2301–2310.

Hanski, I., Kuussaari, M. and Nieminen, M. (1994) Metapopulation structure and migration in the butterfly *Melitaea cinxia*. *Ecology* **75**, 747–762.

Hassell, M.P., Southwood, T.R.E. and Reader, P.M. (1987) The dynamics of the viburnum whitefly (*Aleurotrachelus jelinekii* Fraunf): a case study on population regulation. *Journal of Animal Ecology* **56**, 283–300.

Holmes, E.E., Lewis, M.A., Banks, J.E. and Veit, R.R. (1994) Partial differential equations in ecology: spatial interactions and population dynamics. *Ecology* **75**, 17–29.

Hunter, L.D. (1975) Phloridzin and apple scab. *Phytochemistry* **14**, 1519–1522.

Hunter, M.D. (1987) Opposing effects of spring defoliation on late season oak caterpillars. *Ecological Entomology* **12**, 373–382.

Hunter, M.D. (1992) A variable insect–plant interaction: the relationship between tree budburst phenology and population levels of insect herbivores among trees. *Ecological Entomology* **17**, 91–95.

Hunter, M.D. and Hull, L.A. (1993) Variation in concentrations of phloridzin and phloretin in apple foliage. *Phytochemistry* **34**, 1251–1254.

Hunter, M.D. and Price, P.W. (1992) Playing chutes and ladders: heterogeneity and the relative roles of bottom-up and top-down forces in natural communities. *Ecology* **73**, 724–732.

Hunter, M.D. and Schultz, J.C. (1993) Induced plant defenses breached? Phytochemical induction protects an herbivore from disease. *Oecologia (Berlin)* **94**, 195–203.

Hunter, M.D. and Schultz, J.C. (1995) Fertilization mitigates chemical induction and herbivore responses within damaged oak trees. *Ecology* **76**, 1226–1232.

Hunter, M.D. and Willmer, P.G. (1989) The potential for interspecific competition between two abundant defoliators on oak: leaf damage and habitat quality. *Ecological Entomology* **14**, 267–277.

Hunter, M.D., Hull, L.A. and Schultz, J.C. (1994) Evaluation of resistance to tufted apple bud moth (Lepidoptera: Tortricidae) within and among apple cultivars. *Environmental Entomology* **23**, 282–291.

Kareiva, P.M. (1994) Space: the final frontier for ecological theory. *Ecology* **75**, 1.

Klingauf, F. (1971) Die Wirkung des glucosids phlorizin auf das wirtswahlverhalten von *Rhopalosiphum insertum* (Walk.) und *Aphis pomi* De Geer (Homoptera: Aphididae). *Zeitschrift fur angewandte Entomologie* **68**, 41–55.

Lande, R. (1987) Extinction thresholds in demographic models of territorial populations. *American Naturalist* **130**, 624–635.

Lincoln, D.E. and Langenheim, J.H. (1979) Variation of *Satureja douglassi* monoterpenoids in relation to light intensity and herbivory. *Biochemical Systematics and Ecology* **7**, 289–298.

Lincoln, D.E. and Mooney, H.A. (1984) Herbivory on *Diplacus aurantiacus* shrubs in sun and shade. *Oecologia* **64**, 173–176.

Louda, S.M. and Rodman, J.E. (1983a) Ecological patterns in the glucosinolate content of a native mustard, *Cardamine cordifolia*, in the Rocky Mountains. *Journal of Chemical Ecology* **9**, 397–421.

Louda, S.M. and Rodman, J.E. (1983b) Concentration of glucosinolates in relation to habitat and insect herbivory for the native crucifer *Cardamine cordifolia*. *Biochemical Systematics and Ecology* **11**, 199–207.

Louda, S.M., Farris, M.A. and Blua, M.J. (1987) Variation in methylglucosinolate and insect damage to *Cleome serrulata* (Capparaceae) along a natural soil moisture gradient. *Journal of Chemical Ecology* **13**, 569–581.

MacDonald, R.E. and Bishop, C.J. (1952) Phloretin: an antibacterial substance obtained from apple leaves. *Canadian Journal of Botany* **30**, 486–489.

Meagher, R.L. and Hull, L.A. (1986) Predicting apple injury caused by *Platynota idaeusalis* (Lepidoptera: Tortricidae). *Journal of Economic Entomology* **79**, 620–625.

Molofsky, J. (1994) Population dynamics and pattern formation in theoretical populations. *Ecology* **75**, 30–39.

Montgomery, M.E. and Arn, H. (1974) Feeding response of *Aphis pomi*, *Myzus persicae*, and *Amphorophora agathonica* to phlorizin. *Journal of Insect Physiology* **20**, 413–421.

Mopper, S. and Simberloff, D. (1995) Differential herbivory in an oak population: the role of plant phenology and insect performance. *Ecology* **76**, 1233–1241.

Mopper, S., Beck, M., Simberloff, D. and Stiling, P. (1995) Local adaptation and agents of mortality in a mobile insect. *Evolution* **49**, 810–815.

Muzika, R.-M. (1993) Terpenes and phenolics in response to nitrogen fertilization: a test of the carbon/nutrient balance hypothesis. *Chemoecology* **4**, 3–7.

Muzika, R.-M. and Pregitzer, K.S. (1992) Effect of nitrogen fertilization on leaf phenolic production of grand fir seedlings. *Trees* **6**, 241–244.

Power, M.E. (1992) Top-down and bottom-up forces in food webs: do plants have primacy? *Ecology* **73**, 733–746.

Preszler, R.W. and Boecklen, W.J. (1994) A three-trophic-level analysis of the effects of plant hybridization on a leaf-mining moth. *Oecologia* **100**, 66–73.

Price, P.W. (1992) Plant resources as the mechanistic basis for insect herbivore population dynamics. In *Effects of Resource Distribution on Animal–plant Interactions* (eds M.D. Hunter, T. Ohgushi and P.W. Price) pp. 139–173, Academic Press, San Diego.

Price, P.W. and Hunter, M.D. (1995) Novelty and synthesis in the current development of population dynamics. In *Population Dynamics: New Approaches and Synthesis* (eds N. Cappuccino and P.W. Price) pp. 389–412, Academic Press, San Diego.

Pulliam, H.R. (1988) Sources, sinks, and population regulation. *American Naturalist* **132**, 652–661.

Raa, J. (1968) Polyphenols and natural resistance of apple leaves against *Venturia inaequalis*. *Netherland Journal of Plant Pathology* **74**, 37–45.

Ricker, W.E. (1954) Stock and recruitment. *Journal of the Fish Research Board of Canada* **11**, 559–623.

Rodman, J.E. and Chew, F.S. (1980) Phytochemical correlates of herbivory in a community of native and naturalized Cruciferae. *Biochemical Systematics and Ecology* **8**, 43–50.

Rosenthal, G.A. and Janzen, D.H. (1979) *Herbivores: Their Interaction with Secondary Plant Metabolites*, Academic Press, New York.

Rosenthal, G.A. and Berenbaum, M.R. (1991) *Herbivores: Their Interactions with Secondary Plant Metabolites*, Vol. 1, 2nd edn, Academic Press, San Diego.

Rossiter, M.C. (1987) Use of secondary hosts by non-outbreak populations of the gypsy moth. *Ecology* **68**, 857–868.

Rossiter, M.C., Schultz, J.C. and Baldwin, I.T. (1988) Relationships among defoliation, red oak phenolics, and gypsy moth growth and reproduction. *Ecology* **69**, 267–277.

SAS Institute, Inc. (1985) *SAS User's Guide: Statistics*, Cary, North Carolina.

Schultz, J.C. (1988) Many factors influence the evolution of herbivore diets, but plant chemistry is central. *Ecology* **69**, 896–897.

Schultz, J.C. (1992) Factoring natural enemies into plant tissue availability to herbivores. In *Effects of Resource Distribution on Animal–plant Interactions* (eds M.D. Hunter, T. Ohgushi and P.W. Price) pp. 175–197, Academic Press, San Diego.

Schultz, J.C. and Baldwin, I.T. (1982) Oak leaf quality declines in response to defoliation by gypsy moth larvae. *Science* **217**, 149–151.

Schultz, J.C., Foster, M.A. and Montgomery, M.E. (1990) Hostplant mediated impacts of a baculovirus on gypsy moth populations. In *Population Dynamics of Forest Insects* (eds A.D. Watt, S.R. Leather, M.D. Hunter and N.A.C. Kidd) pp. 303–313, Intercept, Andover.

Sjögren Gulve, P. (1994) Distribution and extinction patterns within a northern metapopulation of the pool frog, *Rana lessonae*. *Ecology* **75**, 1357–1367.

Speight, M.R. and Wainhouse, D. (1989) *Ecology and Management of Forest Insects*, Oxford University Press, Oxford.

Stewart-Oaten, A. and Murdoch, W.W. (1990) Temporal consequences of spatial density dependence. *Journal of Animal Ecology* **59**, 1027–1045.

Stiling, P. and Rossi, A.M. (1995) Coastal insect herbivore communities are affected more by local environmental conditions than by plant genotype. *Ecological Entomology* **20**, 184–190.

Strauss, S.Y. and Karban, R. (1994) The significance of outcrossing in an intimate plant–herbivore relationship. 1. Does outcrossing provide an escape from herbivores adapted to the parent plant? *Evolution* **48**, 454–464.

Tallamy, D.W. and Raupp, M.J. (1991) *Phytochemical Induction by Herbivores*, John Wiley and Sons, Inc., New York.

Taper, M.L. and Case, T.J. (1987) Interactions between oak tannins and parasite community structure: unexpected benefits of tannins to cynipid gall-wasps. *Oecologia* **71**, 254–261.

Taper, M.L., Zimmerman, E.R. and Case, T.J. (1986) Sources of mortality for a cynipid gall-wasp (*Dryocosmus dubiosus* (Hymenoptera: Cynipidae)): the importance of the tannin/fungus interaction. *Oecologia* **68**, 437–445.

Thomas, C.D., Singer, M.C. and Boughton, D. (1996) Catastrophic extinction of population sources in a butterfly metapopulation. *American Naturalist* **148**, 957–975.

Thompson, J.N. (1988) Coevolution and alternative hypotheses on insect–plant interactions. *Ecology* **69**, 893–895.

Tilman, D. (1994) Competition and biodiversity in spatially-structured habitats. *Ecology* **75**, 2–16.

Varley, G.C., Gradwell, G.R. and Hassell, M.P. (1973) *Insect Population Ecology. An Analytical Approach*, Blackwell Scientific Publications, Oxford.

Watkinson, A.R. and Sutherland, W.J. (1995) Sources, sinks and pseudosinks. *Journal of Animal Ecology* **64**, 126–130.

Watt, A.D., Leather, S.R., Hunter, M.D. and Kidd, N.A.C. (1990) *Population Dynamics of Forest Insects*, Intercept, Andover.

Weis, A.E. and Campbell, D.R. (1992) Plant genotype: a variable factor in insect–plant interactions. In *Effects of Resource Distribution on Animal–plant Interactions* (eds M.D. Hunter, T. Ohgushi and P.W. Price) pp. 75–111, Academic Press, San Diego.

Whitham, T.G. (1989) Plant hybrid zones as sinks for pests. *Science* **244**, 1490–1493.

Williams, A.H. (1964) Dihydrochalcones; their occurrence and use as indicators in chemical plant taxonomy. *Nature* **202**, 824–825.

Zangerl, A.R. and Berenbaum, M.R. (1990) Furanocoumarin induction in wild parsnip: genetics and populational variation. *Ecology* **71**, 1933–1940.

Zangerl, A.R. and Berenbaum, M.R. (1993) Plant chemistry, insect adaptations to plant chemistry, and host plant utilization patterns. *Ecology* **74**, 47–54.

FOREST STRUCTURE AND THE SPATIAL PATTERN OF PARASITOID ATTACK

Jens Roland, Phil Taylor and Barry Cooke

7.1 INTRODUCTION

The study of forest insect population dynamics has taken three general approaches: (1) long-term intensive study of population processes at a single site, such as that of the winter moth (Varley and Gradwell, 1960); (2) shorter-term study of several widely scattered sites, such as those of the gypsy moth (Campbell and Sloan, 1977); and (3) the monitoring of population response to experimental manipulation of abundance – for example, that of the gypsy moth (Gould *et al.*, 1990).

In the first approach, the impact of population processes such as fecundity, parasitism and predation are estimated over time as population densities rise and fall. The result is detailed knowledge of processes at that particular site, but generalization to other populations is difficult as it is not known how representative the results are of dynamics at a larger spatial scale. In addition, because only a single site is studied, the impact of processes such as dispersal can only be inferred from changes in density not explained by fecundity and mortality (e.g. Royama, 1984).

In the second approach, general population patterns are inferred from a composite of the dynamics at multiple sites. For example, Liebhold (1992) demonstrated double-equilibrium dynamics for gypsy moth populations on patterns of population change pooled from 83 widely separated populations. The assumption of this type of analysis is that populations at all sites pass through similar trajectories of density, fecundity and mortality. Although such studies are able to account for some of the variation in dynamics seen among sites, they obscure fine-scale processes acting at any one of them. This is perhaps not surprising as local dynamics will differ depending on site characteristics and the status of neighbouring populations. If multiple study sites are spaced widely enough for density estimates to be independent, there is no opportunity to estimate one of the most important processes influencing population dynamics – that of movement. Being able to estimate movement between two adjacent sites (a desirable goal in any population study) means by definition that the two are not independent of each other.

The third approach uses experimental manipulation of abundance and it measures the response to this manipulation by processes such as attack by natural enemies (e.g. Gould *et al.*, 1990; Ferguson *et al.*, 1994). These studies are powerful in evaluating the impact of population processes locally but, as their authors point out, their drawback is the limited spatial scale (typically < 1 ha) at which the manipulations are logistically feasible. For example, if host increase normally occurs over a large spatial scale rather than in small foci (eg. Liebhold and McManus 1991), then parasitoids would not aggregate in response to variation in host abundance (because there is very little variation at the scale over which a parasitoid would search); the experimental creation of a 'hot spot' of host abun-

---

Forests and Insects. Edited by A.D. Watt, N.E. Stork and M.D. Hunter. Published in 1997 by Chapman & Hall, London. ISBN 0 412 79110 2.

dance might give a false indication of the strength of the parasitoid response.

An understanding of forest insect dynamics at a large scale, and which takes into account habitat variability, habitat configuration and movement of both herbivores and natural enemies, might best be done using a landscape approach.

7.2 A LANDSCAPE APPROACH

In addition to the methods described above, a landscape approach to population dynamics examines the way in which fine-scale processes such as parasitism, predation and disease transmission are affected by landscape structure at the local scale (the scale at which individual insects move over their lifetime) and at the regional scale (the scale of populations rather than of individuals). This is different from traditional studies of habitat use which are done at the stand level without concern for factors such as stand adjacency and the configuration of the mosaic of stands. Analytical tools such as spatial statistics and geographical information systems (GIS) make a landscape approach more feasible. Rather than analysing processes at each site independently, analyses can include effects of each variable at different spatial scales (Heads and Lawton, 1983), the effect of stand adjacency, dynamics of nearby populations (Liebhold and Elkinton, 1989), and of habitat connectivity (Taylor et al., 1993). By studying populations that are close to each other (within the movement range of the host and its natural enemies) there is the added potential for insights into the importance of movement between subpopulations. Discovery of plants by herbivores (Solbreck, 1995), discovery of hosts and aggregation by parasitoids and predators (Heads and Lawton, 1983; Kareiva, 1987) and transmission of disease through host populations (Dwyer, 1992) all depend on movement of organisms.

Spatial patterns of mortality have been studied for many insects, but with very few exceptions (e.g. Liebhold and Elkinton, 1989) such analyses are done in a purely statistical manner. For example, the response of parasitoids to host density is assessed by the use of regression analysis (Walde and Murdoch, 1988; Stiling, 1988; Dempster, 1983), but no consideration is given to the spatial pattern over which host abundance varies. When gypsy moth populations were sampled intensively over a grid of census points with 25–50 m resolution (Liebhold and Elkinton, 1989), densities were found to be highly variable both among sites and between the different life stages. Unfortunately the patterns of mortality and movement which might have caused these changes in densities were not recorded. A spatially process-oriented view of predation and parasitism is useful because the ability of natural enemies to respond to host abundance is governed in part by their ability to move through a mosaic of different habitats in both agricultural (Kareiva, 1987; Kruess and Tscharntke, 1994) and forested (Rogers and Williams, 1993) landscapes. Such fine-scale effects of habitat on natural enemy movement have been argued as one reason for the obscuring of density-dependent patterns of mortality when averaged over a larger area (Hassell, 1985; Liebhold and Elkinton, 1989), and for the preponderance of density-vague regulation of populations (Strong, 1986).

Laboratory studies (Huffaker, 1958) and very fine-scale field studies (Kareiva, 1987; Kruess and Tscharntke, 1994) have examined the effect of habitat structure on dynamics. The importance of habitat structure in each of these fine-scale studies is through its effect in altering movement of the host or predator. Despite several fine-scale examples of altered predator–prey dynamics, Taylor (1991) has pointed out that there are virtually no examples of large-scale effect of habitat structure on predator–prey systems.

7.3 HOW CAN LANDSCAPE AFFECT FOREST INSECT POPULATIONS?

7.3.1 FECUNDITY

Fecundity of forest insects can be affected by landscape either directly, by distribution of hosts plants, or by the effect of landscape on plant quality. Examples include the effect of the incidence of flower cones on jackpine budworm

(*Choristoneura pinus pinus* Freeman), which tends to be much higher along forest margins than in forest interior (Vince Nealis, Canadian Forest Service, Sault Ste Marie, Ontario, personal communication) and which strongly affects larval growth and survival. The result of this pattern is that margins of jackpine stands adjacent to clearings often have higher density of *C. pinus* and more severe defoliation.

7.3.2 MORTALITY

Parasitism is spatially variable and (for species attacking forest defoliators) can be affected by physical structure of the forest. For example, *Compsilura concinnata* (Mg.) parasitizes hosts at a lower rate on isolated trees than on trees in large clumps (Schwenke, 1958). The egg parasitoid *Ooencyrtus kuwanai* (Howard) attacks host egg masses at a higher rate in stand interiors than near or in clearings. Parasitism of forest tent caterpillar cocoons by *Arachnidomyia aldrichi* is lower at the forest edge along lake margins than in the forest interior (Batzer, 1955). The opposite pattern is seen for the chalcid wasp *Brachymeria intermedia* (Nees), which attacks gypsy moth at a greater rate in open sunny areas than in shaded areas (Leonard, 1971). Part of the explanation for these spatial patterns result from microclimatic preferences of the attacking parasitoids, particularly with respect to humidity and insolation (Weseloh, 1976). These two variables exhibit strong gradients along forest edges to a distance of up to 100 m into the forest. Such fine-scale behavioural patterns of natural enemies, when combined with regional patterns of forest structure, could have dramatic effects on regional herbivore dynamics. For example, egg masses of the gypsy moth tend to be more abundant along forest edges compared with the forest interior (Bellinger *et al.*, 1989), as are the eggs of some butterfly species (Courtney and Courtney, 1982). The effect of such an increase in spatial variation in abundance might affect the ability of natural enemies to discover hosts because they are more patchily distributed; or local high abundance could swamp natural enemies, thereby reducing rates of attack per individual.

7.3.3 MOVEMENT

Rates of movement of herbivores (Turchin, 1991; Crist and Wiens, 1995; Chapter 18, this volume), predators (Taylor and Merriam, 1995), parasites (Rogers and Williams, 1993; Sheppard, 1994) and other insect natural enemies (Landis and Haas, 1992) can all be limited (or enhanced) by the spatial configuration of habitat mosaic. Some population models have shown that differential dispersal of host and parasitoid create very different dynamics (Reeve, 1988; Hassell *et al.*, 1991). For example, enhanced dispersal of the host can permit 'escape' from its natural enemies even if the latter are effective in suppressing the host locally once it is discovered. Fine-scale empirical studies, both in the laboratory (Huffaker, 1958) and in the field (Kareiva, 1987; Kruess and Tscharntke, 1994) clearly show that habitat structure can affect the movement of herbivores relative to that of their natural enemies and as a result affect dynamics of both.

7.4 FOREST FRAGMENTATION AND THE DYNAMICS OF FOREST TENT CATERPILLAR

The forest tent caterpillar, *Malacosoma disstria* Hbnr., is one of the dominant forest defoliators in the boreal forests of North America. Populations exhibit periodic outbreaks over large areas of boreal forest with a frequency of 9–14 years (Sippell, 1962; Daniel, 1990; Myers, 1993). Forest tent caterpillars are subject to parasitism by a suite of parasitoid species. The same parasitoid species are dominant in Ontario (Sippell, 1957), Minnesota (Witter and Kulman, 1972, 1979) and Alberta (Parry, 1995). The two most dominant parasitoids are the tachinid fly *Patelloa pachypyga* and the sarcophagid fly *Arachnidomyia aldrichi*. *P. pachypyga* oviposits on caterpillar-damaged trembling aspen foliage, and the eggs are subsequently ingested by the feeding host caterpillar. The fully fed maggot emerges after the host has spun a cocoon in foliage. In contrast to the tachinid, *A. aldrichi* lays eggs directly on the host cocoon. Fully developed *A. aldrichi* maggots emerge well after the host has pupated (Hodson, 1939). Although there are phenological differences in the timing of attack by each of these species (Sippell, 1957; Parry, 1995), they are both abundant in collections of recently formed pupae. Rates of parasitism by *A. aldrichi* have been recorded as being the same in the canopy and understorey (Sippell, 1957), or higher in the understorey (Parry, 1995). There is no consistent difference in parasitism by *P. pachypyga* in the canopy vs. the understorey (Parry, 1995).

In at least parts of its range, the forest tent caterpillar exhibits longer outbreaks in forests that have been fragmented because of clearing, compared with outbreaks in continuous forests (Roland, 1993). Among individual townships (100 km² each), for each additional kilometre of forest edge per km² (as a measure of fragmentation) outbreaks last about a year longer (Roland, 1993). Because viral epizootics (Stairs, 1966; Clark, 1958; Myers, 1993) and parasitism (Sippell, 1957; Witter and Kulman, 1979; Parry, 1995) are strongly implicated in causing population collapse, the altered dynamics seen in fragmented forests may result from the interaction between one or both of these sources of mortality and habitat structure (Roland, 1993; Roland and Kaupp, 1995; Roland and Taylor, 1995, 1997).

We have studied both variation in tent caterpillar abundance and variation in the processes that affect abundance in a current outbreak of forest tent caterpillar near Edmonton, Alberta, Canada. This work is done at two spatial scales. The first is a fine-scale grid of 107 population sample points (Figure 7.1) with 50 m spacing between adjacent sites over an area of 32 ha. This grid is characterized by mixed trembling aspen (*Populus tremuloides*) and balsam poplar (*Populus balsamifera*) averaging about 13 m in height. The grid has a large clearing along the south edge, and three smaller clearings around water bodies within the grid. A larger grid (420 km²) of 127 sample points with 1.8 km spacing between adjacent points was also sampled (see map in Roland and Taylor, 1995) but is not discussed here. In this chapter we present preliminary results on the fine-scale effect of forest structure on the spatial pattern of parasitism of this forest defoliator.

7.4.1 METHODS AND MATERIALS

Abundance of tent caterpillar and parasitism by *P. pachypyga* and *A. aldrichi* were estimated from cocoons collected between June 28 and July 4, 1995, at each of the 107 sampling sites. Abundance was estimated from timed collections of cocoons from the understorey; the time taken to collect 50 cocoons, to a maximum of 15 minutes, was recorded. If 50 cocoons were collected in less than 15 minutes then the number which would have been collected in 15 minutes was estimated. Additional cocoons were collected at each site to improve the estimates of parasitism. The number of *P. pachypyga* and *A. aldrichi* produced from each collection was counted and the rate of parasitism was estimated for each of the 107 sites. For each parasite species, the effect of host density (log-transformed) and two landscape features were used as explanatory variables in a multiple regression of parasitism (arc-sine square-root transformed). The two landscape features were: (1) the amount of forest within 100 m of each site and (2) the amount of forest edge per unit forest (expressed as km/km²) within 100 m of each site. Landscape features were estimated from a classified air-photo mosaic made from 1:20 000 scale false-colour infra-red photography of the 420 km² study. The digitally scanned version of this mosaic had a resolution of 5 m per pixel. Landscape features around each population sample point were estimated using SPANS geographical information system software (Intera Tydac Technologies, 1993).

In addition to the grid studies described above, earlier studies at Kakabeka Falls, Ontario, in 1992 estimated percentage parasitism along transects from forest edge to forest interior. Collections of late larvae (for tachinid parasites) and cocoons (for *A. aldrichi*) were made at intervals from the forest edge (0 m) to 130 m into the forest along four separate transects. Rates of parasitism were estimated, proportions were arc-sine square-root transformed and analysed using analysis of covariance with transect as a main factor and distance into the forest as a covariate.

7.4.2 RESULTS

(a) *Patelloa pachypyga*

Parasitism by *P. pachypyga* varied from 0 to 89% across the grid. Parasitism increased with host

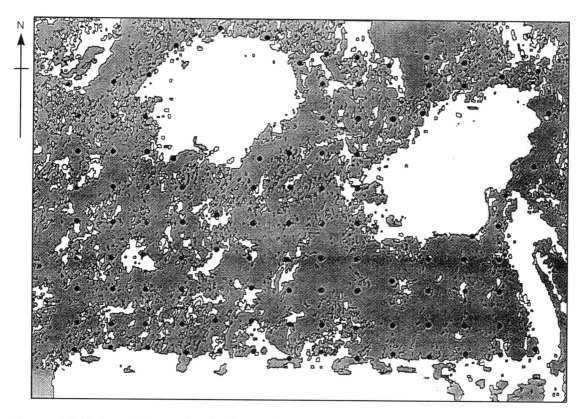

Figure 7.1 Grid of population sample points in aspen forests near Cooking Lake, Alberta. Stippled areas, aspen forest; white areas, clearings. Sample points are separated by approximately 50 m.

abundance, but was more strongly influenced by the amount of forest at each collection site (Table 7.1). Tent caterpillars at sites surrounded by large forest stands exhibited higher rates of parasitism by *P. pachypyga* compared with sites with more fragmented forests. Across the grid, *P. pachypyga* caused greater rates of parasitism in the central large section of forest and reduced rates along the large forest edge at the south end of the study area and around the three large clearings at the ponds (Figures 7.1 and 7.2a). This pattern of reduced parasitism with increased forest edge is consistent with fine-scale transect studies of parasitism by tachinids in Ontario boreal forest (Figure 7.3), and in Alberta forests (Parry, 1994) where rates of parasitism were lower within 50 m of the forest edge compared with that in the forest interior. Similarly, large-scale patterns of parasitism by *Patelloa* in Alberta in 1994 (Roland and Taylor, 1995) showed reduced tent caterpillar parasitism by *P. pachypyga* in fragmented forests compared with that in continuous forests.

(b) *Arachnidomyia aldrichi*

Parasitism by the sarcophagid fly *A. aldrichi* varied across the grid from 0 to 73%. As with *P. pachypyga*, parasitism by *A. aldrichi* increased with increase in host density. Unlike the tachinid, however, there was no effect of forest structure across the grid on rates of parasitism by *A. aldrichi* (Table 7.2); parasitism was similar (Figure 7.2b) regardless of the amount of forest around each point, and regardless of the amount of forest edge. There was no significant reduction in parasitism along the large forest edge at the south end of the study area

Table 7.1 Multiple regression analysis for rate of parasitism (arc-sine square-root transformed) by the tachinid fly *Patelloa pachypyga* as a function of host density (log-transformed), amount of forest (km^2) within 100 m, and km of forest edge per km^2 within 100 m of each of 107 sample points in the grid ($r^2 = 0.17$)

Variable	Coefficient	t-Value	P
Host density (log)	0.138	2.469	0.015
Forest	0.429	2.199	0.030
Edge/km^2 of forest	−0.001	−0.145	0.885

nor along the edges of the forest adjacent to ponds (Figures 7.1 and 7.2b). Transects from the forest interior to the forest edge (at sites in Ontario) showed no reduction in *A. aldrichi* parasitism (Figure 7.4), unlike the reduction seen for *P. pachypyga*. These patterns are consistent with larger-scale patterns of parasitism by A. aldrichi (Roland and Taylor, 1995) which were unrelated to the degree of forest fragmentation.

7.4.3 DISCUSSION

Two dominant parasitoids attacking forest tent caterpillar during population decline are the sarcophagid fly *Arachnidomyia aldrichi* and the tachinid fly *Patelloa pachypyga*. These two species dominate tent caterpillar populations in Ontario (Sippell, 1957), Minnesota (Witter and Kulman, 1979; Hodson, 1939), and in much of Alberta (Parry, 1995). *P. pachypyga* is less efficient in attacking tent caterpillar in forests that are fragmented than in continuous forests. It is not known whether this pattern results from habitat preferences by this fly or from altered search behaviour at the forest edge compared with in the forest interior. Other tachinids, such as *Compsilura concinnata* (Mg.) attacking the noctuid *Acronycta aceris* L., also exhibit reduced parasitism on individual trees and higher parasitism in clumps of trees (Schwenke, 1958). The egg parasitoid *Ooencyrtus kuwanai* (Howard) caused higher rates of parasitism in forest interiors than in or adjacent to clearings (Weseloh, 1972). Part of the explanation for high parasitism in forest interiors has been attributed to the effect of microclimate preferences of the parasitoids (Weseloh, 1976). Preference by *Ooencyrtus kuwanai* for cool humid conditions result in its abundance in the forest interior and scarcity along forest edges and in open forests (Weseloh, 1976). The opposite pattern – preference for warm dry microclimates – has been documented for other species (Weseloh, 1976), a pattern which is likely to result in higher host parasitism along forest edges than in the forest interior.

Longer outbreaks of forest tent caterpillar in fragmented forests may in part be the result of reduced abundance or reduced efficiency of at least part of the parasitoid community. In addition to parasitism, population processes such as virus

Table 7.2 Multiple regression analysis for rate of parasitism (arc-sine square-root transformed) by the sarcophagid fly *Arachnidomyia aldrichi* as a function of host density (log-transformed), amount of forest (km^2) within 100m, and km of forest edge per km^2 within 100 m of each of 107 sample points in the grid ($r^2 = 0.22$)

Variable	Coefficient	t-Value	P
Host density (log)	0.252	5.195	0.000
Forest	0.166	0.992	0.323
Edge/km^2 of forest	0.001	0.795	0.428

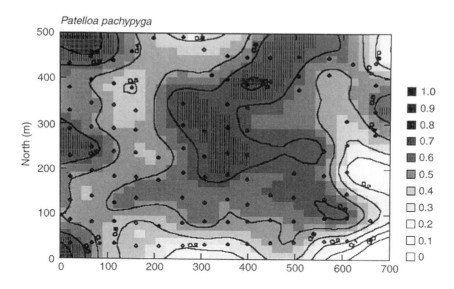

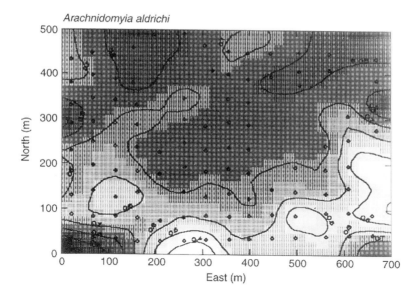

Figure 7.2 Spatial pattern of parasitism by (a) tachinid fly *Patelloa pachypyga* and (b) sarcophagid fly *Arachnidomyia aldrichi* across 32 ha grid. Smoothed estimates are distance-weighted least squares. (Wilkinson *et al.*, 1992.)

transmission, may be reduced along forest edges (Roland and Kaupp, 1995).

Other forest insect species appear to exhibit landscape-dependent dynamics. Mott (1963) noted that population outbreaks of spruce budworm were more severe in large forests than in those which were fragmented. Similarly, the western spruce budworm, *Choristoneura occidentalis* Freeman, shows more severe, longer-lasting and more synchronous outbreaks in the southwestern United

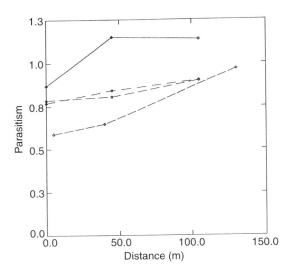

Figure 7.3 Parasitism by tachinids along four transects from forest edge to forest interior (in metres from edge) at Kakabeka Falls, Ontario, 1992 ($N = 25$ larvae per point). Parasitism is reduced near forest edge ($F = 19.63$, $P = 0.003$).

States, over the same period that forests have become more contiguous (a result of favourable climate and fire suppression) (Swetnam and Lynch 1993). The two budworm species show a pattern opposite to that for forest tent caterpillar, suggesting that although landscape may be an important component of forest insect dynamics, the patterns of interaction between population process and landscape are likely to be species dependent.

The data presented here were collected in a single year of widespread host abundance. The real impact of reduced parasitism in fragmented forests will not be evident until the current outbreak begins to collapse. As the outbreak declines it is predicted that pockets of outbreak should linger in the more fragmented parts of the 420 km² study area, and should decline more rapidly in the contiguous forests. Lingering pockets of tent caterpillar in fragmented forests would be predicted to be parasitized at a lower rate than in comparable pockets in continuous forests. Preliminary results here suggest that this may be most strongly manifest through parasitism by tachinids. We feel that large-scale spatial studies of population processes, interacting with habitat, carried out over the course of population rise and collapse will provide valuable insight into the role of natural enemies on host dynamics.

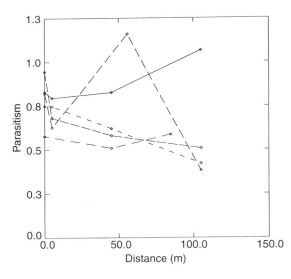

Figure 7.4 Parasitism by *Arachnidomyia aldrichi* along four transects from forest edge to forest interior (in metres from edge) at Kakabeka Falls, Ontario, 1992 ($N = 100$ cocoons per point). There is no effect of distance on parasitism ($F = 1.57$, $P = 0.232$).

REFERENCES

Batzer, H.O. (1955) Some effects of defoliation of aspen, *Populus tremuloides* Michx., stands in northern Minnesota by the forest tent caterpillar, *Malacosoma disstria* Hbn., with notes on parasitism of cocoons by *Sarcophaga aldrichi* Park. and cocooning habits of the host. MSc Thesis, University of Minnesota, St Paul. 66 pp.

Bellinger, R.G., Ravlin, F.W. and McManus, M.L. (1989) Forest edge effects and their influence on gypsy moth (Lepidoptera: Lymantriidae) egg mass distributions. *Environmental Entomology* **18**, 840–843.

Campbell, R.W. and Sloan, R.J. (1977) Natural regulation of innocuous gypsy moth populations. *Environmental Entomology* **6**, 315–322.

Clark, E.C. (1958) Ecology of the polyhedrosis of tent caterpillars. *Ecology* **39**, 132–139.

Courtney, S.P. and Courtney, S. (1982) The 'edge-effect' in butterfly oviposition: causality in *Anthocharis cardamines* and related species. *Ecological Entomology* **7**, 131–137.

Crist, T.O. and Wiens, J.A. (1995) Individual movements and estimation of population size in darkling beetles (Coleoptera: Tenebrionidae). *Journal of Animal Ecology* **64**, 733–746.

Daniel, C.J. (1990) Climate and outbreaks of the forest tent caterpillar in Ontario. MSc Thesis, University of British Columbia, Vancouver.

Dempster, J.P. (1983) The natural control of butterflies and moths. *Biological Reviews* **58**, 461–481.

Dwyer, G. (1992) On the spatial spread of insect pathogens: theory and experiment. *Ecology* **73**, 479–494.

Ferguson, C.S., Elkinton, J.S., Gould, J.R. and Wallner, W.E. (1994) Population regulation of gypsy moth (Lepidoptera: Lymantriidae) by parasitoids: does spatial density dependence lead to temporal density dependence? *Environmental Entomology* **23**, 1155–1164.

Gould, J.R., Elkinton, J.S. and Wallner, W.E. (1990) Density dependent suppression of experimentally created gypsy moth, Lymantria dispar (Lepidoptera: Lymantriidae), populations by natural enemies. *Journal of Animal Ecology* **59**, 213–234.

Hassell, M.P. (1985) Insect natural enemies as regulating factors. *Journal of Animal Ecology* **54**, 323–334.

Hassell, M.P., Comins, H.N. and May, R.M. (1991) Spatial structure and chaos in insect population dynamics. *Nature* **353**, 255–258.

Heads, P.A. and Lawton, J.H. (1983) Studies on the natural enemy complex of the holly leaf-miner: the effects of scale on the detection of aggregative responses and the implications for biological control. *Oikos* **40**, 267–276.

Hodson, A.C. (1939) Sarcophaga aldrichi Parker as a parasite of *Malacosoma disstria* Hbn. *Journal of Economic Entomology* **32**, 396–401.

Huffaker, C.B. (1958) Experimental studies on predation: dispersion factors and predator–prey oscillations. *Hilgardia* **27**, 343–383.

Intera Tydac Technologies (1993) *SPANS GIS Reference Manual*, Intera Tydac Technologies, Nepean, Ontario.

Kareiva, P. (1987) Habitat fragmentation and the stability of predator–prey interactions. *Nature* **326**, 388–390.

Kruess, A. and Tscharntke, T. (1994) Habitat fragmentation, species loss, and biological control. *Science* **264**, 1581–1584.

Landis, D.A and Haas, M.J. (1992) Influence of landscape structure on abundance and within-field distribution of European corn borer (Lepidoptera: Pyralidae) larval parasitoids in Michigan. *Environmental Entomology* **21**, 409–416.

Leonard, D.E. (1971) *Brachymeria intermedia* (Hymenoptera: Chalcidae) parasitizing gypsy moth in Maine. *Canadian Entomologist* **103**, 654–656.

Liebhold, A.M. (1992) Are North American populations of gypsy moth (Lepidoptera: Lymantriidae) bimodal? *Environmental Entomology* **21**, 221–229.

Liebhold, A.M. and McManus, M.L. (1991) Does larval dispersal cause the expansion of gypsy moth outbreaks? *Northern Journal of Applied Forestry* **8**, 95–98.

Liebhold, A.M. and Elkinton, J.S. (1989) Use of multidimensional life tables for studying insect population dynamics. In *Estimating Insect Populations* (eds L. McDonald, B. Manly, J. Lockwood and J. Logan) pp. 360–369, Springer-Verlag, Berlin.

Mott, D.G. (1963) The forest and the spruce budworm. In *The Dynamics of Epidemic Spruce Budworm Populations* (ed. R.F. Morris) *Memoirs of the Entomological Society of Canada* **31**, 189–202.

Myers, J.H. (1993) Population outbreaks in forest Lepidoptera. *American Scientist* **81**, 240–251.

Parry, D. (1994) The impact of predators and parasitoids on natural and experimentally created populations of forest tent caterpillar, *Malacosoma disstria* Hübner (Lepidoptera: Lasiocampidae). MSc Thesis, University of Alberta, 91 pp.

Parry, D. (1995) Larval and pupal parasitism of the forest tent caterpillar, *Malacosoma disstria* Hübner (Lepidoptera: Lasiocampidae), in Alberta, Canada. *Canadian Entomologist* **127**, 877–893.

Reeve, J. (1988) Environmental variability, migration, and persistence in host–parasitoid systems. *American Naturalist* **132**, 810–835.

Rogers, D.J. and Williams, B.G. (1993) Monitoring trypanosomiasis in space and time. *Parasitology* **106**, S77–S92.

Roland, J. (1993) Large-scale forest fragmentation increases the duration of tent caterpillar outbreaks. *Oecologia* **93**, 25–30.

Roland, J. and Kaupp, W.J. (1995) Reduced transmission of forest tent caterpillar NPV at the forest edge. *Environmental Entomology* **24**, 1175–1178.

Roland, J. and Taylor, P.D. (1995) Herbivore–natural enemy interactions in fragmented and continuous forests. In *Population Dynamics: New Approaches and Synthesis* (eds N. Cappuccino and P.W. Price), pp. 195–208, Academic Press, San Diego.

Roland, J and P.D. Taylor (1997) Insect parasitoid species respond to forest structure at different spatial scales. *Nature* **386**, 710–713.

Royama, T. (1984) Population dynamics of the spruce budworm *Choristoneura fumiferana*. *Ecological Monographs* **54**, 429–462.

Schwenke, W. (1958) Local dependence of parasitic insects and its importance for biological control. *Proceedings of the Tenth International Congress of Entomology* **4**, 851–854.

Sheppard, D.C. (1994) Dispersal of wild-captured, marked horn flies (Diptera: Muscidae). *Environmental Entomology* **23**, 29–34.

Sippell, W.L. (1957) A study of the forest tent caterpillar *Malacosoma disstria* Hbn., and its parasite complex in Ontario. PhD Thesis, University of Michigan, Ann Arbor. 147 pp.

Sippell, W.L. (1962) Outbreaks of the forest tent caterpillar, *Malacosoma disstria* Hbn., a periodic defoliator of broad-leaved trees in Ontario. *Canadian Entomologist* **94**, 408–416.

Solbreck, C. (1995) Long-term population dynamics of a seed-feeding insect in a landscape perspective. In *Population Dynamics: New Approaches and Synthesis* (eds N. Cappuccino and P.W. Price), pp. 279–301, Academic Press, San Diego.

Stairs, G.R. (1966) Transmission of virus in tent caterpillar populations. *Canadian Entomologist* **98**, 1100–1104.

Stiling, P. (1988) Density-dependent processes and key factors in insect populations. *Journal of Animal Ecology* **57**, 581–593.

Strong, D.R. (1986) Density-vague population change. *Trends in Ecology and Evolution* **1**, 39–42.

Swetnam, T.W. and Lynch, A.M. (1993) Multicentury, regional-scale patterns of western spruce budworm outbreaks. *Ecological Monographs* **63**, 399–424.

Taylor, A. (1991) Studying metapopulation effects in predator–prey systems. In *Metapopulation Dynamics: Empirical and Theoretical Investigations*, pp. 305–323, Academic Press, London.

Taylor, P.D. and Merriam, G. (1995) Wing morphology of a forest damselfly is related to landscape structure. *Oikos* **73**, 43–48.

Taylor, P.D., Fahrig, L., Henein, K. and Merriam, G. (1993) Connectivity is a vital element of landscape structure. *Oikos* **68**, 571–573.

Turchin, P. (1991) Translating foraging movements in heterogeneous environments into the spatial distribution of foragers. *Ecology* **72**, 1253–1266.

Varley, G.C. and Gradwell, G.R. (1960) Key factors in population studies. *Journal of Animal Ecology* **45**, 313–325.

Walde, S.J. and Murdoch, W.W. (1988) Spatial density dependence in parasitoids. *Annual Review of Entomology* **33**, 441–466.

Weseloh, R.M. (1972) Spatial distribution of gypsy moth (Lepidoptera: Lymantriidae) and some of its parasitoids within a forest environment. *Entomophaga* **17**, 339–351.

Weseloh, R.M. (1976) Behavior of forest insect parasitoids. In *Perspectives in Forest Entomology* (eds J.F. Anderson and H.K. Kaya), pp. 99–110, Academic Press, New York.

Wilkinson, L., Hill, M., Miceli, S. *et al.* (1992) *SYSTAT for Windows: Graphics, Version 5 Edition*, Evanston, Illinois. 636 pp.

Witter, J.A. and Kulman, H.M. (1972) A review of the parasites and predators of tent caterpillars (*Malacosoma* spp.) in North America. *Minnesota Agriculture Experimental Station, Technical Bulletin 289*, 48 pp.

Witter, J.A. and Kulman, H.M. (1979) The parasite complex of the forest tent caterpillar in northern Minnesota. *Environmental Entomology* **8**, 723–731.

PART THREE
INSECTS IN FOREST ECOSYSTEMS

TERMITES AS MEDIATORS OF CARBON FLUXES IN TROPICAL FOREST: BUDGETS FOR CARBON DIOXIDE AND METHANE EMISSIONS

8

David E. Bignell, Paul Eggleton, Lina Nunes and Katherine L. Thomas

8.1 INTRODUCTION

8.1.1 FOREST INSECTS AND THE GLOBAL CARBON CYCLE

Forests contain more organic carbon than all other terrestrial ecosystems and presently account for approximately 90% of the annual carbon flux between the atmosphere and the Earth's land surface (Groombridge, 1992). Perturbations to forest cover therefore have the potential to influence the global carbon cycle significantly and may be crucial in promoting (or mitigating) climate change. Insects, although having a high diversity in tropical forests, might be supposed to have insufficient biomass density to influence carbon fluxes significantly (e.g. Burghouts *et al.*, 1992). This chapter examines whether this proposition is true in the case of termites, making use of data from assemblages in the Mbalmayo Forest Reserve of southern Cameroon, which have been characterized and quantified under a range of disturbance conditions. Until recently forest ecologists have paid relatively little attention to the role of species complexes in ecosystem processes, particularly the transformation and flux of energy and matter (Carney, 1989; Likens, 1992; Lawton, 1994). New estimates of the contribution of a species-rich group of invertebrates to C-fluxes may therefore help to clarify the relationship between ecosystem functions and the species richness of communities, and to address the issue of whether losses of diversity amongst insects are likely to matter functionally, beyond the level of mere academic interest. A correlation between species diversity and the biological stability of ecosystems has been proposed by May (1973), but it is unclear whether this relationship is causal, or merely circumstantial.

Social insects are often assumed to be amongst the most abundant insects in tropical ecosystems (e.g. Wilson, 1993), although little reliable comparative data for different insect groups exists. Of the social insects, ants and termites are thought to be the most abundant. Data for leaf-litter ants in Ghana indicate abundances of $117/m^2$ (Belshaw and Bolton, 1994) and in southern Cameroon $100–200/m^2$ (including canopy ants, Watt *et al.*, 1995). The termite literature now contains estimates of abundance and biomass density from each of the three global blocks of tropical forest which suggest that they may be an order of magnitude more abundant than ants (Eggleton and Bignell, 1995). Any comparison with non-social insects confirms the high relative density of social insects: for example, estimates of beetles at about $0.5 \text{ g}/m^2$ (Rodriguez,

Forests and Insects. Edited by A.D. Watt, N.E. Stork and M.D. Hunter. Published in 1997 by Chapman & Hall, London. ISBN 0 412 79110 2.

1992) and all non-social insects at 3 g/m² (Stork and Brendell, 1993) are well below the corresponding data for termites in similar ecosystems (Wood and Sands, 1978). The highest densities reported for termites, 50–100 g/m² in southern Cameroon forest (Eggleton *et al.*, 1996), are clearly greater than any other component of the invertebrate biota and may constitute as much as 95% of all soil insect biomass (Chapter 15). If direct carbon fluxes from invertebrates are of any significance, by far the greatest contribution will be made by termites.

8.1.2 ECOSYSTEM FUNCTIONS AFFECTED BY TERMITES

Termites are frequently dominant invertebrates in tropical soils (Wood and Sands, 1978; Eggleton and Bignell, 1995) and, compared with many other invertebrate groups, are well known and tractable in the sense that a well advanced taxonomy is available (Pearce and Waite, 1994). The roles of termites as direct mediators of organic decomposition and as important components of tropical soil faunas influencing humification, soil conditioning, nitrogen fixation, water dynamics, aggregate binding and the formation of clay-mineral complexes are widely recognized (Lee and Wood, 1971; Collins, 1983; Wood, 1988; de Bruyn and Connacher, 1992; Martius, 1994; Brussaard and Juma, 1996). The role of termites as keystone species is discussed by Lawton *et al.* (1995). A secondary feature of the ecological impact of termites is the production of methane from anaerobic microsites within the gut (Zimmerman *et al.*, 1982). Methane is one of the principal greenhouse gases, contributing about 18% to the radiative forcing of climate (Anon., 1992).

8.1.3 THE BASIS OF TERMITE DIGESTION AND RESPIRATORY METABOLISM

The traditional view that termites must form obligate associations with microorganisms to degrade cellulosic materials has recently been challenged by demonstrations that endogenous cellulases are produced by a variety of termite species (wood- or grass-feeders) across a broad taxonomic spectrum (e.g. Kovoor, 1970; Potts and Hewitt, 1974; O'Brien *et al.*, 1979; Schulz *et al.*, 1986; Rouland *et al.*, 1988; Veivers *et al.*, 1991). More detailed investigations of cellulose dissimilation are few, but in one or two African species of the fungus-associated genus *Macrotermes*, some evidence exists that endogenously produced cellulase is sufficient to support the notional glucose requirement of the termites (Veivers *et al.*, 1991; Bignell *et al.*, 1994). There is thus no apparent involvement of either fungal or bacterial partners in the dissimilatory process, and glucose appears to be the principal substrate supporting respiration in a manner similar to that of other aerobic eukaryotes. However, it is still the case that all termites lose viability, either immediately or eventually, after separation from some or all of their microbial partners (Cleveland, 1925; Sands, 1969; Eutick *et al.*, 1978) and even in those species where an endogenous cellulase is demonstrable, there is still an enlarged hindgut containing an abundant and heterogeneous community of prokaryotes. Acetate and some other short-chain fatty acids which are the typical endproducts of fermentative microbial metabolism are known to be produced and accumulated in termite hindguts (Anklin-Mühlemann *et al.*, 1995) and can be utilized by termites tissue to support oxidative respiration (Mauldin, 1982; Odelson and Breznak, 1983; Hogan *et al.*, 1988). There is therefore a conflict of evidence concerning the mutualistic relationships of termites which needs to be resolved and which is of obvious relevance to any examination of their role in C mineralization, especially since methane (CH_4) is not a metabolic end-product of eukaryotes and can only be produced by certain varieties of bacteria. Explanations of the relationship which hold that microbes have no more than a casual role or that endogenous cellulase is produced to effect a marginal improvement in the digestion of plant fibres are not consistent with the design of the termite digestive system and, arguably, would provide insufficient selective advantage to explain their evolution in diverse termite groups.

A novel explanation of the termite–microbe relationship in the Termitidae (higher termites: the group dominant in tropical forests) is suggested by the observation of O'Brien and Breznak (1984) that termite tissues are unable to decarboxylate pyruvate, apparently because they lack an active form of the key respiratory enzyme, pyruvate decarboxylase (PDH). This new proposal is illustrated in Figure 8.1, which shows a scheme for carbohydrate dissimilation in termites in which the hindgut microbiota participate principally as decarboxylating agents for pyruvate (and thus generating rather less than one third of the CO_2 produced by the overall respiratory process), but with important secondary roles in methanogenesis, acetogenesis and (probably) nitrogen fixation (Breznak and Brune, 1994; Tayasu et al., 1994). The significance of the secondary reactions is that they provide a sink for reducing equivalents (hydrogen), which preserves the redox balance of the system as a whole and in the case of acetogenesis adds as much as one third to the total of acetate accumulated (Breznak and Switzer, 1986; Breznak, 1994). It is clear that the balance between acetogenesis and methanogenesis varies in a consistent way between species and between trophic groups, with soil-feeding forms producing more methane (per unit weight of termite) than others (Brauman et al., 1992). This again has clear implications for assemblage C-flux budgets, since populations dominated by soil-feeders or in which one or two soil-feeding forms have very high biomass density (e.g. Eggleton et al., 1996) will be larger gross producers of methane (see below).

It is likely that more evidence to corroborate these new concepts of termite–microbe symbiosis will emerge soon. For example, in *Nasutitermes walkeri*, it has now been demonstrated that the hindgut bacteria produce most acetate in the presence of pyruvate, rather than glucose or trehalose (P. Veivers and M. Slaytor, unpublished). Further, enzymatic assays have shown that at least two pathways are available for pyruvate metabolism, one of which (PDH pathway, Figure 8.1) is aerobic and leads to the production of CO_2 within the gut community by the tricarboxylic acid cycle. This implies that oxygen is available to at least some of the intestinal organisms, a conclusion strongly and elegantly supported by the recent microelectrode

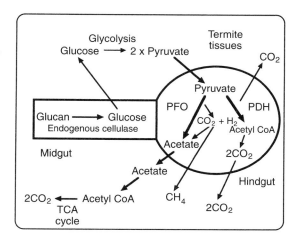

Known reductive reactions of gut	ΔG^0 (kJ/e$^-$)
$4H^+ + 4e^- + O_2 \longrightarrow 2H_2O$	−119
$8H^+ + 8e^- + CO_2 \longrightarrow CH_4 + 2H_2O$	−16.3
$8H^+ + 8e^- + 2CO_2 \longrightarrow CH_3COOH$	−13.1
$8H^+ + 8e^- + N_2 \longrightarrow 2NH_3 + H_2$	−5.5

Figure 8.1 Simplified outline of carbohydrate metabolism in wood- and litter-feeding higher termites. Cellulase is produced in midgut and salivary glands without assistance of microorganisms and in sufficient quantity to generate the glucose which supports respiratory metabolism. However, decarboxylation of pyruvate, an essential step before any net gain of energy can be achieved, can only be carried out by intestinal microbiota, as appropriate enzyme (pyruvate dehydrogenase) is not active in termite tissues. At least two pathways, with differing electron acceptors, are available to microorganisms, depending on prevailing redox conditions (M. Slaytor, personal communication). Pathways are mediated by pyruvate dehydrogenase (PDH) and pyruvate ferredoxin oxidoreductase (PFO), respectively. Reoxidation of electron acceptors (and therefore continuation of carbohydrate dissimilation) can be obtained by at least four reductive reactions (lower set of equations; ΔG^0 is associated free energy change at 298°K and 1 atms), one of which results in production of methane. Balance between these reactions, and consequently rates of methane efflux, varies between termite species. End-product of dissimilation is acetate (and some other short-chain fatty acids), which can be oxidized aerobically by termite host, or acetyl-CoA, oxidized by bacteria in aerobic regions of gut. Scheme would not apply in lower termites where cellulolytic flagellate protozoans are present in hindgut.

study of Brune *et al.* (1995). Hydrogen gas is also produced by termite guts (e.g. Hungate, 1946; Odelson and Breznak, 1983; Anklin-Mühlemann *et al.*, 1995), but there are insufficient data to show whether production varies with trophic group.

Almost all the information available on termite respiratory metabolism is from wood- and litter-feeding forms where it could not be seriously disputed that the degradation of cellulose (perhaps with other polysaccharides), by whatever mechanism, was the central digestive process. By contrast, virtually nothing is known about soil-feeders or intermediate forms feeding at the wood/soil interface, where polysaccharide might not be the principal resource utilized (Bignell, 1994). However, some idea of the principal dissimilation which supports respiration may be obtained from the respiratory quotient (RQ, the molar ratio of CO_2 produced to O_2 consumed). RQs close to or above 1.0 would indicate the dissimilation of carbohydrate, especially if accompanied by methane production. If CO_2, H_2 and CH_4 effluxes are known, it is possible to calculate the amount of oxygen that would be needed to oxidize completely a carbohydrate substrate with the general formula $(CH_2O)_n$, i.e. to predict a theoretical RQ, which can then be compared with measured values. Our work in the Mbalmayo Forest Reserve has enabled us to make experimental determinations of RQ, along with measurements of CO_2 and CH_4 effluxes, in termites of different trophic groups.

8.1.4 TROPHIC SPECIALIZATIONS OF TERMITES

Accurate information on the natural history and feeding habits of termites is scarce, especially for predominantly subterranean species, amongst which in forests many may be soil-feeders. Accordingly, it is not yet possible to assign species unambiguously to functional groups that are ecologically robust (Eggleton and Bignell, 1995). The recent literature recognizes four broad trophic categories to which most termites can be allocated empirically by reference to site of discovery, colour of abdomen and known biology (Martius, 1994; de Souza and Brown, 1994; Eggleton *et al.*, 1995, 1996). The categories are as follows.

1. **Soil Feeders**. Termites apparently feeding predominantly on the upper mineral soil horizons. The gut contents of such termites are characteristically heterogeneous, with as many as 10 different categories including fine roots (Sleaford *et al.*, 1996), but the most abundant organic material ingested is soil organic matter. Some fraction of this is assumed to be utilized but the identity of the resource and the mechanism of its digestion are unknown (Bignell, 1994).
2. **Wood-soil feeders**. Termites apparently feeding only or predominantly within soil under logs, within soil plastered on the surface of rotting logs or within highly decayed (friable) wood. This group includes termites whose gut contents are similar to those of soil-feeders, including the presence of much mineral material, but with more recognizable plant tissue and/or wood fibres in various states of degradation (Sleaford *et al.*, 1996).
3. **Wood feeders**. Termites feeding on wood and excavating galleries in larger items of woody litter. The digestive mechanisms of these termites are the best characterized and there is clear evidence that cellulose and some other polysaccharides are digested while, simultaneously, atmospheric nitrogen is fixed by the gut flora. It is unclear whether lignin is degraded to any significant extent (except Macrotermitinae: Breznak and Brune, 1994).
4. **Litter and grass foragers**. Termites that forage for leaf litter, dry standing grass and small woody litter, cutting these items where they occur. These termites are assumed to be degraders of cellulose and other structural plant polysaccharides.

Other trophic categories are known (for example, dung feeders and lichen feeders), but have only localized importance. Categories 3 and 4 include termites (Macrotermitinae) having epigeal or subterranean nests in which fungus gardens are cultivated. In these forms the forage is composted by the fungus, and fungal tissue is subsequently eaten by the termites, but the termites themselves have the ability to degrade cellulose efficiently and the real basis of the association is unclear (Veivers *et al.*, 1991; Bignell *et al.*, 1994; Anklin-Mühlemann *et al.*, 1995). In Macrotermitinae the metabolic activity of the cultivated fungus (and especially the production of CO_2) is substantial and its metabolic rate exceeds that of its termite partners (Wood and Sands, 1978).

The relevance of this trophic diversity to any attempt to define or quantify the carbon fluxes associated with termites is that many termites are feeding well down the humification gradient on organic material that is already partially degraded and physically modified. Therefore it cannot necessarily be assumed that all respiratory metabolism is based on a simple digestion of cellulose, with glucose becoming the principal substrate, without an experimental determination of gas exchanges. Any departure from the classic cellulose/glucose dissimilation scheme is unlikely to affect weight-specific metabolic rates directly (as these are largely determined by physiological scaling constants), but will have implications for hydrogen sinks and nitrogen metabolism (Breznak, 1994; Breznak and Brune, 1994).

8.2 CURRENT INFORMATION ON THE ROLE OF TERMITES

8.2.1 QUANTIFICATION OF CARBON FLUXES IN SAVANNAS AND FORESTS

The importance of termites as decomposers in savanna and desert systems has been confirmed by quantitative studies (Lepage, 1974, Wood and Sands, 1978; Collins, 1981; Holt, 1987, 1990; Jones and Nutting, 1989; Jones, 1990; Whitford *et al.*, 1991). The main approaches have been: (1) the estimation of litter and wood consumption by termites, which are then compared with measured rates of litter production and/or net primary production, and (2) respiration measurements, which can be used to determine the rate at which carbon is mineralized directly to the atmosphere by termites (and, where appropriate, their fungal associates). Wood and Sands (1978) reported population respiration rates of 11–118 kg C/ha per annum for West African savannas (the most thoroughly studied), and argued that termites could consume up to 55% of surface litter (about 20–25% of the grass standing crop). Averaging savanna ecosystems, termite consumption is about 20% of the grass standing crop – a little less than mammalian herbivores and about 36% of the net primary production lost to fire (Wood and Sands, 1978). Clearly, these figures support the thesis that termites are a major, though not necessarily predominant, factor in savanna ecosystem processes. Estimates of C mineralization by savanna termites as a proportion of all CO_2 production are few, but published assessments range up to 20% (Holt, 1987).

By contrast, few data are available for tropical forests, but pilot studies (reviewed by Martius,1994) suggest that here, too, termites may consume up to a third of the litter. Martius makes the telling point that even at a mineralization rate of 1–2% of net primary production, termites would make the same relative contribution to the decomposition process as all macroarthropods combined in temperate forests. However, there are large uncertainties in the measurement of termite biomass, consumption and respiration which make it possible that the direct contribution of forest termites to C mineralization may be much larger. Indirect effects through interactions with other organisms are entirely unquantified. These include the stimulation of microbial activity by the conditioning of soil, the supply of carbon to predators and the fixation of nitrogen, permitting subsequent carbon sequestration.

8.2.2 ASSEMBLAGE CHARACTERIZATION IN THE MBALMAYO FOREST RESERVE: SMALL-SCALE DATA FROM LIGHTLY AND HEAVILY DISTURBED FORESTS

The present study has arisen from the UK Natural Environment Research Council TIGER Programme (Terrestrial Initiative in Global Environmental Research). The TIGER work is an extensive (5-year) study of termite biodiversity in the Mbalmayo Forest Reserve of southern Cameroon, encompassing a number of sites where long-term trials of land-use systems related to forest simplification, clearance and afforestation are in place. Basic laboratory facilities are available to us, located within a large and diverse forest block, permitting not only an exhaustive taxonomic and ecological characterization of the assemblage, but also the easy collection of live termites in sufficient number and fresh condition for several physiological measurements to be attempted.

The biodiversity, abundance and biomass density of the Mbalmayo termite assemblage are documented in detail by Eggleton *et al.* (1995, 1996). The data will not be repeated in full here (a summary of biomass densities, averaged across two years of sampling, is given later in Table 8.7), but they are assumed to be at least partially representative of African tropical forests as a whole and therefore form the basis of our estimates of gas effluxes from termites in this biogeographical block.

Data from the Mbalmayo Forest Reserve (and other studies, reviewed by Eggleton and Bignell, 1995) show that severe disturbance produces a large drop in species richness and biomass density, especially amongst soil-feeding termites. However, lighter overall disturbance, such as conversion to plantation or in late successional secondary forest, produces more subtle changes. Biomass density may drop slightly, but abundance remains about the same. The significant effect is a change in the proportions of soil- and wood-feeding forms, with soil-feeders predominating in near primary and old secondary forest, while wood-feeders become more common in young plantations. The changes have an effect on gas effluxes, as demonstrated in sections 8.3 and 8.4. One important feature of the Mbalmayo Reserve (in its current state) is that savanna termite species are not present to colonize disturbed sites (Eggleton *et al.*, 1995) and our data therefore genuinely reflect the responses of forest forms to changes in habitat character.

8.2.3 ROLE OF TERMITES IN TRACE GAS PRODUCTION

Although the estimated contribution of methane released by termites to the global atmospheric methane budget has been revised downwards in recent years (e.g. Fraser *et al.*, 1986; Khalil *et al.*, 1990; Martius *et al.*, 1993), methane production varies considerably between species (Brauman *et al.*, 1992). While the termite assemblages of tropical forests remain incompletely characterized, the possibility that termites are an important net source of methane cannot be excluded. We present (below) new extrapolations for global budgeting and discuss some of the uncertainties that surround the scaling-up calculation.

Zimmerman *et al.* (1982) were the first to suggest that termites might be important global sources of CH_4, CO_2 and H_2. Their extrapolations were based upon laboratory studies using two species of wood-feeding termites (*Reticulitermes tibialis* Banks and *Gnathamitermes perplexus* Banks). Weight-specific productions of trace gases measured in the laboratory were treated as typical of gas production from termite species globally. Estimates of biomass density (in fact rates of consumption, based on data in Wood and Sands, 1978) in various vegetational regions were then used to sum the potential contribution of termites from each area. Their final estimates for methane are shown in Table 8.1. Zimmerman *et al.* went on to suggest that the destruction of forests would lead to a great increase in savanna-type habitats (especially wet savannas) which they believed had the highest termite biomass densities, and thus in turn to a great increase in trace gas production by termites.

The first workers to question these estimates seriously were Rasmussen and Khalil (1983), who presented laboratory-based data for another wood-feeding termite (*Zootermopsis angusticollis* Hagen) and reassessed the projections of consumption by termites. They concluded that the estimate of global methane production by termites made by Zimmerman *et al.* was far too high and, on the basis of the efflux from *Zootermopsis angusticollis*, could only be achieved if termites in nature produced 3.5 times the methane they produced in the laboratory. Consequently, Rasmussen and Khalil reduced the estimate of the contribution by termites to global methane production from *c.* 40% to less than 15%. Further criticisms of both previous estimates were made by Collins and Wood (1984). They pointed out that there was (1) a failure to take into account soil-feeding termites, estimated by them as roughly one-third of all tropical forest termites (Collins and Wood assumed that soil-feeders

Table 8.1 Estimates of annual global methane production by termites made since 1982

Authors	Estimated global production, Tg	Comments
Zimmerman et al. (1982)	75–310	One species measured; global biomass estimates inaccurate
Rasmussen and Khalil (1983)	10–90	One additional species measured; reassesses the above
Collins and Wood (1984)	10–30	Better biomass data; separates soil- and wood-feeders, but assumes soil-feeders produce little CH_4
Seiler et al. (1984)	2–5	Based on fluxes from mound-nests; concluded that CH_4 oxidation in soil > production
Fraser et al. (1986)	6–42	Only wood-feeders studied
Martius et al. (1993)	5–36	Based on production by Neotropical *Nasutitermes* mound-nesters; global biomass estimates not made
Hackstein and Stumm (1994)	6–51	Very few species studied; unwarranted assumptions made about biomass density.
This study	3–96	See text for method of extrapolation

would produce less methane than wood-feeders) and (2) incorrect calculation of consumption rates, partly because the termites used in laboratory tests of methane output were amongst the very largest known and consequently unrepresentative. Collins and Wood concluded that the estimate of global methane production by termites by Zimmerman et al. was at least an order of magnitude too large (Table 8.1).

Seiler et al. (1984) were the first to gather significant field data on termite production of trace gases, from studies of six genera of mound-building, wood-feeding termites in South Africa. The conclusion, for CH_4, was that the amount of oxidation in the soils was likely to be greater than the small efflux of CH_4 from termite mounds. Khalil et al. (1990) obtained field data from mound-building, wood-feeding Australian termites, and in addition to methane also examined the fluxes of CO_2, $CHCl_3$, N_2O, CO, H_2 and light hydrocarbons. Again, the conclusion was that termites were not important sources of any of these gases. They also showed that the soil around the mounds was consuming methane due to the presence of methylotrophic bacteria. The general conclusions of this careful study were that the trace gas production by termites was likely to be too small to justify the major effort that would be required to remove uncertainties. A low estimate of the termite contribution to regional methane budgets (0.5–1.5 Mt for tropical Africa) was also made by Delmas et al. (1991, 1992), again based on data from mound-building forms, but this time including soil-feeders. It was estimated that contributions from swampy forest soils and biomass burning (conditions arguably incompatible with high termite biomass densities) greatly exceeded effluxes from termite mounds.

While the series of papers cited above concentrated on scaling-up calculations to assess the global significance of termite methane production, contributions by Brauman et al. (1992) and by Rouland et al. (1993) examined the variations of methane flux between species and particular feeding groups (24 species from Africa and North America were investigated). From their data sets, soil-feeding forms were shown to produce more methane than wood-feeders, thus overturning the key assumption of Collins and Wood (1984), although there was a great deal of variation within both groups. Interestingly, termites with low rates of methane production showed higher rates of acetogenesis (from $H^{14}CO_3$)

in vitro, confirming that methane production is not an inevitable consequence of termite hindgut fermentation, as alternative mechanisms of achieving overall redox balance are available.

The sole contribution to the study of termite methane effluxes based on South American species (the hectarage of Amazonian tropical forest is 60% of the global total) is that of Martius *et al.* (1993), which presents data from the wood-feeding, nest-building *Nasutitermes* species that dominate Neotropical termite faunas. They concluded that Amazonian termites are producing more methane than other termites (on the basis of comparisons with the 11 species in Khalil *et al.*, 1990), but their extrapolations suggested a low figure for the termite contribution to global methane production (Table 8.1). Although no data for soil-feeding forms were presented (such forms are probably important components of the Amazonian fauna), Martius *et al.* did point out that the mounds of soil-feeders were known to contain methylotrophic bacteria and that the net contribution from these termites was therefore very uncertain. Recently, Hackstein and Stumm (1994) reviewed methane production by terrestrial arthropods. Their conclusion concerning termites was that the most they could produce globally is 50.7 Tg, based on an assumption that biomass density nowhere exceeded 6 g/m^2, and on methane flux measurements from four species of wood-feeding termites that produce relatively little methane. The review is, however, illuminating in that it shows that the methane effluxes of termites are, on a weight-specific basis, not necessarily exceptional.

8.3 PHYSIOLOGICAL STUDIES UNDER FIELD CONDITIONS

We have chosen constant-volume Warburg manometry for the determination of xCO_2 (i.e. the rate of CO_2 efflux on a weight specific basis), metabolic rate (MR) and RQ in 24 species of termite that are common or locally abundant in the Mbalmayo Forest Reserve (protocols given in Figure 8.2). The advantage of this method, in addition to its simplicity, accuracy and reliability, is that power is required only to maintain a water bath at a constant temperature (28°C, roughly 2°C above ambient). The principal difficulty is that the number (biomass) of termites that can be incubated is limited by the small volume of the flask (12–15 ml) and the social interactions that take place between individuals may not be typical of those within the mound, nest or other natural habitats. Further, it is impractical, as well as confusing, to add mound material, soil or wood to the flask, and so the termites are not in contact with their natural substrata and both their behaviour and their physiology may be affected. Generally xCO_2, and therefore RQ, declines slowly during the incubation (normally 3 hours). The reasons for this are unclear, but the first 30 minutes of incubation are assumed to be an adjustment period for the insects and the RQs reported here are therefore averaged over the period from 30 minutes to 180 minutes following the start of the incubation.

The primary determinations of xCH_4 were also made by the incubation of isolated groups of termites in closed containers (Figure 8.3), but because of the need to remove quite large volumes of headspace gas (20 ml) for gas analysis, a minimum flask volume of 100 ml was necessary. However, wherever possible approximately the same biomass of termites was employed for Warburg manometry and methane measurement. Again, the advantage of closed incubation with headspace gas sampling is the simplicity of the method (there is no requirement for electrical power) and the accuracy and sensitivity that can be achieved by subsequent gas chromatography (GC) analysis of the samples. Little or no methane escapes from gas sampling vials if these are correctly closed, making it possible to survey a large number of termites in the field and to replicate measurement before removing the vials to a fully equipped laboratory for gas analysis. The interaction between termites and their mound materials or the soil column may be important for methane budgets because of possible local oxidation of methane. This interaction was investigated using a field-portable mass spectrometer (Figures 8.4 and 8.5), which was able to provide an immediate (and continuous) measurement of methane concentrations in headspace gases and thus faciliate the design of manipulative experiments.

Physiological studies under field conditions 117

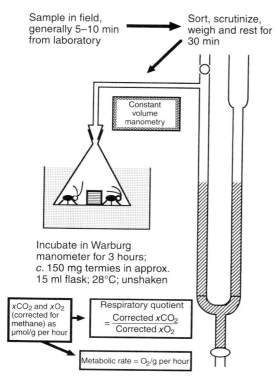

Figure 8.2 Protocol for determination of carbon dioxide efflux (xCO_2) and respiratory quotient (RQ) under field conditions. Sampled termites were sorted on moistened filter paper to remove damaged individuals and weighed before being enclosed in single side-arm Warburg manometer flasks. Eight flasks were employed for each determination in two arrays of four flasks on either side of rectangular perspex water bath of 8 l capacity, maintained at 28°C and stirred with a Bioblock Thermostatic Heater. Each array comprised thermobarometer (containing 1 ml H_2O), experimental flask (containing termites and 0.1 ml 1M KOH in centre well) and two control flasks (containing termites only, and termites plus 0.1 ml H_2O in the centre well, respectively). Flasks and manometers were paired, standardized gravimetrically using mercury and water, and tested for gas-tightness by hydrazine/ferricyanide method of Umbreit *et al.* (1964). Kreb's fluid was employed in the manometers, which were adjusted to constant volume at 30-minute intervals during incubation. Values of xCO_2 and xO_2 were adjusted for methane production (determined separately for each species). Hydrogen effluxes were assumed to be negligible and no correction was applied. Termite mortality was determined at the completion of the incubation and the results from excessively stressed groups of termites were discarded.

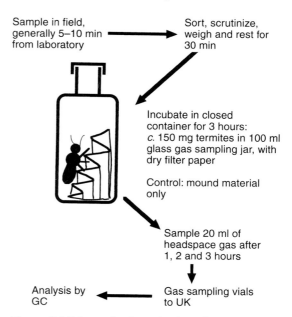

Figure 8.3 Scheme for determination of methane effluxes from isolated worker caste termites. Termites were sorted on moistened filter paper to remove damaged individuals and weighed before being placed in 100 ml amber glass gas sampling bottle (Chromatography Services Ltd), sealed by Chrompack aluminium crimp cap (20 mm diameter) fitted with teflon-lined butyl rubber washer. With a hypodermic syringe, 20 ml of headspace gas was withdrawn at intervals of 1, 2 and 3 hours and transferred to similarly sealed 50 ml Chrompack gas sampling vials from which 20 ml of air had previously been withdrawn by syringe. After each sampling, air was bled back into headspace through syringe needle to equalize pressures. Methane field standards were established at the same time by injecting 20 ml of a 100 ppm mixture of methane in nitrogen (Phase Separations Ltd) into identical vials. Gas sampling vials were analysed for methane in the UK by method of Anklin-Mühlemann *et al.*, (1995). Termite mortality was determined at conclusion of incubation, and results from groups where individuals had died were discarded.

Termites chosen for physiological studies were representative of the major trophic groups and, wherever possible, those which contributed most to the overall biomass density of the assemblage (Eggleton *et al.*, 1996). Termites which form large epigeal mounds or make colony centres within larger logs are easy to obtain in large numbers but may

118 *Termites as mediators of carbon fluxes*

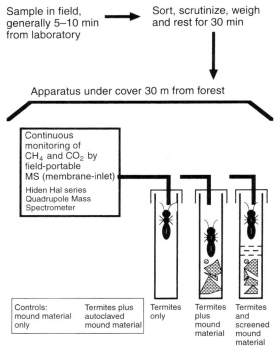

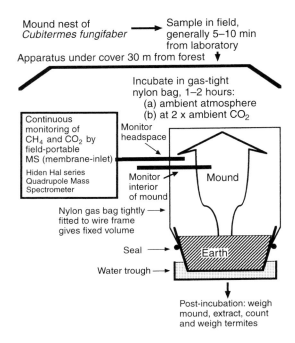

Figure 8.4 Protocol for examining possible interaction between termite methane effluxes and soil or mound materials by membrane inlet mass spectrometry, which allows both carbon dioxide and methane to be determined simultaneously and continuously. Sorted termites were placed in small sealed glass tubes (maximum volume 45 ml, with moistened filter paper) fitted with silicon covered inlet probe (1.56 mm o.d., 0.5 mm i.d., 1 m length). The m/z ratios used to determine CH_4 and CO_2 concentrations were 15 (CH_3^+ fragment ion) and 44, respectively, minimizing interference from O_2 and H_2O vapour. N_2O was monitored at m/z = 30, but was not detected in any experiment. Mass spectrometer was powered from diesel generator and calibrated using standard gas mixtures (EDT Instruments Ltd). Termites were incubated alone or in presence of approximately 1 g of mound material, in contact with termites or separated by wad of gauze. In control incubations mound materials and autoclaved mound materials (heated in steam for 30 minutes at ambient pressure) were used. Forest soil was substituted for mound materials in some experiments.

not be major contributors to gas effluxes (for example, *Macrotermes muelleri*, *Cephalotermes rectangularis* and *Termes hospes*), whereas other forms abundant within the soil but at the same time dis-

Figure 8.5 Scheme for determining net methane and carbon dioxide fluxes from intact mound nests of *Cubitermes fungifaber* by mass spectrometry. Mound was enclosed in nylon gas bag (EDT Instruments Ltd) of known volume (approximately 30 l), sealed around the inlet probe – which was left free in headspace (1–2-hour incubations) or inserted through mound wall and sealed at point of entry (short incubations of 10–30 minutes). For internal gas measurements, end of probe was covered by gas-permeable steel gauze to prevent termites destroying silicon rubber membrane.

persed in diffuse gallery systems (particularly the Apicotermitinae) are difficult to obtain in sufficient numbers for satisfactory replication of gas measurements. The actual list of species examined is therefore not precisely representative of the assemblage as a whole, but provides us with a sufficient variety of forms to validate at least some of the assumptions inherent in the scaling-up calculations.

8.3.1 MEASUREMENT OF CARBON DIOXIDE PRODUCTION AND RQ

A summary of gas flux data at 28°C for 24 readily sampled species is given in Table 8.2. There is a wide range of sizes in termites within both soil-feeding and non-soil-feeding trophic groups (mean worker fresh weights in the Mbalmayo assemblage vary from 0.5 mg for species of *Microtermes* to 26.8 mg for the very large soil-feeder *Labidotermes* sp. nov. 1). Figure 8.6 shows that, as expected, there is a general trend for termites with greater weight (i.e. larger body size) to show a lower weight specific rate of respiration (and hence efflux of carbon as CO_2). A similar trend in termites was shown by Wood and Sands (1978), using data from the existing literature and, where necessary, using an assumed RQ of 1.0 to derive estimates of O_2 consumption from the measured output of CO_2. Our data show this assumption to be generally valid, though a number of species show RQs well above 1.0. Under constant conditions the same biomass of termites will therefore vary in its contribution to carbon mineralization, according to the number of individuals and the proportions of different species that make it up. It is necessary to remember that termites may respire for much of the time inside nests or within the soil column where both temperature and CO_2 concentration are raised (Lee and Wood, 1971). No experimental data are yet available to assess the effect of elevated CO_2 on termite respiration.

There appears to be a difference in the rate of respiration between functional groups, with soil-feeding forms showing lower rates of O_2 uptake (and CO_2 efflux) per unit weight than wood-feeding or litter-feeding forms of approximately the same size (Figure 8.6). In assemblages dominated by wood-feeding forms the same biomass of termites might therefore be expected to mineralize C at a greater rate than those in which soil-feeders are most abundant, as many soil-feeders are often larger forms. Disturbance of old growth or primary forest causes precisely such a shift in assemblage structure, favouring wood-feeding species, although overall termite biomass density is reduced by the most severe disturbance (Eggleton *et al.*, 1995, 1996). Figure 8.6 shows that it is possible to fit separate regression lines to the log/log data for soil-feeding and non soil-feeding species, but in fact the two slopes are not significantly different ($P > 0.05$) and it is not possible to conclude that trophic or behavioural groups have their own weight-specific respiration relationships (cf. Wood and Lawton, 1973). Soil-feeders have relatively large

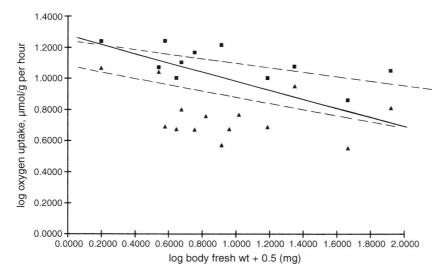

Figure 8.6 Log/log plot of metabolic rate (oxygen uptake per unit weight, from Warburg manometry) against fresh body weight in 23 species of termites from Southern Cameroon. For graphical convenience, 0.5 is added to the log fresh weight. The overall regression equation (all termites combined, continuous line) is: $y = -0.2857x + 1.2486$. Figure shows separate regressions (dashed lines) for non-soil feeders (■) and soil feeders (including wood/soil forms, ▲), though slopes of these two lines are not significantly different ($P > 0.05$).

Table 8.2 Gas exchanges of 24 species of higher termites from the Mbalmayo Forest Reserve, ranked by xCO_2

Termite	Feeding type	Worker weight (mg)	xCO_2 $\mu mol/g^{-1}/h^{-1}$ (corrected)[a]	xO_2 $\mu mol/g^{-1}/h^{-1}$ (corrected)[a]	RQ[b] (corrected)	xCH_4 $\mu mol/g^{-1}/h^{-1}$	xCH_4/xCO_2 molar ratio (corrected)
Jugositermes tuberculatus	Soil	10.6	4.20	3.60	1.17	0.229	0.054
Astalotermes sp. nov. 14	Soil	3.3	4.42	4.92	0.90	0.697	0.015
Cubitermes heghi	Soil	18.2	5.05	4.73	1.07	0.302	0.060
Labidotermes sp. nov.2	Soil	6.5	5.27	4.92	1.07	0.189	0.036
Ophiotermes grandilabius	Soil	4.9	5.58	3.77	1.48	0.147	0.026
Labidotermes sp.nov.1	Soil	26.8	5.65	5.93	1.02	0.265	0.046
Procubitermes arboricola	Soil	3.8	5.70	4.74	1.20	0.325	0.057
Crenetermes albotarsalis	Soil	13.3	6.03	4.73	1.27	0.171	0.028
Apilitermes longiceps	Soil	14.0	6.58	5.77	1.14	0.167	0.025
Macrotermes muelleri minors[c]	Litter (F)	14.7	7.64	7.31	1.05	0.191	0.025
Thoracotermes macrothorax	Soil	11.5	7.73	6.53	1.18	0.242	0.031
Coxotermes boukokoensis	Soil	4.6	8.33	6.37	1.30	0.274	0.032
Cubitermes fungifaber	Soil	6.8	8.38	9.00	0.93	0.263	0.031
Protermes prorepens	Wood (F)	1.4	10.05	10.08	1.00	0.030	0.003
Astalotermes quietus	Soil	2.9	10.62	11.07	0.96	0.229	0.021
Pericapritermes amplignathus	Soil	2.1	11.16	11.71	0.95	0.319	0.028
Nasutitermes latifrons	Wood	4.9	11.67	10.11	1.15	0.163	0.014
Macrotermes muelleri majors[c]	Litter (F)	26.5	11.79	11.30	1.04	0.069	0.006
Pseudacanthotermes militaris	Litter (F)	7.1	12.97	12.02	1.08	0.253	0.020
Microcerotermes parvus	Wood	1.5	13.50	12.77	1.06	0.040	0.003
Amalotermes phacocephalus.	Wood/soil	1.8	15.34	14.80	1.04	0.052	0.003
Cephalotermes rectangularis	Wood/soil	1.2	18.54	17.56	1.06	0.252	0.014
Microtermes sp.	Wood (F)	0.5	18.58	17.42	1.07	0.061	0.003
Termes hospes	Wood/soil	2.6	20.27	16.55	1.22	0.915	0.045
Acanthotermes acanthothorax	Wood (F)	1.1	23.31	11.84	1.97	0.029	0.001

(F) = species with fungal associates.
[a] Averaged over 3 hours ($n = 4$), fluxes corrected for CH_4, but not H_2.
[b] Averaged over 12 × 1-hour intervals (four replicates of 3 hours incubation).
[c] Separate worker castes of *Macrotermes muelleri*.

guts containing substantial quantities of (presumably inert) mineral material (Bignell, 1994), which may explain the tendency towards lower weight-related respiratory rates, but the data are too variable to establish this as a clear difference.

The range of RQs observed (Table 8.2) is clearly consistent with carbohydrate dissimilation in all trophic groups, but values in excess of about 1.10 (10 of the species examined) cannot be explained solely by the release of some C as CH_4, but may imply the production of large amounts of hydrogen. Another explanation is that there is some dissimilation by anaerobic bacteria utilizing pathways whose redox balances differ from the classical scheme of carbohydrate fermentation (Gottschalk, 1986). The exisiting literature on termite RQ has been reviewed by Peakin and Josens (1978), but very few measurements on forest species, especially soil-feeding forms, have been reported. Hébrant's (1970) suggestion that the RQs of *Cubitermes exiguus* and *Cubitermes sankurensis* are in the region of 0.7 is not supported by the present study, though the gradual decline of RQ with time of incubation described by Hébrant is in accordance with our observations. Rouland *et al.* (1993) give the following RQs for soil-feeding forms: *Noditermes* sp., 0.64; *Thoracotermes macrothorax*, 0.64; *Astratotermes* sp., 0.67. The RQ value of 1.97 given in the present study for *Acanthotermes acanthothorax* is almost certainly spurious. This termite is notably prone to high mortality during incubation, but is included in the data set as it is commonly observed in woody litter in Mbalmayo forests.

8.3.2 MEASUREMENT OF METHANE PRODUCTION BY ISOLATED TERMITES

Table 8.2 shows that weight-specific CH_4 efflux is generally greatest in soil-feeding forms and wood/soil forms feeding in the immediate vicinity of decaying wood. The range of production, from less than 0.1 μmol/g/h (*Microtermes* sp., *Microcerotermes parvus*, *Protermes prorepens* and *Acanthotermes acanthothorax*) to 0.915 μmol/g/h (*Termes hospes*) is broadly in agreement with the surveys of African forest and savanna species made by Fraser *et al.* (1986), Brauman *et al.* (1992) and Rouland *et al.* (1993). Comparisons with other recent published studies (for example, Seiler *et al.*, 1984; Kahlil *et al.*, 1990; Martius *et al.*, 1993) are more problematical, as these authors addressed the net effluxes from whole mounds. While such an approach has the apparent advantage of leaving the termites undisturbed in a natural habitat and allows any normal oxidation processes in mound walls to continue, our contention from extensive quantitative sampling of forest sites is that the vast majority of termite individuals (and therefore most of the biomass density) are located in the soil column or in larger decaying woody items and not in mounds (Eggleton and Bignell, 1995; Eggleton *et al.*, 1996). The critical determinations are therefore the rates of CH_4 efflux from isolated termites and the interactions of the gas with soil (and mound) methylotrophs.

The proportion of C efflux from termites as CH_4 rather than as CO_2, (i.e. the molar ratio $CH_4:CO_2$) varies from close to zero to some 5% or 6%. Again, this proportion is generally higher in soil-feeding forms, but an assemblage dominated (in biomass terms) by active wood-feeding and litter-foraging species with high empirical rates of respiration might nevertheless be responsible for a greater absolute amount of CH_4 evolution than one in which soil-feeders were most abundant.

In the case of a (hypothetical) termite producing approximately 6% of its total C efflux as CH_4 (actually 6.25% in the closest balanced equation given below), the stoichiometry of classical fermentation (i.e. where a carbohydrate substrate is completely degraded to CH_4, CO_2 and H_2O only) would be:

$$16C_6H_{12}O_6 + 84O_2 \rightarrow 6CH_4 + 90CO_2 + 84H_2O$$

The theoretical RQ for this reaction is 1.07, but of the three species producing more than 5% of their C efflux as CH_4 (*Jugositermes tuberculatus*, 5.2%; *Procubitermes arboricola*, 5.4%; *Cubitermes heghi*, 5.6%), only *C. heghi* has a measured RQ (manometric data corrected for the accumulation of CH_4 in the flasks) at the predicted value. The RQs of the other two species are higher (1.17 and 1.20, respectively), while a number of other species show RQs above or well above 1.10 while producing less than 5% of their C efflux as CH_4. In these cases the production of methane alone cannot account for the observed RQs.

8.3.3 USE OF THE MASS SPECTROMETER IN THE FIELD

(a) Interactions between termites and mound materials

Simultaneous determinations of CH_4 and CO_2 effluxes over short incubations (up to 3 hours) of termite species, selected to represent the assemblage, freshly removed from mounds or rotting wood were made with a mass spectrometer (MS) (Table 8.3, treatments 2–5). Opportunities for replication of measurements on some species are limited under field conditions and therefore only those efflux data that are means ($n = 3$) are shown. A comparison of treatments 1 and 2 shows that measurements of gas effluxes by MS are broadly comparable to those made by the more sensitive GC and Warburg systems (treatment 1, data abstracted from Table 8.2) and the rank order of species is identical for CH_4. Adding mound material to the incubation vessel (treatment 3) increased CH_4 production in three of five species (*Thoracotermes macrothorax*, *Cubitermes fungifaber* and *Termes hospes*) and CO_2 production in four (these species plus *Microcerotermes parvus*). In *Thoracotermes macrothorax* and *Cubitermes fungifaber* the enhancement of CH_4 production by mound materials occurred whether termites were in contact with the materials or separated from them by a gauze (data not shown). A similar effect was obtained in the presence of soil (also not shown). In *Thoracotermes macrothorax*, *Cubitermes fungifaber* and *Termes hospes*, enhancement of CH_4 efflux was also obtained in the presence of autoclaved mound material (treatment 4). CO_2 efflux was also greater than when termites were incubated alone. A similar effect could be obtained with autoclaved soil (not shown). Mound material incubated without termites (treatment 5) did not produce any detectable CH_4 and negligible CO_2.

These limited data suggest that important interactions can take place between termites and their mound materials, which will affect net gas effluxes. Two possible interactions (not mutually exclusive) are (1) that metabolic rate is stimulated in the presence of mound materials and soil (or stress reactions are reduced) and (2) that a proportion of the CH_4 released by the termites is oxidized by adjacent mound materials (or soil). A detailed argument supporting the latter proposition is given by Kahlil *et al.* (1990).

(b) Gas effluxes of intact mounds

When intact mounds of *Cubitermes fungifaber* and *Thoracotermes macrothorax* were incubated within a gas-tight bag, both CH_4 and CO_2 were evolved. Similar effluxes were observed from aerial purse nests of *Astalotermes quietus*. The production of CH_4 and CO_2 by an intact mound of *Cubitermes fungifaber* was greater than the predicted efflux based on a post-incubation determination of worker biomass and extrapolation from the *in vitro* data presented in Table 8.2, but the same as predicted from MS measurement of isolated termites (Table 8.4). CO_2 efflux was more than the prediction based on GC work, but less than that extrapolated from MS determinations. The mounds and nests are net sources of both gases. Internal CO_2 was about 10 times the ambient level and internal CH_4 was elevated by about 30–40 times, compared with forest air samples. Internal CH_4 concentrations in soil-feeder mounds have not been previously reported, but are broadly in line with those reported by Khalil *et al.* (1990) for the Australian savanna species *Coptotermes lacteus* (maximum 0.002%) and *Amitermes laurensis* (maximum 0.005%). P. Zimmerman and J. Darlington (1996, unpublished observations) give the internal CH_4 concentration in the mound of the Kenyan species *Cubitermes ugandensis* as 0.002%, and in the vented outflow from mounds of *Macrotermes jeanneli* as up to 0.001%. The effect of elevated internal CO_2 concentration on rates of methanogenesis is unknown, but we have shown in preliminary experiments that the rate of production by isolated workers of *Thoracotermes macrothorax* is increased by about 25% when CO_2 levels are increased by about 10 times (data not shown).

8.4 SCALING-UP AND ITS PITFALLS

Although there have been several attempts to estimate global carbon fluxes due to termites (especially fluxes of CH_4, listed in Table 8.1), none of these studies has adequately emphasized the enormous difficulties of such an exercise. In this section we review the uncertainties and show how previous studies are deficient when judged against them, and conclude that sufficient data do not yet exist to allow confident estimates to be made. We do, however, give a range of probable global values for CO_2 and CH_4 for discussion purposes and to illustrate the magnitude of the uncertainties. The particular difficulty of scaling-up is that it involves a nested set of assumptions, each supported by extremely patchy data and so subse-

Table 8.3 Comparison of methane (CH_4) and carbon dioxide (CO_2) emissions (μmol/g per hour) by termites and mound material under differing conditions of incubation *in vitro*. (mean of three determinations shown for each gas). Details of gas detection and measurement are given in Figures 8.4 and 8.5.

Termite species	Treatment (1)		Treatment (2)		Treatment (3)		Treatment (4)		Treatment (5)	
	CH_4 μmol g⁻¹h⁻¹	CO_2 μmol g⁻¹h⁻¹	CH_4 μmol g⁻¹h⁻¹	CO_2 μmol g⁻¹h⁻¹	CH_4 μmol g⁻¹h⁻¹	CO_2 μmol g⁻¹h⁻¹	CH_4 μmol g⁻¹h⁻¹	CO_2 μmol g⁻¹h⁻¹	CH_4 μmol g⁻¹h⁻¹	CO_2 μmol g⁻¹h⁻¹
Thoracotermes macrothorax	0.242	7.73	0.207	5.09	0.288	12.30	0.439	10.87	n.d.	n.d.
Cubitermes fungifaber	0.263	8.38	0.340	11.8	0.41	22.5	0.53	14.8	n.d.	0.06
Termes hospes	0.915	20.27	1.00	19.95	1.41	38.39	1.18	34.11	n.d.	–
Astalotermes quietus	0.697	10.62	0.53	28.02	0.41	25.33	0.38	14.35	n.d.	–
Microcerotermes parvus	0.040	13.50	n.d.	20.3	n.d.	24.55	n.d.	17.37	n.d.	n.d.

n.d. = not detected

Treatment descriptions
(1) 100 ml vessel; measurement by GC (termites only)
(2) 45 ml vessel; measurement by MS (termites only)
(3) 45 ml vessel; measurement by MS (termites + mound)
(4) 45 ml vessel; measurement by MS (termites + autoclaved mound)
(5) 45 ml vessel; measurement by MS (mound material only)
Incubation vessels contained 0.15–0.40 g termites (Treatment 1), 0.3–1.0 g termites (Treatments 2, 3, 4) and 4–5 g mound materials (Treatment 5).

Table 8.4 Emissions of carbon dioxide and methane from intact mounds of two soil-feeding termites (*T. macrothorax* and *C. fungifaber*) and an aerial purse-nest of *A. quietus*, determined by mass spectrometry. Mounds and purse nests were freshly sampled from mature plantation within the Mbalmayo Forest Reserve of Southern Cameroon and measured under ambient conditions in an adjacent laboratory. Incubations were carried out for 24 hours after sampling. Predicted gas effluxes (for *C. fungifaber* only) are based on post-incubation determinations of termite biomass and extrapolated from determinations of *in vitro* emissions. Methodology is illustrated in Figure 6. Measurement of gas emissions was continuous; actual rates of efflux are therefore averaged over the whole period of incubation

Termite species	CH_4 efflux from mould (μmol/h)	Predicted CH_4 efflux (μmol/h)	CO_2 efflux from mould (μmol/h)	Predicted CO_2 efflux (μmol/h)	Internal CH_4 concentration of mound (%)	Internal CO_2 concentration of mound (%)
Cubitermes fungifaber	11	8.5* 11.0**	675	379.8* 1019.7**	0.003	0.11
Thoracotermes macrothorax	18	–	208	–	0.007	0.33
Astalotermes quietus	7.5	–	103	–	0.005	0.132

* from GC analysis in 100 ml vessels over a 3 h incubation, $n = 5$ determinations.
** from MS analysis in 45 ml vessels over a 3 h incubation, $n = 3$ determinations.

quent errors are both unpredictable and cumulative.

8.4.1 DIFFICULTY OF ESTIMATING POINT ASSEMBLAGE DATA FOR TERMITES

The determination of termite biomass density presents both logistical and statistical problems and even the most accurate estimates have very wide confidence limits, due to the aggregated distribution of termite abundance and large variations between seasons (Eggleton and Bignell, 1995). Previous scaling-up estimates have generally been based on rather dated termite assemblage population data (the best summary of such data is Wood and Sands, 1978), which tended to take into account only a small proportion of any given termite assemblage. The most usual error is to concentrate population sampling on nests, on the assumption that most termites are found there. This is much less true of tropical forest than of tropical savanna ecosystems, with the consequence that point termite biomass density may be underestimated (Eggleton and Bignell, 1995). In some cases (e.g. Eggleton *et al.*, 1996) mound populations make up less than 10% of total termite biomass.

The spatial patchiness and temporal unpredictability of termite occurrence seems to be particularly marked for some larger soil-feeding Termitinae and Apicotermitinae (Eggleton *et al.*, 1996). For example, in the Bilik near-primary forest investigated by Eggleton *et al.* (1996), *Labidotermes* sp. nov. 1 (worker fresh weight, 27 mg) made up 94% of all termite biomass density during the short dry season of 1992, but was not found again on the site until 1995. Absolute termite biomass was 25 times greater in 1992 than in 1993, when a replicate sampling was made. In terms of gross methane production by the assemblage, *Labidotermes* sp. nov. 1 would, when present, still be producing 7 times as much CH_4 and 6 times as much CO_2 as the rest of the assemblage put together, even if these other termites were producing the gases at the maximum observed rate for any species (i.e. equivalent to the rates of *Termes hospes* and *Acanthotermes acanthothorax*, respectively; Table 8.2).

Unfortunately, the scale of sampling required to characterize completely the termite assemblage of even one area of forest has never been available and, arguably, never will be within the likely limits of funding available to termite scientists working in non-arable ecosystems (Eggleton and Bignell, 1995). Therefore, strictly speaking, the data scarcely exist to estimate local termite biomass density to within one order of magnitude. This basic fact alone makes all existing (and subsequent) estimates of termite global carbon fluxes unreliable, but does at least indicate the scale on which future sampling of termite populations needs to be executed to generate acceptable data.

8.4.2 THE PROBLEM OF ESTIMATING WEIGHT-SPECIFIC GAS PRODUCTION BY TERMITES

The construction of a carbon-flux budget for any termite assemblage (beyond those where a mere handful of species are present) necessarily involves the selection of feeding types for investigation *in vitro* (i.e. isolated workers not in contact with their food, mound or gallery materials) or measurements of fluxes from accessible mounds (or both approaches, as in the present study). In only two cases (Brauman *et al.*, 1992, and the present study, though arguably also Seiler *et al.*, 1984) can the selection of species for investigation be considered representative of natural assemblages. Even here forms which have a high frequency of occurrence in soil cores but which do not occur in sufficiently large aggregations to permit easy collection (for example, *Pericapritermes* spp.) may be excluded, even though their contribution to biomass density is likely to be high. Conversely, species such as *Termes hospes*, which are spectacular CO_2 and CH_4 producers on a weight-specific basis and easy to collect in large numbers, may not in reality have any dominant ecosystem role (this species occurred in only two of 10 forest sites sampled quantitatively by Eggleton *et al.*, 1996, where it made up less than 4% of the individuals collected). Similarly, one species with a high visual impact in the assemblage because of its large, densely populated mounds and its active foraging over the surface of the ground (*Macrotermes muelleri*) did not occur in any randomized qualitative or quantitative sample (Eggleton *et al.*, 1995, 1996).

The difficulty of knowing whether termite gas exchanges *in vitro* are similar to those that occur in natural situations, where the insects are undisturbed and where social interactions are unhindered, has been mentioned above. Further, our calculations have ignored the contributions of immature termites and alates, whose biomass densities we have considered small in comparison with fully differentiated workers (Eggleton *et al.*, 1996). Of equal concern is the nature and physical organization of the inert materials which separate termites in their natural habitats within mounds and galleries from the atmosphere. Forms living in highly decayed wood at or above the surface of the ground, or in carton nests in similar locations, might be supposed to inhabit relatively porous structures where the accumulation of high concentrations of CO_2 would be less pronounced and the escape of CH_4 to the subcanopy atmosphere unrestricted. In contrast, forms in sound wood, within hard earthen mounds or within subterranean galleries may experience quite different ambient conditions from those in which *in vitro* measurements are made, and much of the CH_4 they produce could be oxidized by methylotrophic bacteria before entering the atmosphere. Further complications arise from the well-established effects of disturbance and water content on the viability of soil methylotrophs (Keller *et al.*, 1990; Castro *et al.*, 1994), and from the foraging patterns of different termite species, where individuals may at one time be located within a comparatively sealed mound or nest and at another be foraging through relatively scattered galleries in the surface layers of the soil or moving through open litter layers. Only where gas exchanges are measured from undisturbed mounds *in situ* can it be said that termite carbon fluxes have been examined under natural conditions, but even here it is not usually the case that mound populations have been subsequently enumerated or weighed. The links established between gas fluxes and termite biomass density may therefore be weak. Our limited data on the interactions between isolated termites and their mound materials and forest soil suggest that a CH_4 oxidation factor of about 0.5 would be appropriate in scaling-up calculations for some soil-feeding species, but a different factor might apply to foraging termites and those soil-feeding forms not constructing mounds (together the majority in forests). Published static chamber experiments (e.g. Tathy *et al.*, 1992, in Congo) support the received wisdom that tropical forest soils are net methane sinks unless wet, but recent work in the Mbalmayo Reserve using similar methodology has provided some evidence that non-inundated soils can be net sources where they contain a large biomass of soil-feeding termites (J.A. MacDonald, P. Eggleton and D.E. Bignell, unpublished observations). Once again the conclusion must be that insufficient data are yet available to permit a reliable estimate of global gas productions by termites.

8.4.3 THE PROBLEM OF UNIVERSALITY OF POINT DATA

Scaling-up at all but the local species pool level is complicated by the high beta-diversity of termites (Eggleton *et al.*, 1994). The dispersal powers of termites appear to be limited and levels of endemism, at both generic and specific level, are therefore high. Thus, point abundance data cannot easily be used to extrapolate to other areas, and it is especially difficult to extrapolate from one biogeographical region to another. Further, there is a clear latitudinal gradient in termite diversity (Collins, 1989; Eggleton 1994), with termite abundance and species richness negligible at high latitudes, which needs to be taken into account when scaling-up. Fortunately, forests at mid to high latitude (i.e. subtropical, temperate and boreal forests) have low to zero biomass density (Eggleton and Bignell, 1995). Therefore the variation between continental land masses in the extent of such forests is unimportant (Table 8.6, below). Studies such as Hackstein and Stumm (1994) appear to have overestimated the biomass density of termites in subtropical and warm temperate regions.

In addition to the latitudinal gradient there is a more complex longitudinal variation across the three main tropical forest blocks (Neotropical, Afrotropical and Indo-Malayan). The order of generic richness, for instance, is Afrotropical > Neotropical > Indo-Malayan (Eggleton *et al.*, 1994) and the same rank order seems to be reflected in abundance and biomass data (Eggleton and Bignell, 1995). By contrast, most scaling-up studies have been based on one or a few model species (for example subtropical or temperate wood-feeding forms) which cannot be considered representative of termites on a global scale, and on biomass data that does not consider the effect of 'diversity anomalies' (*sensu* Ricklefs and Latham, 1993) between major tropical forest blocks. A large-scale species-level survey of termite methane production encompassing both taxonomic and biogeographical variation is still needed.

8.5 PROBABLE RANGE OF GLOBAL CARBON DIOXIDE AND METHANE FLUXES FROM TERMITES

This section suggests an approach that will permit the approximate limits of global termite carbon fluxes to be derived from the meagre empirical evidence and that takes into account the difficulties and caveats discussed above. In one sense the figures should not be taken as yet another estimate but as a heuristic exploration of ways of getting to a better estimate. As additional data are accumulated (for example, we are at present collecting assemblage and carbon flux data from tropical rain forest in Malaysia) we hope that both the method of estimation and the data will improve.

8.5.1 METHANE

The methods to extrapolate global production rates of both methane and carbon dioxide were essentially the same, as outlined below, and the ranges of production rates estimated for both gases are shown later, in Table 8.6. All of our figures are gross estimates and do not take into account the effects of methane oxidation, although they are partly based on efflux measurements from a small selection of individual species (described above) that examined the interaction between methane-emitting termites and soil or their mound materials.

(a) Land types and areas

These were chosen to represent areas with particular vegetational, latitudinal and biogeographical factors that are important in affecting termite species richness, biomass density and functional group structure. Area estimates are taken from Groombridge (1992) and are based on FAO data for 1990.

(b) Biomass density

These estimates are based on subjective extrapolations from the existing data (Wood and Sands, 1978; Eggleton and Bignell, 1995; and especially Eggleton *et al.*, 1996). The biomass density figures for African tropical forest are based on the minimum and maximum values for the Mbalmayo (Cameroon) assemblage in old secondary and near-primary plots obtained in two successive sampling years (1992 and 1993) by Eggleton *et al.* (1996), and should be regarded as the most accurate. Biomass estimates for Indo-Malayan (South-East Asian) and South American forests are extrapolated from existing point data listed by Eggleton and Bignell (1995), using the Cameroon results to calibrate the maximum values (e.g. the scaling of total populations relative to nest populations). As few quantitative data exist for South America, our extrapolations for this forest block may be highly inaccurate. In particular, soil-feeder biomass densities have never been adequately estimated in South America.

(c) Soil-feeding:non-soil-feeding ratio

Feeding group structure has been recognized as an extremely important predictor of methane efflux (e.g. Brauman *et al.*, 1992; Rouland *et al.*, 1993), the clear implication of their data being that assemblages dominated by soil-feeders will have a larger gross production of methane per unit of biomass. To reinforce this conclusion we have taken all available data on methane effluxes and combined them (Table 8.5). The combined data set shows that soil-feeders have a reported methane output a little more than twice that of wood-feeders. For any vegetational/biogeographical region, a scaling factor can be derived from the relative biomasses of soil-feeding and non-soil-feeding termites (the latter category comprising both wood and litter feeders) and this can be used to estimate overall methane production per unit of termite biomass. These constants are shown in Table 8.6. It should be noted that the most accurate estimates of biomass density for soil-feeding forms are almost certainly those for Africa, and hence in part the high scaling ratio for methane production in this region, although it is also true that soil-feeders are genuinely predominant here, especially in forests. Estimates for proportions of feeding groups outside Africa are based on regional generic and species richness rather than on true biomass density data. Nevertheless we are confident in our general assumption that Neotropical and Indo-Malayan forests will have, relatively, a lower biomass density of soil-feeding termites. Figures for savannas may be underestimates if litter-feeding termites (especially Macrotermitinae in African and Oriental regions, which appear to produce more methane than wood-feeders) have high biomass densities there.

(d) Accounting for interactions with soil and/or mound materials

The figures for methane production in both soil- and wood-feeders have been adjusted upwards by 20% for the estimate of maximum efflux (only). This is to take into account the generally higher methane effluxes observed in termites that have contact with soil or mound material (see above). A second adjustment for methane oxidation (a reduction) will also be eventually necessary once an understanding of the importance of this process for each trophic group has been obtained.

(e) Conclusions for methane and anthropogenic effects

Table 8.6 shows that we estimate the limits of the contribution made by termites to global methane production to the atmosphere as 3–19% (17–96 Tg) of a total production from all sources of 500 Tg. All previous scaling-up studies (except Zimmerman *et al.*, 1982) produced estimates towards the lower end of this range. Since we ascribe the greater part of gross methane production to soil-feeding forms, and some part of this may be oxidized by methylotrophic bacteria within the soil or mound matrix, the lower limit of our estimate (3% of global production) could be further reduced. The true impact of termites on the global methane budget is therefore still uncertain.

Zimmerman *et al.* (1982) argued that anthropogenic conversion of tropical forest to derived savanna would lead to a great increase in termite biomass density and therefore to a concomitant increase in methane generation. This conclusion is at odds with the existing data on biomass density (reviewed by Eggleton and Bignell, 1995), which shows that termite biomass is, in general, less in savannas than in forests, and that savanna populations contain relatively fewer soil-feeders. However, it is still highly pertinent to ask how disturbances to forest which fall short of complete clearance (for example simplification by logging, conversion to plantation or the shortening of fal-

Table 8.5 Reported gross methane production (mean ± 1SD) of 68 termite species (worker caste) grouped by feeding habit (main data sources: Brauman *et al.*, 1992; present study (Table 8.2) and other references in Table 8.1).

Feeding group	No. of species	Methane production ($\mu mol/g/h$)	SD
Soil	21	0.416	0.274
Wood/soil	3	0.406	0.452
Litter	9	0.354	0.292
Wood	34	0.176	0.234
Grass	1	0.021	–

Table 8.6 Summary of estimated maximum and minimum global carbon dioxide and methane productions by termites

Region	Area $(m^2 \times 10^6)$	Biomass density (g/m^2)	Scaling factor (per system) SF : NSF	CO_2 production Tg/y	CH_4 production Tg/y
Tropical forest (African)	6.0	24–130	0.9	231–1249	8.1–50
Tropical forest (Neotropical)	8.4	10–60	0.6	161–964	3.5–24.5
Tropical forest (Oriental)	2.8	5–20	0.3	31–125	0.47–2.1
Temperate forest (wet/dry)	2.1	1–3	0.0	6–17	0.1–0.2
Savanna (wet/dry)	24.6	5–20	0.2	281–1124	4.2–18.4
Temperate grassland	6.7	1–3	0.0	18–54	0.2–0.6
Cultivated land	1.2	0.1	0.0	3–9	0–0.1
Desert scrub	1.8	1–3	0.0	5–14	0.1–0.2
Estimated annual production by termites				735–3557	17–96
Estimated terrestrial production by biota				60000.00*	n.a.
Estimated total annual production (all sources)				156900.00*	500**

*figures from Schimel (1995)
**figures from Houghton et al. (1990)
n.a. = not available

Extrapolations are calculated from land surface area of tropics and subtropics inhabited by termites divided into eight gross vegetational/biogeographical regions. In each region termite biomass density is estimated from our own and other data (for listing, see Eggleton and Bignell, 1995), then apportioned between soil-feeding (SF) and non soil-feeding (NSF) forms. Averaged values for gas effluxes in each trophic group (there is assumed to be a difference) are then used to compute annual CO_2 and CH_4 production for each region, using biomass density, feeding group proportion and size of region. Regional productions are summed to give a global total.

low/regrowth periods in the slash-and-burn cycle) might affect the potentially large methane fluxes from their termite populations. Site-specific data from Cameroon (Table 8.7, derived from Eggleton et al., 1995) suggest that as forests are simplified (conversion to arable use) both total termite biomass and the relative contribution of soil-feeding and wood/soil-feeding forms decline. On this basis, extrapolations of local methane fluxes would predict the production of less gas in simplified systems compared with mature (and therefore relatively diverse) forest. However, this conclusion ignores the loss of methane oxidizing potential in soil that may accompany disturbance (Keller et al., 1990), and until this factor is better quantified we are in no position to make confident assertions. Other factors that influence methane oxidation by microorganisms include soil texture (Dow et al., 1993), soil moisture content (Castro et al., 1994) and nitrogen inputs (Steuder et al., 1989). Termites themselves cause considerable physical disturbance to soils, but it is unclear whether microbial activities are directly affected by this perturbation (Lee and Wood, 1971; Brussaard and Juma, 1996).

If instead of using the 1990 FAO estimates of global forest cover, we substitute the 1980 figures (Groombridge, 1992) into the scaling-up calculation, assuming maximum termite biomass densities and that all forest becomes savanna when cleared, the estimated maximum 1980 figure for termite

Table 8.7 Efflux of carbon dioxide and methane from termite assemblages in an undisturbed forest site and sites subjected to disturbance and clearance

Year	Site	Termite biomass (g/m^2)	Contribution by soil-feeding and wood/soil-feeding forms (% total abundance)	Estimated gross CO$_2$ production by termites (mg/m^2/h)	CO$_2$ production (as % of total soil respiration)	Estimated gross CH$_4$ production by termites (mg/m^2/h)
1	Complete clearance	19.89	82	2.6	1.6	0.064
	Weeded bush fallow	1.74	5	< 0.01	< 0.1	0.004
	Young plantation	35.13	43	5.2	1.9	0.129
	Mature secondary	114.16	82	10.5	3.0	0.415
	Near primary	140.59	93	10.4	3.2	1.118
2	Complete clearance	0.11	81	< 0.01	< 0.1	< 0.001
	Weeded bush fallow	2.09	4	< 0.1	< 0.1	0.005
	Young plantation	10.98	22	0.8	0.5	0.032
	Mature secondary	39.11	77	3.8	1.1	0.109
	Near primary	8.31	54	0.7	0.2	0.027

Biomass and the proportion contributed by soil-feeding and wood/soil-feeding forms are shown in five one-hectare sites within the Mbalmayo Forest Reserve in southern Cameroon, sampled in two successive years (July 1992 and July 1993; Eggleton *et al.*, 1996). The sites represent a disturbance gradient from complete clearance (very severe) to near primary (light or undisturbed). The biomass data are then used to estimate gross gas effluxes, using the measurements of CO$_2$ and CH$_4$ production by representative species in the laboratory. Soil respiration was measured on each of the sites by a portable IRGA system, using closed circulation through a static chamber.

methane production is 104 Tg/y (*c.* 21% of the total from all sources). This suggests that methane production by termites may have dropped by 8% in the period 1980–1990, as the result of the loss of forest during this period. This will presumably have been partially or completely offset by the increase in methane effluxes due to biomass burning, ungulate farming, urbanization and other human-induced activity.

The geographical limits of tropical forests have fluctuated widely during the Quaternary period. At the beginning of the Eocene, for example, tropical forests were present from palaeolatitudes 32° N to 32° S (Creber and Chaloner, 1985) and presumably termites had a concomitantly wider range at this time. If biomass density was at a similar level to that of the present, the total methane production by termites would have been several times greater than today, and at those levels would surely have had a significant effect on the heat exchanges of the atmosphere. However, the soil-feeding forms that (on present day evidence) contribute most to methane fluxes may have evolved too recently to have been a significant component of assemblages in Eocene forests (Noirot, 1992; Eggleton *et al.*, 1994). For further comments on methane production by termites during the recent Quaternary, see Petit-Marie *et al.* (1991).

8.5.2 CARBON DIOXIDE

Calculations of global CO$_2$ production by termites have been made by essentially the same methods employed for CH$_4$ (Table 8.6). Our estimates are 0.3–1.0 Gt C/y from a total of *c.* 60 Gt C/y from global net respiration (i.e. 0.2–2.0% of the total terrestrial natural efflux; Schimel, 1995). This figure is an order of magnitude smaller than that suggested by Zimmerman *et al.* (1982), but although small it represents up to 2% by a group with only *c.* 0.01% of the terrestrial global species richness. This 2% exceeds estimates of a number of other elements in the global carbon cycle that attract serious study – for example, C uptake by northern hemisphere regrowth forests is *c*. 0.5 Gt C/y (Houghton *et al.* 1990). Our estimate that the direct contribution made by termites to global CO_2 production is small is consistent with the views of some soil biologists that invertebrate animals as a whole make a small contribution to carbon fluxes in terrestrial ecosystems (Swift *et al.*, 1979). This does not exclude the possibility that there are important indirect effects through physical and chemical processes and by interactions with other organisms (Brussaard and Juma, 1996).

8.6 OVERALL CONCLUSIONS AND FUTURE WORK

Our data and global extrapolations suggest that the direct impact of termite gas exchanges on atmospheric chemistry might be much greater for methane than for carbon dioxide. This conclusion is in part a consequence of the relative and absolute abundance of methane-producing forms in tropical forests (not adequately considered in previous published studies of global methane production by termites), but also reflects the simple fact that such forests remain extensive in area (especially Amazonia) and have a large influence on the terrestrial carbon cycle. The many uncertainties associated with scaling-up calculations are far from being resolved and we still do not have the data required to show beyond doubt that termites have a significant role in global carbon fluxes. However, we suggest that the true figures for global gas production by termites in tropical forests lie in the range 423–2338 Tg/y for CO_2 and 12–77 Tg/y for CH_4. The upper values represent 1.5% and 15%, respectively, of the estimated annual production of CO_2 and CH_4, respectively, from all sources. Disturbance of tropical forest, including simplification and conversion to plantation, should reduce gross CH_4 production, but the effects on CO_2 are less certain. To make scaling-up estimates more accurate, the most pressing needs are for good quantitative data on the termite assemblages of Neotropical forests and further information on the oxidation of termite-produced methane in soils. In order to reduce the range in our current estimates of gas production by termites, seasonal and annual changes in termite biomass density must also be determined.

ACKNOWLEDGEMENTS

We thank members and guests of the TIGER 1.5 consortium (John Lawton, Dieudonné Nguele, Bill Sands, Tom Wood, Brian Waite, Mike Hodda, Eileen Wright, David Jones, Nick Mawdsley, Melanie West, Patrick Meir, Corinne Rouland, Alan Brauman, Pam Veivers, Ichiro Tayasu, Jannette MacDonald, Graham Lawton and Nicholas Bignell) who have contributed to the work described in this chapter. In addition we thank David Lloyd for the loan of the mass spectrometer and Michael Slaytor for many illuminating discussions of termite biochemistry. We are grateful to the Government of Cameroon (ONADEF), the Overseas Development Administration (UK) and International Institute of Tropical Agriculture for permission to work and logistical assistance in Cameroon. The work was funded by the Natural Environment Research Council through its Terrestrial Initiative in Global Environmental Research, award no. GST/02/625.

REFERENCES

Anklin-Mühlemann, R., Bignell, D.E., Veivers, P.C. *et al.* (1995) Morphological, microbiological and biochemical studies of the gut flora in the fungus-growing termite *Macrotermes subhyalinus*. *J. Insect Physiol.* **41**, 929–940.

Anon. (1992) *Climate Change (1992) Supplementary report to the Intergovernmental Panel on Climate Change scientific assessment*, Cambridge University Press, New York, 200 pp.

Belshaw, R. and Bolton, B. (1994) A survey of the leaf litter ant fauna in Ghana, West Africa (Hymenoptera: Formicidae) *J. Hymenoptera Res.* **3**, 5–16.

Bignell, D.E. (1994) Soil-feeding and gut morphology in higher termites. In *Nourishment and Evolution in Insect Societies* (eds J.H. Hunt and C.A. Nalepa), pp. 131–158, Westview Press, Boulder.

Bignell, D.E., Slaytor, M., Veivers, P.C. *et al.* (1994) Functions of symbiotic fungus gardens in higher termites of the genus *Macrotermes*: evidence against the acquired enzyme hypothesis. *Acta Microbiologia et Immunologica Hungarica* **41(1)**, 391–401.

Brauman, A., Kane, M.D., Labat, M. and Breznak J.A. (1992) Genesis of acetate and methane by gut bacteria of nutritionally diverse termites. *Science* **257**, 1384–1387.

Breznak, J.A. (1994) Acetogenesis from carbon dioxide in termite guts. In *Acetogenesis* (ed. H.L. Drake), pp. 303–330, Chapman & Hall, New York.

Breznak, J.A. and Brune, A. (1994) Role of microorganisms in the digestion of lignocellulose by termites. *Ann. Rev. Entomol.* **39**, 453–487.

Breznak, J.A. and Switzer, J.M. (1986) Acetate synthesis from H_2 plus CO_2 by termite gut microbes. *Appl. Environ. Microbiol.* **52**, 623–630.

Brune, A., Emerson, D. and Breznak, J.A. (1995) The termite gut flora as an oxygen sink: microelectrode determination of oxygen and pH gradients in guts of lower and higher termites. *Appl. Environ. Microbiol.* **61**, 2688–2695.

Brussaard, L. and Juma, N.G. (1996) Organisms and humus in soils. In *Humic Substances in Terrestrial Ecosystems* (ed. A. Piccolo), pp. 329–359, Elsevier, Amsterdam.

Burghouts, T., Ernsting, E., Korthals, G. and de Vries, T. (1992) Litterfall, leaf litter decomposition and litter invertebrates in primary and selectively logged dipterocarp forest in Sabah, Malaysia. *Phil. Trans. Roy. Soc. Lond.*, **B335**, 407–416.

Carney, H.J. (1989) On the competition and the integration of population, community and ecosystem studies. *Funct. Ecol.* **3**, 637–641.

Castro, M.S., Melillo, J.M., Steuder, P.A. and Chapman, J.W. (1994) Soil moisture as a predictor of methane uptake by temperate forest soil. *Canadian Journal of Forest Research* **24**, 1805–1810.

Cleveland, L.R. (1925) The effects of oxygenation and starvation on the symbiosis between the termite *Termopsis*, and its intestinal flagellates. *Biol. Bull.* **48**, 309–326.

Collins, N.M. (1981) The role of termites in the decomposition of wood and leaf litter in the southern Guinea savanna of Nigeria. *Oecologia* **51**, 389–399.

Collins, N.M. (1983) The utilization of nitrogen resources by termites (Isoptera) In *Nitrogen as an Ecological Factor* (eds J.A. Lee, S. McNeill and I.H. Rorison), pp. 381–412, Blackwell Scientific Publications, Oxford.

Collins, N.M. (1989) Termites. In *Ecosystems of the World: Tropical Rain Forest Ecosystems* (eds H. Leith and M.J.A. Werger), pp. 455–472, Elsevier, Amsterdam.

Collins, N.M. and Wood, T.G. (1984) Termites and atmospheric gas production. *Science (Washington)* **224**, 84–86.

Creber, G.T. and Chaloner, W.G. (1985) Tree growth in the Mesozoic and Tertiary and the reconstruction of palaeoclimates. *Palaeogeography, Palaeoclimatology, Palaeoecology* **52**, 35–60.

De Bruyn, L.A.L. and Conacher, A.J. (1990) The role of termites and ants in soil modification: a review. *Aus J. Soil Res.* **28**, 55–93.

Delmas, R.A., Marenco, A., Tathy, J.P. *et al.* (1991) Sources and sinks of methane in the African savanna. CH_4 emissions from biomass burning. *J. Geophys. Res.* **96**, 7287–7299.

Delmas, R.A., Servant, J., Tathy, J.P. *et al.* (1992) Sources and sinks of methane and carbon dioxide exhanges in mountain forest in equatorial Africa. *J. Geophys. Res.* **97**, 6169–6179.

Dow, H., Kedwell, L. and Levin, I. (1993) Soil texture parameterization and the methane uptake in aerated soils. *Chemosphere* **26**, 697–713.

Eggleton, P. (1994) Termites live in a pear-shaped world: a response to Platnick. *J. Nat. Hist.* **28**, 1209–1212.

Eggleton, P. and Bignell, D.E. (1995) Monitoring the response of tropical insects to changes in the environment: troubles with termites. In *Insects in a Changing Environment* (eds R. Harrington and N.E. Stork), pp. 434–497, Academic Press, London.

Eggleton, P., Williams, P.H. and Gaston, K.J. (1994) Explaining global termite diversity ... productivity or history? *Biodiversity and Conservation* **3**, 318–330.

Eggleton, P., Bignell, D.E., Sands, W.A. *et al.* (1995) The species richness of termites (Isoptera) under differing levels of forest disturbance in the Mbalmayo Forest Reserve, Southern Cameroon. *J. Trop. Ecol.* **11**, 85–98.

Eggleton, P., Bignell, D.E., Wood, T.G. *et al.* (1996) The diversity, abundance and biomass of termites under differing levels of disturbance in the Mbalmayo Forest Reserve, Southern Cameroon. *Phil. Trans. R. Soc. Lond. B.* **251**, 51–68.

Eutick, M.L., Veivers, P.C., O'Brien, R.W. and Slaytor, M. (1978) Dependence of the higher termite *Nasutitermes exitiosus* and the lower termite *Coptotermes lacteus* on their gut flora. *J. Insect Physiol.* **24**, 363–368.

Fraser, P.J., Rasmussen, R.A., Creffield, J.W. *et al.* (1986) Termites and global methane – another assessment. *J. Atmosph. Chem.* **4**, 295–310.

Groombridge, B. (ed.) (1992) *Global Biodiversity. Status of the Earth's Living Resources*, Chapman & Hall, London.

Gottschalk, G. (1986) *Bacterial Metabolism*, Springer-Verlag, Berlin.

Hackstein, J.H.P. and Stumm, C.K. (1994) Methane production in terrestrial arthropods. *Proc. Natnl. Acad. Sci. (USA)* **91**, 5441–5445.

Hébrant, F. (1970) Étude du flux énergétique chez deux espèces du genre Cubitermes Wasmann (Isoptera, Termitinae), termites humivores des savanes tropicales de la région Éthiopienne. Thèse de Docteur en Sciences, Université Catholique de Louvain, Louvain.

Hogan, M., Veivers, P.C., Slaytor, M. and Czolij, R.T. (1988) The site of cellulose breakdown in higher termites (*Nasutitermes walkeri* and *Nasutitermes exitiosus*) *J. Insect Physiol.* **34**, 891–899.

Holt, J.A. (1987) Carbon mineralization in semi-arid Northeastern Australia: the role of termites. *J. Trop. Ecol.* **3**, 255–263.

Holt, J.A. (1990) Carbon mineralization in semi-arid tropical Australia: the role of mound-building termites. *Aust. J. Ecol.* **15**, 133–134.

Houghton, J.T., Jenkins, G.J. and Ephraums, J.J. (1990) *Climate Change: the IPCC Scientific Assessment*, Cambridge University Press.

Hungate, R.E. (1946) The symbiotic utilization of cellulose. *J. Elisha Mitchell Sci. Soc.* **62**, 9–24.

Jones, J.A. (1990) Termites, soil fertility and carbon cycling in dry tropical Africa – a hypothesis. *J. Trop. Ecol.* **6**, 291–305.

Jones, S.C. and Nutting, W.L. (1989) Foraging ecology of subterranean termites in the Sonoran Desert. In *Special Biotic Relationships in the Arid Southwest* (ed. J.O. Schmidt), pp. 79–106, University of New Mexico Press, Albuquerque.

Keller, M., Mitre, M.E. and Stallard, R.F. (1990) Consumption of atmospheric methane in soils of Central Panama: effect of agricultural development. *Global Biogeochemical Cycles* **4**, 21–27.

Khalil, M.A.K., Rasmussen, R.A., French, J.R.J. and Holt, J.A. (1990) The influence of termites on atmospheric trace gases: CH_4, CO_2, $ChCl_3$, N_2O, CO, H_2, and light hydrocarbons. *J. Geophys. Res.* **95**, 3619–3634.

Kovoor, J. (1970) Presence d'enzymes cellulolytique dans l'intestin d'un termite supérieur, *Microcerotermes edentatus* (Was). *Ann. Sci. Natn. Zool.* **12**, 65–71.

Lawton, J.H. (1994) What do species do in ecosystems? *Oikos* **71**, 1–8.

Lawton, J.H., Bignell, D.E., Bloemers, G.F. *et al.* (1995) Carbon flux and diversity of nematodes and termites in Cameroon forest soils. *Biodiversity and Conservation* **5**, 261–273.

Lee, K.E. and Wood, T.G. (1971) *Termites and Soils*, Academic Press, New York.

Lepage, M. (1974) Les termites d'une savanne sahélienne (Ferlo Septentrional, Sénégal): peuplement, population, consommation, role dans l'écosysteme. Doctoral Thesis, University of Dijon, Dijon.

Likens, G.E. (1992) *The Ecosystem Approach: Its Use and Abuse*, Ecology Institute, Oldendorf/Luhe.

Martius, C. (1994) Diversity and ecology of termites in Amazonian forests. *Pedobiologia* **38**, 407–428.

Martius, C., Wassmann, R., Thein, U. *et al.* (1993) Methane emission from wood-feeding termites in rain forests of Amazonia. *Chemosphere* **26(1–4)**, 623–632.

Mauldin, J.K. (1982) Lipid synthesis from (^{14}C)-acetate by two subterranean termites, *Reticulitermes flavipes* and *Coptotermes formosanus*. *Insect Biochem.* **12**, 193–199.

May, R.M. (1973) *Stability and Complexity in Model Ecosystems*, Princeton University Press, New Jersey.

Noirot, C. (1992) From wood- to humus-feeding: an important trend in termite evolution. In *Biology and Evolution of Social Insects* (ed. J. Billen), pp. 107–119, Leuven University Press, Leuven, Belgium.

O'Brien, R.W. and Breznak, J.A. (1984) Enzymes of acetate and glucose metabolism in termites. *Insect Biochem.* **14**, 639–643.

O'Brien, G.W., Veivers, P.C., McEwen, S.E. *et al.* (1979) The origin and distribution of cellulase in the termites *Nasutitermes exitiosus* and *Coptotermes lacteus*. *Insect Biochem.* **9**, 619–625.

Odelson, D.A. and Breznak, J.A. (1983) Volatile fatty acid production by the hindgut microbiota of xylophagous termites. *Appl. Environ. Microbiol.* **45**, 1602–1613.

Peakin, G.J. and Josens, G. (1978) Respiration and energy flow. In *Production Ecology of Ants and Termites* (ed. M.V. Brian), pp. 111–163, Cambridge University Press.

Pearce, M.J. and Waite, B. (1994) A list of termite genera with comments on taxonomic changes and regional distribution. *Sociobiology* **23**, 247–263.

Petit-Marie, N., Fintugne, M. and Rouland, C. (1991) Atmospheric methane ratios and environmental changes in the Sahara and Sahel during the last 130 years. *Palaeogeography, Palaeoclimatology, Palaeoecology* **86**, 197–204.

Potts, R.C. and Hewitt, P.H. (1974) The partial purification and some properties of the cellulase from the termite *Trinervitermes trinervoides* (Nasutitermitinae) *Comp. Biochem. Physiol.* **47B**, 317–326.

Rasmussen, R.A. and Khalil, M.A.K. (1983) Global production of methane by termites. *Nature (Lond.)* **301**, 700–702.

Ricklefs, R.E. and Latham, R.E. (1993) Global patterns of diversity in Mangrove floras. In *Species Diversity in Ecological Communities* (eds D. Schulter and R.E. Ricklefs), pp. 215–229, The University of Chicago Press, Chicago.

Rodriguez, J.M.G. (1992) Abundancia e distribuicao de coleopteros do solo em capoeira de terra firma na regiao de Manaus – AM Brazil. *Acta Amazonica* **22**, 323–333.

Rouland, C., Brauman, A., Labat, M. and Lepage, M. (1993) Nutritional factors affecting methane emission from termites. *Chemosphere* **26**, 617–622.

Rouland, C., Civas, A., Renoux, J. and Petek, F. (1988) Purification and properties of cellulases from the termite *Macrotermes mulleri* and from its symbiotic fungus *Termitomyces* sp. *Comp. Biochem. Physiol.* **91B**, 459–465.

Sands, W.A. (1969) The association of termite and fungi. In *Biology of Termites* (eds K. Krishna and F.M. Weesner), pp. 495–524, Academic Press, London.

Schimel, D.S. (1995) Terrestrial ecosystems and the carbon cycle. *Global Change Biology* **1**, 77–91.

Schulz, M.W., Slaytor, M., Hogan, M. and O'Brien, R.W. (1986) Components of cellulase from the higher termite *Nasutitermes walkeri*. *Insect Biochem.* **16**, 929–932.

Seiler, W., Conrand, R. and Scharfe, D. (1984) Field studies of methane emission from termite nests into the atmosphere and measurements of methane uptake by tropical soils. *J. Atmos. Chem.* **1**, 171–186.

Sleaford, F., Bignell, D.E. and Eggleton, P. (1996) A pilot analysis of gut contents in termites from the Mbalmayo Forest Reserve, Cameroon. *Ecological Entomology* **21**, 279–288.

de Souza, O.F.F. and Brown, V.K. (1994) Effects of habitat fragmentation on Amazonian termite commuities. *J. Trop. Ecol.* **10**, 197–206.

Steuder, P.A., Bowden, R.D., Melillo, J.M. and Aber, J.D. (1989) Influence of N fertilization on the methane uptake in temperate forest soils. *Nature (Lond.)* **341**, 314–315.

Stork, N.E. and Brendell, M.J.D. (1993) Arthropod abundance in lowland rain forest of Seram. In *The Natural History of Seram, Maluka, Indonesia* (eds I.D. Edwards, A.A. MacDonald and J. Proctor), pp. 115–130, Intercept, Andover.

Swift, M.J., Heal, O.W. and Anderson, J.M. (1979) *Decomposition in Terrestrial Ecosystems*, Blackwell Scientific Publications, Oxford.

Tathy, J.P., Delmas, R., Marenco, B. *et al.* (1992) Methane emission from flooded forest in central Africa. *J. Geophys. Res.* **97**, 6159–6168.

Tayasu, I., Sugimoto, A., Wada, E. and Abe, T. (1994) Xylophagous termites depending on atmospheric nitrogen. *Naturwissenschaften* **81**, 229–231.

Umbreit, W.W., Burris, R.H. and Stauffer, J.F. (1964) *Manometric Techniques*, Burgess Publishing Company, Minneapolis.

Veivers, P.C., Muhlemann, R., Slaytor, M. *et al.* (1991) Digestion, diet and polyethism in two fungus-growing termites: *Macrotermes subhyalinus* and *Macrotermes michaelseni*, Sjostedt. *J. Insect Physiol.* **37**, 675–682.

Watt, A., Stork, N., McBeath, C. *et al.* (1995) *Ecology of Insects in Cameroon Plantation Forests*. Final report to UK Overseas Development Agency (ODA)/GOC Office National de Développment de Forêts (ONADEF): Forest Management and Regeneration Project.

Whitford, W.G., Ludwig, J.A. and Noble, J.C. (1991) The importance of subterranean termites in semi-arid ecosystems in southeastern Australia. *J. Arid Environ.* **22**, 87–92.

Wilson, E.O. (1993) *The Diversity of Life*, Harvard University Press, Harvard.

Wood, T.G. (1988) Termites and the soil environment. *Biol. Fert. Soils.* **6**, 228–236.

Wood, T.G. and Lawton, J.H. (1973) Experimental studies on the respiratory rates of mites (Acari) from beech-woodland leaf litter. *Oecologia* **129**, 169–191.

Wood, T.G. and Sands, W.A. (1978) The role of termites in ecosystems. In *Production Ecology of Ants and Termites* (ed. M.V. Brian), Cambridge University Press.

Zimmerman, P.R., Greenberg, J.P., Wandiga, S.O. and Crutzen, P.J. (1982) Termites, a potentially large source of atmospheric methane. *Science (Wash.)* **218**, 563–565.

HERBIVORY IN FORESTS – FROM CENTIMETRES TO MEGAMETRES

Margaret D. Lowman

> To know the forest, we must study it in all aspects, as birds soaring above its roof, as earth-bound bipeds creeping slowly over its roots.
>
> Alexander F. Skutch,
> *A Naturalist in Costa Rica* (1971)

9.1 INTRODUCTION

Herbivory, or the consumption of plant material, is a subject of great economic as well as biological importance (review: Barbosa and Schultz, 1987; Price *et al.*, 1991; Hunter *et al.*, 1992; Bernays and Chapman, 1994). **Herbivory** is the term applied to the consumption of any plant material, and ranges from 0% up to 100% (also called **defoliation** at this level of severity) leaf surface area removed. This chapter follows the common convention of using herbivory to indicate folivory (i.e. applies only to foliage). The most abundant herbivores are insects, although birds and mammals also play important roles in certain forest types (Morrow, 1977; Lowman, 1995; review: Perry, 1994).

When walking through a forest or woodland, we usually focus our observations on a narrow band of green foliage, from ground level to perhaps two metres in height. This represents at most only 10% of the foliage in a stand, with the majority of leaves located high above our heads and consequently beyond our observations. Since the majority of plant–herbivore relationships occur where the foliage is situated, it is obvious that herbivory has remained relatively understudied. Only recently have ecologists begun to recognize the importance and also the severity of herbivory as a canopy process.

The impact of herbivory on trees is very poorly understood. There are three main reasons to account for this:

1. Most experimental studies of herbivory have been restricted to annuals and other small plants, for obvious logistical reasons. The impact of herbivory on a tree ranges from the possible stimulation of new growth in instances of moderate grazing (e.g. Lowman, 1982) to entire stand dieback in cases where insect epidemics have repeatedly defoliated entire crowns (e.g. Lowman and Heatwole, 1992), but it is difficult to manipulate whole trees experimentally for quantified and statistically rigorous measurements.
2. Until recently, the methods of measurement of herbivory were extremely variable, with some techniques being utilized that may not yield accurate results (reviewed in Lowman, 1984a). The logistical challenges of accurately sampling throughout this architecturally complex array of foliage contribute to our inability to measure herbivory accurately in a tree crown.
3. The herbivores themselves – usually insects – are also difficult to study, due to their relatively small body size as well as to their cryptic habits

---

Forests and Insects. Edited by A.D. Watt, N.E. Stork and M.D. Hunter. Published in 1997 by Chapman & Hall, London. ISBN 0 412 79110 2.

136 *Herbivory in forests*

in a large three-dimensional matrix of foliage (review: Lowman and Moffett, 1993).

From the perspective of a tree, there exists an evolutionary roulette of avoiding leaf damage and thereby maintaining one's photosynthetic tissue intact. From a herbivore's perspective, leaves represent a challenge, both physically (e.g. locating and chewing) and chemically (e.g. detoxifying and digesting). These interactions have led to a plethora of protective mechanisms produced by the trees in defence of their foliage. The herbivores have responded with consequential adaptations to accommodate toxins in addition to extracting nutrition, as well as improvements in their chewing mouthparts to overcome the physical barriers (review: Bernays and Chapman, 1994). Despite the fact that forest canopies may look like homogeneous green arrays of foliage, there exist many different cohorts or populations of leaves, each with its own mechanism for avoiding consumption.

In the life of a leaf, it is critical to survive the vulnerable weeks of foliar expansion without being eaten (Coley, 1983; Lowman, 1985). Thereafter, different defences such as toughness, stinging hairs, or perhaps even overwhelming abundance, may provide protection from herbivory. Conversely, a herbivore confronts a complex world of choices in the canopy. Different bites of leaf tissue must be recognized: soft versus tough; nutritious versus non-nutritious; toxic versus benign; moist versus dry; old versus young; apparent versus non-apparent; rare versus common; and probably other variables that have not yet been recognized by biologists (review: Lowman, 1995). All of these factors contribute to the process of herbivory, which – in forest canopies – is much more complex than previously assumed when the observations were limited to small, rapid, ground-level-only leaf samples.

This chapter reviews the spatial variation in forest and woodland herbivory, using data from forests in Australia (New South Wales and Queensland), Central America (Belize and Panama), South America (Peru), Africa (Cameroon) and North America (Massachusetts and New York). Data from forests has been compared where possible, though land-use practices in Australia and many tropical regions have reduced the original forests to woodland tracts and, in extreme cases, to fragmented stands of isolated trees. The variability in herbivory is quantified with respect to several spatial factors along increasing scales of size: leaf, branch, canopy height and light, individual crowns, species, stands, forest types and continents. Using a case study from Australian dry sclerophyll woodlands, the extreme variability in herbivory with respect to human perturbations is examined. Finally, I attempt to evaluate the challenges of understanding herbivory and other forest canopy processes in conservation and management.

9.2 HISTORY OF STUDIES OF HERBIVORY IN FORESTS

Historically, most studies of herbivory have been conducted in an over-simplistic fashion. Samples of leaves were collected at one point in time in temperate deciduous forests and usually from the lower canopy (i.e. within easy reach of the sampler). Consequently, herbivory levels reflected the amounts of damage inflicted upon understorey, deciduous shade leaves for one period of time, and frequently averaged 3–10% leaf surface area loss (reviews: Bray and Gorham, 1964; Landsberg and Ohmart, 1989; Lowman and Heatwole, 1992) (Table 9.1). Because of the widespread use of this method of sampling (termed **discrete sampling**), herbivory was calculated as a discrete, snapshot event (Diamond, 1986) and failed to account for spatial or temporal variability.

In recent years, the advent of experimental ecology has led to more rigorous and accurate sampling within forest ecosystems. Forest herbivory is no exception. The traditional methods of measuring herbivory by destructive sampling of small quantities of leaves are usually considered inadequate for addressing this complex canopy process, and more comprehensive methods have been employed (review: Lowman, 1984a; Landsberg and Ohmart, 1989). Sampling designs require careful replication with respect to branches, canopy heights, individual crowns and different forest sites (Figure 9.1).

Table 9.1 Levels of herbivory in different forests of the world, as sampled by different authors using a variety of sampling techniques over the past three decades (herbivory expressed as percentage leaf area removed)

Site	Study	% Grazed Annually
Australia		
Cool temperate	Lowman, 1984	26
Subtropical	Lowman, 1984	22
Warm temperate	Lowman, 1984	24.6
Dry eucalypt forest	Lowman and Heatwole, 1992	7–15
Fragmented eucalypt stands	Lowman and Heatwole, 1992	15–300
Dry forest	Ohmart et al., 1983	3–5
New Guinea		
Tropical	Wint, 1982	9–12
Americas		
Costa Rica, ferns	Hendrix and Marquis, 1983	5.5–9.9
Costa Rica, trees	Stanton, 1975	7.5 (young) 30 (old)
Panama, tropical	Wint, 1983	13
Barro Colorado, tropical	Leigh and Smythe, 1978	8
	Odum and Ruis-Reyes, 1970	7.8
	Coley, 1983	21 (up to 190)
Peru, tropical	Lowman and Wittman (unpublished)	13.7
Puerto Rico, subtropical	Leigh and Windsor, 1982	15
	Benedict, 1976	5.5–16.1
Massachusetts, temperate	Lowman (unpublished)	7–15
	Bray, 1961	5–9
Africa		
Cameroon, tropical	Lowman, Moffet and Rinker, 1993	8.5

Not surprisingly, when herbivory was measured more comprehensively and accurately, the resulting levels of grazing reported were more variable and also usually higher than results obtained with discrete sampling. The comprehensive methods that have been developed, called **long-term sampling** (Lowman, 1984a), incorporate spatial and temporal factors into the sampling, thereby accounting for the dynamic aspects of herbivory as a forest process. Such factors include temporal factors (e.g. phenology of the leaves and herbivores, leaf age, age of trees) and spatial factors (e.g. branch height, light levels, tree species, site, aspect, forest type, and even such large-scale factors as continent). Long-term measurements serve to quantify accurately the variability in herbivory for many forests, both temporal and spatially, though very few such studies have been conducted (but see Coley, 1983; Lowman, 1985, 1992; Brown and Ewel, 1987, 1988; Lowman and Heatwole, 1992).

The results of long-term sampling have presented a new perspective on the role of herbivores in forests. When herbivory was measured over longer durations (> 1 year) and included a wider range of leaf cohorts throughout tree canopies, high levels of grazing were reported. Instead of 3–10% leaf area losses annually, herbivory averaged 15–30% when measured by long-term sampling, though it ranged from as low as 3% (Ohmart et al., 1983) to as high as 300% foliage removal (the latter occurring when three successive flushes were eaten from trees in fragmented woodlands) (Lowman and

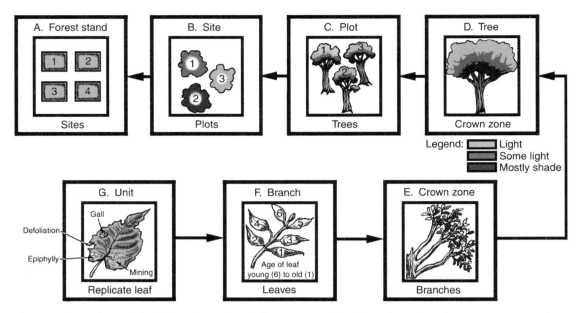

Figure 9.1 Experimental depiction of sampling design to measure herbivory throughout forest canopy, expanding in spatial scale from leaf to branches, to crown zone (heights, light), to individual trees, to site, to forest stands.

Heatwole, 1992). Insect pests can have catastrophic effects on trees (review: Barbosa and Schultz, 1987), though outbreaks are considered more common in temperate forests than in tropical forests (but see Wong *et al.*, 1991).

9.3 METHODS USED TO MEASURE HERBIVORY

The traditional methods of measuring foliage consumption by herbivores were to harvest leaves from tree branches, and measure their hole area either by visual estimates of damage (e.g. Wint, 1983) or by actual graph paper calculations of holes in leaf surfaces (e.g., Bray, 1961; Odum and Ruiz-Reyes, 1970; Leigh and Smythe, 1978; Lowman, 1984b). This discrete sampling is also termed 'reach-and-grab' sampling (Figure 9.2).

More recently, leaves are usually marked permanently and observed over time for herbivore activities. Leaves can be traced and this trace returned to the laboratory for measurement with an area meter (Lowman, 1992; Lowman and Heatwole, 1992) or measured against a clear acetate template that is marked with a grid of square centimetres (Coley, 1983). Most recently, a computer training course, called 'The Herbivory Game', was developed so that students can learn to estimate herbivory visually with accurate results (Lowman and Bell, unpublished).

For this review, long-term data from previous studies were compared (Lowman, 1985, 1992; Lowman and Heatwole, 1992). For leaf observations and measurements, several canopy access techniques were used, including single-rope techniques, cherry pickers, a construction crane, a raft and dirigible apparatus, and canopy walkways (Figures 9.3, 9.4, 9.5 and 9.6). In total, more than 15 000 leaves were used to obtain the proportions of leaf area removed. Long term sampling results were used for all comparisons, except in cases where sampling techniques were being evaluated. In all cases, proportions of leaf surface area missing were recorded (rather than actual centimetres missing) because this created comparable data among species of varying leaf sizes. Pilot studies on the proportional changes in hole sizes with leaf expansion illustrated that proportions missing remain consistent throughout leaf expansion (Lowman, 1984a).

Figure 9.2 Traditional method of sampling for herbivory, called 'reach and grab', whereby sampler harvests leaves that are easily within reach of trails at ground level. Because of this mode of harvest, earlier measurements of herbivory did not account for leaves in upper canopy, where herbivory levels can be entirely different.

The extent of replication of leaf samples within a tree crown was determined by pilot studies (Lowman, 1985), or in some cases by our ability to reach different portions of the canopy (e.g. Belize). Between 1980 and 1995, trees were measured for herbivory in a range of forest and woodland types (temperate deciduous, temperate dry sclerophyll, montane rain forest, tropical rain forest, subtropical rain forest, warm temperate rain forest and seasonal dry forest) and locations (Australia, Africa and North, Central and South America).

9.4 HERBIVORY WITH RESPECT TO SPATIAL SCALES

9.4.1 LEAVES AND BRANCHES

Herbivory on individual leaves ranged from 0% to 100%, representing the full possible extent of variation and illustrating the importance of utilizing adequate sample sizes. Some species typically suffered only 1–5% leaf surface area loss (e.g. *Bernoullia flammea* in the Belizean tropical rain forests: Appendix 9A; *Clusia* sp. and *Philodendron* sp. on the Peruvian tropical forests: Appendix 9B). In cases of low and homogeneous herbivory, smaller sample sizes were adequate to provide accurate results. In other species, herbivory ranged from negligible damage to entire defoliation of whole crowns (e.g. *Eucalyptus nova-anglica* in Australian woodlands). In these cases where herbivore attack was intermittent, larger sample sizes were required to reduce the standard error.

Sampling design was a very important factor in terms of obtaining accurate results. When a small number of leaves of *Eucalyptus blakelyi* were collected to estimate herbivory, the measurements greatly underestimated the level of damage. At least

Figure 9.3 Most commonly utilized method of canopy access: single rope technique, or SRT. The author scales *Ceiba pentandra* in Peru.

Table 9.2 Variation in levels of herbivory at small spatial scales: differences in proportions of leaf surface area eaten between leaves on several branches of *Eucalyptus nova-angelica* in eucalypt woods of eastern Australia (site information and data from Lowman and Heatwole, 1992)

Number of leaves	Herbivory %	Standard error
3	15.0	10.2
5	30.0	18.5
10	39.5	11.4
30	50.3	5.8
50	59.3	5.1
100	60.5	3.4

30–50 leaves, comprising approximately 3–10 branches, were required to provide an accurate estimate of herbivory for this species with an acceptable level of statistical confidence (Table 9.2).

9.4.2 CANOPY HEIGHTS AND LIGHT LEVELS

Herbivory varies significantly at different heights in tree crowns, which may be due to the differences in light levels as well as a direct result of the height of the leaves (Table 9.3). Some insects may prefer leaves closer to ground level as a consequence of where the insects enter their life cycle as herbivores. In rain forests, herbivory is significantly higher in the understorey where the shaded leaves were softer, larger and lower in tannins and phenolics than in the canopy (Lowman and Box, 1983). For example, *Doryphora sassafras* averaged 27.8% leaf area lost in the understorey leaves of warm temperate rain forests in Australia, compared with 17.6% herbivory in the upper canopy (Lowman, 1992).

In the Australian woodlands, eucalypts show an even greater variation in levels of insect damage with respect to height and light. *Eucalyptus blakelyi* ranged from 6% to 47% between upper and lower canopy, respectively, at one site, but it only ranged from 1% to 5% leaf area losses in another site (Table 9.4). (In contrast, it suffered complete defoliation in sites further south: Landsberg, 1990.) Presumably, herbivory in the lower canopy is a consequence of the behavior of some herbivores that feed heavily in the foliage near ground level, thereby avoiding bird predation above. In contrast, some herbivores (e.g. scarab beetles) were observed to feed gregariously in the upper canopy, moving down as they consumed the foliage.

9.4.3 INDIVIDUAL TREES

Spatial variability in herbivory with respect to different trees requires extensive replication and has not been well studied. In our limited measurements of at least three trees of each species per site in Australian dry sclerophyll forest, we found that significant differences exist in the levels of insect damage between some trees but not others (Table

Figure 9.4 Canopy cranes provide excellent access to small tracts of forest within reach of the crane arm, and cranes have been established in four forest types to date.

9.4). For example, in *E. blakelyi* trees at Newholme (NSW, Australia), herbivory ranged from 46% to 60.8% leaf area loss annually, but at Eastwood (NSW, Australia) two trees averaged similar levels of 4.7% and 4.9% leaf area losses (Table 9.4; see also Lowman and Heatwole, 1992).

9.4.4 TREE SPECIES

Trees of different species exhibited enormous variability in herbivory levels, with their physical and chemical defence attributes obviously contributing to these differences. In the dry sclerophyll forests, herbivory differed most significantly among species, with *E. nova-anglica* losing up to 300% leaf surface area per year, and *E. blakelyi* as little as 4–10% (Figure 9.7) (but see Landsberg, 1990). Similarly, in Australian rain forests, damage levels showed significant statistical differences, with some species ranging from as low as 3–5% for *Toona ciliata* to as high as 42% for *Dendrocnide excelsa* (Lowman, 1992).

Table 9.3 Variation in levels of herbivory at moderate spatial scales: differences in proportions of leaf surface area eaten between different heights, light levels and sites of eucalypt canopies in eastern Australia (site information and data from Lowman and Heatwole, 1992)

Site	Height (m)	Herbivory %	Standard error
Wood Park	2	47	3.9
	7	28	4.8
	15	6	2.0
Eastwood	2	5	2.3
	6	3	1.8
	10	1	0.3

142 *Herbivory in forests*

Figure 9.5 Unique mode of canopy access whereby hot-air dirigible (*Radeau des Cimes*) is pulling sled or skimmer through treetops for sampling. The author samples herbivores from one corner of sled in lowland tropical rainforest in Cameroon, Africa.

9.4.5 WITHIN AND BETWEEN STANDS

Between three different stands of montane or cool temperate rain forest in Australia, herbivory averaged 38%, 39.4% and 18.8% (Selman and Lowman, 1983). In contrast, dry sclerophyll woodlands appeared to have even higher variability of herbivory levels between sites, with ranges of 9%, 60% and 98.5% (Figure 9.7 and Table 9.5). Other researchers have found herbivory levels as low as 3% for eucalypt forests (Landsberg and Ohmart, 1989). Further investigation to explain the variabil-

Table 9.4 Variation in levels of herbivory at moderate spatial scales: differences in proportions of leaf surface area eaten between different trees, species and sites of eucalypt canopies in eastern Australia (site information and data from Lowman and Heatwole, 1992)

Species	Tree	Site		
		Eastwood	Wood Park	Newholme
Eucalyptus blakelyi	1	4.7	29.5	46.0
	2	4.9	36.5	50.4
	3	16.8	34.5	60.8
Eucalyptus viminalis	1	10.0	38.1	25.9
	2	19.1	20.4	29.8
	3	7.5	31.3	–

Herbivory with respect to spatial scales 143

Figure 9.6 Walkways and bridges provide a moderately priced, permanent base for canopy studies. The author uses a bridge in temperate deciduous forest in Massachusetts, USA, to measure herbivory on *Quercus rubra*.

ity in herbivory of eucalypt canopies revealed that the land-use history of a stand could significantly affect the outbreak patterns of beetles (section 9.5).

9.4.6 FOREST TYPES

Herbivory levels throughout entire forests were averaged when the majority of species were mea-

Table 9.5 Variation in levels of herbivory at moderately large spatial scales: differences in proportions of leaf surface area eaten between different sites and stands of eucalypt woodlands in eastern Australia (site information and data from Lowman and Heatwole, 1992)

	% Herbivory sampling methods		
	Visual	Discrete	Long-term
Rural dieback stand			
Wood Park	3	30.6	98.5
Newholme	3	33.9	59.9
Undisturbed forests			
Walcha Woods	1	14.7	10.6
Eastwood	1	15.8	9.0

144 *Herbivory in forests*

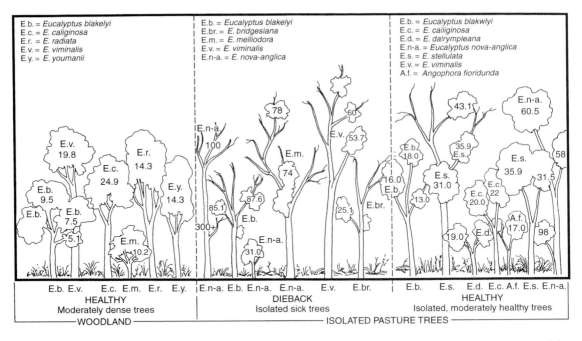

Figure 9.7 A depiction of herbivory in the dry sclerophyll forests of Australia, illustrating variation at the spatial scales of height in crown, individual trees, species and site.

sured, or when the species of greatest abundance had been measured. Seven forest and woodland types were examined, ranging from 7 to 300% leaf surface area lost annually (Table 9.6). In general, temperate forests had lower herbivory than the tropical forests, but more extensive measurements are required with careful replicated sampling within and between forest types. Australian eucalypts, situated in the southern temperate regions, represent

Table 9.6 Variation in levels of herbivory at large spatial scales: differences in proportions of leaf surface area eaten between different forest canopies throughout the world

Forest type	% Herbivory
Temperature deciduous	15
Australian dry forest	15–300
Cloud forest	26
Subtropical	16
Warm temperate	21
Tropical	12–30

an anomaly because they had moderate herbivory in some stands and extremely high levels in others. Because of these relatively high levels of herbivory (i.e. as many as three successive flushes were defoliated per year for some eucalypts), this ecosystem is examined in greater detail in section 9.5.

9.4.7 CONTINENTS

Similarities and differences in ecosystem structure and function between continents is the subject of great debate and has even been the theme of an international symposium (Dodson and Westoby, 1985; Ohmart, 1985). One such controversy occurred over the levels of herbivory in Australian ecosystems, whereby some scientists suggested that insect damage is higher in forests of this continent than in other continents (Fox and Morrow, 1983; Lowman, 1987; but see Ohmart *et al.*, 1983). In our data collection, this apparent anomaly still exists – Australia exhibited much higher herbivory than any other continent (Table 9.7). Different explanations have been suggested, such as the existence of proportionally higher numbers of herbivores in Australian ecosystems, more palatable leaf tissue on that continent, or even simply artefacts of sampling. The latter explanation is illustrated in Table 9.5, where three different sampling methods (visual, long-term and discrete), when performed on the same trees at the same sites, yielded significantly different levels of damage. Such controversial results jeopardize confidence in pre-existing studies. Rigorous measurements with adequate replication are required to test carefully for the differences between different variables, and to minimize the interactions possible when examining a complex factor in a complex ecosystem.

9.5 HERBIVORY IN FORESTS – ABOVE- AND BELOW-GROUND COMPONENTS

The herbivory levels for Australian eucalypt woodlands were an order of magnitude higher than other forest types. Over the past century, eucalypts have been susceptible to a widespread dieback syndrome – hypothesized to be insect-related – whereby trees on millions of acres of rural countryside are threatened by crown decline and ultimate mortality (reviewed in Heatwole and Lowman, 1986). This dieback phenomenon has been reported in other forests of the world, where causes have ranged from insect outbreaks to acid rain (review: Mueller-Dombois, 1986; Huettl and Mueller-Dombois, 1993), but nowhere does it appear as severe as in Australia (Mueller-Dombois, personal communication).

A multitude of factors has been suggested to explain the eucalypt dieback syndrome, including frost, salinity, excessive grazing of trees by livestock, invasion of non-native grasses that alter the soil chemistry and exclude trees, fluctuations in bird populations, changes in the watertable, and agricultural practices that result in fragmentation and reduction in tree cover. Even incidents such as koalas feeding in eucalypt crowns were implicated as possible causes of dieback. (In fact, in some rural regions, land owners proposed culling koalas to control the eucalypt dieback.) The severity of dieback is most evident where land use has been intense, with heavy grazing, clearing, aerial fertilizer applications and establishment of non-native grasses. Insects are considered a major component of the dieback syndrome, but very few studies had quantified the severity of herbivory in eucalypt crowns.

Long-term herbivory measurements on rural eucalypts in Australia, conducted during 1983–1989, revealed that herbivory levels were variable at every spatial scale – among leaves, branches, individual trees and sites (Lowman and Heatwole, 1992). However, in some cases (e.g. *Eucalyptus nova-anglica*), herbivory was fairly homogeneous, with complete defoliation occurring throughout all crowns in all stands. Insect herbivores have been implicated as a major cause of the dieback syndrome in Australia, with casual reports of outbreaks of scarab beetles over the past century (Norton, 1886; Old *et al.*, 1984). Despite the belief that insects were a factor in dieback, very little quantified measurement of insect damage or of insect population dynamics occurred before 1980.

The life cycle of the scarab beetle (genus *Anoplognathes*), responsible for a major portion of crown damage in eucalypts, includes a larval phase, in which the grubs feed on roots underground. Further measurements of dieback trees

Table 9.7 Variation in levels of herbivory at the largest spatial scale: differences in proportions of leaf surface area eaten between different continents of the world (sampling techniques are also compared between discrete – 'reach and grab' – and long-term measurements, illustrating that the latter show higher levels of herbivory because they are more comprehensive in scope)

Location	% Herbivory Sampling methods	
	Discrete	Long term
Belize (Central America)	7.0	–
Panama (Central America)	7.9	30
Peru (South America)	13.7	–
Cameroon (Africa)	8.5	–
Australia		
Tropical	7.3	20
Subtropical	6.9	16
Montane	7.9	26
Dry forest	7.34	15–300

revealed that their root systems were severely reduced by insect larvae, although the species remains unknown (Lowman et al., 1987). In a harvesting episode, entire trees of E. nova-anglica were felled; their root systems were carefully extracted from the ground (including root remains where larvae had tunnelled) and weighed; and their foliage was clipped and measured for insect damage. The root biomass of a healthy and a dieback eucalypt lost 5% and 19%, respectively, to larval consumption (Table 9.8). In essence, some eucalypts actually received a double dose of herbivory whereby insect larvae fed on the root systems of trees and then emerged as adults to consume the foliage. Some beetle larvae are thought to have a higher survival rate in farmland that has non-native grasses, application of fertilizers, and increased compaction of soil around tree bases from livestock seeking shade (Heatwole and Lowman, 1986). As the problem accelerates, the numbers of trees decline and the survival of beetles around the (fewer) remaining trees increases. Some sites of Australian eucalypt stands appear to suffer from the highest herbivory ever recorded, including both foliage and woody tissue consumption.

9.6 IMPACTS OF HERBIVORY ON FORESTS

What do these relatively high levels of herbivory suggest in terms of the dynamics of the canopy community? In many forests, by virtue of their high levels of herbivory, insect consumption represents a major pathway for the transfer of primary productivity (i.e. leaf material) to other organisms. It may even approach the significance of secondary consumption (i.e. the decomposition pathway) which has traditionally been considered the major means of transfer of leaf material back into the forest ecosystem.

In forests with relatively low levels of herbivory, the importance of insects for removal and transfer of foliage may be less significant. But in many forests, the variability of herbivory between years makes it difficult to quantify the importance of herbivory. In some cases (e.g. gypsy moths in the American northeast: Barbosa and Schultz, 1987), the cycles of herbivores have greatly altered the original forest composition, albeit without any predictable patterns. This variability in the role of herbivores in the forest nutrient dynamics is of obvious importance to our understanding of the conservation and management of forest habitats (Bernays and Chapman, 1994). Some insects, that are otherwise cryptic and restricted to the upper canopy, may be important to the structure and function of the ecosystem. Like pollinators, herbivores are essential to the maintenance of trees in forest ecosystems and may contribute to species diversity by virtue of their selectivity between different tree species (Mattson and Addy, 1973). The importance of this relationship has not been well documented, due to the challenges of scale for such field experimentation, but such efforts are currently underway in Australia (J. Connell, personal communication) and in Panama (S. Hubbell, personal communication).

The unusually high fluctuations in herbivory – even outright defoliation – of dry sclerophyll forests indicates a strong tolerance to stress, probably a consequence of many generations of adaptation to physical as well as biological limitations. Eucalypts are renowned for their ability to tolerate periods of drought, temperature extremes and other physical conditions, such as poor soils and flooding (review: Saunders et al., 1990). Conversely, in the adjacent Australian rainforests, the levels of herbivory are much more homogeneous at the scale of the entire community compared with their dry forest counterparts. But in the rainforests, herbivory levels may vary more significantly at a smaller spatial scale, such as within crowns. Changes in height, light and species all result in different susceptibilities of foliage to herbivores. Further comparative studies

Table 9.8 Herbivory of eucalypts including both the foliage removal and root removal components of insect attack, as compared between healthy and unhealthy trees (site information and data from Lowman et al., 1987)

Component	Dieback tree		Healthy tree	
	%	Biomass (g)	%	Biomass (g)
Foliage	51	3 191	11.1	1 134
Roots	19	151 248	5	52 029

at larger spatial scales – such as between forest types and continents – will provide further information on trophic structures of forests, especially relative proportions of herbivores and foliage.

As habitat destruction continues to threaten the world's forests, in both tropical and temperate regions, canopies will become more reduced and fragmented in area, thereby eliminating species. It is believed that many canopy organisms have already disappeared before they were ever scientifically described, and most of them are presumed to be insects (Erwin, 1982, 1991; Wilson 1992), including many herbivores. Understanding the maintenance of species diversity in tropical habitats is still an urgent priority (Connell, 1978; Connell and Lowman, 1989). The complex interactions between canopy foliage and herbivores represent a potential site for ecological change as human activities continue to alter forest ecosystems. Many insect pest outbreaks in forests are a result of human perturbation (e.g. gypsy moths in the temperate deciduous forests of North America, reviewed by Elkinton and Liebhold, 1990). Another startling example is the scarab beetle implicated in tree dieback over millions of acres the rural landscape in Australia (Heatwole and Lowman, 1986). Although pest outbreaks are still regarded as relatively rare events in forests, it is possible that such events were not well documented in the past due to the logistic difficulties of observing defoliation in the canopies of tall forests. Now, with better techniques for canopy access, it is possible to measure the natural processes regulating tree–herbivore interactions to understand better the implications of imbalances that result from human impacts.

REFERENCES

Barbosa, P. and Schultz, J.C. (1987) *Insect Outbreaks*, Academic Press, New York.
Bernays, E.A. and Chapman, R.L. (1994) *Host Plant Selection by Phytophagous Insects*, Chapman & Hall, New York.
Bray, J.R. (1961) Primary consumption in three forest canopies. *Ecology* **45**, 165–167.
Bray, J.R. and Gorham, E. (1964) Litter production in forests of the world. *Adv. Ecol. Res.* **2**, 101–157.
Brown, B.J. and Ewel, J.J. (1987) Herbivory in complex and simple tropical successional ecosystems. *Ecology* **68**, 108–116.
Brown, B.J. and Ewel, J.J. (1988) Responses to defoliation of species-rich and monospecific tropical plant communities. *Oecologia* **75**, 12–19.
Coley, P.D. (1983) Herbivory and defensive characteristics of tree species in a lowland tropical forest. *Ecol. Monogr.* **53**, 209–233.
Connell, J.H. (1978) Diversity in tropical rain forests and coral reefs. *Science* **199**, 1302–1310.
Connell, J.H. and Lowman, M.D. (1989) Low diversity tropical and subtropical forests. *American Naturalist* **134**, 88–119.
Diamond, J. (1986) Overview: laboratory experiments, field experiments and natural experiments. In *Community Ecology* (eds J. Diamond and T.J. Case), Harper and Row, New York.
Dodson, J.R. and Westoby, M. (1985) *Are Australian Ecosystems Different? Proceedings of Special Symposium of the Ecological Society of Australia*, Vol. 14.
Elkinton, J.S. and Liebhold, A.M. (1990) Population dynamics of gypsy moth in North America. *Annu. Rev. Entomol.* **35**, 571–596.
Erwin, T.L. (1982) Tropical forests: their richness in Coleoptera and other arthropod species. *Coleopt. Bull.* **36**, 74–75.
Erwin, T.L. (1991) How many species are there? Revisited. *Conservation Biology* **5**, 330–333.
Fox, L.R. and Morrow, P.A. (1983) Estimates of damage by insect grazing on *Eucalyptus* trees. *Austr. J. Ecol.* **8**, 139–147.
Hairston, N.G., Smith, F.E. and Slobodkin, L.B. (1960) Community structure, population control, and competition. *Amer. Nat.* **94**, 421–425.
Heatwole, H. and Lowman, M.D. (1986) *Dieback – Death of an Australian Landscape*, Reed Books, Sydney.
Huettl, R.F. and Mueller-Dombois, D. (1993) *Forest Decline in the Atlantic and Pacific Region*, Springer-Verlag, Germany.
Hunter, M.D., Ohgushi, T. and Price, P.W. (1992) *Effects of Resource Distribution on Animal–Plant Interactions*, Academic Press, New York.
Kitching, R.L., Bergelsohn, J.M., Lowman, M.D. *et al.* (1993) The biodiversity of arthropods from Australian rainforest canopies: General introduction, methods, sites and ordinal results. *Austr. J. Ecol.* **18**, 181–191.
Landsberg, J. (1990) Dieback of rural eucalypts: does insect herbivory relate to dietary quality of tree foliage? *Austr. J. Ecol.* **15**, 73–87.
Landsberg, J. and Ohmart, C.P. (1989) Levels of defoliation in forests: patterns and concepts. *Trends in Ecol. Evol.* **4**, 96–100.

Leigh, E. and Smythe, N. (1978) Leaf production, leaf consumption, and the regulation of folivory on Barro Colorado Island. In *The Ecology of Arboreal Folivores* (ed. G.G. Montgomery), Smithsonian Press.

Lowman, M.D. (1982) The effects of defoliation of coachwood (*Ceratopetalum apetalum*) on growth and mortality of seedlings. *Austr. J. Bot.* **10**, 190–200.

Lowman, M.D. (1984a) An assessment of techniques for measuring herbivory: is rainforest defoliation more intense than we thought? *Biotropica* **16**, 264–268.

Lowman, M.D. (1984b) Grazing of *Utethesia pulchelloides* larvae on its host plant, *Argusia argentea*, on coral cays of the Great Barrier Reef. *Biotropica* **16**, 14–18.

Lowman, M.D. (1985) Temporal and spatial variability in insect grazing of the canopies of five Australian rainforest tree species. *Austr. J. Ecol.* **10**, 7–24.

Lowman, M.D. (1987) Insect herbivory in Australian rain forests – is it higher than in the neotropics? In *Are Australian Ecosystems Different? Proceedings of Special Symposium of the Ecological Society of Australia*, Vol. 14, pp. 109–121.

Lowman, M.D. (1992) Leaf growth dynamics and herbivory in five species of Australian rain-forest canopy trees. *J. Ecol.* **80**, 433–447.

Lowman, M.D. (1995) Herbivory in tropical forest canopies. In *Forest Canopies* (eds M.D. Lowman and N. Nadkarni), Academic Press, California.

Lowman, M.D. and Box, J.D. (1983) Variation in leaf toughness and phenolic content among 5 species of Australian rain forest trees. *Austr. J. Ecol.* **8**, 17–25.

Lowman, M.D. and Heatwole, H. (1992) Spatial and temporal variability in defoliation of Australian eucalypts. *Ecology* **73**, 129–142.

Lowman, M.D. and Moffett, M. (1993) The ecology of tropical rain forest canopies. *Trends Ecol. Evol.* **8**, 103–108.

Lowman, M.D., Higgins, W.D. and Burgess, A.D. (1987) The impact of herbivory on the root and foliage biomass of a New England eucalypt (*Eucalyptus nova-anglica*). *Austr. J. Ecol.* **12**, 38–47.

Lowman, M.D., Moffett, M. and Rinker, H.B. (1993) A new technique for taxonomic and ecological sampling in rain forest canopies. *Selbyana* **14**, 75–79.

Mattson, W.J. and Addy, N.D. (1973) Phytophagous insects as regulators of forest primary production. *Science, NY* **190**, 515–522.

Morrow, P. (1977) Host specificity of insects in a community of three codominant *Eucalyptus* species. *Austr. J. Ecol.* **2**, 89–106.

Mueller-Dombois, D. (1986) Perspectives for an etiology of stand-level dieback. *Ann. Rev. Ecol. Systematics* **17**, 221–243.

Norton, A. (1886) On the decadence of Australian forests. *Proc. R. Soc. Qld.* **3**, 15–22.

Odum, H.T. and Ruiz-Reyes, J. (1970) Holes in leaves and the grazing control mechanism. In *A Tropical Rain Forest* (eds H.T. Odum and R.F. Pigeon) pp. I-69–I-80, US Atomic Energy Commission, Oak Ridge, Tennessee.

Ohmart, C.P. (1985) Is insect defoliation in eucalypt forests greater than that in other temperate forests? *Proc. Ecol. Soc. Australia* **14**, 121.

Ohmart, C.P., Stewart, L. and Thomas, J.R. (1983) Leaf consumption by insects in three Eucalyptus forest types in southeastern Australia and their role in short-term nutrient cycling. *Oecologia* **59**, 322–330.

Old, K.M., Kile, G.A. and Ohmart, C.P. (1984) *Eucalypt Dieback in Forests and Woodlands*, CSIRO, Australia.

Perry, D.A. (1994) *Forest Ecosystems*, Johns Hopkins University Press, Maryland.

Price, P.W., Lewinsohn, T.M., Fernandes, G.W. and Benson, W.W. (1991) *Plant–Animal Interactions, Evolutionary Ecology in Tropical and Temperate Regions*, John Wiley and Sons, New York.

Saunders, D.A., Hopkins, A.J.M. and How, R.A. (eds) (1990) Australian ecosystems – 200 years of utilization, degradation, and reconstruction. *Proc. Ecol. Soc. Australia*, Vol. 16, Surrey Beatty & Sons, Sydney.

Selman, B. and Lowman, M.D. (1983) The biology and herbivory rates of *Novacastria nothofagi* Selman (Coleoptera: Chrysomelidae), a new genus and species on *Nothofagus moorei* in Australian temperate rain forests. *Austr. J. Zool.* **31**, 179–191.

Wilson, E.O. (1992) *The Diversity of Life*, Harvard University Press, Massachusetts.

Wint, G.R.W. (1983) Leaf damage in tropical rain forest canopies. In *Tropical Rain Forest: Ecology and Management* (eds S.L. Sutton, T.C. Whitmore and A.C. Chadwick), Blackwell Scientific Publications, England.

Wong, M., Wright, S.J., Hubbell, S.P. and Foster, R.B. (1991) The spatial pattern and reproductive consequences of outbreak defoliation in *Quararibea asterolepis*, a tropical tree. *J. Ecol.* **78**, 579–588.

Appendix 9A Herbivory levels in the forest canopy of a tropical rain forest at Blue Creel Preserve, southern Belize

Genus	Height	Light	% Defoliated
Aechmea	15	shade	0.00
Androlepis	10	shade	0.00
Anthurium	10	shade	16.33
Bernoullia no. 1	25	sun	4.55
Bernoullia no. 2	29	sun	3.67
Bernoullia no. 3	30	sun	3.64
Bernoullia no. 4	32	sun	3.92
Calophyllum no. 1	22	sun	6.10
Calophyllum no. 2	28	sun	6.73
Calophyllum no. 3	33	sun	3.58
Chrysopetalum	4	shade	5.87
Clavija	2	shade	6.60
Clusia	12	shade	0.65
Dracaena	5	sun	0.56
Grias	10	shade	3.96
Inga	2	shade	8.29
Lonchocarpus	27	shade	0.75
Orbignya	2	shade	0.57
Pachira	20	sun	11.29
Psychotria	2	shade	10.84
Pterocarpus	15	shade	18.63
Randia	2	shade	15.16
Rondoletia	4	shade	11.18
Sebastiana no. 1	20	shade	2.17
Sebastiana no. 2	25	shade	0.51
Sebastiana no. 3	15	shade	0.42
Sloanea	2	shade	5.82
Terminalia no. 1	18	shade	5.32
Terminalia no. 2	27	shade	6.37
Terminalia no. 3	30	sun	35.64
Terminalia no. 4	32	sun	15.63

Appendix 9B Herbivory levels in the forest canopy of a tropical rain forest at ACEER in northwestern Peru

Genus	% Defoliated
Brosimum	6.7
Cecropia	7.6
Clusia	2.4
Couratari	14.8
Cynometra	14.8
Erythrina	17
Eugenia	4.9
Grias	20
Inga	28.2
Otoba	10.2
Philodendron	1.8
Piper	4.8
Rinorea	5.8
Spondias	20.6
Terminalia	46

PART FOUR
FOREST PESTS

COMPARATIVE ANALYSIS OF PATTERNS OF INVASION AND SPREAD OF RELATED LYMANTRIIDS

Pedro Barbosa and Paul W. Schaefer

10.1 INTRODUCTION

One of the questions that most interest ecologists is: 'Why do some invading species succeed and others fail?' (Leston, 1957; Elton, 1958; Baker and Stebbins, 1965; Sailer, 1978, 1983; Wilson and Graham, 1983; Parsons, 1983; Mooney and Drake, 1986; Pimentel, 1986; US Congress OTA, 1993). Why, for example, has the European cabbage butterfly *Pieris rapae* L. invaded North America, Bermuda, New Zealand, Hawaii and other Pacific Islands over the last century, but other European *Pieris* such as *P. mannii* and *P. erganae* have not? Even when two related species invade, the ability of their populations to grow and spread may differ dramatically. The house sparrow (*Passer domesticus*) has occupied all of the United States in about 50 years, whereas the tree sparrow (*Passer montanus*), 90 years after its introduction, still has a relatively restricted distribution around the area of St Louis, Missouri (Ehrlich, 1986).

Being able to determine the causes for the success or failure of most, if not all, invasions is highly improbable, particularly when invasions have failed. Usually, little or no data are available on populations in which individuals fail to reproduce, or in which females do reproduce but populations perish nevertheless. Even researchers who believe that invasions can be assessed *ex posteriori*, may not believe that the results of such experimental analyses are likely to be meaningful. This view is expressed in the concluding statement in Simberloff (1986). He stated:

> I remain convinced that the reasons for success or failure of any attempt can be determined by extensive field study, but that these reasons will reside in aspects of the particular species and system that are so idiosyncratic that they will defeat any attempt at concise generalization. In fact, the variety of insect introductions is so great that I think any general theory with relatively few parameters will make incorrect predictions about success and failure many times.

Whether or not Simberloff's admonishment is valid or invalid is still a subject of intense debate. Regardless of the merits of Simberloff's caveat, we suggest that there is an equally interesting and related question that offers more opportunities for experimental inquiry and greater insights into invasive species (Andow *et al.*, 1993). That is: of those invading species that successfully establish populations, why do some readily reach outbreak levels and/or spread rapidly from the point of their initial invasion, while others do not?

Inquiries about the ecology of invasions are complicated by the lack of consensus and by vagueness in the terms and concepts currently used

---

Forests and Insects. Edited by A.D. Watt, N.E. Stork and M.D. Hunter. Published in 1997 by Chapman & Hall, London. ISBN 0 412 79110 2.

to discuss this topic. Even a question as basic as, 'What is or is not an invader?' is not always clear or obvious. For example, the invasion of North America by three of the lymantriids discussed in this chapter is a matter of record. However, although most consider the fourth species – the rusty tussock moth – to be an indigenous species, there are a few (rare) reports suggesting it is an invader.

Several terms commonly used to describe the phases of invasion are clarified below to facilitate comprehension of our discussion. An invasion is an attempt by a species to colonize an area beyond its current distribution. Such attempts may fail or may result in establishment of a reproducing population (i.e. in colonization). The result of an invasion is a discontinuous expansion of a species' distribution, i.e. there is an area between the source population and the invaded area that is uninhabited by the invader. In contrast, invading species often increase their distribution subsequent to invasion. We refer to this phenomenon as spread. Spread describes the continuous, relatively uninterrupted expansion of a species distribution. A successfully spreading invader is one that is capable of expanding its geographical range beyond the point of invasion, regardless of the density that the population reaches while doing so.

Some invading species do routinely reach outbreak levels (Barbosa and Schultz, 1987), although the densities reached by one acknowledged outbreak species (whether an invader or not) may be significantly lower than those reached by another outbreak species. For an invading species it may be difficult to establish what constitutes an outbreak when compared with another species in the invaded habitat. Nevertheless, invading species may reach levels far above those typical of most unrelated species, or of congeners in the same habitat. Invading species, however, may successfully colonize a new habitat without reaching outbreak levels. We suggest that the designation of an outbreak species should be made relative to the abundance of other species (which are ecological analogues) in the same type of habitat.

We consider that three out of the four lymantriids discussed here are outbreak species, as their populations typically far exceed the generally low abundance of other forest macrolepidoptera. Many, if not most, macrolepidoptera occur at low densities. Data on the macrolepidoptera collected on two common eastern US forest trees, *Acer negundo* L. (box elder) and *Salix nigra* Marsh (black willow), over a period of four years, illustrate the typical abundance at which macrolepidoptera occur (Barbosa et al., unpublished data). Only 1–5 larvae per year were collected for an average of 67% and 52% of the species on box elder and willow, respectively, in each of the four years (Figures 10.1, 10.2). The number of larvae per species was small even though larvae were collected from 10 trees per species in each of four sites, and were sampled weekly for four months each year. The defoliation produced by the entire species assemblage on each of these trees was so insignificant that it would have been difficult to measure rigorously (personal observation). These data on abundance suggest that most species of tree feeding macrolepidoptera in these habitats occur at relatively low densities. Thus, populations of native or invading species need not be extraordinarily large before being considered outbreak populations.

This chapter compares the biology and ecology of four species within the lepidopteran family Lymantriidae. Three species (*Lymantria dispar* L., the gypsy moth; *Euproctis chrysorrhoea* L., the browntail moth; and *Leucoma salicis* (L.), the satin moth) invaded North America in the late nineteenth and early twentieth centuries. In contrast, the fourth species (*Orgyia antiqua* (L.), the rusty tussock moth) is an indigenous holarctic species (Ferguson, 1978) though it has been cited as an introduced species by Baker (1972). Because these four species are taxonomically related, comparisons of their life histories (Table 10.1) are less likely to be confounded by phylogenetic constraints. In addition, comparisons are facilitated because the introductions of the three species occurred within a relatively short period of each other. The gypsy moth, browntail moth and satin moth were introduced into North America in 1868/69, 1897 and 1920, respectively. The patterns of population increases and declines and of the

Introduction 155

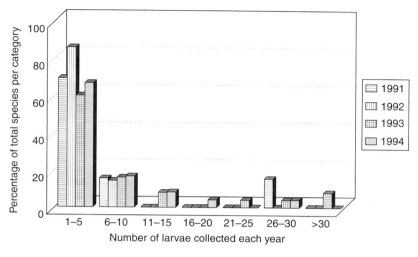

Figure 10.1 Relative abundance of macrolepidoptera collected from *Acer negundo* (box elder) between 1991 and 1994; annual collection from a total of ten trees in each of four sites, sampled once a week for four months.

spread of populations of these three invading species provide fascinating contrasts and insights into the ecology of invasions.

Gypsy moth and browntail moth populations increased rapidly subsequent to their introductions and both became pest species. For example, invasion by the gypsy moth quickly led to the complete defoliation of trees in a 603 km^2 area around Boston, Massachusetts, within 23 years of its introduction (Fernald, 1892). By the 1920s, its distribu-

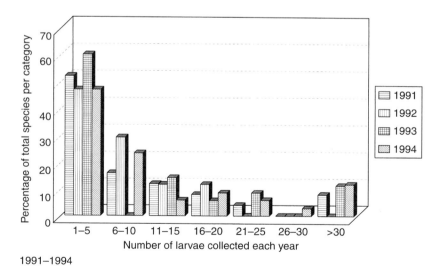

Figure 10.2 Relative abundance of macrolepidoptera collected from *Salix nigra* (black willow) between 1991 and 1994; annual collection from a total of ten trees in each of four sites, sampled once a week for four months.

Table 10.1 Comparison of life history traits of *Orgyia antiqua*, *Leucoma salicis*, *Euproctis chrysorrhoea* and *Lymantria dispar*

Species	Feeding behaviour	Distribution	Periodicity	Condition of ♀	Type of dispersal	Phagy	Over-wintering stage	Average Fecundity per ♀	Number of generations	Larval feeding phenology
Orgyia antiqua[1] (rusty tussock moth)	Feed individually. Early surface epidermis feeding. Later whole leaf.	Northern North America and Eurasia to Iceland.[2]	Diurnal[3]	Wingless[4] and flightless.	First instar ballooning. Flightless females.	Polyphagous, deciduous and coniferous trees. Orchard pest[5].	One layer of exposed eggs, not covered with hair/froth.	300–400 eggs (usually laid on cocoon).	One generation in US.	May to mid summer in US.
Leucoma salicis[6] (satin moth)	Instars 1–3 = surface epidermis feeders; 4–7 whole leaf.[7]	Introduced to New World (1920) from temperate Eurasia.[8]	Adults nocturnal. Both sexes attracted to light.	Winged, ♀♀, capable of flight.	Adult flight and 1st instar ballooning (speculated).	Mostly *Populus* spp., some *Salix*.[9]	3rd instar inside web (4 × 2mm) spun by 2nd instar.[10]	316–516: covered with hardened white secretion.[11]	One generation in North America and most of Europe.[12]	1st and 2nd instar mid July to late summer; 3rd–7th instar April to June.
Euproctis chrysorrhoea[13] (browntail moth).	Feed gregariously. Early instars feed on leaf surface. Form silk shelter (pre-webs)[14]. Late instars eat whole leaf.	Introduced (1897) from Europe. Currently restricted to few coastal areas of New England.	Both sexes nocturnal (attracted to light). Most feeding by larvae nocturnal.	Winged, ♀♀, capable of flight.	Adult flight; orient to host plant. No first instar ballooning. Late instars wander from defoliated hosts.[15]	Common rosaceous trees favoured: apple, pear, cherry, etc. Will feed on oak, willow, etc.[16]	Communal webs with many larvae (3rd instars).[17]	Hair covered clusters averaging 200–500 eggs on underside of leaf.	One generation.	1st–2nd instar early August to early September; 3rd to final instar April to late June.
Lymantria dispar[18] (gypsy moth).	Feed individually. Early surface feeding. Later whole leaf.	Introduced from Europe in 1868/69.	Larval feeding usually nocturnal. Adults crepuscular or nocturnal.	Winged, flightless ♀♀.	First instar ballooning. Human transport. Late instar wandering.	Most polyphagous of 4 spp. Oaks preferred but feeds on deciduous trees and conifers.	Layered eggs in mass covered with hair.	About 500–800 eggs/mass.	One generation	May (N. England) or April (mid Atlantic) to mid June.

[1] In the past known as *Notolophus antiqua*. In North America occurs as three subspecies; *nova* (most of what is collected in the USA), *badia* (from the Pacific northwest from British Columbia to California), *argillacea* (from one population in Alaska).

[2] In North America occurs south to the middle Atlantic states and northern California. Widespread throughout the British Isles.

[3] Unlike most other species of *Orgyia* in North America.

[4] Wings described in British reference (Carter, 1984) as 'vestigial, covered with greyish white scales'.

[5] Most commonly noted on tree species typical of young hardwood stands and/or relatively tolerant understorey tree species. Hosts include species of fir, spruce, larch, pine, hemlock, birch, alder, willow, poplar, maple, elm, apple and cherry.

[6] Previously known as *Stilpnotia salicis* (L.).

[7] Over 90% of consumption occurs in the spring period.

[8] Became established in North America in two widely separated populations: one in New England and the other in British Columbia. The east *L. salicis* occurs from Massachusetts to the gulf of St. Lawrence region (in all Atlantic provinces) and southward into New York. The western population occurs from Vancouver Island (to the Okanagan Valley) in the southern interior of British Columbia, southward through western Washington and Oregon to Modoc County, California. In the east it moved from Boston to the Magdalen Islands, Canada in 17 years.

[9] On the North American continent has shown a preference for ornamental or shade poplar trees. Feeding on aspen and willow occurs apparently in absence of preferred hosts; however, reports exist of infestations in natural stands of aspen (Canada).

[10] Although in North America and most of Europe the species overwinters as larvae there appear to be numerous European reports of overwintering in the egg stage.

[11] Egg masses with over 1000 eggs have been recorded in the field.

[12] Two generations are reported in southern Europe.

[13] Also known by synonym *Nygmia phaeorrhoea* Donovan.

[14] These so-called pre-webs are reinforced to form the overwintering webs, within which larvae construct silken hibernacula. Diapausing larvae may go through 2nd to 4th stadia but most emerging spring larvae are 3rd instars. Late instars are not gregarious.

[15] Schaefer (1974) did not observe behaviour suggesting 1st instar dispersal; however, he notes that Curtis (1782) had an illustration showing the 'dropping' behaviour of 1st instars. It is not known whether European browntail moth larvae balloon. Long-range dispersal of moth adults is by wind transport; short-range adult dispersal is oriented around host plants.

[16] Host plant preferences changed among populations at the time of peak populations and spread to those of current populations. Although *Prunus maritima* is a prominent host in current coastal habitats, it is unlikely that it played a major role for populations in the Boston area in 1915. Preferences may have changed from area to area. For example, in Nova Scotia and New Brunswick *Pyrus malus* (apple) was the common host of the browntail moth.

[17] The number of larvae in overwintering webs vary from site to site and depending on the host plant on which they occur. Sites in Maine and Massachusetts average 317 and 560 larvae per web (Schaefer, 1974) but in certain years, sites and plants (such as *Prunus maritima*) can average 862 larvae/web.

[18] Previously referred to as *Porthetria dispar*.

Table is based on data from Rogers and Burgess (1910), Burgess and Crossman (1927,1929), Glendenning (1932), Reeks and Smith (1956), Baker (1972), Schaefer (1974), Wagner (1977), Wagner and Leonard (1979a,b, 1980), Ferguson (1978), Doane and McManus (1981), Barbosa *et al.* (1989), Sterling and Speight (1989), West (1993).

tion comprised much of eastern New England, corresponding quite closely to the distribution of the browntail moth. The gypsy moth has persisted for well over 100 years, periodically reaching outbreak levels and continually spreading (Figure 10.3). Within about 18 years of its introduction the distribution of the browntail moth encompassed all of New England (Figure 10.4), east of the Connecticut River, north to Maine, and east to New Brunswick, Canada, an area of about 150 000 km^2 (Schaefer, 1974). The difference was that, while gypsy moth populations continued to expand, those of the browntail moth receded (Figures 10.5, 10.6). At the end of its decline, browntail moth populations were restricted to isolated islands of Casco Bay, Maine and coastal sand dune habitats of Cape Cod, Massachusetts.

The later invasion by the satin moth (*L. salicis*), in 1920, occurred in two widely separate areas: one in New England (the northeast corner of the United States) and the other in British Columbia, Canada (north of the northwest corner of the United States) (Figure 10.6). Not only does this species have life cycle and life history traits that are quite similar to those of the browntail moth (Table 10.1) but also its population dynamics, subsequent to its invasion, were similar. As with the gypsy moth and the browntail moth, the satin moth quickly reached outbreak levels and pest status (Burgess and Crossman, 1927; Reeks and Smith, 1956). Indeed, within 6–9 years of its introduction heavy defoliation and tree mortality were recorded in both eastern United States and Canada (Wagner, 1977). The satin moth was of sufficient concern in the United States that early attempts at introducing biological control agents against the gypsy and browntail moths also included efforts aimed at the satin moth. In contrast, today the satin moth is considered an infrequent pest, only occasionally reaching outbreak levels, for relatively short periods (rarely more than three to four years) and causing little damage (Wagner, 1977). Today the satin moth, just like the browntail moth, occurs at low population densities. Unlike the browntail moth, the satin moth is widely distributed throughout New England. It feeds on the few *Populus* species that are planted as ornamentals, although occasionally it is observed in natural forests such as in aspen stands of Maine and Canada (Wagner and Leonard, 1979b).

This chapter first discusses several alternative hypotheses which propose a suite of traits and/or interactions which differentiate species that successfully invade and spread from those that rarely, if ever, do so successfully (Leston, 1957; Elton, 1958; Baker and Stebbins, 1965; Sailer, 1978, 1983; Wilson and Graham, 1983; Parsons, 1983; Mooney and Drake, 1986). These hypotheses are ones most likely to provide viable explanations for the patterns of abundance and spread discussed above. The alternatives discussed include the following hypotheses:

1. Species that successfully invade and spread exhibit greater genetic polymorphism than non-invaders.
2. Invaders are characterized by specific weedy species traits which facilitate successful invasion and spread.
3. The success or failure of invading species is mediated by human activities.
4. The composition of the natural enemy complex affecting a species, and its effectiveness, determines the success or failure of invasion and spread.

Although these hypotheses do not comprise all possible alternatives, they do represent commonly discussed hypotheses (see references above) or hypotheses that may have relevance because of the unique characteristics of the lymantriids discussed. Finally, we present a comparison of the ecology and biology of three invasive and one indigenous lymantriid as suggestive evidence which leads us to propose the hypothesis that **host plant availability and quality (as a consequence of either inter- or intra-plant species variation) is a major driving force determining the spread and abundance of invading species, subsequent to their invasion**. Clearly, any biological phenomenon such as the spread and abundance of invading species cannot be explained solely by a single factor. Nevertheless, certain factors may be keystone causal mechanisms. We believe that host plant quality is such a factor in the ecology of invasive species.

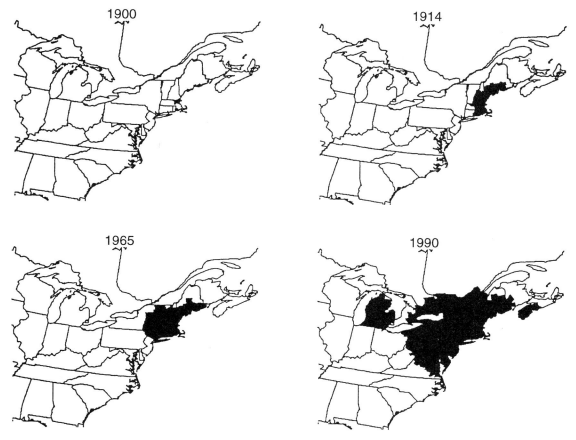

Figure 10.3 Historical pattern of spread of gypsy moth populations in the United States. (From Liebhold *et al.*, 1995.)

10.2 ALTERNATIVE EXPLANATIONS FOR HISTORICAL PATTERNS OBSERVED IN THESE INVASIVE LYMANTRIIDS

10.2.1 GENETIC VARIANCE

The level of genetic polymorphism with regard to one or more adaptive traits has been proposed as a key characteristic of successful invasive and outbreak species (Parsons, 1983; Kim and McPheron, 1993). Indeed, contrary to common assumptions that a population bottleneck (experienced by some invading species) reduces genetic variance, Goodnight (1988), Leberg (1992), Bryant and Meffert (1995) and others have found that bottlenecks do not necessarily reduce variability and under certain conditions can lead to increased additive genetic variance. Thus, one might ask: is the degree of genetic variation greater in invasive species for traits that enhance the likelihood of population growth and spread compared with those in non-invasive and non-outbreak species? Another related hypothesis states that successful establishment and spread depends on the area, within the source population, from which invaders originate (Remington, 1968; Parsons, 1983). A prediction of the hypothesis is that source area determines the genetic structure and life history traits of invaders (Scorza, 1983; Brussard, 1984). However, opinions vary on whether individuals from core areas or the margins of source populations are the most genetically variable and the most likely to successfully

160 *Comparative analysis of lymantriid invasion and spread*

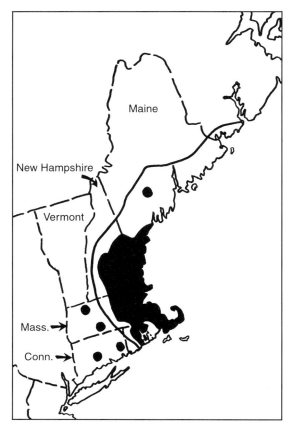

Figure 10.4 Distribution of browntail moth and gypsy moth populations in the United States in 1910. The black area is that infested with gypsy moth; browntail moth occupied the area to the right of the black line. (From Rogers and Burgess, 1910.)

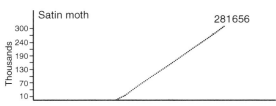
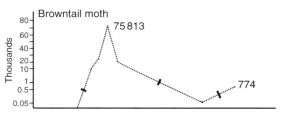
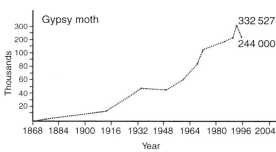

Figure 10.5 Changes over time in area infested by browntail, gypsy and satin moths in the United States. (Based on data from Wallner, 1989; Ferguson, 1978.)

invade and colonize. One view is that invaders originating from subpopulations near the core of the species' native geographical distribution (generally the most favourable environment) will be large and consist of highly heterozygous individuals that are well adapted to a wide range of environmental conditions (i.e. individuals having a great number of alleles at each locus). A corollary of this view is that the core area population may consist of a great variety of individual genotypes specifically adapted to different areas within the core. In either case, these core individuals would be expected to be highly successful at invading the area of introduction and spreading beyond it. In contrast, subpopulations near the margin are presumed to be under intense selection for adaptation and are typically small and highly homozygous, and less likely to be successful invaders and/or outbreak species (Remington, 1968; Parsons, 1983).

In general, the proponents of the genetically based explanation of success of invasive and/or outbreak species might hypothesize that the gypsy, browntail, satin and rusty tussock moths differ significantly in genetic variance of traits of importance in reaching outbreak levels or spreading

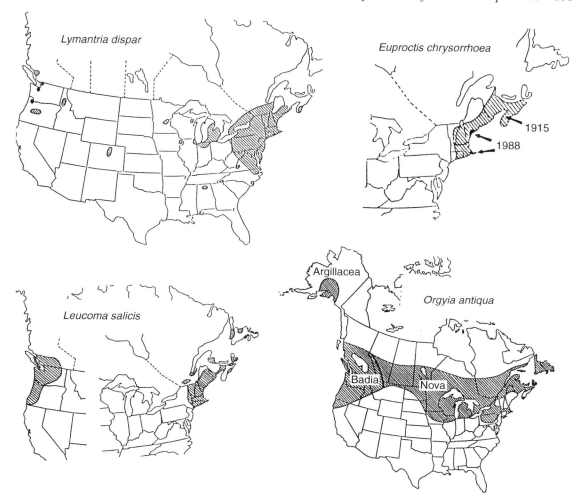

Figure 10.6 Distributions of browntail, gypsy, misty tussock and satin moths in North America. (Based on data from Wallner, 1989; Ferguson, 1978.)

rapidly. More specifically, a prediction of this hypothesis would be that the greatest to lowest levels of genetic variability would be found in the gypsy, browntail, satin and rusty tussock moths, respectively. Unfortunately, this hypothesis and its predictions remain to be tested.

What little we do know about the genetics of lymantriids is limited to the gypsy moth. The genetic diversity of North American populations is, of course, a function of the original diversity of the genotypes in the invading population. There is evidence that there exists significant genetic variation in a variety of traits in the gypsy moth (Rossiter et al., 1990, 1993; Rossiter, 1991). However, analyses of allozyme variation at 20 loci have found extremely low levels of genetic variation in North American gypsy moths, compared with European populations (Harrison et al., 1983). A survey of restriction fragment length polymorphisms (RFLPs) has indicated that North American and French gypsy moths cannot be discriminated whereas moths from Hokkaido (Japan) and China are distinct from North American moths, as well as distinct from each other (Harrison and Odell,

1989). The low level of genetic variability of the successful invader and outbreak species, the gypsy moth, is perhaps unexpected based on the predictions of theory outlined above. This does not mean that, though low, the degree of genetic variance may not nevertheless be greater than that of the other lymantriids. Since we have no comparable data for the other species we cannot determine whether gypsy moth variability is greater than that of the other lymantriids discussed here.

10.2.2 INVADERS AS WEEDY SPECIES

The term 'weed' has become less of a botanical category and more of a psychological category for the description of invasive organisms. Thus, some have speculated that invasive species can be characterized by certain 'weedy traits', perhaps tantamount to those attributed to r-selected versus K-selected species (Dobzhansky, 1965; Crovello and Hacker, 1972; Ziegler, 1976; Parsons, 1983). These so-called weedy species might be characterized (Zimdahl, 1983) as species that have:

- special adaptations for long and short distance dispersal;
- discontinuous germination (or its equivalent, e.g. extended hatch period);
- great longevity of an initial stage (such as a seed or first instar);
- lack of an environmental trigger for germination (or hatch);
- rapid growth of initial stages;
- tolerance of climatic and edaphic variations;
- great competitive ability;
- early age to reproduction;
- ease of mating;
- reproductive output under a variety of environmental conditions, as well as continuous high reproduction under favourable conditions;
- minimum losses during non-growing season (i.e. high overwintering or aestivation survival);
- several generations.

It is difficult to attribute these traits in a rigorous fashion to Lepidoptera species. Not only do few species exhibit these so-called weedy species traits but Lepidoptera, in general, appear to be inherently poor invaders – there is a disparity in their representation among successfully invading species. The proportion of successfully invading Lepidoptera species is not a random sample; it is far smaller than their proportion among known insect taxa (Simberloff, 1986). In addition, Lepidoptera that are invaders or outbreak species do not appear to exclusively exhibit 'weedy species characteristics', when compared with non-invaders or non-outbreak species. For example, the four species discussed here have many similarities in their life history traits even though they differ dramatically in their ability to increase their population size and spread beyond their current distribution. Nevertheless, there is simply no information which directly addresses this issue in the four species under consideration. Indeed, playing devil's advocate one might suggest that – until we have a clear understanding of the traits that facilitate invasion, population increase and population spread in these and other species – no valid comparison can be made among invasive, non-invasive, outbreak, non-outbreak, rapidly spreading and non-spreading populations or species.

10.2.3 ANTHROPOGENIC DISTURBANCES

Have anthropogenic (human) disturbances or activities (such as vehicle transport of egg masses, transported eggs on plant materials, ship ballast, etc.) facilitated invasion and spread (Pimentel, 1993; US Congress OTA, 1993)? The answer is clearly yes but human activities may have just as easily impeded, as well as assisted, the invasion, growth and spread of populations (Kim, 1983). A major effort to control the browntail moth was initiated in the early part of this century (1900–1920) and involved the removal and destruction of over-wintering webs (Kirkland, 1906). For example, 67 000 webs were destroyed in Nova Scotia, Canada, between 1906 and 1914. By 1905 more than 700 bushels of webs (about 57 million larvae) were destroyed in Massachusetts. About 1.2 million webs were destroyed in Maine in 1907, and another 30 million from 1933 to 1936 (Schaefer, 1974). This effort is believed by some to have contributed to the decline of browntail moth populations.

However, equally impressive and manually intensive control efforts were initiated against the gypsy moth in the late part of the nineteenth century which involved the removal of all egg masses from trees in infested areas, as well as the outright removal of the trees themselves (Kirkland, 1906). Yet the latter effort clearly failed. In contrast, there is little doubt that the increased mobility of the American public has had a major impact in the spread of the gypsy moth.

It is difficult to believe that these human inputs played a key role in distinguishing the population dynamics of these lymantriids or of the satin moth. Nor are they likely explanations for the dramatic differences in the ultimate distribution and size of their populations. Finally, little can be said about the most widespread of human activities, insecticide spraying – if only because we have no evidence of any differential susceptibility of the lymantriids discussed that might be responsible for subsequent differences in distribution. In addition, in absolute terms it would appear that the vast majority of the amount of insecticides applied in the northeast occurred well after the dichotomous changes in the pattern of spread and ultimate distribution of the lymantriids.

10.2.4 NATURAL ENEMIES

Although there are many examples of predator–prey interactions in which natural enemies impede the development of outbreak populations (Price, 1987), such is not the case for many introduced species. Invading species are usually introduced without any of their natural enemies. This and the failure of native natural enemies to regulate the invading species is often the rationale for the introduction of natural enemies from the area of origin of the invader. However, in general, neither native nor introduced natural enemies appear to have had any significant impact on any of the invasive lymantriids discussed here (Howard and Fiske, 1911). Nor do they appear to have played any obvious role in changes in their rate of spread.

For example, the browntail moth is attacked by about 65 parasitoid species world wide (Thompson, 1946, in Schaefer, 1974) and by about 46 parasitoid species in Europe (Wolff and Krausse, 1922, in Schaefer, 1974). In North America, 10 parasitoid species of the browntail moth have been recovered or become established. Predators, parasitoids and pathogens have been cited as the cause of population declines of the browntail moth (Schaefer, 1974) but direct evidence of a cause-and-effect relationship does not exist. For example, rarely have levels of parasitism in sampled cohorts of browntail moths exceeded 30%; they have tended to range between 10 and 30% (Schaefer, 1974). Life tables generated by Schaefer (1974) indicate that the largest proportion of generational mortality occurs among first and second instars. The causal agents for that mortality were 'all factors' (including predation, parasitism, disease, etc.) 'combined'. Thus, it is unclear how large a component natural enemies were in this mortality.

Only a small proportion of the natural enemies released against the gypsy moth ever became established. It appears that an equilibrium level of about 10 species was reached by about 1930 (Tallamy, 1983). These have been unsuccessful at preventing the spread of the moth, or its ability to reach outbreak levels.

The gypsy and the browntail moths colonized New England in the absence of their European natural enemies (Burgess and Crossman, 1929). However, natural enemies of the satin moth were resident in New England prior to its invasion because they had already been released against the gypsy moth (e.g. *Blepharipa scutellata*, now *pratense*; *Compsilura concinnata*; *Apanteles melanoscelus*, now *Cotesia melanoscela*; and *Calosoma sycophanta*, in 1906), against the browntail moth (e.g. *Eupteromalus nidulans*) or against both (e.g. *Meteorus versicolor*). Although *C. sycophanta* was not considered to be a significant source of mortality, *B. pratense* was viewed as having potential value and *C. concinnata* was readily and widely reared from field-collected satin moth larvae. Nevertheless, stage-specific mortality of satin moth by its parasitoids rarely exceeded 20% (Wagner, 1977; Wagner and Leonard, 1980). In addition, although Burgess and Crossman

(1927) noted the presence of fungal disease, at relatively insignificant levels (comprising 2.6% of the mortality), they also noted the presence of the fungus *Spicaria* sp. in British Columbia, causing 90% mortality of larvae. Overwintering mortality has been associated with the fungi *Spicaria canadensis*, *Isaria* sp. and *Beauveria globulifera*. However, in general, mortality due to natural enemies does not appear to play a significant role in the regulation of satin moth populations.

10.3 HOST PLANT QUALITY: MEDIATOR OF THE ABUNDANCE AND SPREAD OF INVADING SPECIES

A key prediction in most theoretical analyses of the rate of spread of invading species is that a species' range should increase asymptotically with time in a homogenous habitat (Andow *et al.*, 1993, and references therein). However, habitats of invading species are rarely homogeneous and, indeed, may vary dramatically over time and space, particularly with regard to the number, type and quality of host plants. In fact, we know that herbivore host plants are quite variable. The potential impact of this variation has become increasingly obvious given the large number of studies demonstrating the significant impacts that host plants can have on herbivore physiology, behaviour and ecology (references in Strong *et al.*, 1984; Fritz and Simms, 1992; Hunter *et al.*, 1992; Stamp and Casey, 1993; Carde and Bell, 1995), as well as on dispersal and migration (Johnson and Birks, 1960; Sutherland, 1969; Denno, 1985).

Although the basic concept – that host plant quality/availability influences the distribution of invading species – may not have been formulated as a hypothesis in other studies, there are various individual studies demonstrating the importance of host plant quality and availability in determining the spread and growth of populations. The movement and expansion of *Euphydryas editha* has been associated, in large part, with host plant availability (Ehrlich *et al.*, 1984; Ehrlich, 1986). Studies of *Epilachna varivestris* suggest that host plant variation is a key factor influencing flight behaviour, local sex ratio, life history traits, reproduction, and over time its evolution (Saks, 1993). Host plant suitability was a key feature of White's (1974, 1978) proposal that the origins of insect outbreaks were associated with plant stress and the effects of subsequent changes in host plants of herbivores. To date, to our knowledge, no experimental tests of our hypothesis have been undertaken.

In the absence of direct tests, we compare the ecology of three invasive and one indigenous species, and specifically the differences in host plant breadth and the differential influence of host plants on the gypsy, browntail, satin and rusty tussock moths. As a result of this analysis we present a hypothesis that may explain the observed patterns of abundance and spread of these species, i.e. that **host plant availability and quality (as a consequence of either inter- or intra-plant species variation) is a major driving force determining the spread and abundance of invading species, subsequent to their invasion**.

10.3.1 GYPSY MOTH

What are the forces that have facilitated the spread of the gypsy moth? Various factors, including the ineffectiveness of native and introduced natural enemies (Doane and McManus, 1981), may have been important. Nevertheless, we speculate that host availability and quality may have been key mediators of the spread of gypsy moth populations. Polyphagy, the ability to disperse via first instar ballooning among undefoliated hosts throughout the habitat, and the ability to achieve high fecundity on an array of host plants are key traits through which the influence of host plant quantity/quality may have been expressed. The gypsy moth can survive on a wide variety of the hosts found in the North American forests (Hough and Pimentel, 1978; Barbosa and Greenblatt, 1979; Barbosa *et al.*, 1983; Gross *et al.*, 1990). Although development, survival and fecundity are reduced on some tree species, the ability of first instars to differentiate host species tends to redistribute larvae from less suitable hosts to more suitable host species (Lance and Barbosa, 1980, 1981; Barbosa *et al.*, 1989).

In recent decades research has shown that defoliation of plants results in significant changes in their suitability (Tallamy and Raupp, 1991). In particular, it has been noted that repeated defolation, over several seasons, produces a dramatic induction of defensive compounds as well as nutritional changes in the foliage of several species (Haukioja and Niemelä, 1979; Wallner and Walton, 1979; Werner, 1979; Haukioja and Neuvonen, 1987; Neuvonen and Haukioja, 1991; Tuomi et al., 1984). Differences in tree quality and tree species composition, and the associated differential survival and fecundity of the moth, in front and behind the leading edge, may have had important consequences for the spread of the gypsy moth. Although the gypsy moth may be tannin adapted (Schultz and Lechowicz, 1986), high levels of phenolics in general, and of tannins in particular, resulting from previous defoliation are likely to limit growth and survival (Schultz and Baldwin, 1982; Rossiter et al., 1988). Chemical changes due to natural defoliation of oaks and other hosts can be sufficient to reduce gypsy moth pupal mass (Valentine et al., 1983) and thus resulting fecundity.

However, plant quality deterioration (associated with defoliation) would be less likely, or less intense, in forested areas in front of the leading edge of spreading gypsy moth populations – areas usually lightly defoliated, if at all. Thus, it is not untenable to assume that populations grow and spread rapidly in these areas, in the absence of detrimental changes in foliage. In contrast, reduction in the number and/or intensity of outbreaks is more likely behind the leading edge where defoliation, and its subsequent effects, are most intense. We suggest that this was and continues to be a mechanism which, at least in part, may explain the patterns of spread of the gypsy. A study of changes in feeding by larvae provides suggestive evidence that trees in front and behind the leading edge differ in suitability. Montgomery (1989) measured consumption by gypsy moths of black oak foliage from defoliated stands versus foliage from trees in undefoliated stands. He found that larvae ate more foliage from the undefoliated stands than from the defoliated stands. The difference in consumption may have been due to the significantly higher level of total phenolics and lower level of free sugars found in defoliated stand foliage compared with that in undefoliated stand foliage (Montgomery, 1989). Further studies are needed that directly assess the differential suitability of foliage from trees behind and in front of leading edge populations, on survival and fecundity.

In addition to short-term changes, over time gypsy moth defoliation and associated tree mortality result in significant changes in forest tree species composition (Campbell and Sloan, 1977). Surviving individuals of favoured species such as oaks are likely to be more tolerant or resistant to defoliation. This does not mean that repeated defoliation is not possible in areas once defoliated; however, it does appear that susceptibility of an area does decline with time (Gansner et al., 1995). Short-term changes are reinforced by long-term changes in defoliation where areas less favoured and suitable understorey tree species, such as maples, increase in abundance as more light penetrates through an open canopy. Gansner et al. (1995) noted that, in Pennsylvania forests, tree species diversity is greater now than 15 years ago.

10.3.2 BROWNTAIL MOTH

We speculate that the differential suitability of potential tree hosts and the species composition of invaded habitats are major factors in the ability of browntail moths to reach outbreak levels and spread successfully. We suggest that as tree species composition changes from area to area so does the ability of species such as the browntail moth to survive and reach outbreak levels. This differential survival and population growth may have been a consequence of host plant-mediated competition for appropriate habitats between the browntail moth and the gypsy moth. Schaefer (1974) speculated that the temporal and spatial coincidence of the introduction of the gypsy and browntail moths, initially, may have accelerated the spread of the browntail moth. The gypsy moth was well established and abundant in the Somerville and Cambridge (Massachusetts) area when the browntail moth was initially introduced. In addition, the gypsy moth is extremely polyphagous (feeding on over 400 species of plants) (Table 10.2) and thus many potential browntail moth hosts (Table 10.2) were likely to have been subject to defoliation by the gypsy moth earlier in the season.

Perhaps the most important ecological interaction explaining changes in the distribution of the browntail moth may have been simply the inability of females to oviposit on already defoliated hosts, or the rejection of small refoliated leaves (on previously defoliated trees) by ovipositing females (Schaefer, 1974). This would have obvious consequences on population growth and spread. Lack of foliage, or of appropriate foliage, is most likely to have been due to defoliation by a persistent outbreak species such as the gypsy moth. Larval phenology of the gypsy moth is such that any browntail moth females, co-inhabiting an area with the gypsy moth, would have to contend with defoliated hosts at about the time they would be prepared to oviposit (i.e. about July). We suggest that constraints resulting from changes in host availability or quality, associated with gypsy moth defoliation, may have induced adult dispersal and encouraged the spread of browntail moth populations during the 13–15 years after the introduction of the browntail moth (Figures 10.4, 10.5).

With time, however, the spread of the gypsy moth eventually outpaced that of the browntail

Table 10.2 Tabulation of published records of feeding by *Euproctis chrysorrhoea*, *Leucoma salicis*, *Lymantria dispar*, and *Orgyia antigua* on host plants[1] (from Schaefer, unpublished. data)

Host plant family	*Euproctis chrysorrhoea*[2]		*Leucoma salicis*[3]		*Lymantria dispar*		*Orgyia antigua*	
Fagaceae	119	(10.4)			505	(13.5)	19	(6.9)
Rosaceae	432	(37.7)			487	(13.0)	78	(28.2)
Pinaceae					387	(10.3)	47	(17.9)
Betulaceae	55	(4.8)			326	(8.7)	36	(13.0)
Salicaceae	68	(5.9)	154	(84.2)	295	(7.9)	29	(10.5)
Aceraceae	43	(3.7)			195	(5.2)		
Ulmaceae	46	(4.0)			108	(2.8)		
Juglandaceae	10	(0.8)			105	(2.8)		
Leguminosae	16	(1.3)			88	(2.3)		
Ericaceae					83	(2.2)	14	(5.1)
Oleaceae	35	(3.0)			66	(1.7)		
Tiliaceae	14	(1.2)			61	(1.6)		
Anacardiaceae	14	(1.2)			54	(1.4)		
Compositae					51	(1.3)		
Gramineae					42	(1.1)		
Caprifoliaceae	30	(2.6)			39	(1.0)		
Hamamelidaceae					37	(0.9)		
Cupressaceae					31	(0.8)		
Moraceae					27	(0.7)		
Polygonaceae	10	(0.8)			24	(0.6)		
Saxifragaceae	15	(1.3)						
Elaeagnaceae	13	(1.1)						
Vitaceae	11	(0.9)						
Comaceae	10	(0.8)						
Hippocastanaceae	10	(0.8)						
Myricaceae	10	(0.8)			25	(0.6)		
Lauraceae					25	(0.6)		
All others	183	(16.9)	29	(15.8)	670	(19.0)	54	(17.5)

[1] Each number represents the number of published citations of feeding on species within each plant family. The value in parentheses is the percentage of the total records that the number represents (Schaefer, unpublished data).
[2] Approximately 12, 8, and 6 % of all published records are of species in the genera *Prunus*, *Quercus* and *Pyrus*, respectively.
[3] Approximately 59 and 25 % of all published records are of species in the genera *Populus* and *Salix*, respectively.

moth (Figures 10.5, 10.6). The browntail moth may have ultimately been distributed within the distribution of the gypsy moth and hence in a 'sea' of defoliated and refoliated host plants, or (given that its host range is narrower than that of the gypsy moth) found itself in habitats dominated by undefoliated but unsuitable hosts (Table 10.2). As the browntail moth spread rapidly into habitats that were climatically and ecologically distinct from those of coastal areas, the availability of suitable host trees may have become increasingly limited. Because the browntail moth overwinters as a third instar, the availability of foliage in spring is critical. Not only do trees selected by females the previous season have to be nutritionally suitable but also they have to be phenologically suitable (i.e. the timing of budbreak is critical). Phenology can be altered by abiotic factors and biotic factors such as defoliation. The phenology of budbreak of trees severely defoliated in the previous season has been shown to shift in the seasons after defoliation (Heichel and Turner, 1976). Even slight shifts in budbreak have been shown to have dramatic impacts on herbivore survival (Barbosa and Wagner, 1989; Hunter and Lechowicz, 1992; Quiring, 1992). We speculate that under the aforementioned conditions populations of the browntail moth began to recede.

Over time the contraction of the browntail moth's distribution may have been fostered by higher levels of tolerance or resistance to herbivory among trees surviving gypsy moth defoliation (Gansner *et al.*, 1995), and thus a greater availability of foliage for the browntail moth . Reduced susceptibility and defoliation may have resulted in greater availability of appropriate (or more) foliage behind the front, on which females could oviposit. Such changes in the habitats behind the leading edge of browntail moth populations were probably in direct contrast to increasing lack of suitability of trees in front of the leading edge, which were likely to have suffered extensive defoliation due to spreading gypsy moth populations. Differential survival of larvae on tree species in habitats may have amplified the changes described above.

The relative differences in nutritional suitability of dominant species in interior forests compared with those of hosts in coastal habitats also may have led ultimately to a constriction of the browntail moth's distribution to coastal habitats, as has been noted in North America and in Europe (Sterling and Speight, 1989). For the browntail moth, even within favoured habitats, the survival, size and number of larvae per overwintering web (all keys to its population dynamics) vary depending on the host plant upon which larvae feed, and on which they create overwintering webs. For example, the weight of larvae emerging from webs on *Myrica pennsylvanica* is only 40% of that of larvae from webs on *Rosa rugosa*. Similarly, the average number of larvae per web can be as high as 600 on *Pyrrus malus* or as low as 275 in webs on *Amelanchier arborea*. Survival of larvae on *R. rugosa* is about 92% whereas that of larvae on *Quercus ilicifolia* is about 60% (Schaefer, 1974).

The exact reasons why the browntail moth is restricted to coastal habitats have yet to be determined, although observations of this tendency are well documented. The regression of the distribution may have continued until the habitats that could maintain low but persistent populations (i.e. the coastal habitats) were populated to the general exclusion of interior habitats. Host plant-mediated low levels of mortality from natural enemies in coastal habitats may have facilitated the restriction of the browntail moth to such habitats.

Might host plants also mediate multitrophic interactions and consequently help determine patterns of distribution? Host plant/habitat differences can affect resident herbivores in a variety of ways. The affinity of the browntail moth for coastal habitats has been recognized for populations in Great Britain (Theobald, 1919), the Netherlands (de Fluiter, 1934; van der Linde and Voute, 1967) and Germany (Gasser and Steiner, 1949; Lubke, 1952), as well as in New England. In the United States, populations in Casco Bay, Maine, occur on islands that range from deciduous forest habitats to shrub habitats, and these in turn may differ from inland forests. In each type of habitat the moth may utilize a different set of host plants (interior: species in the genera *Crataegus*, *Quercus*, *Populus* and *Prunus*; vs. coastal: species in *Rosa*, *Myrica*, *Amelanchier*, *Rubus* and *Ribes*). Habitats in Cape Cod,

Massachusetts, are also distinct from those in Maine. The Cape Cod habitat consists of sand dunes inhabited by relatively few plants and the browntail moth relies on *Prunus maritima* and *Rosa rugosa*. We speculate that host plants may mediate important ecological interactions between the browntail moth and its natural enemies. We have found no evidence to suggest that the effectiveness of natural enemies of the four lymantriids discussed here was sufficient to explain differences in the abundance of their populations. However, host plants may nevertheless mediate other interactions so as to restrict these species to certain habitats, or influence their rate of spread.

Residence on one host plant may subject a herbivore to more or less mortality than it would be exposed to on another host. Preliminary results from a study of macrolepidopterans on box elder and black willow (Barbosa *et al.*, unpublished data) suggest that host plant-associated differences can lead to significant differences in levels of parasitism among a variety of herbivores. When a species occurs on box elder it suffers higher levels of parasitism than when the same species occurs on willow (Table 10.3). Equally interesting are differences in total parasitism of species that occur on both box elder and black willow. In this two-year data set, species that occur only on box elder were compared with their congeners on willow (such as those in the genera *Zale* and *Acronicta*) in three out of four cases. Levels of total parasitism of congeners on box elder were significantly greater than those of the larvae of the congeners on black willow (Table 10.4). This occurs even though the total number of species and larvae collected from willow is always greater than on box elder (Figures 10.7, 10.8).

A similar phenomenon may occur in the browntail moth. Levels of parasitism of the browntail moth by the polyphagous tachinid *Compsilura concinnata* and the monophagous *Townsendiellomyia nidicola* are significantly higher in larvae on *P. maritima* compared with those on *R. rugosa* (Schaefer, 1974). Unfortunately, few direct comparisons have been made of the population dynamics of the four lymantriids discussed here to determine the potential for differential parasitism on different tree hosts. Nevertheless, the greater effectiveness of these natural enemies, and perhaps others, against larvae on certain tree species may help keep coastal populations low and decrease the likelihood of spread. Host plant effects may also be indirect. Scarce resources

Table 10.3 Comparison of percentage parasitism for selected Macrolepidoptera on black willow and box elder*

	Box elder			Willow		
	N	Parasites (no.)	Parasitism (%)	N	Parasites (no.)	Parasitism (%)
1991						
Halysidota tessellaris	5	2	40.0	5	0	0.0
Hyphantria cunea	6	5	83.3	10	4	40.0
Orgyia leucostigma	1	1	100.0	8	4	50.0
Automeris io	3	3	100.0	12	0	0.0
Total	15	11	73.3	35	8	22.8
1993						
Halysidota tessellaris	3	1	33.3	15	1	6.7
Hyphantria cunea	13	3	23.1	23	5	21.7
Orgyia leucostigma	25	11	44.0	34	7	20.6
Eutrapela clemataria	11	2	18.2	19	2	10.5
Other Species	30	2	6.7	44	2	4.5
Total	71	19	26.8	135	17	12.6

*In both years the differences in total percentage parasitism was statistically significant according to Fisher's Exact Test (a = 0.05).

Table 10.4 Comparison of percentage parasitism between macrolepidopteran congeners on black willow and box elder in 1991 and 1993

Tree species	Caterpillar species	1991		1993	
		N	Parasitism (%)	N	Parasitism (%)
Box elder	*Zale galbanata*	17	17.6	43	46.5
Black willow	*Zale lunata*	41	19.5	32	6.3 *
Box elder	*Acronicta americana*	25	48.0	22	36.4
Black willow	*Acronicta oblinita*	11	9.1 *	16	6.3 *

*Represents significant differences according to Fisher's Exact Test (a = 0.05).

in a habitat (such as the current habitats of the browntail moth) may lead to a significant reduction in the overall number of resident species compared with that found in a resource-rich habitat (e.g. inland forests). A consequence of that species reduction may be that maintaining a population in a depauparate (harsh) habitat may be facilitated for a herbivore whose natural enemies are polyphagous. The density and effectiveness of its natural enemies are likely to depend on the presence and abundance of alternate hosts. Important natural enemies of the browntail and satin moths such as *Apanteles lacteicolor*, *A. melanoscelus*, *C. concinnata* and *Trichogramma minutum* require alternate hosts, which may not be available in harsh, species-poor habitats. The parasitoid *A. lacteicolor* was not found in larvae collected from Cape Cod although it was present in other habitats (Schaefer, 1974). Similarly, vertebrate predators persist and are most effective in habitats where prey are readily available. Schaefer (1974) noted that there was a lower diversity of birds in the coastal dune habitats of Cape Cod, which in turn may have resulted in reduced intensity of predation in those habitats. Thus, differences in host plants in different habitats might play a major role in determining the success or failure of spread of populations, and in their distribution.

10.3.3 SATIN MOTH

The satin moth is similar to the browntail moth in its feeding behaviour, phenology, overwintering

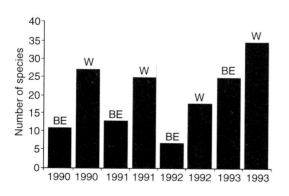

Figure 10.7 Species abundance of macrolepidoptera from black willow (W) and box elder (BE).

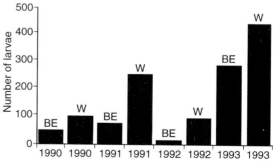

Figure 10.8 Number of macrolepidoptera reared from black willow (W) and box elder (BE). Individuals that died as larvae without yielding parasitoids were excluded.

stage, and in other aspects of its life history (Table 10.1). However, the satin moth differs from the other three lymantriids in one very important characteristic: its host plant range is relatively limited (Table 10.2). Unlike the other polyphagous lymantriids in North America, the satin moth restricts its feeding mostly to species of *Populus* (Wagner and Leonard, 1979a), although on occasion in North America and world-wide it is recorded as feeding on some *Salix* species (Table 10.2). This species was introduced in 1920 in suburban areas (both in Canada and in the eastern United States) where there was an abundance of ornamental poplars such as Lombardy poplar (*Populus nigra* L. *italica*) and balm of Gilead (a hybrid of *P. balsimifera* L.), or native poplars (e.g. *P. deltoides* Battr. ex. Marsh) that often grow in recently disturbed areas. We speculate that the widespread abundance of poplars and the monophagy of the satin moth provided conditions that facilitated rapid population growth and spread, shortly after its introduction. However, the subsequent major reduction in the abundance of poplars may have made the satin moth an infrequently outbreaking species, and one which typically occurs at relatively low densities.

In North America the satin moth has adapted to novel, native species, but does not do well on these hosts over extended periods. For example, decades after its introduction Wagner and Leonard (1979a) reported satin moth defoliation of two forest-inhabiting native poplars, *P. grandidentata* Michx. and *P. tremuloides* Michx. However, satin moth development on the latter two hosts is poor. Dry-weight gain of larvae of the satin moth is greater on widespread species that grow in urban and suburban areas, such as Lombardy poplar (*P. nigra italica*) and cottonwood (*P. deltoides*), than it is on *P. tremuloides*. Weight gain is poorest on bigtooth aspen (*P. grandidentata*) (Wagner and Leonard, 1979a). These differences should translate into differences in fecundity and population growth since satin moths may lay as few as 250 but may lay over 500 eggs per female (Table 10.1) (Glendenning, 1932).

The differences in suitability of hosts trees may be particularly critical if the abundance of the most suitable hosts declines significantly. Following the outbreaks, and rapid spread that occurred after its introduction, changes in tree species composition in urban and suburban habitats (which occur over a large proportion of New England) probably directly and indirectly reduced the likelihood of maintaining outbreak levels in the satin moth's expanded geographical range. We speculate that these changes probably had a devastating effect on the size of satin moth populations. Changes in the tree species composition of the satin moth's habitat were very likely due to several factors. Although a host such as Lombardy poplar is the most suitable for the satin moth, it is also a species with a relatively short life span. Thus, unless regenerated or replanted, over time it will be unavailable. If host availability decreases sufficiently it is likely to have an impact on outbreak populations. In addition, because poplars in general are highly susceptible to a large array of arthropods and pathogens, the number of newly planted poplars has declined sharply since the satin moth's introduction. Similarly, established poplars have been culled at a greater rate than other ornamental, street and park trees. Changes in community preferences for poplars were surely influenced by the thinking and recommendations of arborists during the period. In the middle of this century basic texts on tree maintenance specifically state that 'poplars are not recommended for street-side plantings' and that, for example, 'Lombardy poplar is rapidly disappearing in the eastern United States because of its extreme susceptibility to the canker disease' (Pirone, 1959). Although poplars were quite common as shade, street and park trees when the satin moth was introduced, they are currently rare in the same area today. All of these factors directly or indirectly may have contributed to a decreased likelihood that outbreak satin moth populations would occur, although sufficient poplars remained to maintain much of its distribution.

10.3.4 RUSTY TUSSOCK OR VAPOURER MOTH

Although exhibiting a life history nearly identical to that of the gypsy moth, the indigenous and wide-ranging species *Orgyia antiqua* has rarely been reported to reach outbreak levels except in Canada (Baker, 1972). This species is similar to the gypsy moth in how it overwinters, its lack of flight in females, its ballooning first instars, etc. However, the rusty tussock moth differs significantly from the gypsy moth in that it rarely exhibits the relatively high fecundity that gypsy moth females achieve (Table 10.1). The average number of eggs per rusty tussock moth female is similar to that produced by the satin moth and the browntail moth, but the rusty tussock moth female is wingless and incapable of flight (unlike the satin moth and the browntail moth) and relies on first instar ballooning for dispersal. Although ballooning is an effective dispersal mechanism, larvae experience high levels of mortality (Mason and McManus, 1981; Dwyer and Elkinton, 1995). In the gypsy moth this mortality is counterbalanced somewhat by its high fecundity but the same may not be true for the rusty tussock moth. We speculate that the inability of the rusty tussock moth to achieve the relatively high fecundity achieved by other ballooning species like the gypsy moth, and the limited number of suitable host trees on which ballooning larvae might settle (Table 10.2), may be key reasons for the infrequency of rusty tussock moth outbreaks.

10.4 CONCLUSION

Although available historical, empirical and experimental data provide support for the hypothesis that **host plant availability and quality (as a consequence of either inter- or intra-plant species variation) is a major driving force determining the spread and abundance of invading species, subsequent to their invasion**, the best evidence comes from direct tests of the hypothesis. These four lymantriids offer a unique and fascinating opportunity to test the proposed hypothesis. For each of the four species the influence of host plant quantity and quality is manifested in different ways and thus the tests of hypothesis for each species differs.

For the gypsy moth the critical prediction is that the performance of gypsy moth populations will differ in front and behind the leading edge of its distribution. Experiments can be conducted testing the suitability of foliage from the same tree species, selected from trees both in front and behind the leading edge. An assessment of growth, development, survival and fecundity (and their impact on population growth and dispersal) would provide important evidence.

Tests of the relative suitability of hosts' plants for the browntail moth in different habitats are also important tests of the hypothesis. For one, the prediction of the hypothesis as it relates to the browntail moth suggests that budbreak phenology of trees (relative to the initiation of feeding by overwintering larvae) in coastal habitats should differ from that of trees in interior mesic forest habitats. This can certainly be tested in an array of host tree species. Similarly a second prediction, that survival, development, fecundity, etc. are better in coastal trees than on closely related species in interior habitats, can be tested. Finally, differences in the natural enemy complex and their effectiveness in coastal and interior habitats, and whether those differences can be attributed to the influences exerted by host trees can also be tested. Clearly, all three tests represent initial attempts to determine what factors may be important causes for the tendency of the browntail moth to be restricted to coastal habitats. In recent years, for the first time since its distribution receded to its current distribution, browntail moth populations have been found on the mainland off coastal islands of Maine. This change in distribution from about 129 to 2005 km^2, in 1995 (Figure 10.5), offers a unique opportunity to study this species, particularly if the latter trend continues. A comparison of the patterns of host plant utilization and other aspects of the ecology of populations on the leading edge and island habitats should provide useful insights into the ecology of invasions.

Our speculation about the influence of the abundance and quality of poplars on the abundance and distribution of the satin moth may be tested by comparing habitats differing in the abundance of poplar species. A prediction of our hypothesis would be that populations in habitats with a greater proportion of favoured poplar species would support significantly higher population densities; habitats with the greater proportion of favoured species would support higher populations even when compared habitats have the same overall proportion of poplar species. A third aspect of the hypothesis could be assessed with a widespread survey of the presence of the satin moth. The hypothesis would be supported if the satin moth distribution was directly related to the presence of poplars.

Finally, the influence of host tree species on the rusty tussock moth was speculated to be based on the inability of the fecundity of females to counteract the mortality resulting from first instar ballooning. Experimental manipulation of the fecundity of the rusty tussock moth by combining two, three or more egg masses on small field trees should be possible. One can then determine if the ultimate density, persistence and spread of the population is a function of initial number of eggs.

Obviously, these are but a few of the many experimental tests that can be initiated to test the hypothesis presented in this chapter. The important point is that the hypothesis is testable and can produce important insights into the ecology of invasive species.

ACKNOWLEDGEMENTS

We thank Caroline White, Dennis McGrevy, Susan Barth, Paul Gross and Alejandro Segarra for their assistance in gathering the data presented in this publication. We also express our appreciation to several anonymous reviewers whose comments and suggestions helped us to convey our ideas more directly and succinctly.

REFERENCES

Andow, D.A., Karieva, P.M., Levin, S.A. and Okubo, A. (1993) Spread of invading organisms: patterns of spread. In *Evolution of Insect Pests. Patterns in Variation* (eds K.C. Kim and B.A. McPheron) pp. 219–242, John Wiley & Sons, New York.

Baker, W.L. (1972) *Eastern Forest Insects*. Misc. Publ. No. 1175, US Department of Agriculture, Forest Service. Superintendent of Documents, US Government Printing Office, Washington, DC. 642 pp.

Baker, H.G. and Stebbins, G.L. (1965) *The Genetics of Colonizing Species*, Academic Press, New York.

Barbosa, P. and Greenblatt, J.A. (1979) Suitability, digestibility, and assimilation of various host plants of the gypsy moth, *Lymantria dispar* (L.). *Oecologia* **43**, 111–119.

Barbosa, P. and Schultz, J.C. (1987) *Insect Outbreaks*, Academic Press, San Diego, CA.

Barbosa, P. and Wagner, M.R. (1989) *Introduction to Forest and Shade Tree Insects*, Academic Press, San Diego, CA.

Barbosa, P., Waldvogel, M., Martinat, P. and Douglass, L.W. (1983) Developmental and reproductive performance of the gypsy moth *Lymantria dispar* (L.) on selected hosts common to mid-Atlantic and southern forests. *Environ. Entomol.* **12**, 1858–1862.

Barbosa, P. Krischik, V. and Lance, D. (1989) Life-history traits of forest-inhabiting flightless Lepidoptera. *Amer. Midl. Nat.* **122**, 262–274.

Brussard, P.F. (1984) Geographic patterns and environmental gradients: the 'central–marginal' model in *Drosophila* revisited. *Annu. Rev. Ecol. Syst.* **15**, 25–64.

Bryant, E.H. and Meffert, L.M. (1995) An analysis of selection response in relation to a population bottleneck. *Evolution* **49**, 626–634.

Burgess, A.F. and Crossman, S.S. (1927) *The satin moth a recently introduced pest*, USDA Bulletin 1469. 22 pp.

Burgess, A.F. and Crossman, S.S. (1929) *Imported insect enemies of the gypsy moth and the brown-tail moth*, USDA Tech. Bull. 86.

Campbell, R.W. and Sloan, R.J. (1977) Forest stand conditions and the gypsy moth. *For. Sci. Monogr.* 19.

Carde, R. and Bell, W.J. (1995) *Chemical Ecology of Insects 2*, Chapman & Hall, New York.

Carter, D.J. (1984) *Pest Lepidoptera of Europe. With Special Reference to the British Isles*, W. Junk Publishers, Dordrecht, Germany.

Crovello, T.J. and Hacker, C.S. (1972) Evolutionary strategies in life table characteristics among feral and urban strains of *Aedes aegypti* (L.). *Evolution* **26**, 185–196.

De Fluiter, H.J. (1934) Over *Nygmia phaeorrhoea* Donovank den astaardsatijavlinder, en de factoren, Welke tijdens de winterrust de gestalsterkte van dit insekt decrmeeren. *Tjdschr. Ned. Heidemij.* **49**, 1–35. (In Schaefer, 1977.)

Denno, R.F. (1985) The role of host plant condition and nutrition in the migration of phytophagous insects. In *The Movement and Dispersal of Agriculturally Important Biotic Agents* (eds D.R. MacKenzie, C.S. Barfield, G.C. Kennedy *et al.*) pp. 151–172, Claitor's Publishing House, Baton Rouge, LA.

Doane, C.C. and McManus, M.L. (1981) *The Gypsy Moth: Research Towards Integrated Pest Management*, USDA Forest Service Technical Bull. No. 1584, Washington, DC.

Dobzhansky, Th. (1965) 'Wild' and 'domestic' species of *Drosophila*. In *The Genetics of Colonizing Species* (eds H.G. Baker and G.L. Stebbins) pp. 533–546, Academic Press, New York.

Dwyer, G. and Elkinton, J.S. (1995) Host dispersal and the spatial spread of insect pathogens. *Ecology* **76**, 1262–1275.

Ehrlich, P.R. (1986) Which animals will invade? In *Ecology of Biological Invasions of North America and Hawaii* (eds H.A. Mooney and J.A. Drake) pp. 79–95, Springer Verlag, New York.

Ehrlich, P.R., Launer, A.E. and Murphy, D.D. (1984) Can sex ratio be defined or determined? The case of a population of checkerspot butterflies. *Amer. Natur.* **124**, 527–539.

Elton, C.S. (1958) *The Ecology of Invasions by Animals and Plants*, Methuen, London.

Ferguson, D.C. (1978) *The Moths of America North of Mexico. Fasicle 22.2. Noctuoidea, Lymantriidae*, E.W. Classey and R.B D. Publications Inc., Middlesex. 110 pp.

Fernald, C.H. (1892) The gypsy moth. Report by the Division of Entomology. *Massachusetts Hatch Experiment Station Bull.* **19**, 109–116.

Fritz, R.S. and Simms, E.L. (1992) *Plant Resistance to Herbivores and Pathogens. Ecology, Evolution, and Genetics*, University of Chicago Press, Chicago.

Gansner, D.A., Quimby, J.W., King, S.L. *et al.* (1995) Tracking changes in the susceptibility of forest land infested with gypsy moth. *Gypsy Moth News* **37**, 3–6.

Gasser, G. and Steiner, P. (1949) Massenvermehrung und Bekampfung des Goldafters auf der Insel Borkum und Juist. *Nachrbl. Biol. Zent. Anst. Braunschw.* **1**, 163–165. (In Schaefer, 1977.)

Glendenning, R. (1932) *The satin moth in British Columbia*, Can. Dept. Agric. Pamph. 50, n.s. rev. 3rd edn. 15 pp.

Goodnight, C.J. (1988) Epistasis and the effect of founder events on the additive genetic variance. *Evolution* **42**, 441–454.

Gross, P., Montgomery, M.E. and Barbosa, P. (1990) Within and among site variability in gypsy moth performance on five tree species. *Environ. Entomol.* **19**, 1344–1355.

Harrison, R.G. and Odell, T.M. (1989) Mitochrondrial DNA as a tracer of gypsy moth origins. In *Proceedings. Lymantriidae: A Comparison of Features of New and Old World Tussock Moths* (eds W.E. Wallner and K.A. McManus) pp. 339–350, Forest Service Gen. Tech. Rept. NE-123, USDA, Broomall, PA.

Harrison, R.G., Wintermeyer, S.F. and Odell, T.M. (1983) Patterns of genetic variation within and among gypsy moth, *Lymantria dispar* (Lepidoptera: Lymantriidae), populations. *Ann. Entomol. Soc. Amer.* **76**, 652–656.

Haukioja, E. and Niemelä, P. (1979) Birch leaves as a resource for herbivores: seasonal occurrence of increased resistance in foliage after mechanical damage of adjacent leaves. *Oecologia* **39**, 151–159.

Haukioja, E. and Neuvonen, S. (1987) Insect population dynamics and induction of plant resistance: testing the hypothesis. In *Insect Outbreaks* (eds P. Barbosa and J.C. Schultz), Academic Press, San Diego, CA.

Heichel, G.H. and Turner, N.C. (1976) Phenology and leaf growth of defoliated hardwood trees. In *Perspectives in Forest Entomology* (eds J.F. Anderson and H.K. Kaya) pp. 31–40, Academic Press, New York.

Hough, J.A. and Pimentel, D. (1978) Influence of host foliage on development, survival, and fecundity of the gypsy moth. *Environ. Entomol.* **7**, 97–102.

Howard, L.O. and Fiske, W.F. (1911) The importation into the United States of the parasites of the gypsy moth and the brown tail moth. *USDA Agric. Bur. Entomol. Bull.* **91**.

Hunter, A.F. and Lechowicz, M.S. (1992) Foliage quality changes during canopy development of some northern hardwood trees. *Oecologia* **89**, 316–323.

Hunter, M.D., Ohgushi, T. and Price, P.W. (1992) *Effects of Resource Distribution on Animal–Plant Interactions*, Academic Press, New York.

Johnson, B. and Birks, P.R. (1960) Studies on wing polymorphism in aphids. I. The developmental process involved in the production of the different forms. *Entomol. Exp. et Appl.* **3**, 327–339.

Kim, K.C. (1983) How to detect and combat exotic pests. In *Exotic Plant Pests and North American Agriculture* (eds C. Wilson and C.L. Graham) pp. 15–38, Academic Press, New York.

Kim, K.C. and McPheron, B.A. (1993) *Evolution of Insect Pests*, John Wiley and Sons, New York.

Kirkland, A.H. (1906) *The gypsy and brown-tail moths*, Office of the Superintendent for Suppressing the Gypsy and Brown-tail Moths, Commonwealth of Massachusetts, Bull. No. 2, Wright & Potter Printing Co., State Printers. 33 pp.

Lance, D. and Barbosa, P. (1980) Dispersal of larval Lepidoptera with special reference to forest defoliators. *The Biologist* **91**, 90–110.

Lance, D. and Barbosa, P. (1981) Host tree influences on the dispersal of first instar gypsy moths, *Lymantria dispar. Ecol. Entomol.* **6**, 411–416.

Leberg, P.L. (1992) Effects of population bottlenecks on genetic diversity as measured by allozyme electrophoresis. *Evolution* **46**, 477–494.

Leston, D. (1957) Spread potential and the colonization of islands. *Syst. Zool.* **6**, 41–46.

Liebhold, A.M., MacDonald, W.L., Bergdahl, D. and Mastro, V.C. (1995) Invasion by exotic forest pests:

a threat to forest ecosystems. *For. Sci. Monogr.* **30**, 1–49.

Lubke, A. (1952) Der Goldafter auf den Nordsee-Inselin. *Z. Pflschutz.* **59**, 221–223.

Mason, C.J. and McManus, M.L. (1981) Larval dispersal of the gypsy moth. In *The Gypsy Moth: Research Toward Integrated Pest Management* (eds C.C. Doane and M.L. McManus), USDA Tech. Bull. 1584, Washington, DC.

Montgomery, M. (1989) Relationship between foliar chemistry and susceptibility to *Lymantria dispar*. In *Proceedings. Lymantriidae: A Comparison of Features of New and Old World Tussock Moths* (eds W.E. Wallner and K.A. McManus) pp. 339–350, Forest Service Gen. Tech. Rept. NE-123, USDA, Broomall, PA.

Mooney, H.A. and Drake, J.A. (1986) *Ecology of Biological Invasions of North America and Hawaii*, Springer Verlag, New York.

Neuvonen, S. and Haukioja, E. (1991) The effects of inducible resistance in host foliage on birch-feeding herbivores. In *Phytochemical Induction by Herbivores* (eds D.W. Tallamy and M.J. Raupp) pp. 277–291, John Wiley & Sons, New York.

Parsons, P.A. (1983) *The Evolutionary Biology of Colonizing Species*, Cambridge University Press, New York.

Pimentel, D. (1986) Biological invasions of plants and animals in agriculture and forestry. In *Ecology of Biological Invasions of North America and Hawaii* (eds H.A. Mooney and J.A. Drake) pp. 148–162, Springer Verlag, New York.

Pimentel, D. (1993) Habitat factors in new pest invasions. In *Evolution of Insect Pests. Patterns in Variation* (eds K.C. Kim and B.A. McPheron) pp. 165–181, John Wiley & Sons, New York.

Pirone, P.P. (1959) *Tree Maintenance*, Oxford University Press, New York.

Price, P.W. (1987) The role of natural enemies in insect populations. In *Insect Outbreaks* (eds P. Barbosa and J.C. Schultz) pp. 287–312, Academic Press, San Diego, CA.

Quiring, D.T. (1992) Rapid change in suitability of white spruce for a specialist herbivore, *Zeiraptera canadensis*, as function of leaf age. *Can. J. Zool.* **70**, 2132–2138.

Reeks, Q.A. and Smith, C.C. (1956) The satin moth, *Stilpnotia salicis* (L.), in the maritime provinces and observations on its control by parasites and spraying. *Can. Entomol.* **88**, 565–579.

Remington, C.L. (1968) The population genetics of insect introduction. *Annu. Rev. Entomol.* **13**, 799–824.

Rogers, D.M. and Burgess, A.F. (1910) *Report on the field work against the gypsy moth and the brown-tail moth*, USDA Bur. Ent. Bull. 87. 81pp.

Rossiter, M.C. (1991) Maternal effects generate variation in life history: consequences of egg weight plasticity in the gypsy moth. *Funct. Ecol.* **5**, 386–393.

Rossiter, M.C., Cox-Foster, D.L. and Briggs, M.A. (1993) Initiation of maternal effects in *Lymantria dispar*: genetic and ecological components of egg provisioning. *J. Evol. Biol.* **6**, 577–589.

Rossiter, M.C., Schultz, J.C. and Baldwin, I.T. (1988) Relationships among defoliation, red oak phenolics, and gypsy moth growth and reproduction. *Ecology* **69**, 267–277.

Rossiter, M.C., Yendol, W.G. and Dubois, N.R. (1990) Resistance to *Bacillus thuriensis* in gypsy moth (Lepidoptera: Lymantriidae): genetic and environmental causes. *J. Econ. Entomol.* **83**, 2211–2218.

Sailer, R.I. (1978) Our immigrant insect fauna. *Bull. Entomol. Soc. Amer.* **24**, 3–11.

Sailer, R.I. (1983) History of introductions. In *Exotic Plant Pests and North American Agriculture* (eds C. Wilson and C.L. Graham) pp. 15–38, Academic Press, New York.

Saks, M.E. (1993) Variable host quality and evolution in the Mexican bean beetle. In *Evolution of Insect Pests. Patterns in Variation* (eds K.C. Kim and B.A. McPheron) pp. 329–350, John Wiley & Sons, New York.

Schaefer, P.W. (1974) Population ecology of the brown-tail moth (*Euproctis chrysorrhoea* L.) (Lepidoptera: Lymantriidae) in North America. PhD Dissertation, University of Maine, Orono, ME.

Schultz, J.C. and Baldwin, I.T. (1982) Oak leaf quality declines in response to defoliation by gypsy moth larvae. *Science* **217**, 149–151.

Schultz, J.C. and Lechowicz, M.J. (1986) Hostplant, larval age, and feeding behavior influence midgut pH in the gypsy moth (*Lymantria dispar*). *Oecologia* **71**, 133–137.

Scorza, R. (1983) Ecology and genetics of exotics. In *Exotic Plant Pests and North American Agriculture* (eds C.L. Wilson and C.L. Graham), Academic Press, New York.

Simberloff, D. (1986) Introduced insects: a biogeographic and systematic perspective. In *Ecology of Biological Invasions of North America and Hawaii* (eds H.A. Mooney and J.A. Drake) pp. 3–26, Springer-Verlag, New York.

Stamp, N.E. and Casey, T.M. (1993) *Caterpillars. Ecological and Evolutionary Constraints on Foraging*, Chapman & Hall, New York.

Sterling, P.H. and Speight, M.R. (1989) Comparative mortalities of the brown-tail moth, *Euproctis chrysorrhoea* (L.) (Lepidoptera: Lymantriidae), in southeast England. *Bot. J. Linn. Soc.* **101**, 69–78.

Strong, D.R., Lawton, J.H. and Southwood, R. (1984) *Insects on Plants. Community Patterns and Mechanisms*, Harvard University Press.

Sutherland, O.R.W. (1969) The role of the host plant in the production of winged forms by two strains of the pea aphid, *Acyrtosiphon pisum*. *J. Insect Physiol.* **15**, 2179–2201.

Tallamy, D.W. (1983) Equilibrium biogeography and its application to insect host–parasite systems. *Amer. Natur.* **121**, 244–254.

Tallamy, D.W. and Raupp, M.J. (1991) *Phytochemical Induction by Herbivores*, John Wiley & Sons, New York.

Theobald, F.V. (1919) Insects of the sea buckthorn. *The Entomologist* **52**, 169–171. (In Schaefer, 1977.)

Thompson, W.R. (1946) *A catalogue of the parasites and predators of insect pests*, Sect. 1, Part 8. The Imperial Parasite Service, Belleville, Ontario, Canada.

Tuomi, J., Niemelä, P., Haukioja, E. et al. (1984) Nutrient stress: an explanation for anti-herbivore responses to defoliation. *Oecologia* **61**, 208–210.

US Congress, Office of Technology Assessment (OTA) (1993) *Harmful Non-Indigenous Species in the United States*, OTA-F-565, US Government Printing Office, Washington, DC. 391 PP.

Valentine, H.T., Wallner, W.E. and Wargo, P.M. (1983) Nutritional changes in host foliage during and after defoliation and their relation to the weight of gypsy moth pupae. *Oecologia* **57**, 298–302.

Van der Linde, R.J. and Voute, A.D. (1967) Das Auftreten des Goldafters (*Euproctis chrysorrhoea* L) in den Neiderlanden und der mogliche Einfluss der Nahrung auf die Schwankungen in der Populationsdichte. *Z. Angew. Ent.* **60**, 85–96. (In Schaefer, 1977.)

Wagner, T.L. (1977) Population ecology of the satin moth, *Stilpnotia salicis* L. (Lepidoptera: Lymantriidae). PhD Dissertation, University of Maine, Orono, ME.

Wagner, T.L. and Leonard, D.E. (1979a) The effects of parental and progeny diet on development, weight, and survival of pre-diapause larvae of the satin moth, *Leucoma salicis* (Lepidoptera: Lymnatriidae). *Can. Ent.* **111**, 721–729.

Wagner, T.L. and Leonard, D.E. (1979b) Aspects of mating, oviposition, and flight in the satin moth, *Leucoma salicis* (Lepidoptera: Lymantriidae). *Can. Ent.* **111**, 833–840.

Wagner, T.L. and Leonard, D.E. (1980) Mortality factors of satin moth, *Leucoma salicis* (Lep.: Lymantriidae) in aspen forests in Maine. *Entomophaga* **25**, 7–16.

Wallner, W.E. and Walton, G.S. (1979) Host defoliation: a possible determinant of gypsy moth population quality. *Ann. Entomol. Soc. Amer.* **72**, 62–67.

Wallner, W.E. (1989) Overview of pest Lymantriidae of North America. In *Proceeding. Lymantriidae: A Comparison of Features of New and Old World Tussock Moths* (Tech. Coords W.E. Wallner and K.A. McManus) pp. 65–79, USDA Northeast Forest Experiment Station. Gen. Tech. Rep. NE-123, Broomall, PA.

Werner, R.A. (1979) Influence of host foliage on development, survival, fecundity, and oviposition of the spear-marked black moth, *Rheumaptera hastata*. *Can. Entomol.* **111**. 317–322.

West, B.K. (1993) *Orgyia antiqua* Ochs. (Lep.: Lymantiidae): voltinism. *Ent. Rec. and J. Var.* **105**, 241–242.

White, T.C.R. (1974) A hypothesis to explain outbreaks of looper caterpillars, with special reference to populations of *Selidosema suavis* in a plantation of *Pinus radiata* in New Zealand. *Oecologia* **16**, 279–302.

White, T.C.R. (1978) The importance of a relative shortage of food in animal ecology. *Oecologia* **33**, 71–86.

Wilson, C. and Graham, C.L. (1983) *Exotic Plant Pests and North American Agriculture*, Academic Press, New York.

Wolff, M. and Krausse, A. (1922) *Die forstlichen lepidopteren*, G. Fischer, Jena. 337 pp. (In Schaefer, 1977.)

Ziegler, J.R. (1976) Evolution of the migration response: emigration by *Tribolium* and the influence of age. *Evolution* **30**, 579–592.

Zimdahl, R.L. (1983) Where are the principal exotic weed pests? In *Exotic Plant Pests and North American Agriculture* (eds C. Wilson and C.L. Graham) pp. 183–217, Academic Press, New York.

THREATS TO FORESTRY BY INSECT PESTS IN EUROPE

Keith R. Day and Simon R. Leather

11.1 INTRODUCTION

A number of insects have achieved pest status in Europe over the last half century since plantation forestry increased in importance. The pest problems vary in severity from country to country but certain pests have had a European-wide impact, in particular the pine sawflies *Diprion pini* and *Neodiprion sertifer*, the pine beauty moth (*Panolis flammea*), the large pine weevil (*Hylobius abietis*), the green spruce aphid (*Elatobium abietinum*), the nun moth (*Lymantria monacha*) and the bark beetles *Ips typographus* and *Dendroctonus micans*. There are several other pests that have achieved local importance in some European countries but those listed above can be regarded as the most important of European forestry today. Some countries have had a number of devastating outbreaks of different pests over the last decade. Poland, where 20% of the country is forested (8.7 million ha and 80% of this in conifers), is an example (Figure 11.1). All nun moth outbreaks in Poland required control using aerial application of insecticides, and so important were they to the country that a stamp commemorating the events was issued. Pest problems in Poland were attributed to a long history of mismanagement, plantation establishment and increasing air and soil pollution from industrial sources (Capecki, 1982; Sierpinski, 1984) and although they emphasize the potential importance of insect pests, their severity and longevity are not typical of problems in most European countries.

In recent years, policy emphasis has been on the multifunctional management of forests to support their protection, development and exploitation. However, while the quality and quantity of timber products remain high priorities, increasing attention is being given to the wider role of insect communities in forest ecosystem functioning. Sometimes the role of insect pests is surprising and profound. *Zeiraphera diniana*, the larch budmoth, can apparently prevent succession and hold subalpine forests perpetually in a subclimax state by periodically suppressing the growth of cembran pine (*Pinus cembra*) in favour of the dominant European larch (*Larix decidua*). This appears to be a natural phenomenon (Baltensweiler, 1975). The extent to which foresters balance the negative effects of insects on timber production with the acceptance of their natural roles hinges crucially on their actual economic costs. More than ever before, modern forestry is driven by the relationship between costs and benefits; where the costs of pests are significant, the prominence of pest species is often used as a measure of the success of silvicultural practices.

The threats from forest insect pests in Europe have a number of origins and causes, which are reviewed here. The intensive management and silviculture of forests with high economic value will almost inevitably dispose them to pest problems from time to time because insects will be exposed to novel combinations of abiotic and biotic condi-

Forests and Insects. Edited by A.D. Watt, N.E. Stork and M.D. Hunter. Published in 1997 by Chapman & Hall, London. ISBN 0 412 79110 2.

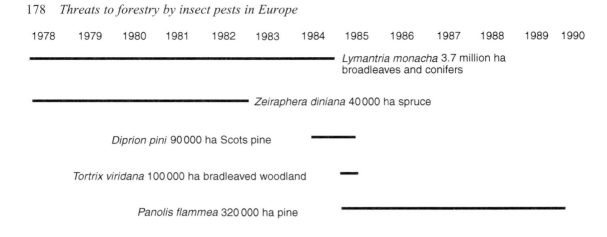

Figure 11.1 Outbreaks of forest insect pests in Poland since 1978.

tions in such forests, and we are rarely able to predict the consequences with any certainty. However, the extent of our knowledge of the effects of silvicultural practice continues to grow. The use of new tree species and the modification of genetic diversity of trees in forests also plays an important role in generating pest problems. The genetic differentiation of pest species which enables the establishment of new ecological relationships with their hosts is a cause for concern, while the movement of pests between regions which leads to extensions of their range can also create threats, but in principle may be easier to predict and regulate. Where pest species are particularly responsive to silvicultural practice there is cause for greater optimism, since a sound ecological understanding of the pests may enable sustainable intervention to form a part of the multifunctional management strategy. The success of this is often underpinned by the development of effective risk estimation systems.

11.2 CHARACTERISTICS OF EUROPEAN FORESTS AND THEIR INSECT PESTS

The extent of forested areas differs greatly between European regions, ranging from 6% of total land area (Ireland) and 12% (UK) to 70% (Sweden) and 80% (Finland). In the European Community 34% of land is forested (150 million ha), though this is still a small proportion of the forest on continental Europe as a whole – 900 million ha in the former Commonwealth of Independent States (CIS), for example (Wainhouse, 1987).

The conifer species in Europe are largely pines and spruces, with a greater representation of larch further east in the former CIS; and the broadleaves are largely oak and beech, with birch being particularly well represented in Scandinavia (Figure 11.2). In central Europe, nine conifer species and 12 broadleaved genera predominate among forest trees and together host around 500 species of insect pests (Table 11.1) (Klimetzek, 1993). Klimetzek (1993) defines pests in the broadest sense as insect herbivores that inflict some definite and sustained injury to the tree. Pine and spruce alone support about 100 species. A fifth of the forest tree pests are generalists; 60 species feed on both broadleaved trees and conifers and a further 15 and 30 species are confined to conifers and broadleaves, respectively. Two-thirds of the pest species are beetles and moths (Coleoptera and Lepidoptera) and the most significant pests of European forests in recent years have been from these orders.

The extent to which the range and abundance of a tree species determines the number of herbivorous insects, and indeed the number of pest species, associated with it has been the subject of several

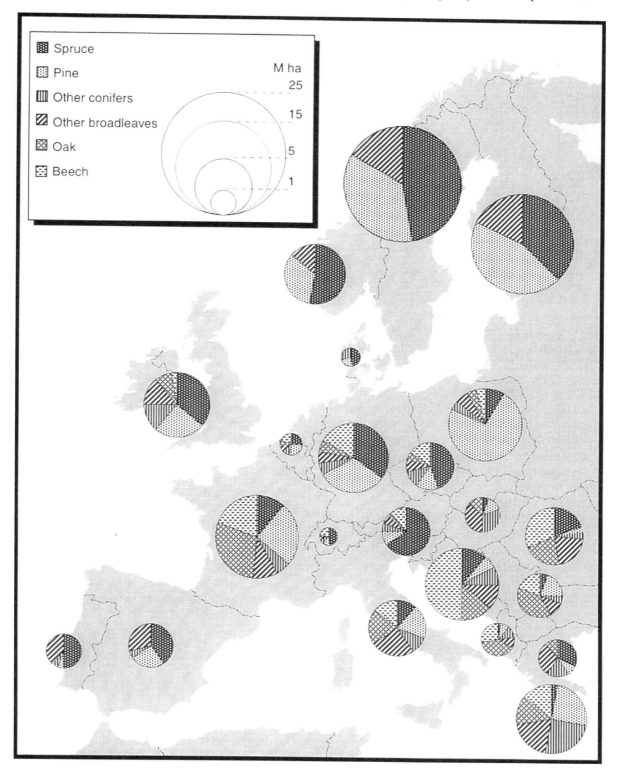

Figure 11.2 Main tree types and their area of occupancy in Europe. (Redrawn from Klimetzek, 1993.)

Table 11.1 Number of European pest insects associated with different tree genera (source: Klimetzek, 1993)

Conifers			Broadleaves		
Genus	Common name	Pests	Genus	Common name	Pests
Pinus	pine	183	*Quercus*	oak	136
Picea	spruce	154	*Populus*	poplar	133
Larix	larch	110	*Salix*	willow	124
Abies	fir	109	*Fagus*	beech	117
Pseudotsuga	Douglas fir	82	*Betula*	birch	110
			Alnus	alder	104
			Ulmus	elm	104
			Fraxinus	ash	101
			Acer	maple	97
			Castanea	chestnut	95
			Carpinus	hornbeam	95
			Tilia	lime	94

studies (Claridge and Evans, 1990) and is mentioned elsewhere in this volume. In general, and using Britain as a model, more widespread species of trees do not necessarily have higher numbers of pest species; there are many other factors, such as the presence of taxonomically related plants in the flora, that also dispose trees to the aquisition of new species. On the other hand, trees whose range and abundance in Europe has recently increased as a result of plantation have clearly acquired new pests, and this point will be developed further when discussing the use of exotic species in contemporary forestry.

Pest problems are often very localized. For example, France is one of Europe's largest countries with arguably the greatest range of forest climates and ecological diversity. In one year (1991) more than 5000 surveyed forests yielded records of damage by 49 pest insect species, but only three of these (*Thaumetopoea pityocampa*, *Pityogenes chalcographus* and *Ips typographus*) were considered a significant problem in more than 4% of the forests (Anon., 1991).

Some regional differences in the type of pest problems experienced are clear. Compared with Scandinavian countries, there have been few problems with bark beetles (Scolytidae) in the British Isles but a number of significant problems with defoliating insects. Where deciduous trees represent a large proportion of the forest area, it is the defoliating Lepidoptera that predominate. In a country like Romania the chief of these are species frequently associated with oaks (*Lymantria dispar* and *Tortrix viridana*), pests which in 1988 jointly attacked a maximum area of forest in that country amounting to 1.5 million ha (Teodorescu and Simionescu, 1994). These authors admit that the extent and duration of such attacks may have been exacerbated by pesticide use.

Trees are long-lived, even in intensively managed forests, and the threats posed by insect pests at different stages in the growth of an even-aged plantation might be expected to vary systematically. One generalized relationship suggests relative resistance to pests during the phase of highest current annual increment (Wainhouse, 1987), a period when tree 'vigour' might be expected to be highest. Whether or not a pest is a problem is a function of the tree's absolute size, its tolerance or constitutive resistance to attack and the abundance of the pests. With this in mind it is possible to envisage, for any particular tree species, a sequence of threats from pests ecologically pre-adapted to cause damage at different stages in a forest rotation (Figure 11.3). For the suite of pests known to be significant threats to Sitka spruce, for example, there are some species-characteristic affiliations with tree age or size, or both.

Elatobium abietinum may be found in equal numbers and producing the same results on trees at all ages, whereas *Hylobius abietis* causes greatest damage to transplanted seedlings because the weevil is attracted to and breeds in clearfelled forest and gains easy and frequent access to foliage close to ground level. In this case the susceptibility of seedling conifers does not necessarily require an explanation based on chemical or physical differences from older plants, though age-related defensive tactics are found in other insect–tree relationships (Wagner, 1988; Leather *et al.*, 1987; Steacy, 1993). Of the other main Sitka spruce pests, *Cinara pilicornis* (Carter and Maslen, 1982) and *Operophtera brumata* are associated with young trees prior to canopy closure, *Gilpinia hercyniae* with pole-stage trees (Billany, 1977) and *Dendroctonus micans* with mature or overmature individuals (Lempérière, 1994; Julien, 1995).

The effect of insects on seed crops of European forest trees represents a particular type of threat. Cone and seed insects received little attention until tree improvement programmes were initiated, but in today's conifer seed orchards they are considered to be of major importance. A range of insects are involved as seed predators of Europe's most important conifers: *Strobilomyia* cone flies, *Megastigmus* seed chalcids, *Dioryctria* cone pyralids, *Pissodes* cone weevils, cone tortricids and seed midges (e.g. *Resseliella*). In all, 184 insect species from six orders are recorded pests of Eurasian cones, and of these 75% are strictly cone-related (rather than feeding generally on conifer plant parts) and 47% are specific, at least to host genera (Turgeon *et al.*, 1994).

The strategy for development of European forests varies a great deal between regions. On mainland Europe the main task has been to convert a natural forest into a managed forest, whereas in Britain and Iceland, for example, it has been to re-establish forest where none has existed in recent times (Hummel, 1991), and to do so within clearly defined economic constraints. For this reason in part, Britain is unique in Europe in having such a large proportion of exotic plantation trees. In other European countries, forests have such an important part to play in watertable management or as a component of landscape (for avalanche regulation in Switzerland, for example) that their role as an economic source of timber could be considered subsidiary or negligible. These are important considerations, relevant to the following discussion because

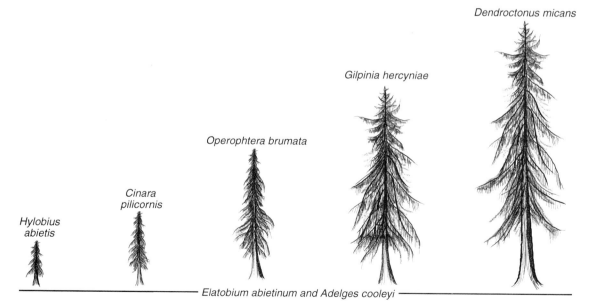

Figure 11.3 Temporal sequence of pest insect associations with Sitka spruce during a typical plantation rotation in the British Isles.

they set the thresholds that determine whether or not the activities of insects in forests can be considered as threats.

There are real difficulties in estimating the profitability of long-term enterprises like forest plantation projects which pivot on discounted cash flow analysis. Economic views in this area remain controversial (Kula, 1988) and forest enterprises in some parts of Europe are still considered economically marginal. Against this background the threat of losing even small levels of productivity to insect pests seems highly significant.

11.3 ASSESSMENT OF THE IMPACT OF PESTS

Insects damage trees in a variety of ways, some more obvious than others. To decide if insects pose an economic threat to forests we require a clear understanding of the loss of forest productivity, preferably integrated over more than one rotation in plantations, and in addition a knowledge of the relationship between the abundance of the pest and the change in tree or forest productivity so that a reliable threshold can be determined. The establishment of meaningful thresholds provides a firm basis for the implementation of effective integrated pest management strategies. Sadly, for most European forest pests little of this information is yet available.

Ips typographus, the spruce bark beetle, has probably been the most destructive pest of spruce in Europe. Between 1977 and 1981 almost 5 million m³ of timber trees were killed by *Ips* in Norway (Bakke, 1991) and 6 million m³ in Sweden (Eidmann, 1992). A similar timber volume was attacked by *Ips* following seasonal gales in 1990 in Belgium (de Proft *et al.*, 1992; Grégoire *et al.*, 1995). More than 50 million m³ have been killed in Europe as a whole since the nineteenth century (Grégoire *et al.*, 1995; Speight and Wainhouse, 1989). The overall economic loss is difficult to quantify since there is a differential cost of attack on trees of different harvested size categories and some of the timber can be salvaged. There is also a significant loss in quality of timber because the bark beetle vectors blue-stain fungi to the timber, but again this problem remains largely unquantified (Grégoire *et al.*, 1995). To put the problem into perspective, in Sweden the volume of trees killed between 1977 and 1981 represented about 20% of the total gross fellings of spruce during the same period (Eidmann, 1992). The fact that *Ips typographus* is a problem on a pan-European scale is illustrated by its current importance in Turkey (Keskinalemdar *et al.*, 1987).

Impact may seem minimal where trees attacked by bark beetles are mature and fungal staining does little to diminish the timber value, but even in these cases the need for further sanitation felling to reduce the spread of the problem has significant economic cost. The great spruce bark beetle (*Dendroctonus micans*) rarely causes mortality of trees in France. Lempérière (1994) surveyed 87 sites in the Limousin, 75% of which were infested by the great spruce bark beetle, but detailed studies of 62 000 trees within sites revealed attacks on only 2.1%. Its presence causes alarm because it attacks mature trees and although timber can be salvaged it pre-empts and compromises normal forest harvesting plans. This in itself is an economic consideration. In addition, *Dendroctonus micans* can significantly affect mature spruce forest whose principal value may lie in soil stabilization on steep slopes (Targamadze *et al.*, 1975).

The greatest cause of mortality of trees in replanted European forests is the pine weevil. *Hylobius abietis* is the only forest pest in Britain against which prophylactic treatment is habitually carried out (Speight and Wainhouse, 1989). Without such protection 30–100% of all tree crops planted on restocked sites in Britain would be lost (Heritage *et al.*, 1989). Normal levels of seedling mortality of Norway spruce in Denmark are 15–30% (Bejer, 1982), and in Finland 10% of Norway spruce and Scots pine (Långström, 1985) are regularly killed. Part of the problem lies in the scale of replanting and the type of felling that precedes it. Ireland faces particular problems in the next 10–20 years because large areas of relatively unproductive lodgepole pine are making way for other conifers, but at the same time pine stumps make particularly good breeding sites for weevils. We have estimated that costs of pesticide application to protect plants in 14 European countries may amount to £40 million per annum. From 1996 pesticide use will be prohibited in Swedish forests and other countries may follow suit, so the threat from the pine weevil may be one of Europe's biggest.

Mortality of trees following attacks by defoliators is a relatively rare event, but larvae of the pine beauty moth (*Panolis flammea*) consume new and old growth foliage and can quickly kill trees where their populations are sufficiently abundant. Death of established crops at several locations in Scotland since 1978 have followed attacks by *Panolis flammea* (Watt and Leather, 1988). The costs of such attacks have been estimated at £540/ha. Over 25 000 ha of pine forest have been treated with insecticide since 1976 (Aegerter, 1995) and as outbreaks are cyclical in nature, this may continue to be a problem for the forseeable future.

Sublethal effects of pests are even more difficult to quantify and the research investment in this area in Europe has often been less than in North America because of the scale and diversity of the pest problems. A common approach is to compare the growth of trees that are attacked by heavy infestations of defoliators with that of unattacked or less heavily attacked trees (Austarå, 1984; Day and McClean, 1991; Day *et al.*, 1991; Thomas and Miller, 1994), or to compare growth of the same trees during periods of high with periods of insignificant pest population density (Austarå *et al.*, 1987; Guryanova, 1988; Straw, 1996; Seaby and Mowat, 1993). These studies share all the problems of having pseudo-controls, but often represent the only available information where pests are unpredictable or controls on large trees are impractical to arrange.

An experimental approach involving simulated defoliation (Britton, 1988; Carter, 1977; Ericsson *et al.*, 1985) serves well to demonstrate the impact of natural defoliation, and lacks only the scale of the real thing and the possibility that insects may provide additional nutrient inputs, etc. which are not duplicated in short-term studies.

Leader extension (and consequently height growth of trees) is very sensitive to defoliation and is the first growth parameter to be affected (Carter and Nichols, 1988; Hayes and Britton, 1986; Seaby and Mowat, 1993). However, it is volume increment that is the most reliable guide to eventual economic productivity of a forest. Incremental growth often shows a delayed response to defoliation (Day and McClean, 1991; Ericsson *et al.*, 1985), again highlighting the need for careful long-term research that integrates changes due to pests over the life of the crop. Some of the studies that estimate increment losses due to European insect pests are listed in Table 11.2. Most refer to the highest levels of defoliation and suggest very significant losses, although the temporal and spatial patchiness of defoliation events such as these may have much less impact over the life of a complete rotation.

Extensive crown damage of larch by the larch budmoth can be seen clearly in the width of annual rings (Schweingruber, 1979) but the increment loss integrated over the long periods in which cycles of insect abundance take place is negligible. This is largely because larch is deciduous and even able to refoliate in the same season. Only 1% of larch standing timber volume is lost by mortality following *Zeiraphera* defoliation in the Alps (Baltensweiler and Rubli, 1984). In parts of Siberia, however, destruction of foliage twice in one season by *Zeiraphera* is thought to affect larch seed crops indirectly (Galkin, 1992).

The impacts of defoliators on incremental growth of broadleaved trees are rarely studied in the context of economic forestry, perhaps because it is assumed that the loss of early season deciduous foliage will be compensated by later season growth. In the Ukraine it is estimated that more than 3 m^3/ha of oak (*Quercus robur*) timber are lost each year to a variety of leaf-gnawing insect pests, and in years of heavy infestation this could reach as much as 10 m^3/ha (Avaramenko *et al.*, 1981). Interpretation of older data on oak growth following caterpillar defoliation suggests an average loss of 50% increment (Gradwell, 1974) but there remains the possibility of substantial compensation for moderate levels of defoliators with no apparent increment loss (Crawley, 1983).

Some insects are, by their nature, low threshold density forest pests. That is to say, the damage they cause devalues the tree beyond the usual measures of growth performance considered above.

Table 11.2 Some examples of studies of European forest insect defoliators and their impact on tree volume increment growth

Insect pest	Forest tree	Degree of defoliation	Source	Increment reduction %
Neodiprion sertifer	*Pinus sylvestris*	Complete (mature)	Baronio *et al.*, 1989	76
		Heavy	Austarå, *et al.*, 1987	33
	Pinus contorta	Complete	Britton, 1988	32
Zeiraphera diniana	*Larix decidua*	Complete (periodic)	Baltensweiler and Rubli, 1984	zero
	Pinus contorta	Heavy (current)	Day *et al.*, 1991	25
Epinotia nanana	*Picea abies*	Heavy (current)	Austarå, 1984	43
Elatobium abietinum	*Picea sitchensis*	Heavy (mature)	Day & McClean, 1991	50
		Heavy (mature)	Thomas and Miller, 1994	50
Tomicus piniperda	*Pinus sylvestris*	Moderate (25% shoots)	Ericsson *et al.*, 1985	16

Rhyacionia buoliana, the pine shoot moth, is a low density pest since it distorts shoots and modifies the growth form of the tree to the detriment of timber value (Haussler, 1990; Kletecka, 1992). Insects that damage plantations of Christmas trees (Donaubauer, 1993) can also be tolerated only at very low levels because the value of the forest product is unusually high. The same is more or less true for the European forest stands that are seed resources.

There are more than 1000 ha of productive seed orchards and 140 000 ha of selected seed stands in Europe. Insect impact seriously affects the number and quality of seeds that can be produced (e.g. Douglas fir). The high value of seeds and the high capital and operating costs of seed production systems (the cost of producing 300 kg seed/ha over 20 years is around £2500/ha per annum) result in a generally low tolerance (economic threshold) to seed and cone pests. In seed orchards, insects often damage 80–100% of the crop, and they may severely reduce the natural regeneration of forests. A loss of 50–100% of the seed crop is common in forests of *Larix decidua*, and in some areas of Greece *Cupressus sempervirens* have very low seed recruitment partly due to insect damage (Turgeon *et al.*, 1994; Roques and Sun, 1995).

11.4 ORIGINS OF PEST PROBLEMS

Insects may reach damaging population levels even in relatively natural forest areas, but the progressive changes in European forest management practice and the quest for further rationalization, more economic returns on investment and standardization of timber products to meet industrial requirements have all contributed from time to time to more frequent, more extensive or completely new pest problems.

There are long records for some forest insect pests. Dendrometric studies give evidence of regular and spectacular infestations of the larch budmoth in the Swiss Valais since 1387, as far back as such records permit (Schweingruber, 1979), but it is clear that the 'problem' is neither attributable to forest management nor likely to be particularly serious in forestry terms (Table 11.2). Generally the more serious problems arise from silvicultural practice.

11.4.1 SITE FACTORS

Removal of forest litter from Bavarian pine forests (the equivalent of 10–20 kg N/ha per annum) was thought to be a major factor causing susceptibility to insects such as *Panolis flammea* and *Lymantria monacha*. Both species have been forest pests in Germany for at least two centuries, and historical records of outbreaks have enabled regional and ecological comparisons to be made (Klimetzek, 1990). Another factor responsible for a decline in pests in the Pfalz region has been the increased planting of hardwoods which grow as mixtures or understorey with pine. Historically, pure stands of pine were more susceptible to pests.

It is almost inevitable that the growing conditions experienced by planted trees will not always favour optimal growth and will sometimes favour pest development. A major research question for forest entomologists in recent years has been whether site factors reduce the defensive capacity of trees or directly improve fertility or survivorship of the insect herbivores. At a time when silvicultural strategies are still developing, it is difficult to avoid pest problems all the time. For example, *Pinus strobus* was recommended for introduction to Italy partly on the basis of its virtual immunity from the pine shoot moth *Rhyacionia buoliana*, but growing in poor, drained or dry soils the trees eventually suffered heavy infestations from this pest (Currado *et al.*, 1983).

Site factors are clearly instrumental in determining patterns of mass outbreaks of the web-spinning sawfly *Cephalcia abietis* in Austria (Schmutzenhofer, 1985). The sawfly causes severe defoliation to older Norway spruce in the central European uplands (Fuhrer and Fischer, 1991) and the problems have increased greatly since the 1970s in eastern Bavaria (Eichhorn and Pausch, 1986). Schmutzenhofer (1985) defines quite specific elevations, rainfall, snow coverage and mean January and July temperatures with which the main foci of attack are associated. These tend to be on lightly water-stressed hilltops. Clearly the pest population growth potential is greatest under these climatic conditions but there is no indication of the indirect role that these might have on plant condition and, hence, indirectly on the insects. In addition, it is of interest that the complex diapause development and highly variable voltinism of *Cephalcia abietis* play a part in the pest problem (Martinek, 1986; 1987). Generation time may be between one and four years and is partly determined by the time at which larvae drop from the tree prior to pupation; those dropping early tend to develop within one year and those dropping later in two years. When larger proportions of the population become univoltine, or the whole nymphal population from several cohorts emerges simultaneously, there is a greater likelihood of outbreaks being generated (Martinek, 1986), and in turn this may be linked to site conditions (Martinek, 1987).

In a similar way, temperature may affect the proportion of pupae of the pine processionary moth (*Thaumetopoea pityocampa*) in prolonged diapause and consequently the size of the emergent adult population. The caterpillars generated by high populations of adult moths severely damage a number of pine species in Mediterranean countries (Speight and Wainhouse, 1989).

As with site type, the effects of weather on pest eruptions may be profound, but it is often unclear if weather directly affects insect performance or if it modifies plant growth and condition or has a differential effect on natural enemies, or indeed, all of these things. A typical level of understanding is for the nun moth (*Lymantria monacha*) which reached outbreak levels during the 1970s in pine and spruce forests in Denmark, Sweden, Germany and Holland (Bejer, 1985). The simultaneous nature of high population densities strongly suggested climatic factors as a dominant release (probably a sequence of hot summers preceded by low spring temperatures), and although sandy soils were a prerequisite, neither local climate nor stand composition was sufficient to predict where the outbreaks actually occurred.

Outbreaks of *Neodiprion sertifer* on pines in the Nordic countries have covered an area of 140 000 ha since the problems became serious in the 1960s (Austarå *et al.*, 1983). Larsson and Tenow (1985) analysed the situation in southern Sweden and concluded that outbreaks were again limited to dry soils and triggered by summer drought. Here, however, the connection between drought and soil type and the physiological state of pine trees, and hence the value of needles as food for *N.sertifer* larvae, is better understood (Larsson *et al.*, 1993). Equally interestingly, the lack of inducible defence responses in the mature foliage of Scots pine may explain why sawfly outbreaks can continue for several years on the same stands (Larsson and Tenow, 1984; Niemelä *et al.*, 1991).

11.4.2 INDIGENOUS SPECIES BECOMING PESTS

Many insect species achieve pest status through changes in silvicultural practice. Two well documented British examples are the pine beauty moth

(*Panolis flammea*) and the winter moth (*Operophtera brumata*), both of which have risen to outbreak levels over the last two decades.

(a) Pine beauty moth (*Panolis flammea*) in Scotland

Panolis flammea is a noctuid moth, native to Britain, where for many years it has been recorded feeding on Scots pine (*Pinus sylvestris*). It has been a major pest of European forestry for several centuries but had never been a problem in Britain until the 1970s (Watt and Leather, 1988), despite being regularly recorded in routine pupal surveys for other pest species, albeit at very low levels (Barbour, 1987). The pine beauty moth has a univoltine life cycle, full details of which are given elsewhere (Watt and Leather, 1988). It is particularly significant that pupae overwinter in the soil/litter interface from August until April.

During the 1950s and 1960s the British Forestry Commission began large-scale plantings of the North American pine species *Pinus contorta*, which was felt to be ideally suited to British silvicultural conditions (Lines, 1966). By the 1970s almost 20% of all trees being planted in Britain were *P. contorta*. Most of these plantings were in the north of Britain, particularly in the north of Scotland where sites tended to be on poorly drained deep peat, regarded as unsuitable for other tree crops. Early establishment appeared to be good, but subsequent silvicultural problems manifested themselves, namely basal sweep in wet soils and snow break in areas subject to heavy snowfall. In 1973, larvae of *P. flammea* were first noticed in a plantation in northern Scotland and the following year an outbreak ensued, to be followed by several further outbreaks in other forest blocks in Scotland requiring large-scale aerial applications of insecticide (Stoakley, 1979). Most outbreaks were confined to the north of Scotland but two outbreaks occurred further south (Watt and Leather, 1988). Outbreaks were linked by the fact that they were associated with *P. contorta* growing largely in monoculture and on poor soil types, mainly deep unflushed peat, though outbreaks did spread to trees growing on other soil types as they progressed (Watt *et al.*, 1989). Foliar and tree growth analyses indicated that the trees growing in these sites were nutritionally stressed and of a lower yield than trees growing in unattacked sites. It was hypothesized that the trees growing on deep unflushed peat sites were stressed *sensu* White (1969, 1974) and that they were therefore more susceptible to insect attack than other less stressed trees (Stoakley, 1979). Subsequent outbreaks, approximately seven years later, again on deep peat sites and on several of the sites previously affected, appeared to support this hypothesis (Barnett, 1987). However, no evidence for tree stress playing a part in the outbreak dynamics was found in either laboratory (Watt, 1986) or field (Leather, 1993) experiments. Rather, it was found that the native host *P. sylvestris* was actually a much better host plant in terms of larval nutrition and survival (Watt, 1989) but that the relative absence of natural enemies (Watt, 1989), in particular predatory beetles (Walsh *et al.*, 1993) and parasitic wasps (Aegerter, 1995), in lodgepole pine forests resulted in the outbreaks seen. Thus, a native species attained pest status through inappropriate silvicultural practice, in this case the planting of an alien crop in virtual monoculture in areas where natural enemies were scarce. Recommendations to prevent further occurrences of this nature have been made in terms of mixed planting and natural enemy enhancing features such as increased floral diversity in rides (Leather, 1992; Leather and Knight, 1997).

(b) Winter moth (*Operophtera brumata*) – a new pest of Sitka spruce in Britain

The winter moth (*O. brumata*) has long been known as a polyphagous pest of broadleaved trees in Britain (Browne, 1968) and has formed the basis of several classic studies in ecology (Varley *et al.*, 1973) as well as being noted as a severe pest of orchards in those countries where it has been introduced (Embree, 1965; Roland, 1990). In Britain the winter moth is commonly seen defoliating oak trees over wide areas in some years, but has not been regarded as a serious pest of British forestry until recently. During the 1950s and early 1960s defoliation of the introduced conifer, Sitka spruce (*Picea sitchensis*), by *O. brumata* was noticed on more than one occasion in northern Scotland. These events were thought to be overspills from outbreaks on heather (*Calluna vulgaris*), a known host of this species which dominated the ground vegetation at these sites (Stoakley, 1985).

More recently, however, widespread damage of Sitka spruce by winter moth has occurred in south and west Scotland (Stoakley, 1985) as well as minor sporadic damage to Sitka spruce crops in northern England (S.R. Leather personal observation). It has been conclusively shown that the latter outbreaks are being caused by winter moth using Sitka spruce as their sole host, completing their entire life cycle on this plant, rather than migrating there when alternative food supplies are exhausted (Stoakley, 1985; Watt and McFarlane, 1991; Watt *et al.*, 1992). The most damaging effects are seen on relatively young trees (thicket stage crops), though the larvae can be found on all growth stages of the crop. On Sitka spruce, winter moth damage is confined to the current year shoots, and in heavy infestations these can be totally destroyed and result in stunting and deformation of the crown and stem.

So far the patterns of winter moth outbreaks remain unexplained. Research has indicated that the outbreaks of winter moth on Sitka spruce are not influenced by tree-nutrient deficiencies, between-tree differences in budburst phenology or differential pupal predation (Hunter *et al.*, 1991). Explanations for winter moth outbreaks on heather which were considered by Kerslake *et al.* (1996) include the existence of parasitoid-free space at high-altitude moorland sites, and the possibility that nutritional adaptations may have recently evolved in winter moth, improving its ability to utilize 'new' hosts such as *Calluna* or *Picea*. In many respects spruce is a suboptimal host for the moth; larvae continue to perform better on oak (Fraser, 1995) and egg positioning on spruce appears to be maladaptive (Watt *et al.*, 1992). However, positive factors are that egg hatch is closely synchronized with budburst (Watt and McFarlane, 1991), and that ballooning first instar larvae are able to disperse widely and increase their exposure to spruce. While there is no evidence yet that *O. brumata* has adapted to spruce, it is certain that the relatively recent increase in range of Sitka spruce has influenced the problem and this is a point we pursue later.

11.4.3 INSECT RANGE EXTENSIONS

Pest problems often originate when insects are able to extend their range. In general, insects that attack seeds and foliage are thought to spread most easily (Bejer, 1981), whereas insects associated with bark and timber have been imported to remoter zones only through trade routes. Insect pests have moved with their hosts during palaeoecological range extensions; for example, there is evidence that *Ips typographus* has followed the biogeography of *Picea abies* since the last glaciation (Stauffer *et al.*, 1992).

Cephalcia lariciphila, the web-spinning larch sawfly, became apparent in Britain for the first time in the 1950s and caused repeated defoliation in plantations of Japanese larch. Like the introduced exotic *Gilpinia hercyniae*, the European spruce sawfly, its spread was followed by natural enemies which successfully regulate populations at low densities (Speight and Wainhouse, 1989). The significance that 'enemy-free space' may have for the threat posed by forest pests echoes the examples of other British pest problems given in the previous section. In quite recent time the root aphids *Pachypappa vesicalis* and *Pachypappella lactea* are species that have extended their European ranges to the British Isles where spruce plantations in suitable soil conditions have provided ideal opportunities for establishment (Carter and Daniellson, 1991).

Dendroctonus micans, the great spruce bark beetle, is a species whose origins are thought to be in Siberia. Spreading further west, it caused widespread damage to plantations of *Picea orientalis* in Georgia in the 1960s, subsequently becoming established in the Carpathian mountains of Romania. Throughout this century there had been records from Scandinavia but it had posed no major problems. In western European countries its spread has followed increasing planting of spruce – particularly in France, where its development as a forest pest is described by Lempérière (1994). The fact that the infested regions in France and Italy (Battisti, 1984) are outside the natural range of *Picea* is crucial; the origins of the problems are based on an extension of the insect's range combined with factors increasing tree susceptibility such as age, certain silvicultural operations, and factors making life for spruce more precarious, such as summer drought. In northern Europe things are a little different since the susceptibility of trees depends more on their weakening by the preceding winter's weather (Bejer, 1988).

The trade-aided movement of *D. micans* to Britain in 1973 was heralded by a risk assessment of the major bark beetles worldwide (Marchant and Borden, 1976), though the beetle was not discovered until 1982. The establishment, monitoring and control of the great spruce bark beetle in Britain is reviewed in Fielding *et al.* (1991), who emphasize the important role played by unrestricted movement of unbarked spruce timber in its spread. Hot summers are significant because they weaken potential host trees through water stress and promote further local dispersion of adults, which fly at the relatively high temperature threshold of 23°C (Fielding *et al.*, 1991).

The budworm *Choristoneura murinana* has been steadily extending its range in a southerly direction from central Europe. For a long time, the insect was thought to be monophagous on silver fir (*Abies alba*) and in recent years it has been the most damaging pest of this tree in France and Italy (Du Merle *et al.*, 1990, 1991), where planting of the host tree has increased. However, new *Abies* species replacing *A. alba* in Greece and the north African conifer, *Cedrus atlantica*, planted in southern France have also become host species. *C. murinana* seems to be quite a generalist on Abietoideae (Pinaceae) and the consequences of the increasing use of exotic Abietoideae species of this subfamily in Europe are that the host range of *C. murinana* is likely to increase further (Du Merle *et al.*, 1992). *Abies alba* is also attacked by the adelgid *Dreyfusia nordmannianae* (Demolis *et al.*, 1991; Binazzi and Covessi, 1991). In its natural range from the Caucasus to Turkey this insect has a complex life history on Nordmann fir (*A. nordmanniana*) and *Picea orientalis* but in central Europe it has become parthenogenetically adapted to the secondary host, silver fir (Alles, 1994). Silver fir grown in the British Isles is so debilitated by the adelgid that it is no longer considered a forestry prospect, but the reasons for this acute level of susceptibility are unknown.

Even species thought to be endemic are with further insight now suspected of having extended their ranges before becoming forest pests. The pine bast scale *Matsucoccus josephi* was discovered in Israel in 1933 where it was thought to be native to Aleppo pine (*Pinus halepensis*). By the late 1980s the scale was infesting all major pine forests in Israel and is now a severe threat to Aleppo pine in all Mediterranean countries. In fact the scale probably originated in Cyprus where it feeds on, but causes no injury to, brutia pine (*Pinus brutia* subsp. *brutia*) (Mendel, 1992).

The risks of introducing exotic pests to Europe and redistributing some that do not yet have pan-European status are sufficiently serious to warrant legislation. In 1977, a directive was issued protecting the European Community from entry of pests and diseases of plants (Phillips, 1978). The great dangers from bark beetles require restrictions on the movement of timber products within Europe, and the UK for example has Protected Zone status against the bark beetle *Ips typographus* (Fielding *et al.*, 1994). Vigilance is maintained against the importation of insect species from North America, though historically only 17 species of Nearctic insects have become established in European forests, compared with a great many more establishments in the other direction (Väisänen, 1995).

11.5 PESTS, EXOTIC TREE SPECIES AND GENETIC DIVERSITY OF TREES

The establishment and cultivation of exotic forest tree species has become a well established strategy in Europe where the performance of native species on difficult and often new forest sites can be exceeded by a well adapted exotic counterpart. For a species like Sitka spruce in Britain, the maximum productivity (a maximum mean annual timber increment of 24 m^3/ha per annum: Malcolm, 1987) even exceeds that recorded in its native North

America. It is no surprise, then, that Sitka spruce represents more than 70% of current forestry planting in the British Isles. Evans (1987) has reviewed the occurrence of phytophagous insect species on this plant since its introduction to Britain some 160 years ago. The 90 or so species are considerably more than would be predicted from the temporal history of the tree in these islands, yet rather less than might be anticipated on the basis of area planted. Evans (1987) concludes that Sitka spruce is likely to continue to accumulate more species in accordance with its area of occupancy. The same range of species and an additional 10 are currently found on Sitka spruce elsewhere in Europe, reflecting substantial transfaunation from native Norway spruce. In general, conifer genera and major geographical regions share a low proportion of pest species (Klimetzek, 1992) and there is therefore a high probability that new European Sitka spruce pests will orginate from *Picea* faunas within continental Europe. Of the sap-sucking guild of aphids on Sitka spruce in Britain, only two species (*Adelges cooleyi* and *Pineus similis*) originate from North America, whereas 11 have transferred from *Picea abies* (Carter, 1990). However, there are still possibilities of transfaunation over significant taxonomic boundaries and geographical distances. For example, *Operophtera brumata* and *Orgyia antiqua* are native insects that have attained pest status on Sitka spruce within the last 20 years (Pinder and Hayes, 1986; Stoakley, 1985) but were formerly associated with hardwoods. *Pissodes strobi* is one of six North American species that could be real threats if accidentally introduced to Sitka spruce in Europe (Evans, 1987; Alfaro, 1989). The expectation of pest problems may be different for exotic hardwoods introduced to Europe, since the evidence is for many more hardwood genera to share pest species (Klimetzek, 1992).

The quest for better quality timber and greater production in forests has led inevitably to the selection of a narrower genetic definition for forest planting stock (Lee, 1990; Mason, 1985). Such is the long-term nature of forestry that estimates of productivity must be made at an early stage from available genotypes (Lee, 1986) and relatively little attention has been given in Europe to the consequences of the lower genetic diversity now found in most modern conifer plantations (Muller-Starck and Zeihe, 1992). Care will always be taken to spread the risk among a number of high-performing genotypes; nevertheless, research is urgently needed to minimize the opportunities open to pests in future. Some resistance characteristics have been realized in near-mature crops (for example, variation in lignification among trees and its relevance to defence against the great spruce bark beetle: Wainhouse *et al.*, 1990) but an uninformed choice of genotypes now, based solely upon insect-free growth performance, may put problems in place for many years hence.

In addition to reducing genetic diversity of trees, plantation monoculture has also changed the degree of genetic patchiness found in natural forest stands. Genetic mosaics in forest stands may result from disturbance so that when gaps appear they allow the establishment of genetically related groups of seedlings (Jones, 1945). Alternatively, a family structure of groups of trees may appear as a consequence of naturally patchy seed dispersal (Linhart, 1989). Either way, the probable outcome is that pest attacks will also follow this genetic mosaic. There is plenty of evidence to suggest that the regional differentiation of tree chemistry in natural forests will provide profound natural barriers to phytophagous insects (Cates and Zou, 1990; Forrest, 1987). Spatial homogenization of genetic structure in European plantation forests of the future may have profound effects on the appearance of pest outbreaks, which may become evident over wider areas rather than in spatially discrete zones.

The variable responses of insects to trees of different provenance (seed origins) are widely recognized. The opportunity to study this level of variation and its consequences for the threat of pests has been afforded by the establishment of seed origin trials for exotic North American species such as *Pinus contorta* and *Picea sitchensis*. Replicated forest field trials of trees grown from seed collected at known North American sites (provenances) has enabled the growth and performance of trees in a wide variety of European climates to be carefully evaluated (Lines, 1976; Ying and McKnight,

1993). Insect infestations have been examined on some of these field trials where the opportunity has arisen.

One example is found in the severity of damage to canopies of lodgepole pine by *Zeiraphera diniana*. A total of 36 provenances planted on deep peat at Shin Forest, Scotland, were examined after an outbreak of the moth that lasted two years, when the trees were 10 years old. Damage was strongly related to tree provenance (Day *et al.*, 1991). The most severely damaged provenances had been included among those recommended for British sites (Lines, 1976). Despite the variability and complex differentiation of lodgepole pine in its natural range (Rehfeldt, 1986), it was found that shoot oleoresin profiles of lodgepole pine corresponded well with damage levels (Figure 11.4) and provided a chemical marker for resistance should this particular forest threat warrant attention in future tree breeding programmes. However, the replication of provenance demonstrations and their usefulness in highlighting possible plant resistance should not dull our appreciation of genotype–environment interactions (Quiring and Butterworth, 1994) which could conspire to alter the expression of resistance in different geographical regions or even from one forest or year to another. Furthermore, it seems that even insects from the same order may not respond in the same way to a similar range of provenances. Patterns of damage caused by *Zeiraphera diniana* among lodgepole pine provenances may not be mirrored by damage caused by the European pine shoot moth *Rhyacionia buoliana* (Esbjerg and Feilberg, 1971) or the pine beauty moth, *Panolis flammea* (Leather, 1985; Steacy, 1993). Other demonstrations of differences in the extent of damage caused by pests to trees of different provenance include European provenances of *Larix decidua* which are differentially susceptible to the larch gall midge *Dasyneura laricis* (Šindelar and Hochmut, 1972).

Special problems may be experienced with exotic or relocated plant species when they are brought into contact with pre-adapted insect species with which they have no coevolutionary history. *Larix kaempferi* (= *leptolepis* Gord.) is significantly more susceptible to infestation by the larch case-bearer (*Coleophora laricella* Hbn.) than European provenances of *Larix decidua* (Šindelar and Hochmut, 1972), while *Pinus contorta* is preferred by the pine shoot moth *Rhyacionia buoliana* Schiff. to Austrian pine (*P. nigra* Arnold) and Scots pine (*P. sylvestris*) (Esbjerg and Feilberg, 1971).

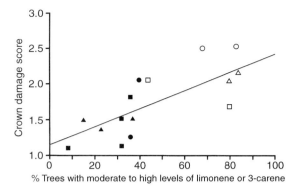

Figure 11.4 Relationship between an index of damage by the moth *Zeiraphera diniana* to the crowns of lodgepole pine (*Pinus contorta*) of different provenance planted in Scotland, and the patterns of shoot monoterpenes associated with each provenance (from Day *et al.*, 1991). Provenances are also assigned to broad regional groups, depending on the geographical origin of the seed.

A graphic case of this is the problem experienced throughout northern and western Europe with the green spruce aphid *Elatobium abietinum*. The potential effects of the insect on forest growth have been mentioned earlier, and the pest problem originates for two principal reasons. The aphid probably had a natural distribution in central and northern Europe which corresponded with the range of Norway spruce, *Picea abies*, but it has become established in more maritime regions of western Europe with the progressive use of spruce species in new forests. In central and eastern Europe the spruce aphid is holocyclic (Kloft *et al.*, 1964) and can resist extreme low temperatures, but in maritime winter climates the temperatures are mild enough to permit anholocyclic populations to reproduce parthenogenetically throughout the winter on the needle leaves of dormant shoots. The different life cycles are not the result of racial differences but of climate (Carter and Austarå, 1994). When warmer spring temperatures coincide with high nutrient levels in spruce shoots prior to budbreak, the normally larger anholocyclic population is able to increase exponentially (Day, 1984a, 1986) whereas a small holocyclic population hatches from overwintered eggs but some time elapses before adults appear and growth becomes exponential. By this time, bud flushing signals a reduced nutrient supply and a window of opportunity for holocyclic spruce aphids is missed. Outbreaks of the aphid are consequently rare in continental Europe (Bejer-Petersen, 1962). In western Europe, the dynamics of annual peak population levels are driven by a suite of winter temperature conditions which, if severe enough, kill aphids outright or do not permit them to develop or reproduce sufficiently rapidly. These are superimposed on overcompensating density-dependent processes affecting populations from one year to the next (Day and Crute, 1990; Thacker, personal communication).

The second reason for the aphid's pest status is the susceptibility of Sitka spruce. Among the range of world *Picea* species on which the performance of the aphid was measured, *P. sitchensis* was one of those providing the greatest potential for aphid population growth (Carter and Nichols, 1985; Nichols, 1987). Comparative population densities on the original host, *P. abies*, are never as high and consequently Norway spruce rarely suffers defoliation by aphids. The host preferences of *Elatobium* are further refined within a species to provenance; earlier studies suggested quite significant differences in the peak population densities reached on a range of Sitka spruce provenances (Day, 1984b) and further work showed differences in the capacity of provenances to recover from attack (Carter and Nichols, 1988). Recent research (Armour, 1995) indicates that British Columbian provenances, favoured by silviculturalists in the British Isles, have a strong affinity for aphids, but the highest rates of aphid growth occurred on Alaskan provenances which are favoured for the developing forest industry in Iceland and which may now explain some of the aphid problems that have been experienced there.

11.6 FOREST INSECT BIOTYPES

Some insect species have extended their range of influence in European forests by exploiting genetic variability in source populations. Larch and pine forms of the budmoth *Zeiraphera diniana* exist in central Europe as host-associated biotypes (Baltensweiler, 1993a). Chief differences between them include larval colouration (Baltensweiler, 1977), female pheromone blend (Guerin *et al.*, 1984), diapause and egg hatch phenology, and larval survival on alternative hosts (Day, 1984c). Phenology and pheromones greatly reduce gene flow and heighten genetic differentiation between populations on larch and pine in many parts of subalpine Europe.

European larch in zones between 1400 and 2000 m above sea level experiences regular and severe defoliation (Baltensweiler *et al.*, 1969) as a result of cyclic fluctuations in populations of the larch biotope of the budmoth. Cembran pine is only damaged when larch-form caterpillars exhaust their original food supply (Baltensweiler, 1975) but the pine host is never defoliated by pine-form caterpillars. By contrast, outside the alpine zones and particularly towards the north and western fringes of Europe it is damage to pine or spruce that is most significant, and it is quite clearly the pine form of the moth that is responsible (Day *et al.*, 1991) while the sympatric larch form of the moth usually remains at low population levels (Day, 1984c).

Zeiraphera diniana is an insect whose eggs require long periods at low temperature for the elimination of diapause (Bassand, 1965; Baltensweiler, 1993b). Mild winter temperature regimes result in inappropriate hatching phenology of eggs and failure of the larvae to coincide with new foliage of hosts. Not surprisingly its distribution in the British Isles is confined to higher altitudes and latitudes in the east (Day, 1984c). In these areas significant damage has been caused to young plantations of lodgepole pine (Day, 1984c, 1994). In general terms, alpine winters favour the development of the larch form and higher latitude winters with greater maritime influence favour the development of the pine form. Underpinning these patterns is the regional availability of host plants. Spruce is damaged regularly by *Zeiraphera diniana* in Poland, eastern Germany and the Czech Republic, where it is one of the major forest species at appropriate altitudes (Cwiklinski and Grodzki, 1989; Theile, 1967; Ždárek *et al.*, 1985). So it is the differences in winter temperature and consequent differences in the phenological relationships of the insect with its host plants that define areas where either of the biotypes will prosper and where different conifer species will be threatened.

Recent research by Emelianov *et al.* (1995) demonstrates genetic differentiation between some populations of the pine and larch budmoth biotypes. They were studied using 24 allozyme loci, some of which clearly separated the two forms regardless of sample locality. The genetic evidence suggests that these might be host races that hybridize, and this would be supported by earlier data on colour morph composition of edge-of-range populations on larch. Caterpillars on larch are normally predominantly dark and those on pine are light in colour; the system generally used to categorize larval colouration has been developed by Baltensweiler (1977). When this was applied to populations on larch in Britain the colour forms were found to be more frequently intermediate, raising the possibility that there is a strong tendency for introgression between typical larch and pine form morphs (Day, 1984c). Further evidence for this comes from studies on sex pheromone polymorphism (Priesner and Baltensweiler, 1987). There is certainly every opportunity for this because larch and pine are frequently associated and the more maritime climate removes some of the temporal isolation between host-specific biotypes. Paradoxically, such introgression might introduce greater variability into populations on some host plants, particularly on larch, and allow greater persistence in a less predictable edge-of-range environment through 'spreading of risk'. Further work on the extent of genetic differentiation between populations on different host plants throughout Europe may or may not confirm this hypothesis. *Zeiraphera diniana* can cause considerable damage to a range of European and exotic conifers in the genera *Larix, Picea* and *Pinus* and at the very least illustrates the options created by taxonomic differentiation that is accompanied by host-association, and the contingent increased threat to forests over a range of climatic zones.

More dramatic instances of the recent evolution and spread of aggressive forest pests are possible. A recent case of an insect pest apparently reinventing itself is the Asian form of the gypsy moth (*Lymantria dispar*) (Schaefer *et al.*, 1984; Winter and Evans, 1994). Females of this form may fly significant distances (3–5 km) and are attracted to light, which at transport depots provides good dispersal opportunities for their egg masses. On trees, the egg masses tend to be deposited on leaves rather than trunks, which means that eggs are eventually protected by snow overwinter (Izhevskii, 1992). The larvae also appear to have a wider host range (some 600 species). The progress of the Asian gypsy moth has been followed with caution in North America (Montgomery, 1995) where previously introduced gypsy moths have an extremely low level of genetic variability. It remains to be seen whether the Asian form has a future in western Europe, where populations are already known to be genetically diverse (Harrison *et al.*, 1983). A threat might be particularly significant in the cork-growing regions of southern Europe, where gypsy moth is already a periodic pest (Luciano *et al.*, 1982).

The existence and significance of biotypes for a wide range of insects are well known (Diehl and Bush, 1984) and should not be ignored in relation to forest insect pests which often have wide geographical distributions and relationships with long-lived hosts. Under such circumstances the origin of allopatric biotypes may have far reaching consequences for the future of European forests.

11.7 PESTS PARTICULARLY RESPONSIVE TO FOREST MANAGEMENT OPERATIONS

Aspects of silviculture involving the harvest of forest products, the recovery from environmental events which destabilize trees and the reforestation of harvested areas present quite critical problems in relation to forest pests. Some insects, mainly those depending on living bark, derive particular benefits when trees are felled or fall through wind or snow damage; the immediate effects of thinning may be negligible and those of windthrow and clearcuts are frequently drastic (Eidmann, 1985) and all will depend upon the timing and area of the events and the amount and distribution of stumps, slash and timber. Contributory to pest development will be post-harvest activities such as site treatments accompanying reforestation and whether timber is left on site.

Windthrow may be a sporadic phenomenon in natural forests but in some areas its effects have been exacerbated by plantation silviculture. In Northern Ireland, for example, newly forested upland areas of first-rotation spruce were often grown on poorly drained peat, where root development was lateral and the maturing trees became very unstable. These plantations were often at risk from high winds and extensive windthrow was the result. The trees were difficult to harvest and it is fortunate that aggressive species of Scolytidae are hitherto absent from the island of Ireland and that significant pest problems did not develop subsequently. Immunity from pest problems following storm damage has not been the case elsewhere.

The spruce bark beetle *Ips typographus* is the most destructive bark beetle in the coniferous forests of Eurasia and good accounts of its ecology and management are given in Berryman (1986) and Berryman and Stenseth (1989). Through pheromonal recruitment, populations are able to build up on damaged or senescent spruce and reach critical local population densities which can overcome the defences of even healthy trees. In effect an increase in beetle population density increases the number of susceptible trees in an area. This contrasts with less aggressive species like *Tomicus piniperda* which depend on dead or dying host material for reproduction, so outbreaks on living trees could not become self-perpetuating, even in severely weakened stands (Långström and Hellqvist, 1993).

Recent outbreaks of spruce bark beetle have all been preceded by combinations of storm damage to trees and drought which are both predisposed to some extent by site type (Eidmann, 1992). Dense or overmature spruce stands may be particularly susceptible, and large continuous areas of spruce are also more likely to be attacked. While it may be impractical to avoid planting trees on sites where there is some future risk of drought or storm damage, one practical way of avoiding major threats from *Ips typographus* has been sanitation to eliminate infested trees and logging debris. There are often forest areas where trees are windthrown on difficult terrain and sanitation has to be delayed. Norway spruce in Switzerland remains attractive for bark beetles for well over a year under such circumstances and if it is not possible to clear all the damaged wood in this period, the best strategy appears to be to deal with the smaller areas and scattered damage first (Forster, 1993).

A suite of additional management practices would include low thinning if a stand is overstocked, shelterwood cutting if a stand is unlikely to respond to thinning, and selective logging of spruce where other species might be encouraged on a dry site (Berryman, 1986). However, thinning must be undertaken sensitively. Opening stands by thinning may subject the remaining trees to sufficient stress to increase, rather than reduce, their susceptibility, or perhaps make them more vulnerable to windthrow (Eidmann, 1985). Construction of dams and ditches which may change forest water levels should also be given considerable thought. The example of the threat of *Ips typographus* emphasizes the importance of a very wide range of forest operations in the management process, from silviculture to the methods of cutting, storage and transport of timber.

Hylobius abietis is truly a pest of plantation forestry since its greatest effects are felt by planted seedlings on clearfelled sites. The weevil is not a significant problem for trees in thinned stands nor for naturally regenerated seedling trees in clearcuts, perhaps because such seedlings are so numerous that weevil-feeding on individual plants is negligible (Eidmann, 1985). The scale of clearfelling has increased substantially in Europe in recent years; a decade ago in Britain the rate of restocking was 6500 ha per annum (Low, 1985). The pine weevil is certain, therefore, to remain an important threat to forests in Europe for many years. Fortunately, and notwithstanding the wide variety of forest conditions where the problem occurs, there is a range of silvicultural operations to which the pest is sensitive. They are summarized in Wilson and Day (1996) and include factors associated with the suitability of breeding sites, factors affecting weevil development rate, the dimensions and treatment of the clearcut in which planting will occur, and the ability of the seedlings to deter or tolerate weevil attack. One factor will serve as an example of how silvicultural treatment of a site will affect damage levels. Removal of most or all of the vegetation on a replanted site by fire or extensive scarification will exacerbate damage because the weevils attracted to the site by the release of tree stump volatiles, and those recruited by breeding on stumps, will have only new seedlings to eat. Partial scarification in spots and stripes will preserve sufficient alternative plant types as food for adult weevils, while discouraging their approach to conifer seedlings over bare ground, so the effect will be less damage (Christiansen and Sandvik, 1974; Soderstrom, 1976; Turchinskaya, 1983). Measures such as these will need to be studied carefully and integrated at the local level with various forms of biological control and other means of insect population suppression or plant protection to minimize the use of pesticides in future.

11.8 THE RISKS TO FORESTS FROM PESTS

The threat of forest insect pests at a local scale can be determined by reference to risk or hazard rating systems. The advantage of recognizing a risk in this way is that action can often be taken in advance of a major problem developing, to minimize or nullify it. Risk rating systems also improve the precision of measures targeted on pest suppression, which give them environmental as well as economic advantages. The terms 'hazard' and 'risk' have been distinguished by Waters *et al.* (1985). **Hazard** is determined by factors (drought, forest age, etc.) that dispose a tree or stand to insect attack. Trees with even high hazard ratings may not be attacked if the pest is absent or rare, so **risk** is said to be a function of the presence and abundance of the pest. Forest insect hazard rating has been popular in North America, where it provides information for managers and proves useful in identifying or locating stands that warrant increased surveillance. It is the kind of approach that suits the scale of North American forestry (Hicks *et al.*, 1987).

Insect attack, or the disposition to insect damage, is the response variable of risk or hazard rating systems, and it may be quantified in various ways and with different levels of precision. It may be interpreted as the need to undertake control measures, it may be measured in discrete damage classes (high, medium and low risk, for example), or it may depend on a complex variable derived from a quantitative measure of damage sustained by a number of sampled trees. In this way the choice of analysis may be a classification model (Pendrel, 1990), discriminant analysis (Belanger *et al.*, 1981), or perhaps linear regression (Stage and Hamilton, 1981). This area has been reviewed by Berryman and Stark (1985). Some of the major reasons for a failure of risk/hazard systems to perform adequately are discussed by Bentz *et al.* (1993).

In Europe there have been a number of attempts to provide advance warning of forest pest problems, and they have been restricted to a few species of greatest economic importance. Znamenskii and Lyamtsev (1983) used discriminant analysis to determine criteria of oak stands that were most likely to indicate if they were reservoirs of gypsy moth populations. Populations of *Ips typographus* can be successfully monitored with pheromone traps, and this has been used as an indication of risk to standing trees in Denmark (Hubertz *et al.*, 1991) and in Norway, Sweden, Finland and Denmark (Weslien *et al.*, 1989). A simple hazard index was developed by Worrell (1983) for forests damaged by *Ips typographus* in Norway; important predictive factors were those variables associated with environmental disturbance (drought) and silviculture (stand composition, growth class) to which the spruce bark beetle is known to be sensitive, but whose quantitative interrelationships are unknown. Risk/hazard rating systems have a known predictive capacity but often a low level of biological understanding. It is therefore not surprising that alternative site characteristics can have equally good predictive value for the same pest; in another study Nef (1994) demonstrated that *Ips typographus* attack intensity on spruce was related to a combination of site factors, soil nutrients (P, C, Ca and Mg) and soil pH.

There are so many site characteristics that are known to affect damage by *Hylobius abietis* (Wilson *et al.*, 1996) and such a poor understanding of how they interact with the population dynamics and prolonged life cycle of the insect, that prospects for developing reliable risk/hazard rating systems may seem equally poor. However, the need for improved targeting of control measures provides a strong encouragement. Short-term forecasts of risk have been evaluated by exposing sections of pine branch to weevil attack and using this to predict seedling damage (Korczynski, 1984). Frequently, weak correlations of weevils sampled on-site with seedling damage (Wilson and Day, 1994) have suggested that predictions of damage might be more reliably tied to hazard criteria, but recent indices of weevil abundance obtained by artificial baits and pitfall traps (Nordlander, 1987) are evidence that risk might in future be distinguished from hazard. Two studies have so far attempted the task of predicting the risk/hazard of pine weevil damage. Nef and Minet (1992) derived an index of risk in 34 forest departments in Belgium from three criteria: damage frequency, an estimation of the importance of damage assessed by forest managers, and the proportion of the forest areas that was treated with weevil control measures. These jointly comprised the response variable whereas potential risk/hazard factors were drawn from information on the annual area of forest planted, two factors relating to the ecological zones where forests were planted, and five further climate or phenological factors. Through stepwise correlation Nef and Minet (1992) revealed that 65% of variation in the risk index could be explained by a temperature factor and the area of the forest planted. Patterns of *Hylobius abietis* damage in Belgium seem to have a strong geographical component which is rather easier to predict than in other parts of Europe, so the model is unlikely to be broadly applicable. In addition the model was unable to resolve risk at the scale of discrete forest plantations, so further work has been necessary.

In Northern Ireland systematic estimates of weevil damage to large samples of seedlings at more than 80 plantations of Sitka spruce were compared with potential risk/hazard variables drawn from a pool of more than 100 recorded at each site (Wilson *et al.*, 1996). A linear response model incorporating four terms explained 53% of the variation in damage at sites in one year. Although apparently low in predictive value, the model was based on a scale of damage assessment and a level of spatial resolution that was rather precise. Using a second year's data from 85 independent plantations, and relaxing the precision of damage prediction to broader, discrete classes, it was found that 52 sites were correctly classified. Fifteen per cent mortality of seedlings at a site is an economic threshold because at this level the whole site is generally replanted. It is of some practical interest, then, that 10 of the 82 sites were high risk in this way, and eight were identified by the model. We might conclude, therefore, that risk and hazard evaluation is a useful way of expressing the threat of forest pests at a local scale and is an important tool that will require development and updating for integrated pest management in forests in future.

11.9 CONCLUSIONS

In conclusion, it appears that forest insect pests in Europe are of major importance, and it is likely that the pressures placed upon the forest industry by restrictions in the use of pesticides Europe-wide, and the extensification of clonal forestry driven by the projection of increasing economic gains, will ensure that the current problems persist well into the next century. In addition, the experiences of forest entomologists in North America leave us in no doubt about the threats posed by accidental importation of key pest insect species from that continent.

ACKNOWLEDGEMENTS

The authors would like to thank the Forestry Commission (UK) and the Northern Ireland Forest

Service (Department of Agriculture) for many years of cooperation on studies that are mentioned in this review. Assistance has been given to KRD by grants from the European Community (AIR3 CT-94-1883) and the Leverhulme Trust.

Appendix 11A Major forest insect pests of Europe

Hemiptera

Adelges abietis (L.)	Pineapple gall woolly aphid
Adelges laricis Vall.	Larch woolly aphid
Aradus cinnamomoeus (Panz.)	Pine bark bug
Cinara piceae (Panz.)	Great black spruce bark aphid
Cinara pilicornis (Htg.)	Brown spruce shoot aphid
Cinara pineae (Mord.)	Large pine aphid
Cryptococcus fagisuga (Lundg.)	Felted beech coccus
Drepanosiphum platanoidis (Schr.)	Sycamore aphid
Dreyfusia piceae (Ratz.)	Balsam woolly aphid
Dreyfusia nordmannianae (Eck.)	Silver fir woolly aphid
Gilletteela cooleyi (Gill.)	Douglas fir woolly aphid
Elatobium abietinum (Walker)	Green spruce aphid
Eucallipterus tiliae (L.)	Lime aphid
Eulachnus agilis (Kalt.)	Spotted pine aphid
Kermes quercus L.	Oak scale or pox
Matsucoccus josephi (Bod. & Harp.)	Pine bast scale
Mindarus abietinus (Horv.)	Balsam twig aphid
Myzus cerasi L.	Cherry blackfly
Pachypappa vesicalis Koch.	Spruce root aphid
Pachypappella lactea (Tullg.)	Spruce root aphid
Phyllaphis fagi (L).	Beech woolly aphid
Pineus pini (Macq.)	Pine woolly aphid
Schizolachnus pineti (Fab.)	Grey pine needle aphid
Stagona pini (Burm.)	Pine root aphid

Lepidoptera

Bupalus piniaria L.	Pine looper moth
Choristoneura murinana (Hbn.)	Fir bud moth
Coleophora laricella Hbn.	Larch case bearer
Cydia conicolana Heyl.	Pine cone moth
Cydia fagiglandana Zell.	Beech seed moth
Cydia strobilella (L.)	Spruce cone tortrix
Dendrolimus pini (L.)	Pine lappet moth
Dioryctria abietella Schih.	Pine knothorn moth
Epinotia nanana (Treitsch.)	Dwarf spruce bell moth
Epinotia tedella (Clerck)	Spruce bell moth
Epirrita autumnata (Bkh.)	Autumnal moth
Euproctis chrysorrhoea (L.)	Browntail moth
Lymantria dispar L.	Gypsy moth
Lymantria monacha L.	Nun moth
Malacosoma neustria (L.)	Lackey moth
Operophtera brumata (L.)	Winter moth
Orgyia antiqua L.	Vapourer moth
Panolis flammea (Den. & Schiff.)	Pine beauty moth
Prays fraxinella Don.	Ash bud moth
Retinia resinella (L.)	Pine resin gall moth

Rhyacionia buoliana (Den. & Schiff.) — Pine shoot moth
Thaumetopoea pityocampa (Den. & Schiff.) — Pine processionary moth
Tortrix viridana (L.) — Green oak tortrix, oak leaf roller moth
Yponomeuta evonymella (L.) — Bird cherry ermine moth
Zeiraphera diniana (Gn.) — Grey larch tortrix, larch bud moth

Diptera
Dasyneura laricis (F. Loew) — Larch bud midge
Phytoagromya populicolia Haliday — Poplar leaf miner
Resseliella picea Seit. — Fir seed gall midge
Thecodiplosis brachyntera (Schwardt.) — Pine needle shortening gall midge

Hymenoptera
Andricus quercuscalicis (Burgs.) — Knopper gall wasp
Cephalcia abietis (L.) — Web-spinning fir sawfly
Cephalcia lariciphila (Wachtl.) — Web-spinning larch sawfly
Diprion pini L. — Large pine sawfly
Gilpinia hercyniae Htg. — European spruce sawfly
Megastigmus pinus Parfitt — Silver fir seed wasp
Megastigmus spermotrophus (Wachtl.) — Douglas fir seed wasp
Nematus melanaspis Htg. — Gregarious poplar sawfly
Nematus pavidus Lepelt. — Lesser willow sawfly
Nematus salicis L. — Large willow sawfly
Neodiprion lecontei (Fitch) — Redheaded pine sawfly
Neodiprion sertifer (Geoff.) — European pine sawfly, fox-coloured sawfly
Pristiphora abietina Christ. — Gregarious spruce sawfly
Pristiphora erichsonii (Htg.) — Large larch sawfly
Pristiphora testacea Jur. — Birch sawfly
Prestiphora wesmaeli Tisch. — Large sawfly
Sirex noctilio Fab. — Steely blue woodwasp
Urocerus gigas (L.) — Giant woodwasp

Coleoptera
Brachyonyx pineti (Payk.) — Pine needle feeding weevil
Chrysomela pipuli (L.) — Large red poplar leaf beetle
Dendroctonus micans Kugelaan — Great spruce bark beetle
Galerucella lineola F. — Brown willow beetle
Hylastes ater Payk. — Black pine beetle
Hylastes brunneus Er. — Black pine beetle
Hylastes cunicularis Er. — Black spruce beetle
Hylobius abietis (L.) — Large pine weevil
Hylobius pinastri (Gyll.) — Small pine weevil
Hylurgopinus rufipes (Eich.) — Canadian elm bark beetle
Ips acuminatus Gyll. — Engraver beetle
Ips cembrae Heer. — Large larch bark beetle
Ips typographus L. — Eight-toothed spruce bark beetle
Otiorhynchus singularis L. — Clay-coloured weevil
Phyllobius pyri L. — Common leaf weevil
Phyllodecta vitellinae L. — Brassy willow beetle
Phyllodecta vulgatissima L. — Blue willow beetle
Pissodes pini L. — Banded pine weevil
Pissodes validirostris Sahl. — Pine cone weevil
Pityogenes bidentatus Herbst. — Two-toothed pine beetle

Plagiodera versicolora (Laich.)	Broader willow leaf beetle
Rhynchaenus fagi L.	Beech leaf miner
Saperda populnea L.	Small polar longhorn beetle
Scolytus multistriatus Marsham	Small elm beetle
Scolytus scolytus Fab.	Large elm beetle
Tomicus minor Htg.	Lesser pine shoot beetle
Tomicus piniperda (L.)	Pine shoot beetle
Xyloterus lineatus (Ol.)	Conifer ambrosia beetle

REFERENCES

Aegerter, J.N. (1995) A three trophic level interaction: pines, pests and parasitoids. Unpublished D.Phil. Thesis, University of York.

Alfaro, R.I. (1989) Stem defects in Sitka spruce induced by Sitka spruce weevil, *Pissodes strobi* (Peck.). In *Insects Affecting Reforestation: Biology and Damage* (eds R.I. Alfaro and S.G. Glover), Canadian Forest Service, Victoria, Canada, 177–185.

Alles, D. (1994) Untersuchungen zum Generationszyklus der Tannenlaus *Dreyfusia nordmannianae* Eckstein (Hom., Adelgidae) in Mitteleuropa. *J. A. Ent.* **117**, 234–242.

Anon. (1991) *La santé des forêts*, Ministère de l'agriculture et de la forêt, Paris, 86 pp.

Armour, H. (1995) Abundance of the defoliating aphid *Elatobium abietinum* on Sitka spruce (*Picea sitchensis*) provenances. *Biology and Environment* **95B**, 145.

Austarå, Ø. (1984) Diameter growth and tree mortality of Norway spruce following mass attacks by *Epinotia nanana*. *Norwegian Forest Research Institute, Research Paper* **10/84**, 1–8.

Austarå, Ø., Annila, E., Bejer, B. and Ehnström, B. (1983) Insect pests in forests of the Nordic countries 1977–1981. *Fauna norv. Ser. B.* **31**, 8–15.

Austarå, Ø., Orlund, A., Svendsrud, A. and Veidahl, A. (1987) Growth loss and economic consequences following two years defoliation of *Pinus sylvestris* by the pine sawfly *Neodiprion sertifer* in West Norway. *Scand. J. For. Res.* **2**, 111–119.

Avaramenko, I.D., Prokopenko, N.I., Mezentsev, A.I. *et al.* (1981) The effect of insects on the formation and productivity of oak stands. *Noveishie dostizheniya lesnoi entomologii eds Aushtilkalnene, A.M.* **4–6**, Vilnius, USSR.

Bakke, A. (1991) Socioeconomic aspects of an integrated pest management program in Norway. Proceedings of a Symposium, 'Integrated Pest Management of Forest Defoliators' (eds A.G. Raske and B.E. Wickman). *For. Ecol. Manage.* **39**, 299–303.

Baltensweiler, W. (1975) Zur bedeutung des Grauen Lärchenwicklers (*Zeiraphera diniana* Gn.) für die lebensgemeinschaft des Lärchen-Arvenwaldes. *Mitt. schweiz. ent. Ges.* **48**, 5–12.

Baltensweiler, W. (1977) Colour-polymorphism and dynamics of larch budmoth populations (*Zeiraphera diniana* Gn., Lep. Tortricidae). *Mitt. schwiez. ent. Ges.* **50**, 15–23.

Baltensweiler, W. (1993a) A contribution to the explanation of the larch budmoth cycle, the polymorphic fitness hypothesis. *Oecologia* **93**, 251–255.

Baltensweiler, W. (1993b) Why the larch budmoth cycle collapsed in the subalpine larch–cembran pine forests in the year 1990 for the first time since 1850. *Oecologia* **94**, 62–66.

Baltensweiler, W. and Rubli, D. (1984) Forestry aspects of the larch budmoth outbreaks in the Upper Engadine, Switzerland. *Mitt. Eidg. Anst. Forst. Versuchs.* **60**, 3–148.

Baltensweiler, W., Giese, R.L. and Auer, C. (1969) The grey larch budmoth, its population fluctuation in optimum and sub-optimum areas. In *Statistical Ecology* (eds G.P. Patil, E.C. Pielou and W.E. Waters), Vol. 2, pp. 401–420, University Park, Penn State University Press.

Baltensweiler, W., Benz, G., Bovey, P. and Delucchi, V. (1977) Dynamics of larch budmoth populations. *A. Rev. Ent.* **22**, 79–100.

Barbour, D.A. (1987) Monitoring pine beauty moth by means of pheromone traps: the effect of moth dispersal. In *Population Biology and Control of the Pine Beauty Moth* (eds S.R. Leather, J.T. Stoakley and H.F. Evans), pp. 7–13, Forestry Commission Bulletin 67, HMSO, London.

Barnett, D.W. (1987) Pine beauty moth outbreaks: associations with soil type, host nutrient status and tree vigour. In *Population Biology and Control of the Pine Beauty Moth* (eds S.R. Leather, J.T. Stoakley and H.F. Evans), pp. 14–20, Forestry Commission Bulletin 67, HMSO, London.

Baronio, P., Faccioli, G. and Butturini, A. (1989) A study of the influence of defoliation by *Neodiprion sertifer* (Geoffr.) (Hym. Diprionidae) on the growth of *Pinus sylvestris* L. in Romagna. *Bolletino dell'Instituto di Entomologia 'Guido Grandi' della Universita degli Studi Bologna* **43**, 17–24.

Bassand, D. (1965) Contribution à l'étude de la diapause embryonnaire et de l'embryogenèse de *Zeiraphera griseana* Hübner (= *Z.diniana* Guénée) (Lepidoptera: Tortricidae). *Rev. suisse Zool.* **72**, 429–542.

Battisti, A. (1984) *Dendroctonus micans* (Kugelann) in Italy (Coleoptera Scolytidae). *Frustula Entomologica* **7–8**, 631–637.

Bejer, B. (1981) Insect risks for introduced and native conifers in Northern Europe, especially in the Nordic countries. *Bulletin Organisation Européene et méditerranéenne pour la Protection des Plantes* **11**, 183–185.

Bejer, B. (1982) The present level of *Hylobius abietis* damage in Denmark. *Dansk Skovforenings Tidsskrift* **67**, 249–256.

Bejer, B. (1985) Nun moth (*Lymantria monacha* L.) outbreaks in Denmark and their association with site factors and climate. In *Site Characteristics and Population Dynamics of Lepidopteran and Hymenopteran Forest Pests* (eds D.Bevan and J.T. Stoakley), Forestry Commission Research and Development Paper 135, pp. 21–26.

Bejer, B. (1988) Sitka spruce and *Dendroctonus micans*. *Dansk Skovforenings Tidsskrift* **73**, 34–42.

Bejer-Petersen, B. (1962) Peak years and regulation of numbers in the aphid *Neomyzaphis abietina* (Walker). *Oikos* **13**, 155–168.

Belanger, R.P., Porterfield, R.L. and Rowell, C.E. (1981) Development and validation of systems for rating the susceptibility on natural stands in the Piedmont of Georgia to attack by the Southern pine beetle. In *Hazard Rating Systems in Forest Insect Pest Management* (eds R.L. Hedden, S.J. Barras and J.E.Coster), Symposium Proceedings, USDA Forest Service General Technical Report WO-27, 79–86.

Bentz, B.J., Amman, G.D. and Logan, J.A. (1993) A critical assessment of risk assessment systems for the mountain pine beetle. *For. Ecol. Manage.* **61**, 349–366.

Berryman, A.A. (1986) *Forest insects: Principles and Practice of Population Management*, Plenum Press, New York, 279 pp.

Berryman, A.A. and Stark, R.W. (1985) Assessing the risk of forest insect outbreaks. *J. Appl. Ent.* **99**, 199–208.

Berryman, A.A. and Stenseth, N.C. (1989) A theoretical basis for understanding and managing biological populations with particular reference to the spruce bark beetle. *Holarctic Ecology*, **12**, 387–394.

Billany, D.J. and Brown, R.M. (1977) The geographical distribution of *Gilpinia hercyniae* Hymenoptera: Diprionidae in the United Kingdom. *Forestry* **50**, 155–160.

Binazzi, A. and Covessi, M. (1991) Contribution to the knowledge of conifer aphids, XII. The genus *Dreyfusia* Boerner in Italy with the description of new species (Homoptera Adelgidae). *Redia* **74**, 233–287.

Britton, R.J. (1988) Physiological effects of natural and artificial defoliation on the growth of young crops of lodgepole pine. *Forestry* **61**, 165–175.

Browne, F.G. (1968) *Pests and Diseases of Forest Plantation Trees*, Clarendon Press, Oxford.

Capecki, Z. (1982) Massive outbreak of *Cephalcia falleni* in the Gorce mountains. *Sylwan* **126**, 41–50.

Carter, C.I. (1977) *Impact of green spruce aphid on growth: can a tree forget its past?* Research and Development Paper 116, Forestry Commission, UK, 27 pp.

Carter, C.I. (1983) Some new aphid arrivals to Britain's forests. *Proceedings and Transactions of the British Entomological and Natural History Society* **16**, 81–87.

Carter, C.I. (1990) Aphid biology and spruce forests in Britain. *Acta Phytopathologica et Entomologica Hungarica* **25**, 393–401.

Carter, C.I. and Austarå, Ø. (1994) The occurrence of males, oviparous females and eggs within anholocyclic populations of the green spruce aphid *Elatobium abietinum* (Walker) (Homoptera: Aphididae). *Fauna norv. Ser. B* **41**, 53–58.

Carter, C.I. and Danielsson, R. (1991) Two spruce root aphids *Pachypappa vesicalis* and *Pachypappa lactea* new to Britain with illustrated keys to the morphs from *Picea* roots. *Entomologist* **110**, 66–74.

Carter, C.I. and Maslen, N.R. (1982) *Conifer Lachnids*, Forestry Commission Bulletin 58, HMSO, London, 73 pp.

Carter, C.I. and Nichols, J.F.A. (1985) Host plant susceptibility and choice by conifer aphids. In *Site Characteristics and Population Dynamics of Lepidopteran and Hymenopteran Forest Pests* (eds D. Bevan and J.T. Stoakley), Forestry Commission Research and Development Paper 135, pp. 94–99.

Carter, C.I. and Nichols, J.F.A. (1988) *The Green Spruce Aphid and Sitka Spruce Provenances in Britain*, Forestry Commission Occasional Paper 19, 7 pp.

Cates, R.G. and Zou, J. (1990) Douglas fir (*Pseudotsuga menziesii*) population variation in terpene chemistry and its role in budworm (*Choristoneura occidentalis* Freeman) dynamics. In *Population Dynamics of Forest Insects* (eds A.D.Watt, S.R. Leather, M.D. Hunter and N.A.C. Kidd), pp. 169–182, Intercept, Andover.

Christiansen, E. (1969) Insect pests in forests of the Nordic countries. *Norsk ent. Tidsskr.* **17**, 153–158.

Christiansen, E. and Sandvik, M. (1974) Damage by *Hylobius abietis* to Scots pine on scarified patches. *Norsk Skogbruk* **20**, 8–9.

Claridge, M.F. and Evans, H.F. (1990) Species–area relationships: relevance to pest problems of British trees. In *Population Dynamics of Forest Insects* (eds A.D. Watt, S.R. Leather, M.D. Hunter and N.A.C. Kidd), pp. 59–69, Intercept, Andover.

Crawley, M.J. (1983) *Herbivory: the Dynamics of Animal–Plant Interactions*, Studies in Ecology 10, Blackwell, Oxford.

Currado, I., Scaramozzino, P.L. and Palenzona, M. (1983) Observations on the behaviour of *Rhyacionia buoliana* Den. & Sch. on *Pinus strobus* L. (Lepidoptera Tortricidae). Preventive note. *Atti XIII Congresso Nazionale Italiano di Entomologia, 1983*, pp. 371–374.

Cwiklinski, L. and Grodzki, W. (1989) The need to control the occurrence of *Zeiraphera griseana* using pheromones. *Las-Polski* **4**, 9–10.

Day, K.R. (1984a) The growth and decline of a population of the spruce aphid *Elatobium abietinum* during a three year study, and the changing pattern of fecundity, recruitment and alary polymorphism in a Northern Ireland forest. *Oecologia* **64**, 118–124.

Day, K.R. (1984b) Systematic differences in the population density of green spruce aphid, *Elatobium abietinum* in a provenance trial of Sitka spruce, *Picea sitchensis*. *Ann. appl. Biol.* **105**, 405–412.

Day, K.R. (1984c) Phenology, polymorphism and insect–plant relationships of the larch budmoth, *Zeiraphera diniana* (Guenée) (Lepidoptera: Tortricidae), on alternative conifer hosts in Britain. *Bull. ent. Res.* **74**, 47–64.

Day, K.R. (1986) Population growth and spatial patterns of spruce aphids (*Elatobium abietinum*) on individual trees. *J. Appl. Ent.* **102**, 505–515.

Day, K.R. (1994) Choice of pupal habitats by a moth, *Zeiraphera diniana* Gn. (Lep., Tortricidae), defoliating lodgepole pine, *Pinus contorta* (Douglas), in northerly, non-alpine zones. *J. Appl. Ent.* **118**, 342–346.

Day, K.R. and Crute, S. (1990) The abundance of spruce aphid under the influence of an oceanic climate. In *Population Dynamics of Forest Insects* (eds A.D. Watt, S.R. Leather, M.D. Hunter and N.A.C. Kidd), pp. 25–33, Intercept, Andover.

Day, K.R. and McClean, S.I. (1991) Influence of the green spruce aphid on defoliation and radial stem growth of Sitka spruce. *Ann. appl. Biol.* **119**, 415–423.

Day, K.R., Leather, S.R. and Lines, R. (1991) Damage by *Zeiraphera diniana* (Lepidoptera: Tortricidae) to lodgepole pine (*Pinus contorta*) of various provenances. *For. Ecol. Manage.* **44**, 133–145.

Demolis, C., Francois, D., Gregy, J.C. *et al.* (1991) Chemical control of *Dreyfusia nordmannianae*. *Bulletin Technique Office National des Forêts, Dole, France* **20**, 41–50.

Diehl, S.L. and Bush, G.L. (1984). An evolutionary and applied perspective of insect biotypes. *Ann. Rev. Ent.* **29**, 471–504.

Donaubauer, E. (1993) Severe damage by insects and fungi in Christmas tree plantations. *Forstschutz Aktuell* **12–13**, 16–19.

Du Merle, P., Avolio, S. and Chambon, J.P. (1990) On the presence of *Choristoneura murinana* and two other tortricid Lepidoptera in the fir forests of Calabria and Lucania. *Italia Forestale e Montana* **45**, 197–212.

Du Merle, P., Chambon, J. and Cornic, J. (1991) Lepidoptera Tortricidae caught with traps baited with synthetic sex pheromone of *Choristoneura murinana*. Data on their distribution in France, with special mention of *Cnephasia communana*. *Bull. Soc. Ent. France* **96**, 155–164.

Du Merle, P., Brunet, S. and Cornic, J.F. (1992) Polyphagous potentialities of *Choristoneura murinana* (Hb.) (Lep., Tortricidae): a 'monophagous' folivore extending its host range. *J. Appl. Ent.* **113**, 18–40.

Ehnström, B., Annila, E., Austarå, Ø. *et al.* (in preparation) Insect pests in forests of the Nordic Countries 1982–1986.

Ehnström, B., Bejer-Petersen, B., Löyttyniemi, K. and Tvermyr, S. (1974) Insect pests in forests of the Nordic Countries 1967–1971. *Ann. Ent. Fenn.* **40**, 37–47.

Eichhorn, O. and Pausch, K.L. (1986) Studies on web-spinning sawflies of the genus *Cephalcia* Panz. (Hym., Pamphiliidae). I. The problem of the generation cycle of *Cephalcia abietis* L. *J. Appl. Ent.* **101**, 101–111.

Eidmann, H.H. (1985) Silviculture and insect problems. *Z. angew. Ent.* **99**, 105–112.

Eidmann, H.H. (1992) Impact of bark beetles on forests and forestry in Sweden. *J. Appl. Ent.* **14**, 193–200.

Embree, D.G. (1965) The population dynamics of the winter moth in Nova Scotia. *Memoirs Entomol. Soc. Canada* **46**, 1–46.

Emelianov, I., Mallet, J. and Baltensweiler, W. (1995) Genetic differentiation in *Zeiraphera diniana* (Lepidoptera: Tortricidae) the larch budmoth: polymorphism, host races or sibling species.? *Heredity* **75**, 416–424.

Ericsson, A., Hellqvist, C., Långström, B. *et al.* (1985) Effects on growth of simulated and induced shoot pruning by *Tomicus piniperda* as related to carbohydrate and nitrogen dynamics in Scots pine. *J. Appl. Ecol.* **22**, 105–124.

Esbjerg, P. and Feilberg, L. (1971) Infestation level of the European pine shoot moth (*Rhyacionia buoliana* Schiff.) on some provenances of lodgepole pine

(*Pinus contorta* Loud.). *The Danish Forest Experiment Station Report* **254**, 345–358.
Evans, H.F. (1987) Sitka spruce insects: past, present and future. *Proc. R. Soc. Edin.* **93B**, 157–167.
Fielding, N.J., Evans, H.F., Williams, J.M. and Evans, B. (1991) The distribution and spread of the Great European Spruce bark beetle *Dendroctonus micans*, in Britain – 1982 to 1989. *Forestry* **64**, 345–358.
Fielding, N., Evans, B., Burgess, R. and Evans, H. (1994) *Protected Zone surveys in Great Britain for* Ips typographus, I. amitinus, I. duplicatus *and* Dendroctonus micans, Forestry Commission Research Information Note 253, London, 6 pp.
Forrest, G.I. (1987) A rangewide comparison of outlying and central lodgepole pine populations based on oleoresin monoterpene analysis. *Biochem. Syst. Ecol.* **15**, 19–30.
Forster, B. (1993) Development of the bark beetle situation in the Swiss storm-damage areas. *Schw. Z. Forst.* **144**, 767–776.
Fraser, S.M. (1995) The colonisation of Sitka spruce by winter moth in Scotland. PhD Thesis, University of London.
Fuhrer, E. and Fischer, P. (1991) Towards integrated control of *Cephalcia abietis*, a defoliator of Norway spruce in central Europe. *Forest Ecology and Management* **39**, 87–95.
Galkin, G.I. (1992) Grey larch leafroller (*Zeiraphera diniana*) in the north of Krasnoyarsk Territory. *Zoologicheskii Zhurnal* **71**, 69–78.
Gradwell, G.R. (1974) The effect of defoliators on tree growth. In *The British Oak* (eds M.G. Morris and F.H. Perring), pp. 182–193, Classey, Faringdon.
Grégoire, J-C., Raty, L. and Defays, E. (1995) Estimating the impact of *Ips typographus* in Southern Belgium – how far can we go? S2.07-05: Integrated control of Scolytid bark beetles. *IUFRO XX World Congress Abstracts, Tampere, Finland: Environmental and Economic Impact of Forest Insect Pests*, pp. 184–185.
Guerin, P.M., Baltensweiler, W., Arn, H. and Buser, H.-R. (1984) Host race pheromone polymorphism in the larch budmoth. *Experientia* **40**, 892–894.
Guryanova, T.M. (1988) Effect of periodic damage to Scots pine by *Neodiprion sertifer* and *Panolis flammea* on the increment of young stands. *Lesnovedenie* **5**, 59–63.
Harrison, R.G., Wintermeyer, S.F. and Odell, T.M. (1983) Patterns of genetic variation within and among gypsy moth, *Lymantria dispar* (Lepidoptera: Lymantriidae) populations. *Ann. Ent. Soc. Am.* **76**, 652–656.
Haussler, D. (1990) *The pine bud-shoot tortricid (*Rhyacionia buoliana *Den. & Schiff.)*, Merkblatt Institut fur Forstwissenschaften Eberswalde **48**, 13 pp.
Hayes, A.J. and Britton, R.J. (1986) Attacks of *Neodiprion sertifer* on *Pinus contorta*. *EPPO Bulletin* **16**, 613–620.
Heritage, S., Collins, S. and Evans, H.F. (1989) A survey of damage caused by *Hylobius abietis* and *Hylastes* spp. in Britain. In *Insects Affecting Reforestation: Biology and Damage* (eds R.I. Alfaro and S.G. Glover), pp. 36–42, Canadian Forest Service, Vancouver, Canada.
Hicks, R.R., Coster, J.E. and Mason, G.N. (1987) Forest insect hazard rating. *J. For.* **85**, 20–25.
Hubertz, H., Larsen, J.R. and Bejer, B. (1991) Monitoring spruce bark beetle (*Ips typographus* (L)) populations under non-epidemic conditions. *Scandinavian Journal of Forest Research* **6**, 217–226.
Hummel, F. (1991) Comparisons of forestry in Britain and mainland Europe. *Forestry* **64**, 141–156.
Hunter, M.D., Watt, A.D. and Docherty, M. (1991) Outbreaks of the winter moth on Sitka spruce in Scotland are not influenced by nutrient deficiencies of trees, tree budburst, or pupal predation. *Oecologia* **86**, 62–69.
Izhevskii, S.S. (1992) New problems with the 'old' gipsy moth. *Zascchita Rastenii Moskva* **11**, 37–39.
Jones, E.W. (1945) The structure and reproduction of the virgin forest of the North Temperate zone. *New Phytol.* **44**, 130–248.
Julien, J. (1995) Aspects écologiques et épidemiologiques des populations d'insectes xylophages statistiques types. Doctorate Thesis, University of Paris VII, 391 pp.
Kerslake, J.E., Kruuk, L.E.B., Hartley, S.E. and Woodin, S.J. (1996) Winter moth outbreaks on Scottish heather moorlands: effects of host plant and parasitoids on larval survival and development. *Bull. Ent. Res* **86**, 155–164.
Keskinalemdar, E., Alkan, S. and Aksu, Y. (1987) Studies on the biology and control of *Ips typographus* L. (Coleoptera: Scolytidae) in Artvin Province. Turkiye I. *Entomoloji Kongresi Bildirileri, Ege Universitesi Izmir*, 737–742.
Kletecka, Z. (1992) Distribution of damage by *Rhyacionia*, *Blastesthia* and *Retinia* tip moths (Lep., Tortricidae). *J. Appl. Ent.* **113**, 334–341.
Klimetzek, D. (1990) Population dynamics of pine-feeding insects: a historical study. In *Population Dynamics of Forest Insects* (eds A.D. Watt, S.R. Leather, M.D. Hunter and N.A.C. Kidd), pp. 3–10, Intercept, Andover.
Klimetzek, D. (1992) Schädlingsbelastung der Waldbäume in Mitteleuropa und Nordamerika. *Forstw. Cbl.* **111**, 61–69.

Klimetzek, D. (1993) Baumarten und ihre Schadinsekten auf der Nordhalbkugel. *Mitt. Dtsch. Ges. Allg. angew. Ent.* **8**, 505–509.

Kloft, W., Kunkel, H. and Erhardt, P. (1964) Weitere Beiträge zur Kenntnis der Fichtenröhrenlaus *Elatobium abietinum* (Walker) unter besondere Berücksichtigung ihrer Weltverbreitung. *Z. angew. Ent.* **55**, 160–185.

Korczynski, I. (1984) Possibilities for predicting damage to Scots pine stands by *Hylobius abietis*. *Sylwan* **128**, 51–56.

Kula, E. (1988). *The Economics of Forestry: Modern Theory and Practice*, Croom Helm, London.

Långström, B. (1985) Damage caused by *Hylobius abietis* in Finland in 1970–1971. Results from the Finnish part of a joint Nordic study. *Folia Forestalia Institutum Forestale Fenniae* **612**, 11 pp.

Långström, B. and Hellqvist, C. (1993) Induced and spontaneous attacks by pine shoot beetles on young Scots pine trees: tree mortality and beetle performance. *J. Appl. Ent.* **115**, 25–36.

Larsson, S. and Tenow, O. (1984) Areal distribution of a *Neodiprion sertifer* (Hym., Diprionidae) outbreak on Scots pine as related to stand condition. *Holarctic Ecology* **7**, 81–90.

Larsson, S. and Tenow, O. (1985) Grazing by needle-eating insects in a Scots pine forest in central Sweden. In *Site Characteristics and Population Dynamics of Lepidopteran and Hymenopteran Forest Pests* (eds D. Bevan and J.T. Stoakley), Forestry Commission Research and Development Paper **135**, pp. 47–55.

Larsson, S., Bjorkman, C. and Kidd, N.A.C. (1993) Outbreaks of diprionid sawflies: why some species and not others? In *Sawfly Life History Adaptations to Woody Plants* (eds M.R. Wagner and K.F. Raffa), pp. 453–483, Academic Press, San Diego, CA.

Leather, S.R. (1985) Oviposition preferences in relation to larval growth rates and survival of the pine beauty moth, *Panolis flammea*. *Ecological Entomology* **10**, 213–217.

Leather, S.R. (1992) *Forest Management Practice to Minimise the Impact of the Pine Beauty Moth*, Forestry Commission Research Information Note 217, 2 pp.

Leather, S.R. (1993) Influence of site factor modification on the population development of the pine beauty moth *(Panolis flammea)* in a Scottish lodgepole pine (*Pinus contorta*) plantation. *Forest Ecology & Management* **59**, 207–223.

Leather, S.R. and Knight, J.D. (1997) Pines, pheromones and parasites: a modelling approach to the integrated control of the pine beauty moth. *Scottish Forestry* (in press).

Leather, S.R., Watt, A.D. and Forrest, G.I. (1987) Insect-induced chemical changes in young lodgepole pine (*Pinus contorta*): the effect of previous defoliation on oviposition, growth and survival of the pine beauty moth, *Panolis flammea*. *Ecological Entomology* **12**, 275–281.

Lee, S. (1986) Tree breeding in Britain. *Forestry and British Timber* **16**, 24–25.

Lee, S. (1990) *Potential Gains from Genetically Improved Sitka Spruce*, Forestry Commission Research Information Note 190, HMSO, London, 6 pp.

Lempérière, G. (1994) Écologie d'un ravageur forestier, *Dendroctonus micans* Kug. (Col. Scolytidae) l'hylésine géant de l'epicéa. *Ecologie* **25**, 31–38.

Lempérière, G. and Malphettes, C.B. (1987) Observations of an infestation of *Pissodes piceae* in silver fir (*Abies alba*) stands in the Limousin region, France. *Rev. For. Francaise* **39**, 39–44.

Lines, R. (1966) Choosing the right provenance of lodgepole pine (*Pinus contorta*). *Scottish Forestry* **20**, 90–103.

Lines, R. (1976) The development of forestry in Scotland in relation to the use of *Pinus contorta*. In Pinus contorta *Provenance Studies*, pp. 2–5, Forestry Commission Research and Development Paper 114.

Linhart, Y.B. (1989) Interactions between genetic and ecological patchiness in forest trees and their dependent species. In *The Evolutionary Ecology of Plants* (eds J.H. Bock and Y.B. Linhart), pp. 383–430, Westview Press, London.

Low, A.J. (1985) *Guide to Upland Restocking Practice*, Forestry Commission Leaflet 8, London, 30 pp.

Löyttyniemi, K., Austarå, Ø., Bejer, B. and Ehnström, B. (1979) Insect pests of forests in the Nordic Countries 1972–1976. *Folia For.* **395**, 1–13.

Luciano, P., Delrio, G. and Prota, R. (1982) Defoliators in forests of *Quercus suber*. *Studi Sassaresi* **29**, 321–365.

Malcolm, D.C. (1987) Some ecological aspects of Sitka spruce. *Proc. R. Soc. Edin.* **93B**, 85–92.

Marchant, K.R. and Borden, J.H. (1976) *Worldwide Introduction and Establishment of Bark and Timber Beetles (Coleoptera: Scolytidae and Platypodidae)*, Pest Management Paper 6, Simon Fraser University, 76 pp.

Martinek, V. (1986) The influence of emergence of the one-year generation on the increase in abundance of adult sawflies of the genus *Cephalcia* Pz. (Hym., Pamphiliidae). *Lesnictvi* **33**, 1057–1074.

Mason, W. (1985) Clones to replace seedlings? *Forestry and British Timber* **14**, 24–26.

Mendel, Z. (1992) Occurrence of *Matsucoccus josephi* in Cyprus and Turkey and its relation to decline of

Aleppo pine (Hompotera: Matsucoccidae). *Entomol. Gener.* **17**, 299–306.

Montgomery, M. (1995) The Asian Gypsy moth (*Lymantria dispar*) – an international story of impacts and cooperation. *IUFRO XX World Congress Abstracts, Tampere, Finland, S2.07-11: Integrated Management of Forest Defoliating Insects*, p. 185.

Muller-Starck, G. and Zeihe, M. (eds) (1992). *Genetic Variation in European Populations of Forest Trees*, J.D. Sauerlander's Verlag, Frankfurt am Main, 271 pp.

Nef, L. (1994) Site characteristics as predictive variables of the vulnerability of Norway spruce stands to *Ips typographus* attacks. *Silva Belgica* **101**, 7–14.

Nef, L. and Minet, G. (1992) Evaluation des risques de dégâts d' *Hylobius abietis* L. dans les jeunes plantations de conifères. *Silva Belgica* **99**, 15–20.

Nichols, J.F.A. (1987) Damage and performance of the green spruce aphid, *Elatobium abietinum* on twenty spruce species. *Ent. exp. et Appl.* **45**, 211–217.

Niemelä, P., Tuomi, J. and Lojander, T. (1991) Defoliation of the Scots pine and performance of diprionid sawflies. *J. Anim. Ecol.* **60**, 683–692.

Nordlander, G. (1987) A method for trapping *Hylobius abietis* L. with a standardized trapping bait and its potential for forecasting seedling damage. *Scandinavian Journal of Forest Research* **2**, 199–213.

Pendrel, B. (1990) *Hazard from the seedling debarking weevil: a revised key to predicting damage on sites to be planted*, Forestry Canada – Maritimes Region, Technical Note 236, 4 pp.

Phillips, D.H. (1978) *The EEC Plant Health Directive and British forestry*, Forest Record 116, 22 pp.

Pinder, P.S. and Hayes, A.J. (1986) An outbreak of the vapourer moth (*Orgyia antiqua* L.: Lepidoptera Lymantridae) on Sitka spruce (*Picea sitchensis* (Bong.) Carr.) in central Scotland. *Forestry* **59**, 97–105.

Priesner, E. and Baltensweiler, W. (1987) Studies on pheromone polymorphism in *Zeiraphera diniana* Gn. (Lep., Tortricidae). 1. Pheromone reaction types of male moths in European wild populations, 1978–85. *J. Appl. Ent.* **104**, 234–256.

de Proft, M., Grégoire, J.C., Pigeon, O. and Bernes, A. (1992) Méthode d'evaluation de l'efficacité des insecticides a l'égard des scolytes en forêt resineuse. *International Symposium on Crop Protection, Gent* **57**, 815–824.

Quiring, D.T. and Butterworth, E.W. (1994) Genotype and environment interact to influence acceptability and suitability of white spruce for a specialist herbivore, *Zeiraphera canadensis*. *Ecol. Ent.* **19**, 230–238.

Rehfeldt, G.E. (1986) *Ecological genetics of Pinus contorta in the Upper Snake river basin of Eastern Idaho and Wyoming*, Research Paper INT-356, USDA, Ogden, 9 pp.

Roland, J. (1990) Interaction of parasitism and predation in the decline of winter moth in Canada. In *Population Dynamics of Forest Insects* (eds A.D. Watt, S.R. Leather, M.D. Hunter and N.A.C. Kidd), pp. 289–302, Intercept Press, Andover.

Roques, A. and Sun, J. (1995) Impact of cone and seed insects on the regeneration potential of conifer forests in Eurasia. *IUFRO XX World Congress Abstracts, Tampere, Finland: Environmental Impact of Forest Insect Pests*, pp. 180–181.

Schaefer, P.W., Weseloh, R.M., Sun, X.L. et al. (1984) Gypsy moth, *Lymantria dispar* (L.) (Lepidoptera: Lymantriidae), in the People's Republic of China. *Environmental Entomology* **13**, 1535–1541.

Schmutzenhofer, H. (1985) Site characteristics and mass outbreaks of *Cephalcia abietis* in Austria. In *Site Characteristics and Population Dynamics of Lepidopteran and Hymenopteran Forest Pests* (eds D. Bevan and J.T. Stoakley), Forestry Commission Research and Development Paper 13, pp. 27–35.

Schvester, D. (1985) Les insectes et la forêt française. *Rev. For. Fr.* **37**, 45–64.

Schweingruber, F.H. (1979) Auswirkungen des Lärchenwicklerbefalls auf die Jahringstruktur der Lärche. Ergebnisse einer Jahrringanalyse mit rontgendensitometrischen Methoden. *Schw. Z. Forst.* **130**, 1071–1093.

Seaby, D.A. and Mowat, D.J. (1993) Growth changes in 20-year-old Sitka spruce *Picea sitchensis* after attack by the green spruce aphid *Elatobium abietinum*. *Forestry* **66**, 371–379.

Sierpinski, Z. (1984) The effect of air pollution on harmful insects in stands of coniferous trees in Poland. *Forstwissenschaftliches Centralblatt* **103**, 83–91.

Šindelar, J. and Hochmut, R. (1972) Variability in the occurrence of some insect pests on various provenances of European larch, *Larix decidua* Miller. *Silvae Genetica* **21**, 86–93.

Soderstrom, V. (1976) The effects of scarification before planting conifers on newly clearfelled areas. *Sveriges Skogsvardsforbunds Tidskrift* **74**, 59–333.

Speight, M.R. and Wainhouse, D. (1989) *Ecology and Management of Forest Insects*, Clarendon Press, Oxford, 374 pp.

Stage, A.R. and Hamilton, D.A. (1981) Sampling and analytical methods for developing risk-rating systems for forest pests. In *Hazard Rating Systems in Forest Insect Pest Management* (eds R.L. Hedden, S.J. Barras and J.E. Coster). Symposium Proceedings, USDA Forest Service General Technical Report WO-27, pp. 87–92.

Stauffer, C., Leitinger, R., Simsek, Z. et al. (1992) Allozyme variation among nine Austrian *Ips typographus* L. (Col., Scolytidae) populations. *J. Appl. Ent.* **114**, 17–25.

Steacy, K. (1993) Resistance to insects in Lodgepole pine: interactions between *Panolis flammea*, *Neodiprion sertifer* and *Zeiraphera diniana*. D.Phil. Thesis, University of Ulster, 182 pp.

Stoakley, J.T. (1979) The pine beauty moth – its distribution, life cycle and importance as a pest in Scottish forests. In *Control of Pine Beauty Moth by Fenitrothion in Scotland* (eds A.V. Holden and D. Bevan), Forestry Commission Occasional Publication 4, pp. 7–12.

Stoakley, J.T. (1985) Outbreaks of winter moth, *Operophtera brumata* L. (Lep. Geometridae) in young plantations of Sitka spruce in Scotland. *Z. angew. Ent.* **99**, 153–160.

Straw, N.A. (1996) The impact of pine looper moth, *Bupalus piniaria* L. (Lepidoptera; Geometridae) on the growth of Scots pine in Tentsmuir forest, Scotland *Forest Ecology and Management* **87**, 209–232.

Targamadze, K.M., Shavliashvili, I.A., Mukhashavria, A.L. et al. (1975) Determining the economic effectiveness of forest pest (European spruce beetle) control expenditure. *VIII International Plant Protection Congress, Moscow* **2**, 23–27.

Teodorescu, I. and Simionescu, A. (1994) Dynamics of defoliating Lepidoptera attacks and the control measures in Romania's forests, 1953–1990. *Ambio* **23**, 2650–2660.

Theile, J. (1967) Zur Massenvermehrung des Grauen Lärchenwicklers, *Zeiraphera diniana* Guénée, in Fichtenbeständen des Erzgebirges (Situation 1966). *Arch. Forstwes.* **16**, 831–835.

Thomas, R.C. and Miller, H.G. (1994) The interaction of green spruce aphid and fertilizer applications on the growth of Sitka spruce. *Forestry* **67**, 329–342.

Turchinskaya, I.A. (1983) The silvicultural and biological basis of prophylactic measures against the pine weevil. *Lesnoe Khozyaistvo* **7**, 50–51.

Turgeon, J., Roques, A. and De Groot, P. (1994) Insect fauna of coniferous seed cones: diversity, host plant interactions, impact and management. *Ann. Rev. Ent.* **39**, 175–208.

Väisänen, R. (1995) Boreal forest ecosystems. *IUFRO XX World Congress, Tampere, Finland, Sub-Plenary Session 3: Biodiversity in Forest Ecosystems.*

Varley, G.C., Gradwell, G.R. and Hassell, M.P. (1973) *Insect Population Ecology*, Blackwell Scientific Publications, Oxford.

Vité, J.P. (1989) The European struggle to control *Ips typographus* – past, present and future. *Holarctic Ecology* **12**, 520–525.

Wainhouse, D. (1987). Forests. In *Integrated Pest Management* (eds A.J. Burn, T.H. Coaker and P.C. Jepson), pp. 361–401, Academic Press, London.

Wainhouse, D., Cross, D.J. and Howell, R.S. (1990) The role of lignin as a defence against the spruce bark beetle *Dendroctonus micans*: effect on larvae and adults. *Oecologia* **85**, 257–265.

Wagner, M.R. (1988) Induced defences in ponderosa pine against defoliating insects. In *Mechanisms of Woody Plant Defences against Insects* (eds W.J. Mattson, J.Levieux and C. Bernard-Dagan), pp. 141–155, Springer-Verlag, New York.

Walsh, P.J., Day, K.R., Leather, S.R. and Smith, A.J. (1993) The influence of soil type and pine species on the carabid community of a plantation forest with a history of pine beauty moth infestation. *Forestry* **66**, 135–146.

Waters, W.E., Stark, R.W. and Wood, D.L. (1985) *Integrated Pest Management in Pine-bark Beetle Ecosystems*, Wiley (Interscience), New York.

Watt, A.D. (1986) The performance of the pine beauty moth on water-stressed lodgepole pine plants: a laboratory experiment. *Oecologia* **70**, 578–579.

Watt, A.D. (1987) The effect of shoot growth stage of *Pinus contorta* and *Pinus sylvestris* on the growth and survival of *Panolis flammea* larvae. *Oecologia* **72**, 429–433.

Watt, A.D. (1989) The growth and survival of *Panolis flammea* larvae in the absence of predators on Scots pine and lodgepole pine. *Ecol. Ent.* **14**, 225–234.

Watt, A.D. and Leather S.R. (1988) The pine beauty moth in Scottish lodgepole pine plantations. In *Dynamics of Forest Insect Populations* (ed. A.A. Berryman), pp. 243–266, Plenum, New York.

Watt, A.D. and McFarlane, A. (1991) Winter moth on Sitka spruce: synchrony of egg hatch and budburst, and its effect on larval survival. *Ecol. Ent.* **16**, 387–390.

Watt, A.D., Leather, S.R. and Stoakley, J.T. (1989) Site susceptibility, population development and dispersal of the pine beauty moth in a lodgepole pine forest in northern Scotland. *J. Appl. Ecol.* **26**, 147–157.

Watt, A.D., Evans, R. and Varley, T. (1992) The egg-laying behaviour of a native insect, the winter moth, *Operophtera brumata* (L.) (Lep., Geometridae), on an introduced tree species, Sitka spruce *Picea sitchensis*. *J. Appl. Ent.* **114**, 1–4.

Weslien, J., Annila, E., Bakke, A. et al. (1989) Estimating risks for spruce bark beetle (*Ips typographus* (L.)) damage using pheromone-baited traps and trees. *Scandinavian Journal of Forest Research* **4**, 87–98.

White, T.C.R. (1969) An index to measure weather-induced stress of trees associated with outbreaks of psyllids in Australia. *Ecology* **50**, 905–909.

White, T.C.R. (1974) A hypothesis to explain outbreaks of looper caterpillars, with special reference to populations of *Selidosema suavis* in a plantation of *Pinus radiata* in New Zealand. *Oecologia* **16**, 279–301.

Wilson, W.L. and Day, K.R. (1994) Spatial variation in damage dispersion, and the relationship between damage intensity and abundance of the pine weevil (*Hylobius abietis* L.). *Int. J. Pest Manage.* **40**, 46–49.

Wilson, W.L. and Day, K.R. (1996) Variation in the relative abundance of the large pine weevil among Sitka spruce plantation sites. *Forestry* **69**, 169–171.

Wilson, W.L., Day, K.R. and Hart, E. (1996) Predicting the extent of damage to conifer seedlings by the pine weevil (*Hylobius abietis* L.): a preliminary risk model by multiple logistic regression. *New Forests* **12**, 203–222.

Winter, T.G. and Evans, H.F. (1994) *The Asian strain of Gypsy moth, Lymantria dispar: a significant threat to trees*, Arboriculture Research and Information Note 124, Forestry Commission, Farnham, 5 pp.

Worrell, R. (1983) Damage by the spruce bark beetle in south Norway 1970–80. A survey and factors affecting its occurrence. *Medd. Nor. inst. Skogforsk* **38**, 1–34.

Ying, C.C. and McKnight L.A. (eds) (1993) *Proceedings of the IUFRO International Sitka Spruce Provenance Experiment, Edinburgh*, British Coumbia Ministry of Forests and the Irish Forestry Board, Ireland, 230 pp.

Ždárek, J., Vrkoč, J. and Skuhravý, V. (1985) Outbreak of the spruce form of the larch bud moth (*Zeiraphera diniana* Guen.) in Czechoslovakia. In *Site Characteristics and Population Dynamics of Lepidopteran and Hymenopteran Forest Pests* (eds D. Bevan and J.T. Stoakley), pp. 56–60, Forestry Commission Research and Development Paper 135.

Znamenskii, V.S. and Lyamtsev, N.I. (1983) Criterion for identifying stands that are reservoirs of *Lymantria dispar*. *Lesnoe Khozyaiustvo* **1**, 60–61.

FOREST PESTS IN THE TROPICS: CURRENT STATUS AND FUTURE THREATS   12

Martin R. Speight

12.1 INTRODUCTION

Undoubtedly, many species of insect cause serious losses to tropical forests, and the current status of these pests is predominantly a function of ecological interactions between insect and host tree, which become biased in favour of the herbivore. In addition, the future of tropical forests is threatened by forest management in various guises which is incapable, or unwilling, to redress this balance.

The definition of 'tropical forest' is problematic; many forests occur outside the strict boundaries of the Tropics of Cancer and Capricorn but they still possess the basic ecological characteristics of their true 'tropical' counterparts. Indeed, afforested land which used to provide a clinal system of subtle habitat change has in reality now become more demic as fragmentation continues apace. Forests are normally thought of either as areas of natural tree cover, which may or may not have been logged at some stage, or as plantations of native or exotic tree species. The term 'forest' may not just imply continuous tree cover; the science and practice of agro- or social (village) forestry are just as important to the end user as any other type of forestry – perhaps more so when the survival of families or whole villages may depend on high yield and high vigour of trees planted within a farming system. Tropical forests occur in both developed and developing countries where economic and social factors differ tremendously.

The potential for pest problems varies considerably with the type of forest, and the perspective of whoever is using it. It is dangerous to predict which type of system is more or less prone to pest damage than another, though simplistic dogma suggests that the more intensive the system, the higher is the risk. After all, an insect species or population is only perceived as a 'pest' when economic damage occurs, or when gross yield is reduced in a subsistence system. In the absence of this concept, large numbers of insects in tropical forests are only of biodiversity interest.

12.2 USES OF TROPICAL FOREST PRODUCTS

From a managerial point of view, tropical forests can be conveniently, though rather artificially, split into two basic types: one where the trees are grown for commercial gain, and the other where they form part of agroforestry or village forestry. In the first system, some net profit is usually expected, often via financial returns from exported products such as pulp, chip or logs, so that the cost of pest management may be offset against the increased value of the product. This increase in value may enable the use of pest management systems that are relatively expensive and/or technically complex, such as treatment with chemicals or pathogens. In reality, the profit margins in tropical forestry are frequently small, such that this sort of management strategy may not be feasible. Another problem is

---

Forests and Insects. Edited by A.D. Watt, N.E. Stork and M.D. Hunter. Published in 1997 by Chapman & Hall, London. ISBN 0 412 79110 2.

that large-scale plantation projects not only demand high yield per unit time from their crops, but also require in many cases a particular shape or structure of the resulting timber. For example, pine plantations in northern Luzon in the Philippines are expected to yield a straight log of a minimum length and girth; severe attacks by shoot- and stem-boring Lepidoptera render the crop unmarketable because the boring activities of the pest result in stunted and bushy trees (Speight and Speechly, 1982a). In the second type of forest system, where local villagers rely mainly on biomass production, any reduction in the gross yield may result in insufficient fuel or fodder to sustain the basics of life; timber quality or shape are of less importance. Intermediate between these two extremes (and beyond the scope of this chapter) comes a host of non-timber products that are derived from tropical forests or form small industries within a farm–forestry system. The list of these products includes plant resins, gums, tannins and dyes, essential oils, insecticides, drugs and pharmaceuticals, fruits, honey and shellac (Hanover, 1988).

12.3 IMPACT OF FOREST PESTS IN THE TROPICS

Most experience in tropical forest pest management derives from plantations, of either native or exotic tree species, though lessons learnt in these rather artificial situations may well be applicable to natural or semi-natural forests under some form of utilization. Any tendency to manipulate a forest ecosystem, whatever the type, which perturbs it away from its evolved norm may provoke increases in insect populations and consequent enhanced damage to foliage, fruits, stems and roots. This does not necessarily imply that natural tropical forests do not exhibit large fluctuations in insect population densities from time to time, but even native tree species grown under abnormal conditions might be expected to be attacked more seriously by indigenous insects than in the natural forest. For example, various indigenous species of mahogany suffer more attacks from shoot-borers when raised in unshaded, pure stands (Newton *et al.*, 1993). The debate comparing the risks from pests between exotic and native tree species continues; managerially, each system has to be evaluated separately.

Whether native or exotic species are used, in plantations or in natural forests, in order to assess the viability of pest management in terms of cost-effectiveness it is vital to have quantitative data on the impact of insect attack and damage to trees. Ideally these data should be convertible to a loss in revenue by the end of the forest rotation, though it may be sufficient to gain an awareness of the consequences of insect damage. For example, heavy defoliation by moth or sawfly larvae may in itself be tolerable as long as the attacks do not persist for many generations of the insect, but if the crop is rendered susceptible by the defoliation to secondary attacks from borers such as cerambycid or ambrosia beetles, then tree death can occur.

This vital impact data is distressingly unavailable for a very large proportion of tropical forest pests. Published work is in short supply, and that which is available tends to be qualitative and anecdotal. Without such impact data, any attempts at active pest management in the tropics will be no more than a 'stab in the dark', lacking in clear goals. For example, a common strategy adopted by many afforestation projects in the tropics, especially in the early stages of growth such as seedling or new transplant, is to adopt routine insurance treatment with pesticides at the first sign of any damage, with no knowledge whatsoever of the likely progress of the potential pest outbreak. One of the most detailed examples of impact comes from tropical Queensland. Forest fire is potentially very damaging, and whilst some tropical forest systems have evolved to use fire as a natural mechanism for regeneration (fire climaxes), most plantation systems are at great risk, especially during the dry season. In 1994, large fires occurred in plantations of *Pinus* species (mainly *P. elliottii*, *P. taeda* and *P. caribaea*), caused it is thought by either sparking power lines or a smouldering woodchip heap. (F.R. Wylie, DPI, Queensland, personal communication). The fire produced enormous numbers of standing trees and logs that were ideal breeding sites for a North American species of bark beetle, *Ips grandicollis*, which was accidentally introduced to Australia some years ago. The most serious knock-on effect involved the degradation of the salvaged timber by a blue-stain fungus introduced by the beetles into fire damaged logs and standing trees. Dr Wylie's data on the fire and subsequent pest outbreak show that nearly 9000 ha were burnt, with 926 000 m^3 of timber being damaged. In study plots, between 66 and 100% of trees were attacked by *Ips*; 150 000 m^3 of timber could not be salvaged before significant blue-stain occurred, rendering it unmarketable, but between 540 000 and 810 000 m^3 were saved by rapid salvage operations. The salvage operation cost around A$10 million, and a log dump covering 34 ha had to be placed under continuous water spraying to deter the beetle. This dump is intended to store the logs for three to five years before being distributed on the market.

Some other examples of impact studies are summarized in Table 12.1; note that both exotic and native tree crops are presented, since both have been found to present problems under certain conditions. Only very few examples provide impact data in terms of monetary loss; it is relatively simple to convert mortality data into lost revenue assuming that a market value can be estimated, but it is much more complex to ascribe such figures to increment losses where the trees lose yield but continue to grow.

12.4 WHY PEST OUTBREAKS OCCUR – PREVENTION RATHER THAN CURE

Forest pest management anywhere in the tropical or temperate world should be most concerned with prevention. In order to obtain guidelines for this prevention, it is useful to consider the causes of insect pest outbreaks in tropical forestry, in the knowledge that if we possess the information about what has gone wrong, we can, if we so choose, avoid the same mistakes in the future. Converting a natural forest to a plantation may upset all sorts of stabilizing factors; abiotic conditions such as wind speed, temperature and humidity may promote herbivore populations at the expense of their predators and parasites, and the clearing of natural vegetation may reduce habitats for natural enemies (though the same action may also remove natural reservoirs for pest invaders). One central philosophy is the concept of stressing agencies, which reduce the vigour of a tree and render it more susceptible to insect attack. Loss of vigour in trees can be attributed to a wide variety of direct or indirect causes, but is usually linked to aspects of a tree's environment (e.g. soil type, climatic regime, altitude) or to its interaction with (1) other trees, either within the same species or non-related species, via competition, or (2) herbivores or pathogens that use the tree as a host. There is a wealth of literature available on the links between insect attack and reductions in host tree vigour (e.g. Speight and Wainhouse, 1989; Foggo and Speight, 1993; White, 1993; Speight, 1995), though some authors are uneasy about this. Watt (1994), for example, is of the opinion that there is very little evidence to support a 'stress hypothesis', at least in the case of foliage feeding insects, but he admits that two-thirds of field studies (very few of which were in the tropics) showed a positive relationship between water stress in trees and increased insect attack. There can be no doubt that in terms of forest management, a preventive pest management strategy must attempt to avoid low vigour in trees and hence reduce the probability of significant damage.

With these concepts in mind, Table 12.2 presents a series of factors involved in promoting outbreaks of tropical forest pests (see also Ivory and Speight, 1993). Each scenario will be considered in detail below, with examples.

12.4.1 NATURAL DISASTERS

Occasionally, phenomena occur which silviculturalists cannot easily avoid. High winds can often cause devastation in plantations, especially in those with heavy thinning or where trees are growing on soils that are very light or shallow, such that root systems are unable to hold the trees in windy conditions. Fire and flood, drought and volcanic activity are classic 'stress-provoking' phenomena, with a high probability of increased and potentially lethal insect (and pathogen) attack.

Table 12.1 Examples of insect damage to tropical trees

Insect species	Insect group	Host tree	Country	Attack mode	Severity of impact	Reference
Dioryctria spp.	Lepidoptera: Pyralidae	*Pinus oocarpa*	Indonesia	Shoot/bark boring	70% mortality of young trees	Intari and Ruswandy, 1986
Heteropsylla cubana	Hemiptera: Psyllidae	*Leucaena leucocephala*	Australia	Sap feeder	Dry matter production reduced by 55% of norm	Palmer *et al.*, 1989
Heteropsylla cubana	Hemiptera: Psyllidae	*Leucaena leucocephala*	Australia	Sap feeder	Reduction in fodder by 52% reduction in timber by 79%	Bray and Woodroffe, 1991
Hyblaea puera	Lepidoptera: Hyblaeidae	*Tectona grandis*	SW India	Defoliation	44% loss of volume increment; 50% loss of height increment	Nair *et al.*, 1985
Hypsipyla grandella	Lepidoptera: Pyralidae	*Swietenia mahogoni*	Florida	Shoot boring	50% of container-grown young trees infested	Howard and Meerow, 1993
Macrotermes spp/ *Microtermes* spp./ *Odontotermes* spp.	Isoptera	*Eucalyptus* spp.	Ethiopia/ Africa/ India	Stem and root severance	Up to 100% of new transplants killed	Cowie and Wood, 1989; Cowie *et al.*, 1989
Neotermes tectonae	Isoptera	*Tectona grandis*	Indonesia (Java)	Stem boring	Timber degrade from construction quality to fuelwood; losses of 8.2 m³/ha, worth £350/ha appr.	Intari, 1990
Oiketicus kirbyi	Lepidoptera: Psychidae	*Rhizophora mangle* (mangrove)	Ecuador	Defoliation	80% of foliage removed over 1200 hectares	Gara *et al.* 1990
Platypus spp.	Coleoptera: Platypodidae	*Acacia crassicarpa*	Sabah	Stem boring accompanying black stain	Up to 50 holes per tree	Thapa, 1992
Pteroma plagiophelps	Lepidoptera: Psychidae	*Paraserianthes falcataria*	SW India	Defoliation	Death of 22% of trees and 17% severely damaged after 2.5 years of repeated defoliation	Nair and Mathew, 1992
Xylosandrus compactus	Coleoptera: Scolytidae	*Khaya* spp.	India	Bark boring in seedlings and young saplings	60–70% infestation	Meshram *et al.*, 1993
Several unnamed		*Albizia lebbek*	India	Seed predation	50–70% seeds damaged in pods	Harsh and Joshi 1993

Table 12.2 Some major reasons for insect pest outbreaks in tropical forestry

Category	Examples
Natural disasters	Fire, windthrow, drought, etc. providing food or breeding sites
Forest management – avoidance of stress and susceptibility	Site choice and tree species matching in the planning stage
	Genetics and variable tree susceptibility
	Nursery management and the production of healthy transplants
	Choices at planting – monocultures
	Stand management – under-thinning promotes suppression
	Post-harvest manipulation
Pest invasions	International – accidental imports from foreign countries
	National/regional – invasions from infested stands
Inefficiency of natural enemies	Absence or breakdown of regulation
Lack of background biology	Little or no knowledge of pest taxonomy, ecology and host-plant relationships
Misuse of control systems	Removal of natural enemies
	Stress caused by phytotoxicity

12.4.2 FOREST MANAGEMENT

Unlike natural disasters, silvicultural practices are under management control. Many of these practices can enhance pest problems. The concept of inappropriate silviculture embraces a multitude of sins which crop up time and again in many guises. From nursery work to stand maintenance, from choice of site and species to post-harvest manipulation, each step in a typical forest project should be examined to question its influence on the probability of pest problems.

(a) Site choice and tree species selection

More than any other single factor, careful matching between the tree species to be planted and the site on which the forest will be established promotes tree vigour and hence resistance to pest attack. However, this seemingly obvious management tactic is ignored or overlooked by tropical silviculturalists and economists in many parts of the world. Put simply, if we assume that tree species or subspecies have evolved over millions of years to perform optimally in terms of growth and reproduction under a specific set of environmental constraints (soil type, climatic regime, geographical range, interactions with their own species, other plant species, herbivores, etc.), then any major departures from any or all of these site characteristics is likely to result in suboptimal tree performance, namely low vigour.

Trees do experience fluctuations in environmental conditions even within their natural ranges, but it is gross divergence beyond normal extremes that is often the critical factor (Newhook, 1989).

Time and again, serious pest problems in tropical forestry are the result of trees being stressed because they have been planted in sites to which they are not well adapted. The literature is full of examples of increased pest damage due to drought, waterlogging, a paucity of soil nutrients, too much or too little shade, altitude, latitude, etc. (Speight, 1995) and Table 12.3 presents some examples. In most countries, more fertile soils are naturally used for agriculture, so that forestry tends to be relegated to poor, degraded and shallow soils. It may therefore be impossible to provide a tree species with exactly the right conditions in which to grow, but it is quite clear that some attempt to promote vigour in trees by site matching is vital to reduce the likelihood of serious insect pest problems. To this end, intimate knowledge of a tree's ecology is required; this is often wanting, especially when fast-growing exotic species are employed for the first time in a new region.

(b) Genetics

It is well known that trees employ batteries of defence systems to combat herbivorous insects. These defences can be physical (such as sap pressure or leaf toughness), reduced nutrient availabili-

Table 12.3 Examples of insect attack to tropical trees resulting from poor site conditions

Tree species	Insect species/family	Country	Site problem	Reference
Casuarina cunninghamiana	*Rhyparida limbatipennis* (Coleoptera: Chrysomelidae)	Queensland, Australia	High stream salinity, intensive land management	Wylie et al., 1993
Eucalyptus deglupta	*Agrilus opulentus* (Coleoptera: Buprestidae)	Papua New Guinea	Low rainfall/poor drainage	Makihara et al., 1987
Leucaena leucocephala	*Heteropsylla cubana* (Hemiptera: Psyllidae)	Philippines	Drought	Braza, 1988
Pinus caribaea var hondurensis	*Dioryctria horneana* (Lepidoptera: Phyticidae)	Cuba	Low soil nutrients	Herrero et al., 1991
Pinus caribaea/kesiya/oocarpa	*Petrova cristata/Dioryctria spp* (Lepidoptera: Tortricidae/ Phyticidae)	Philippines	Low soil nutrients, low altitude	Speight and Speechly, 1982a
Prosopis glandulosa	*Oncideres cingulata* (Coleoptera: Cerambycidae)	Texas, USA	Dry soils	Ansley et al., 1990
Prosopis juliflora Swietenia macrophylla	Coleoptera: Cerambycidae *Hexacolus guyanensis* (Coleoptera: Scolytidae)	Kenya Guadeloupe	Servere drought in semi-desert Water stress	Speight, 1995 Comic, 1978
Tectona grandis	Coleoptera: Buprestidae	Nepal	Waterlogging because of thick clay soil	Speight, 1995

ty (such as using tannins to complex organic nitrogen and hence prevent animals using it as food) and chemical toxins and antifeedants (such as resins and terpenes). It is, therefore, not surprising to find significant differences in the ability of an insect to utilize a tree, not only within a tree species but between genotypes, and also between other species. Examples of insects responding to differential resistance mechanisms in their host trees include leaf skeletonizers on teak (Muktar-Ahmad and Ahmad, 1991), defoliators on mangroves (Robertson, 1990), psyllids on *Leucaena* (Suresh-Babu *et al.*, 1992), shoot moths on pines (Speight and Speechly, 1982a), cypress aphids on Cupressaceae (Obiri, 1994), longhorn beetles (*Phoracantha*) on eucalypts (Hanks *et al.*, 1995), and shoot-borers on mahoganies (Newton *et al.*, 1993). One of the most notorious groups of pests of recently planted tropical forests is root-feeding termites (Mitchell, 1989). A large number of strategies have been suggested to combat these insects, but in the final analysis it is thought that the use of resistant tree species and the development of resistant varieties offers the only long-term solution (Cowie *et al.*, 1989).

Superficially, therefore, it seems an extreme folly to plant large areas of the tropics with tree genotypes known to be susceptible to indigenous or introduced insects. However, merely identifying a resistant variety or species is not an easy task; tropical silviculture routinely sets up elaborate provenance trials to investigate growth rate, form, and sometimes, general suitability to a locality, but these trials may be only rather cursorily examined and scored for insect or pathogen susceptibility. Once resistance has been detected, the non-susceptible genotypes may be unavailable – because, for example, the seed is in short supply, as was the case for varieties of *Pinus caribaea* resistant to pine shoot moths in South-East Asia (Speight and Speechly, 1982b) – or they may have undesirable silvicultural or economic characteristics, such as slow growth or bad form.

(c) Nursery management

Tropical forest nurseries exhibit close similarities to agriculture systems, in that the young trees are likely to be at their most nutritious and vulnerable to insects. They are planted in large single-species blocks, and are often surrounded by areas of semi-natural vegetation from which generalist and specialist herbivorous insects can migrate with ease. On the other hand, a nursery system is the only real stage of a tropical forest's life when pest control can easily be accomplished, mainly via the use of pesticides. The liberal use of these compounds, very prevalent in developing countries, has to be offset against the likely impact of the pest; heavy treatment as soon as one cricket or butterfly is seen must be avoided. In general terms, it must be the case that poor quality transplants, resulting from pest or disease attack in the nursery, or stress induced during the planting out phase, will produce non-vigorous plantation trees, especially in the first few months after establishment. Such damage to young plants in the nursery can also produce non-vigorous trees in later life, which is well illustrated by patch dying of 8-year-old *Acacia mangium* in Sabah (northeast Borneo) observed in 1993 (Speight, 1995). These trees, which should still have been growing actively at that age, were heavily infested by two types of bark and wood-boring beetles, a cerambycid and a buprestid. The larvae of these typical secondary pests had attacked the trees and killed them by girdling. Since these pests almost invariably only attack stressed trees, evidence of low vigour needs to be looked for. This detective work included the examination of the roots of dead trees in the stands, which revealed marked deformities, such that there was no true tap root, and the roots that were present were twisted and bent. Termite attack had begun in the crooks of the twisted roots. Back in the nursery, over 50% of *A. mangium* seedlings showed twisted, curled or coiled root systems, resulting from mishandling during the pricking out stage. This occurred because the cheap labour force employed had received payment on a piece-work basis; the more they planted per unit time, the more money they received.

(d) Monocultures

The climax forests indigenous to temperate climates, such as oak or lime in parts of Europe and spruce in the Pacific northwest of the United States, could be described as a type of monoculture, in that these forests are mainly dominated by one species. However, they exhibit not only a relatively broad genetic diversity by virtue of sexual reproduction, but also a varied age structure from regenerated saplings to over-mature trees in the same locality. A true monoculture can be thought of as a crop that consists exclusively of an even-aged population of a single plant species, with little or no genetic variation between individuals; specialist herbivores are in theory able to feed and spread without hindrance through the whole stand. In contrast with this, natural tropical forests are often (though not always) immensely diverse botanically – a 2 ha plot in Sabah (northeast Borneo) can contain up to 198 tree species (Paijmans, 1976) – and hence a stand of trees containing but one species, which until now has been standard silvicultural practice in tropical and temperate regions, is distinctly different and ecologically alien. The dogma stating that risks from pest attack increase markedly in monocultures of all sorts of crops is well supported by the literature on tropical forestry. However, Wormald (1992) compares the benefits and drawbacks between mixed and pure stands of tropical trees, and concludes that the situation is rather confusing, mainly because of incomplete ecological knowledge. He suggests that the greatest risk in monoculture plantations is to be expected from exotic, introduced insect pests. Although it is dangerous to generalize, and whilst there are clear economic and silvicultural reasons for planting monocultures in the tropics, trends towards enrichment or line planting instead of pure plantations may reduce risks.

(e) Stand management – thinning

In tropical countries, forest stands may need thinning within a few years after planting; competition between trees in a stand for light, soil nutrients and/or water can quickly become intense, especially on degraded sites where most tropical trees are grown. Even in line plantings of natives such as dipterocarps within a rain forest, this type of competition is to be expected. Suppression of some or all trees in a plantation then ensues, resulting in reduced vigour. Links between thinning regimes and insect attack are fairly well documented from temperate forests (Sartwell and Dolph, 1976); mortality of pines from bark beetle attacks declines rapidly as stands are progressively thinned. As with many aspects of tropical forestry, quantified data of this nature are harder to find. The buprestid beetle, *Agrilus opulentus*, is reported to kill more eucalyptus trees in Papua New Guinea as the stand density increases (Makihara *et al.*, 1987), and Chinese work suggests that unsuitable thinning is one of the factors that increase risks to tropical forest from insect pests (Liu 1992), but most management systems base their thinning strategies on conventional wisdom rather than accurate impact data. Thin stands would seem to be important in helping to prevent pest outbreaks, though there are exceptions. There is a suggestion from India, for example, that seedlings of various broadleaf species suffered greater mortality due to insect herbivore feeding when in open or sparse stands (Khan and Tripathi, 1991); certainly, over-thinning may increase the insolation, which can in itself stress trees, especially if they have evolved to be shade-loving, and as mentioned above there may also be an increased risk of windthrow.

12.4.3 PESTS FROM ELSEWHERE

Quarantine systems operate in many countries to prevent insect species invading from overseas and becoming serious pests. Since many tree species planted in the tropics are exotic, they may for a while be immune to insect attack, before either indigenous pests take them over as new hosts, or their own herbivores manage to get into the new country. One major route for invasion is via trade in timber and other wood products, and developed countries in the tropics now have elaborate and effective plant health regulations governing imports from overseas. Less developed countries may not yet possess the infrastructure to carry out such safeguards; however, since most such countries export their timber rather than import it, they are possibly running fewer risks than the importer. Despite all possible precautions, tropical pests do manage to get to new regions where their host species was thought to be free of pests. One classic example is that of *Leucaena leucocephala*, a multi-purpose legume originating in Central America, and very widely planted in a large number of countries from Fiji to Kenya. The sap-feeding psyllid *Heteropsylla cubana*, also from Central America, has managed to colonize its host in virtually every country and region of the world (Withington and Brewbaker, 1987), with the result that many subsistence systems pinning their hopes on the tree have met with great disappointment. Another well known example is that of the cerambycid beetle *Phoracantha semipunctata*, an extremely serious pest of eucalypts. This species is native to Australia, but is now causing high tree mortalities from California to Zambia, Malawi to Morocco, and has become well established in Portugal and Spain (Selander and Bubala, 1983; Scriven *et al.*, 1986). Overall, attempts to grow exotic tree species free from insect pests may work for a while, but management tactics that grow exotic, susceptible species on unsuitable sites will not flourish indefinitely. It would be much better to grow really vigorous trees, on well suited sites, so that if and when potentially serious pests do arrive, the forests will be best able to limit the damage done.

In some cases, damaging insects may not have to come from overseas to attack newly planted forests, nor are exotic tree species the only ones to run the risk of pest attack. Natural forests or older plantations may harbour reservoirs of insect species that migrate to the new, vulnerable monoculture stands with great success. Serious outbreaks of pine-shoot moths in the Philippines resulted from the planting of large areas of degraded upland sites with various tropical pine species in close proximity (a mere few metres in some cases) to much older stands of indigenous *Pinus* (Speight and Speechly, 1982a). The vigour of the new stands was likely to be low because of unsuitable site conditions (Speight 1995), and the damage done by shoot and stem boring was so severe that the whole project failed. If the insect species is fairly generalist, then primary or secondary forest or even different crops around a forest plantation can act as pest reservoirs. Even non-related tree species may act as pest reservoirs; it has been suggested that oil palm plantations in the vicinity of *Acacia. mangium* stands in the Far East can act as reservoirs of bag worms (Lepidoptera : Psychidae) (Sajap and Siburat 1992). More likely is that oil palm will provide new pests for the burgeoning rattan (also a palm) industry in parts of Malaysia and Borneo.

12.4.4 INEFFICIENCY OF NATURAL ENEMIES

The roles of natural enemies (predators, parasitoids and pathogens) in the regulation of tropical forest pests are rarely quantified. Large numbers of species are recorded in the literature as having been reared from forest pests, but this information is of relatively little use without quantitative data on percentage parasitism and the ability of the surviving pests to recover. Low levels of parasitism are unlikely to be of use in the regulation of pest epidemics before considerable economic damage has been done; this is especially true of 'low density' pests, where, for example, one shoot- or bark-borer may so disfigure a tree as to make it worthless. The tropics certainly possess a richer and more diverse parasitoid fauna than temperate regions (Gauld, 1991, discussing the situation in Costa Rica), though their ability to regulate pest populations at high densities is less well known. There are suggestions, however, that habitat change, such as when forest plantations are established, may adversely influence the abundance and host-finding ability of predators and parasitoids. The plant density, species diversity and structural complexity of a crop are all thought to reduce the efficiency of natural enemies (Coll and Bottrell, 1994), presumably by removing their alternative hosts, or host plants for the adults. Microclimatic changes may also render the new habitats less suitable; crops such as tea intercropped with various tree species have fewer pests and more of their natural enemies due to more optimal conditions of temperature, light and humidity (Wang, 1994). Finally, habitat fragmentation and/or isolation in new stands may result in reduced effects of natural enemies, releasing herbivorous insects from parasitism (Kruess and Tscharntke, 1994).

Natural enemies may be absent from a forest when exotic tree species are colonized by exotic pests. One of the reasons why the leucaena psyllid *Heteropsylla cubana* is thought to be such a serious pest of *Leucaena leucocephala* in Asia is because no significantly effective parasitoids or predators exist as yet in the region (Showler, 1995).

216 Forest pests in the tropics

12.4.5 CONCLUSIONS

The above section has illustrated diverse ways in which insect pest problems can be exacerbated, and in summary, a general preventive strategy for tropical forest pests is illustrated in Figure 12.1. The key to the interactions presented here is the three-way link between the trees, the insects and the site characteristics. In order to best understand and then manipulate these interactions to the detriment of insect pests, it is vital to consider all three components as equally important, instead of the normal situation where forest entomology (and pathology) is ignored until things actually go wrong, whereupon it is very often too late to do anything about the problem. The factors discussed above should be considered at the planning stage of a forestry project in the tropics, with equal emphasis as for silviculture, soil science and economics.

12.5 MANAGEMENT STRATEGIES AFTER PLANTING

One difficulty concerns the fact that tropical forestry encompasses both developed and developing worlds, with a large midway component where certain tropical countries are rapidly becoming industrialized. It is therefore impossible to generalize about pest management tactics that are appropriate in all tropical forest situations. Instead, each scenario needs to be viewed in the context of its specific scientific, economic and practical capabilities. Nowhere in the tropical world is forestry blessed with profit margins as high as those enjoyed by perennial cash crops such as cocoa, coffee or oil palm, and so management tactics have to be cost effective and cheap. An extra drawback to effective pest management in tropical forests is the lack of an advisory network, either because the knowledge base of pest/tree dynamics and impact data as mentioned earlier is simply not available, and/or because the socio-economics of the region or the dictates of aid agencies mean that such advice often has to come from expensive, short-term expatriate consultancies. Temperate forest pest management has been pushing forward with the practical implementation of integrated pest management (IPM) for some time (Speight and Wainhouse, 1989), and many examples of IPM in commercial use are now available in Europe, North

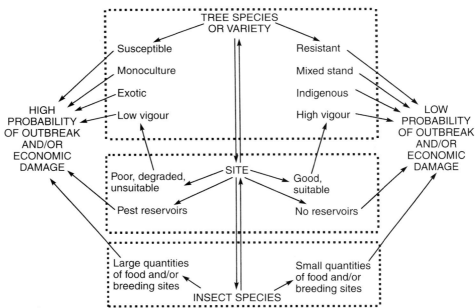

Figure 12.1 Summary of some major factors that may interact to promote insect pest outbreaks in tropical forestry. Many of these factors may be manipulated to reduce the probability of damage.

and South America and parts of Australasia. There is no doubt at all that IPM in forestry is based on prevention rather than cure (control) if at all possible; the extra components of an IPM programme come into play if these preventive systems fail, and they predominantly involve the uses of chemical and/or biological control.

12.5.1 BIOLOGICAL AND CHEMICAL CONTROL IN TROPICAL FOREST PEST MANAGEMENT

Detailed appraisals of biological and chemical control strategies are available elsewhere (e.g. Ivory and Speight, 1993). However, a few general points should be borne in mind. The use of predators, parasitoids and pathogens to regulate insect pest population densities is in general rather hard to achieve, especially in situations where the pests have become a problem due to increased food supply in a favourable climate. Biological control has produced some spectacular success stories, but more often than not the system has failed for a variety of reasons. Single species of parasitoids, and predators in particular, are generally unable to cope with high pest densities due to their fundamental host-finding and manipulation behaviour; pathogens, especially viruses, do not suffer from the same limitations as predators and parasitoids, and so may have a greater potential for pest control in tropical forestry, at least in certain circumstances. A true IPM approach would be either to prevent pest populations reaching high levels or, if they do, to use another control tactic to reduce these densities to a level where biological control can mop up the remainder.

There are few examples of successful biological control on a large scale in tropical forestry so far, though various attempts have been made and are still being investigated. The literature is full of lists of natural enemies recorded from a large number of tropical pest species. Most of this published material merely reports the occurrence of various enemies, with some measure of the percentage mortality caused in the pest population (e.g. teak: Sudheendrakumar, 1986; dipterocarps: Messer *et al.*, 1992; *Leucaena*: Villacarlos and Robin, 1992); there is little or no information about the reductions in pest density and subsequent damage to the trees. Often, there would appear to be little effective reduction of pest attacks; thus, predatory beetles released in Java against *Heteropsylla cubana*, a serious pest of *Leucaena*, had no effect on the damage caused by the sap feeder (Hardi, 1989). However, there is the likelihood of sequential attacks on a single pest species, as its life cycle progresses. For example, *Agrilus sexsignatus* (Coleoptera: Buprestidae), a wood-borer of fast-growing trees such as eucalyptus, is known to be attacked by parasitoids in both the egg and larval stages in the Philippines (Braza, 1989). Average egg parasitism was 47% (maximum 57%), whereas average larval parasitism was 12% (maximum 50%). Combining these two effects, the average overall parasitism within the generation works out as 53% (maximum 86%). The major unknown — typical of many reports describing parasitism and predation on tropical pests — concerns the longer-term effects of this regulation on the pest population density, i.e. will it be brought below the economic threshold and maintained there, thus rendering it effectively harmless?

Optimism is still to be found. *Cinara cupressi* is a serious aphid pest of cypresses in many parts of Africa, including Kenya, Malawi and Tanzania (Ciesla, 1991; Obiri *et al.*, 1994). Unusually, this species also occurs in temperate regions, including the UK, though little harm is done, probably because of the colder climate. The aphid invaded parts of Africa between 1968 and 1986, and because it is an exotic pest, it is possible that natural enemies can be imported from Europe (Mills 1990). *C. cupressi* aggregates on certain individuals of its host tree, *Cupressus lusitanica*, in Kenya (Memmott *et al.*, 1995), which may increase the effectiveness of natural enemies. Unfortunately this tree, which is widely grown as a plantation species, is heavily attacked; a long-term complement to biological control may involve the use of more resistant species, or even hybrids (Obiri, 1994). The International Institute of Biological Control (IIBC) has launched the African Forest Pest Management Programme, which is developing biological control for not only *C. cupressi* but also *H. cubana* and

Phoracantha (S. Murphy, personal communication). The latter pest species also occurs, for example, in southern California, where IPM programmes are being developed that will combine the management of tree stress, resistance selection and forest hygiene with the introduction of egg and larval parasitoids (Paine *et al.*, 1995). The egg parasitoid *Avetianella* sp. (Hymenoptera: Encyrtidae), originally from Australia, has been introduced into South Africa, where over 70% parasitism of *Phoracantha* eggs has been recorded (Tribe, 1992), and IIBC has been releasing the same parasitoid in Zambia during 1995 (Allard, 1995).

Pathogens such as protozoa, nematodes, bacteria, fungi and viruses commonly occur naturally as disease organisms in insects (Sudheendrakumar *et al.*, 1988; Kanga and Fediere, 1991., Misra, 1993). Their spread through a population is heightened as the density of the host insect increases, and hence they have, in theory, great potential to regulate epidemic pest outbreaks. One drawback to the use of protozoa, bacteria and viruses is that they have to be ingested by the host, so that sap feeders and borers that never forage on the outside of the plant will not normally come into contact with the host, even if pathogenicity were possible. The latter is made even less likely since most viruses which are usable, such as the nuclear polyhedrosis viruses (NPVs), are usually host specific, which is important in terms of environmental impact, but unless cross-infective viruses from related species of insect will also kill the target host, then exhaustive searches wherever the pest occurs will be necessary. There is also the problem of application: pathogens would normally be applied to the forest using standard insecticidal technologies, such as spinning disk machines or, more commonly in the tropics, knapsack sprayers. Aerial application is usually impossible for pathogens or chemicals because of the enormous cost and the lack of appropriate technology.

Some viruses crop up naturally in forest pest populations; in the Côte d'Ivoire, for instance, a notodontid moth, *Epicerura pergrisea*, whose larvae defoliate *Terminalia* spp., was found to undergo natural epizootics of an endemic RNA virus (Kanga and Fediere, 1991), whilst larval populations of the pierid butterfly *Eurema blanda*, which severely defoliate *Paraserianthes* (= *Albizia*) in Sabah, have been found to crash at the end of an outbreak due to an unidentified virus infection (Speight, unpublished). The teak defoliator *Hyblaea puera* (Lepidoptera: Hyblaeidae) in southwestern India has long been known to have an effective NPV occurring naturally, and this virus is now being developed as a commercial biocontrol system (K.S.S. Nair, personal communcation; H.F. Evans, personal communication). Virus control of the mahogany shoot-borer, *Hypsipyla* spp., is being actively pursued in both Central America and West Africa (Hauxwell and Speight, unpublished).

The use of synthetic insecticidal chemicals is of course a world-wide popular tactic in all forms of insect pest management, but these chemicals have little real value in tropical forestry. Spraying large stands of maturing trees from the air is out of the question in most tropical forests; even if it was economically or technologically viable, its efficiency is likely to be very limited because of the nature of dense, closed-canopy stands, which prevent the chemical penetrating beyond the surface layer of foliage (Wooster *et al.*, 1989). Chemical control is only appropriate in nurseries and in certain postplanting situations. Even then, the indiscriminate use of insecticides has to be avoided; nursery managers or workers who cannot afford to fail in the production of transplants naturally tend to overspray. This results in all manner of side effects, such as run-off and pollution of drinking water and aquaculture ponds, phytotoxicity of small trees, and a general wasting of severely limited funds. It is vital that simple but dependable impact studies of insect (and fungal) pests in tropical forest nurseries are made available to the end user. Perhaps the most appropriate (and unavoidable) use of insecticides concerns the protection of young transplanted trees from root attack by termites. Many recipes for preventing this sort of damage are available, from the use of resistant, indigenous tree species, to the use of mulches that provide termites with alternative food, to initial overplanting so that some plants can be sacrificed whilst still leaving a commercial stocking density after the attacks (Wardell and Wardell, 1990). In most cases, such as the establishment of large eucalypt forests in Africa, India and South-East Asia, the only dependable system involves the treatment of each planting hole with a persistent insecticide. Until fairly recently, the notorious cyclodeine organochlorines performed this task efficiently and cheaply, but these compounds are, in the main, no longer available and have been replaced in some regions by controlled-release granular formulations of carbosulfan (Marshall suSCon – copyright Incitec Ltd, Brisbane, Australia). Unfortunately, this new system is extremely expensive, so that the cost of the chemical treatment per tree may be several times as much as the value of the transplant itself.

12.6 THE FUTURE OF TROPICAL FOREST PEST MANAGEMENT

There are several key requirements for pest management in tropical forestry, which are summarized in Table 12.4. Some of these requirements have been described earlier, such as the need for in-depth knowledge of the biology and ecology of the pest and its impact on trees. Quantitative surveys of pest incidence on tropical trees are rare so far in the developing world, either because local people do not have the training or education to collect data in an analysable format, or because expatriate consultants have not had the opportunity to set up sampling protocols. One exception involves a potentially serious pest of neem (*Azadirachta indica*), the so-called neem scale insect, *Aonidiella orientalis*. The scale is implicated in a phenomenon called neem decline, where the tops of trees in towns and some shelterbelts exhibit bare branches with small terminal tufts of leaves (Boa, 1994). In Nigeria, protocols for wide-ranging surveys of the decline and also of the scale insect have been provided, so that local villages or districts can assess their own pest problems. Each region is provided with a standard set of instructions and a comprehensive questionnaire which they return to a central agency, the Afforestation Project Co-ordinating Unit (APCU), where analysis and graphical interpretation of the survey data will be carried out (Boa, 1994). In this way, it should be possible to monitor the progress of neem decline, and to study the impact of neem scale on tree health.

Table 12.4 Components required for efficient tropical forest pest management in the future

Research ⟶	Support ⟶	Industry
Basic ecology, taxonomy and impact of insects, as well as pathogens	Databases of pest biology and impact; literature retrieval	Recognition of the equal importance of entomology and pathology in tropical forestry, relative to economics and silviculture
Provenance trials and resistance selection of trees, promotion of indigenous species in high-yield silviculture.	Extension or advisory services readily available to commercial and subsistence growers, using well trained local expertise	Recognition of the importance of prevention rather than cure in pest management, with a willingness to alter forest practices accordingly
Economically viable, appropriate technology systems for pest management	Incorporation of entomolgy and pathology in international aid schemes	
Easy collaboration within and between research workers in tropical countries	Enhanced funding for R and D and the provision of support systems	Consultations with entomologists and pathologists at the planning stages of new and continuing afforestation projects

Clearly, it is of little use to obtain such data without a reliable system for its dissemination to planners, economists and silviculturalists. Perhaps the starting point for this type of system is the publication of insect checklists which provide information on the insect species that cause damage to trees in particular countries. Very few are so far available; the excellent book on pests and diseases of tropical forest trees by F.G. Browne (1968) is sadly long out of date, and in desperate need of revision, but the checklist of forest insects in Thailand (Hutacharern and Tubtim, 1995) provides a good example of what can be made available to initiate searches into forest insect pests and their ecology. Certain tropical countries have produced extremely useful books describing their own forest pests, such as Ghana (Wagner *et al.*, 1991) and the Solomon Islands (Bigger, 1988), but on a broader geographical scale, information retrieval can be enhanced by database maintenance, based initially on simple IBM-compatible computers and latterly being developed on CD-ROM. In the early 1980s, INSPIRE (Interactive Species Information Retrieval) was produced as an aid to tropical forestry in selecting appropriate tree species for their planting sites and also for the particular use of the product, such as pulp, veneer, fuel, etc. (Webb *et al.*, 1984). Although somewhat limited in scope, this computer program provided entomologists and silviculturalists with a quick method of predicting whether or not a particular tree species would be well suited to a site; in other words, was it likely to grow vigorously? In a large number of cases of pest outbreaks in tropical forestry, INSPIRE was able to predict that the chosen trees were not at all suited to the environment in which they had been planted (Ivory and Speight, 1993; Speight, 1995). Unfortunately, this very helpful tool is now out of print and its data are somewhat old. Plans to provide it with a modern interface and up-to-date data have yet to receive substantial funding, though the CAB International Forestry Compendium intends to provide such a product (C. Ison, personal communication). This lack of basic information is clearly alarming, since tropical forestry proceeds apace, leaving behind important decision-making tools for entomologists and silviculturalists. A more recent system, PROSPECT (Programmed Retrieval Of Species by the Property and End-use Classification of their Timbers), is designed expressly to encourage the use of more species of tropical timbers (Smith *et al.*, 1994). This system assumes that monoculture forests are not desirable, and that by using more species of timber tree in the same locality, the diverse nature of a tropical forest can at least be approached. As with INSPIRE, PROSPECT does not directly address forest insects, except where they may affect timber quality. Thus it provides information on a species susceptibility to pinhole borers (ambrosia beetles), powder post beetles, termites and even marine borers, which is thus a useful tool in post-harvest forest protection.

Three other databases may alleviate the problem of the acquisition of background knowledge. CAB International already market TREE-CD and CABPEST-CD which provide a wealth of abstracted literature, describing insect (and other pest) problems and, less frequently, their solutions. In this way, though one-to-one collaboration may be difficult due to lack of funds for international travel, workers may familiarize themselves with the findings of others in a similar situation (assuming that the work has been published in an abstracted journal). The CAB Crop Protection Compendium provides much more detail on selected pest species, including taxonomic keys (C. Schotman, personal communication). The first part of the Compendium, dealing with South-East Asia, became available in early 1997 and, commendably, the price has been set at a realistic figure for developing countries.

Another problem for the efficient and appropriate implementation of tropical forest pest management concerns socio-economic or educational status of the country or region. Some years ago, it was pointed out that, in South-East Asia at least, research activities in forest insect biology and control were inadequate in comparison with the pest problems in the region (Suratmo, 1982). This was due to insufficient financial support, lack of opportunities for collaboration, and the unavailability of trained personnel. In India, where most of the forests are owned by the government, forest pest control has been accorded low priority relative to the more pressing pest problems in agriculture (Nair, 1991). The success of future IPM tactics in the forests will depend on coordinated research and administrative efforts. In 1993, an International Union of Forest Research Organizations (IUFRO) symposium was held in Kerala, southern India, to consider the impact of diseases and insect pests in tropical forests (Nair and Verma, unpublished). It was clear that basic biology and impact data of forest pests was woefully inadequate, and one of the recommendations of the symposium was the urgent need for increased international funding to help this research, and to encourage collaboration within the tropical world and also between temperate and tropical regions. Very few, if any, of the recommendations seem to have yet been implemented. So far, then, the picture appears bleak.

The underlying root of inadequate knowledge, expertise, technology and education lies in the lack of funding for research, development and extension. There is some hope on the horizon, with the development of multidisciplinary approaches to integrated pest management in tropical forestry. One example will illustrate how this system might work.

12.7 THE GOAL OF MULTIDISCIPLINARY PEST MANAGEMENT – EXAMPLE OF MAHOGANY SHOOT-BORER

For a very long time, Lepidoptera in the genus *Hypsipyla* have been known to be serious pests of trees in the Meliaceae. Certainly, pre-war experiences in Latin America pointed to the reductions in yield and tree quality caused by the boring activities of these moth larvae (Anon., 1940, DeLeon, 1941). After the war, large-scale trials with a variety of mahogany species in many countries showed that the moths were polyphagous and very destructive (e.g. Ardikoesoema and Dilmy, 1956). By the 1970s it was apparent that two species of mahogany shoot-borer – *Hypsipyla robusta* and *Hypsipyla grandella* – were problematic in more than 25 tropical countries, from both the old and new worlds (Browne, 1968; Grijpma, 1973; Ivory and Speight, 1993). Detailed examinations of larvae and adults as well as their behaviour and ecology now strongly suggest that there is a complex of species within the genus, only some of which are serious threats (Hauxwell and Speight, unpublished); it is hoped that DNA analysis will clear up some of this confusion. The peak of scientific work directed against the pest appeared in the 1970s (Newton *et al.*, 1993), without much long-term success, and very little mahogany is being planted anywhere. Now, new tactics for control which integrate various strategies are being considered.

The host range of tree species on which the moth larvae feed is wide; most species within the Meliaceae are suitable, including *Swietenia macrophylla* (mahogany), *S. mahagoni*, *Cedrela odorata*, *C. fissilis*, *Toona ciliata* (toona), and many species of *Khaya* (African mahogany) (Browne, 1968; Akanbi, 1973; Baksha, 1990; Yamazaki *et al.*, 1990). The susceptibility of host species varies; for example, *Swietenia* spp. are reported to be less susceptible than *Cedrela* spp. (Dourojeanni-Ricordi, 1963; Yamazaki *et al.*, 1990). Even within a genus such as *Khaya*, some species planted at the same location are more heavily attacked than others (Ardikoesoema and Dilmy, 1956), though the genetic basis for these observations is not always replicable, and environmental parameters could be more important. If the situations in Costa Rica and Australia are compared, it seems that *Toona ciliata*, indigenous to Australia, is more heavily attacked than the exotic *Cedrela odorata* in the same locality; the reverse is true in Costa Rica, where the indigenous *Cedrela* is badly attacked and exotic *Toona* is less attacked (Grijpma, 1973; Hauxwell and Speight, unpublished). Recently, extensive field trials, also in Costa Rica, have shown significant differences in attack rates and damage levels on different genotypes of *S. macrophylla* and *C. odorata* (Newton *et al.*, 1996), and there appears to be a lot of scope for research and development in this area.

Seasonality in attack intensity varies; in some countries, there are marked links between seasonal reflushing of trees at the start of the wet season and the increasing intensity of shoot-borer attacks (Morgan and Suratmo, 1976; Yamazaki, 1987), though in Colombia no clear effects of wet and dry seasons could be detected (Vega-Gonzalez, 1987). During outbreaks, pest generations overlap, and all stages in the life cycle can be found at the same time. A typical *Hypsipyla* life cycle starts when the adult female lays her eggs on leaves or young shoots (though other parts of the tree may be used) (Browne, 1968); between 200 and 300 eggs per female may be laid (Grijpma, 1973). The larvae then tunnel in shoots, though flowers, fruits or even bark may also be attacked. They reappear at intervals, either to colonize a new shoot, or to feed temporarily on external plant tissues. Pupation occurs either in the tunnels or elsewhere in a concealed place on the tree, and the life cycle is completed in approximately 23 to 33 days (Achan, 1968), though colder or more seasonal climates will prolong this period. In heavy attacks, as many as 20 to 40 separate wounds have been observed in terminal stems (Wagner *et al.*, 1991). Although the host tree is rarely killed, terminal buds are damaged and this results in forking, the development of crooked stems and permanent stunting. Progressive attacks produce typical bushy tops to trees (Morgan and Suratmo, 1976) and considerable losses of seed are also reported (Browne, 1968). *Hypsipyla* is normally a pest of young trees; attacks are most severe in the first 2 to 6 years (Morgan and Suratmo, 1976; Vega-Gonzalez, 1987). In Indonesia, 90% of trees of this age, 2–3 m high, were heavily attacked, whereas fewer than 5% of trees older than 14 years or taller than 13 m are damaged (Suratmo, 1977). Quantitative impact data are rather sparse, but Newton *et al.* (1993) provide some information, suggesting that *Hypsipyla* is probably the most important constraint on the commercial planting of various mahogany species in the moist tropics. If it were possible to prevent pest attacks to young trees up to 6 to 8 years, a viable final crop might well be achievable, and it is to this end that a multidisciplinary approach is being directed by collaborative research teams from several countries, including the U.K. and Australia (see Floyd and Hauxwell, 1997).

Figure 12.2 summarizes management tactics being pursued. The commercial use of insecticides, except perhaps in small nursery systems, can be ruled out. Effective control using these chemicals would require repeated sprayings throughout the oviposition period (Ramirez-Sanchez, 1966).

Silvicultural control involves modifications of the forest ecosystem at planting, either by mixing tree species or by altering tree spacing to influence shading on the young crop. Trees growing in sub-optimal sites may be favoured by *Hypsipyla* because of differential shoot quantity and quality (Entwistle, 1967), and the links between trees and site have yet to be established. Mixed planting of mahoganies is being assessed in the Ivory Coast, (Dupuy and Koua, 1993); this might reduce the ability of host-plant-seeking adult moths to find suitable trees, and although mixed line plantings did not reduce pest problems in Peru (Yamazaki *et al.*, 1990), Brunck and Mallet (1993) feel that mixing could form part of an integrated pest management tactic for *Hypsipyla*. Plant resistance has been mentioned above, and shows promise (Newton *et al.*, 1995). Research at Oxford is looking at host plant chemistry in *Cedrela* and *Toona* (Wright, 1996), and if resistant clones can be reliably identified, then shoot tip explant culture techniques are available to propagate them (Maruyama *et al.*, 1989). Host-finding and oviposition behaviour of adult moths is being investigated in Australia (D. Spolc *et al.*, CSIRO, unpublished).

Biological control, especially employing *Hypsipyla*-specific parasitic wasps and flies, has received the highest profile for a long time. Large numbers of species of parasitic Hymenoptera have been reared from both species and most life stages of *Hypsipyla* (e.g. de Sentis, 1973; Hidalgo *et al.*, 1973), and Newton *et al.* (1993) suggest that nearly 50 parasitoid species have been identified so far. However, because of the prolific nature of the pest, coupled with its concealed habits which prolong the time needed for enemies to find and deal with their hosts, little commercial potential for this type of biological control seems to exist at the moment. This form of management is particularly problematic since *Hypsipyla* is a low density pest, so that only one attack on a susceptible tree will destroy its marketability.

Finally, there is potential control available using insect-pathogenic micro-organisms. Two fungi – *Beaveria* spp. and *Metarrhizium* sp. – have been shown to kill *Hypsipyla* larvae in the laboratory (Berrios *et al.*, 1971; Hidalgo-Salvatierra and

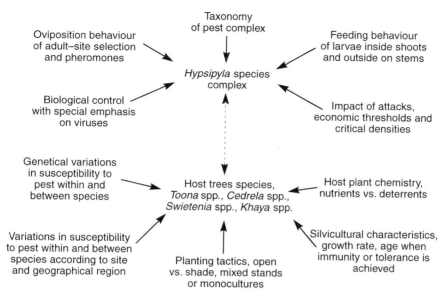

Figure 12.2 Multidisciplinary approach to the research and development of the management of mahogany shootborer, *Hypsipyla* spp.

Palm, 1972), and *Beauveria bassiana* has been isolated from dead *Hypsipyla* larvae from field collections in India (Misra, 1993). Use of these pathogens on a commercial scale has not been attempted. Pathogenic viruses specific to *Hypsipyla* have yet to be identified, but if and when found, they would present an important new tool in the control of the pest. To this end, research based in Oxford and funded by the Forest Research Programme of ODA is presently exploring various sites around the tropical world, such as Costa Rica, Ghana and Sri Lanka. It has already been established that *Hypsipyla* larvae are susceptible to NPVs from other lepidopteran species (Hauxwell, Cory and Speight, unpublished).

No single solution is ever likely to solve the problems of a pest as serious and widespread as *Hypsipyla*. Whilst this system has a long way to go before optimal combinations of pest management strategies can be recommended to foresters, it provides a framework for the future of the management of many other serious forest insect pests in tropical forestry. All that is required is a substantial increase in global concern and investments in research and development funding, with the vital extension of information so derived to silviculturalists and entomologists in the field.

ACKNOWLEDGEMENTS

The author is very grateful to various colleagues who provided unpublished information, and/or critically reviewed the manuscript. Special thanks go to Andy Foggo and Carrie Hauxwell (Department of Zoology, University of Oxford), Ross Wylie (Queensland Forest Service, Brisbane, Australia), Chris Ison, David Nicholson and Charles Schotman (CAB International, Wallingford, UK), Hugh Evans (Forestry Commission, UK), K.S.S. Nair (Kerala Forest Research Institute, India), Adrian Newton (Department of Forestry, University of Edinburgh), Sean Murphy (International Institute of Biological Control, UK), Eric Boa (International Mycological Institute, UK) and, last but not least, Allan Watt (ITE, Scotland). Some of the work described is funded by the Forestry Research Programme, Overseas Development Administration, UK.

REFERENCES

Achan, P.D. (1968) Preliminary observations on the development of an artificial diet for *Hypsipyla robusta* Moore. *Bulletin of Commonwealth Institute of Biological Control* **10**: 23–26.

Akanbi, M.O. (1973) *The Major Insect Borers of Mahogany in Nigeria: a review towards their control*, Research Paper Forest Series No. 16, Federal Department of Forest Research, Nigeria, 8pp.

Allard, G.B. (1995) Report on a visit to Zambia to carry out a biological control programme for the eucalypt borer, *Phoracantha* spp. Unpublished report, International Institute of Biological Control.

Anonymous (1940) Report of the Chief of the Bureau of Entomology and Plant Quarantine, United States.

Ansley, R.J., Meadors, C.H. and Jacoby, P.W. (1990) Preferential attraction of the twig girdler, *Oncideres cingulata texana* Horn, to moisture-stressed mesquite. *Southwestern Entomologist* **15**(4), 469–474.

Ardikoesoema, R. and Dilmy, A. (1956) Some notes on mahogany, especially *Khaya*. *Rimba Indonesia* **5/6**, 266–328.

Baksha, M.W. (1990) *Some Major Forest Insect Pests of Bangladesh and their Control*, Bulletin No. 19, Forest Entomology Series, Forest Entomology Institute, Chittagong, 19 pp.

Berrios, F. and Hidalgo-Salvatierra, O. (1971) Studies on the shoot borer *Hypsipyla grandella*. VI, Susceptibility of the larvae to the fungus *Metarrhizium anisopliae*; VIII, Susceptibility of the larvae to the fungi *Beauveria bassiana* and *B. tenella*. *Turrialba* **21**(2), 214–219; **21**(4), 451–454

Bigger, M. (1988) *The Insect Pests of Forest Plantation Trees in the Solomon Islands*, Solomon Islands' Forest Record No. 4, 190 pp.

Boa, E.R. (1994) *Survey of Neem Scale Infestations in Nigeria: a guide to methods and procedures*, FAO Report TCP/NIR/4452(e).

Bray, R.A. and Woodroffe, T.D. (1991) Effect of the leucaena psyllid on yield of *Leucaena leucocephala* cv Cunningham in south-east Queensland. *Tropical Grasslands* **25**(4), 356–357.

Braza, R.D. (1988) Control of varicose borer in PICOP's bagras plantation. *Canopy International* **14**(2), 7–9.

Braza, R.D. (1989) Parasitoids of immature stages of *Agrilus sexsignatus* (Fisher) (Coleoptera: Buprestidae) attacking *Eucalyptis deglupta* Blume in Surigao del Sur. *Philippine Entomologist* **7**(5), 479–483.

Browne, F.G. (1968) *Pests and Diseases of Forest Plantation Trees*, Clarendon Press, Oxford.

Brunck, F. and Mallet, B. (1993) Problems relating to pests attacking mahogany. *Bois et Forêts des Tropiques* **237**, 9–28.

Ciesla, W.M. (1991) Cypress aphid, *Cinara cupressi*, a new pest of conifers in eastern and southern Africa. *FAO Plant Protection Bulletin* **39**(2–3), 82–93.

Coll, M. and Bottrell, D.G. (1994) Effects of non-host plants on an insect herbivore in diverse habitats. *Ecology* **75**(3), 723–731.

Corbett, A. and Plant, R.E. (1993) Role of movement in the response of natural enemies to agroecosystem diversification: a theoretical evaluation. *Environmental Entomology* **222**(3): 519–531.

Cornic, J.F. (1978) Influence of soil water content on the swarming of *Hexacolus guyanensis* Schedl., mahogany (*Swietenia macrophylla*) bark beetle, in Guadeloupe: preliminary study. *Nouvelles Agronomiques des Antilles et de la Guyane* **4**(3/4), 325–337.

Cowie, R.H. and Wood, T.G. (1989) Damage to crops, forestry and rangeland by fungus-growing termites (Termitidae: Macrotermitinae) in Ethiopia. *Sociobiology* **15**(2), 139–153.

Cowie, R.H., Logan, J.W.M. and Wood, T.G. (1989) Termite (Isoptera) damage and control in tropical forestry with special reference to Africa and Indo-Malaysia: a review. *Bulletin of Entomological Research* **79**(2), 173–184.

DeLeon, D. (1941) Some observations on forest entomology in Puerto Rico. *Caribbean Forestry* **2**, 160–163.

De Sentis, L. (1973) A new neotropical microgasterine (Hymenoptera: Braconidae) parasite of the larvae of *Hypsipyla grandella*. In *Proceedings of the First Symposium on Integrated Control of* Hypsipyla, Turrialba (ed. P. Grijpma), pp. 71–73, IICA Misc. Publ. No. 101.

Dourojeanni-Ricordi, M. (1963) The shoot borer of cedrela and mahogany, *Hypsipyla grandella*. *Agronomia, La Molina* **30**(1), 35–43

Dupuy, B. and Koua, M. (1993) The African mahogany plantations: their silviculture in the tropical rain forest of the Côte d'Ivoire. *Bois et Forêts des Tropiques* **236**, 25–42.

Entwistle, P.F. (1967) Current situation on shoot, fruit and collar borers of the Meliaceae. *Proc. 9th Brit, Commonwealth Forestry Conference*, C.F.I., Oxford, 15 pp.

Floyd, R.B. and Hauxwell, C. (1997) Proceedings of the ACIAR/ODA International Workshop on *Hypsilyla* shootbonners in the Meliaceae. Australian Centre for International Agricultural Research Proceedings Series, Canberra (in press).

Foggo, A. and Speight, M.R. (1993) Root damage and water stress: treatments affecting the exploitation of the buds of common ash *Fraxinus excelsior* by the larvae of the ash bud moth *Prays fraxinella*. *Oecologia* **96**(1), 134–139

Gara, R.I., Sarango. A. and Cannon, P.G. (1990) Defoliation of an Ecuadorian mangrove forest by the bagworm, *Oiketicus kirbyi* Guilding (Lepidoptera: Psychidae). *Journal of Tropical Forest Science* **3**(2), 181–186.

Gauld, I. (1991) The Ichneumonidae of Costa Rica : 1 Introduction, keys to subfamilies, and keys to species of the lower pimpliform subfamilies Rhyssinae, Pimplinae, Poemeniinae, and Cylloceriinae. *Memoirs of the American Entomological Institute* **47**, 1–581.

Grijpma, P. (1973) (ed.) *Proceedings of the First Symposium on Integrated Control of* Hypsipyla, *Turrialba*, IICA Misc. Publ. No. 101.

Hanks, L.M., Paine, T.D., Millar, J.G. and Horn, J.L. (1995) Variation among Eucalyptus species in resistance to eucalyptus long-horned borer in southern California. *Entomologia Experimentalis et Applicata* **74**(2), 185–194.

Hanover, J.W. (1988) *Feasibility Study on Small-farm Production of Gums, Resins, Exudates and Other Non-wood Products*, Multipurpose Tree Species Network Research Series, Paper No. 4, 47 pp.

Hardi, T.W.T. (1989) The dispersal of the leucaena psyllid predator *Curinus coeruleua* in Banyumas Forest District, Central Java. *Buletin Penelitian Hutan* **518**, 23–31.

Harsh, N.S.K. and Joshi, K.C. (1993) Loss assessment of *Albizia lebbek* seeds due to insect and fungus damage. *Indian Forester* **119**(11), 932–935.

Herrero, G., Valdes, E., Rengifo, E. and Gonzales, S. (1991) Preliminary study on the relation between mineral fertilizer treatment and the incidence of *Dioryctria horneana* in *Pinus caribaea*. *Revista Baracoa* **21**(1), 47–58.

Hidalgo, O. and Salvatierra, L.R. (1973) *Trichogramma* sp., an egg parasite. In *Proceedings of the First Symposium on Integrated Control of* Hypsipyla, *Turrialba* (ed. P. Grijpma), pp. 49–50, IICA Misc. Publ. No. 101.

Hidalgo-Salvatierra, O. (1972) Studies on the shoot borer *Hypsipyla grandella*. XIV, Susceptibility of first instar larvae to *Bacillus thuringiensis*. *Turrialba* **22**(4), 467–468.

Howard, F.W. and Meerow, A.W. (1993) Effect of mahogany shoot borer on growth of West Indies mahoganies in Florida. *Journal of Tropical Forest Sciences* **6**(2), 201–203.

Hutacharern, C. and Tubtim, N. (1995) *Check Lists of Forest Insects in Thailand*, Office of Environmental Policy and Planning, Thailand, Volume 1, 392 pp.

Intari, S.E. (1990) Effects of *Neotermes tectonae* Damm attack on the quality and quantity of teak timber in the Kebonharjo Forest Division, Central Java. *Buletin Penelitian Hutan* **530**, 25–35.

Intari, S.E. and Ruswandy, H. (1986) Preferences of *Dioryctria* sp. for three species of pines. *Buletin Penelitian Hutan* **479**, 44–48.

Ivory, M. and Speight, M.R. (1993) Pests and diseases. In *Tropical Forestry Handbook*, Vol. II (ed. L. Pancel), Section 19, Springer Verlag.

Kanga, L. and Fediere, G. (1991) Towards integrated control of *Epicerura pergrisea* (Lepidoptera: Notodontidae), defoliator of *Terminalia ivorensis* and *T. superba*, in the Côte d'Ivoire. Proceedings of 18th International Congress of Entomology, Towards Integrated Pest Management of Forest Defoliators (eds A.G. Raske and B.E. Wickman). *Forest Ecology and Management* **39**(1–4), 73–79.

Khan, M.L. and Tripathi, R.S. (1991) Seedling survival and growth of early and late successional tree species as affected by insect herbivory and pathogen attack in sub-tropical humid forest stands of north-east India. *Acta Oecologica* **12**(5), 569–579.

Kruess, A. and Tscharntke, T. (1994) Habitat fragmentation, species loss, and biological control. *Science* **264**, 1581–1584.

Liu, Y. (1992) Discussion of tropical forest insects on outbreak characteristics. *Forest Research* **5**(5), 570–573.

Makihara, H., Kawamuro, K. and Roberts, H. (1987) Infestation of kamerere plantations attacked by *Agrilus opulentus* Kerramans (Coleoptera: Buprestidae). *Tropical Forestry* **10**, 22–28.

Maruyama, E., Ishii, K., Saito, A. and Migita, K. (1989) Micropropagation of cedro (*Cedrela odorata* L.) by shoot tip culture. *Journal of the Japanese Forestry Society* **71**(8), 329–331.

Memmott, J., Day, R.K. and Godfray, H.C.J. (1995) Intraspecific variation in host plant quality: the aphid *Cinara cupressi* on the Mexican cypress, *Cupressus lusitanica*. *Ecological Entomology* **20**(2), 153–158.

Meshram, P.B., Husen, M. and Joshi, K.C. (1993) A new report of ambrosia beetle, *Xylosandrus compactus* Eichoff. (Coleoptera: Scolytidae) as a pest of African mahogany, *Khaya* sp. *Indian Forester* **119**(1), 75–77.

Messer, A.D., Wanta, N.N. and Sunaya, X. (1992) Biological and ecological studies of *Calliteara cerigoides* (Lepidoptera: Lymantriidae), a polyphagous defoliator of Southeast Asian Dipterocarpacaea. *Japanese Journal of Entomology* **60**(1), 191–202.

Mills, N.J. (1990) Biological control of forest aphid pests in Africa. *Bulletin of Entomological Research* **80**(1), 31–36.

Misra, R.M. (1993) *Beauveria bassiana* (Balsamo) Vuillemin a fungal pathogen of *Hypsipyla robusta* Moore (Lepidoptera: Pyralidae). *Indian Journal of Forestry* **16**(3), 236–238.

Mitchell, M.R. (1989) Susceptibility to termite attack of various tree species planted in Zimbabwe. In *Trees for the Tropics. Growing Australian Multipurpose Trees and Shrubs in Developing Countries* (ed. D.J. Bolan), ACIAR Monograph 10, pp. 215–227.

Morgan, F.D. and Suratmo, F.G. (1976) Host preferences of *Hypsipyla robusta* (Lepidoptera: Pyralidae) in West Java. *Australian Forestry* **39**(2), 103–116.

Mukhtar-Ahmad and Ahmad, M. (1991) Natural resistance in teak clones to leaf skeletoniser, *Eutectona machaeralis* (Lepidoptera: Pyralidae) in south India. *Indian Journal of Forestry* **114**(3), 228–231.

Nair, K.S.S. (1991) Social, economic and policy aspects of integrated pest management of forest defoliators in India. Proceedings of 18th International Congress of Entomology, Towards Integrated Pest Management of Forest Defoliators, Congress of Entomology, Canada (eds A.G. Raske and B.E. Wickman). *Forest Ecology and Management* **39**(1–4), 88–96.

Nair, K.S.S. and Mathew, G. (1992) Biology, infestation characteristics and impact of the bagworm, *Pteroma plagiophelps* Hamps., in forest plantations of *Paraserianthes falcataria*. *Entomon* **17**(1,2), 1–13.

Nair, K.S.S., Sudheendrakumar, V.V., Varma, R.V. and Chako, K.C. (1985) *Studies on the Seasonal Incidence of Defoliators and the Effect of Defoliation on Volume Increment of Teak*, Research Report No. 30, Kerala Forest Research Institute, 78 pp.

Newhook, F.J. (1989) Keynote address: Indigenous forest health in the South Pacific – a plant pathologist's view. *New Zealand Journal of Forest Science* **19**(2/3), 231–242.

Newton, A.C., Baker, P., Ramnarine, S., Mesen, J.F. and Leakey, R.R.B. (1993) The mahogany shoot borer: prospects for control. *Forest Ecology and Management* **57**(1–4), 301–328

Newton, A.C., Cornelius, J., Hernandez, M. and Watt, A.D. (1996) Mahogany as a genetic resource. *Botanical Journal of the Linnaean Society* (in press).

Ngulube, M.R. (1989) Polythene tube sizes for raising eucalyptus seedlings for dryland afforestation programmes in Malawi. *Journal of Tropical Forestry* **5**(1), 30–35.

Obiri, J.F. (1994) Variation of cypress aphid (*Cinara cupressi*) (Buckton) attack on the family Cupressaceae. *Commonwealth Forestry Review* **73**(1), 43–46.

Obiri, J.F., Giathi, G. and Massawe, A. (1994) The effect of cypress aphid on *Cupressus lustitanica* in Kenya and Tanzania. *East African Agricultural and Forestry Journal* **59**(3), 227–234.

Paijmans, K. (ed.) (1976) *New Guinea Vegetation*, CSIRO and ANU Press.

Paine, T.D., Millar, J.G. and Hanks, L.M. (1995) Integrated program protects trees from eucalyptus longhorned borer. *California Agriculture* **49**, 34–37.

Palmer, B., Bray, R.A., Ibrahim, T.M. and Fulloon, M.G. (1989) The effect of the leucaena psyllid on the yield of *Leucaena leucocephala* cv. Cunningham at four sites in the tropics. *Tropical Grasslands* **23**(2), 105–107.

Ramirez-Sanchez, J. (1966) Notes on the control of *Hypsipyla grandella* with insecticides. *Boletin de Instituto Forestal Latino-Americano* **22**, 33–37.

Robertson, A.I. (1990) Plant–animal interactions and the structure and function of mangrove forest ecosystems. *Australian Journal of Ecology* **16**(4), 433–443.

Sajap, A.S. and Siburat, S. (1992) Incidence of entomogenous fungi in the bagworm, *Pteroma pendula* (Lepidoptera: Psychidae), a pest of *Acacia mangium*. *Journal of Plant Protection in the Tropics* **9**(2), 105–110.

Sartwell, C. and Dolph, R.E. Jr (1976) *Silvicultural and Direct Control of Mountain Pine Beetle in Secondgrowth Ponderosa Pine*, USDA Forest Service Research Note, Pacific Northwest Forest and Range Experimental Station PNW-268.

Scriven, G.T., Reeves, E.L. and Luck, R.F. (1986) Beetle from Australia threatens eucalyptus. *California Agriculture* **40**(7–8), 4–6.

Selander, J. and Bubala, M. (1983) *A Survey of Pest Insects in Forest Plantations in Zambia*, Research Note No. 33, Division of Forest Research, Forest Department, Zambia, 33 pp.

Showler, A.T. (1995) Leucaena psyliid, *Heteropsylla cubana* (Homoptera: Psyllidae) in Asia. *American Entomologist* **41**(1), 49–54.

Smith, J.P., Plumptre, R.A., Brazier, J.D. and Dorey, C.E. (1994) *'PROSPECT' for Improved Use of Tropical Timbers*, Tropical Forestry Papers 28, Oxford Forestry Institute, 62 pp.

Speight, M.R. (1995) Host tree stresses and the impact of insect attack in tropical forest plantations. In *International Union of Forest Research Organisations (IUFRO) Symposium*, Kerala Forest Research Institute, Peechi, India (unpublished).

Speight, M.R. and Speechly, H.T. (1982a) Pine shoot moths in S.E. Asia I: Distribution, biology and impact. *Commonwealth Forestry Review* **61**(2), 121–135.

Speight, M.R. and Speechly, H.T. (1982b) Pine shoot moths in S.E. Asia II: Potential control methods. *Commonwealth Forestry Review* **61**(3), 203–211.

Speight, M.R. and Wainhouse, D. (1989) *Ecology and Management of Forest Insects*, Oxford University Press, 335 pp.

Sudheendrakumar, V.V. (1986) *Studies on the Natural Enemies of the Teak Pests,* Hyblaea puera *and*

Eutectona machaeralis, Research Report No. 38, Kerala Forest Research Institute, 23 pp.

Sudheendrakumar, V.V., Ali, M.I.A. and Varma, R.V. (1988) Nuclear polyhedrosis virus of teak defoliator, *Hyblaea puera*. *Journal of Invertebrate Pathology* **51**(3), 307–308.

Suratmo, F.G. (1977) Infestation of the leading shoots of mahogany (*Swietenia macrophylla*) by *Hypsipyla robusta* in West Java, Indonesia. In *BIOTROP – Procs. Symposium on Forest Pests and Diseases in S.E. Asia, Bogor, Indonesia*, pp. 121–132.

Suratmo, F.G. (1982) Pest management in forestry. *Protection Ecology* **49**(3), 291–296.

Suresh-Babu, K.V., Mohankumar, B. and Mathew, T. (1992) Field testing of leucaena germplasm for their relative susceptibility to infestation by the psyllids. *Agricultural Research Journal of Kerala* **30**(2), 135–137.

Thapa, R.S. (1992) Studies of the wide-spread attack of ambrosia beetle, *Platypus* sp., on provenance trials of *Acacia crassicarpa* in Sipitang district of Sabah. *Breeding Technologies for Tropical Acacias. Proceedings of an International Workshop* (eds L.T. Carron and K.M. Aken), pp. 31–34, ACIAR Proceedings Series No. 37.

Tribe, G.D. (1992) New parasitoid for eucalyptus borer discovered. *Bulletin of the Plant Protection Research Institute of South Africa, Plant Protection News* **30**, 16.

Vega-Gonzalez, L.E. (1987) *Growth of Cedrela odorata Managed within a Secondary Shrub Vegetation, or in Initial Association with Agricultural Crops*, CONIF-Informa No. 10, 18 pp.

Villacarlos, L.T. and Robin, N.M. (1992) Biology and potential of *Curinus coeruleus* Mulsant, an introduced predator of *Heteropsylla cubana* Crawford. *Philippine Entomologist* **8**(6), 1247–1258.

Wagner, M.R., Atuahene, S.K.N. and Cobbinah, J.R. (1991) *Forest Entomology in West Tropical Africa: Forest Insects of Ghana*, Kluwer Academic Publishers.

Wang, H.J. (1994) Tea and trees: a good blend from China. *Agroforestry Today* **6**(1), 6–8.

Wardell, A. and Wardell, D.A. (1990) The African termite: peaceful coexistence or total war? *Agroforestry Today* **2**(3), 4–6.

Watt, A.D. (1994) The relevance of the stress hypothesis to insects feeding on foliage. In *Individuals, Populations and Patterns in Ecology* (eds S.R. Leather, A.D. Watt, N.J. Mills and K.F.A. Walters), pp. 53–70, Intercept, Andover.

Webb, D.B., Wood, P.J., Smith, J.P. and Henman, G.S. (1984) *A Guide to Species Selection for Tropical and Sub-tropical Plantations*, Commonwealth Forestry Institute Tropical Forestry Papers no. 15, 256 pp.

White, T.C.R. (1993) *The Inadequate Environment – Nitrogen and the Abundance of Animals*, Springer-Verlag, Berlin, 425 pp.

Whitmore, J.L. (1976) Studies on the shoot borer *Hypsipyla grandella* Lep. In *Pyralidae*, Volume II, Misc. Publication IICA 101, 137 pp.

Whitmore, J.L. (1978) *Cedrela Provenance Trial in Puerto Rico and St. Croix: Establishment Phase*, USDA Forest Service Research Note No. 1TF-16, Institute of Tropical Forestry, Puerto Rico, 11 pp.

Withington, D. and Brewbaker, J.L. (eds) (1987) *Proceedings of a Workshop on Biological and Genetic Control Strategies for the Leucaena Psyllid, November 3–7, 1986, Molokai and Honolulu, Hawaii*, Leucaena Research Reports, 1987, 7(2).

Wooster, M.T., Need, J.T., DuBose, L.A. and Stevenson, H.R. (1989) Malathion penetration of a rain forest canopy following aerial ULV application. *Journal of the Florida Anti Mosquito Association* **60**(2), 62–65.

Wormald, T.L. (1992) *Mixed and Pure Forest Plantations in the Tropics and Sub-tropics*, FAO Forestry Paper no. 103, FAO, Rome, 152 pp.

Wright, G. (1996) The relationship between herbivory and secondary plant compounds in *Cedrela odorata* and *Toona ciliata* in plantation. In *Proceedings of 9th International Symposium on Insect–Plant Relationships, Gwatt* (eds E. Stadler, M. Rowell-Rahier and R. Baur), Kluwer Academic Press, reprinted from *Ent. Exp. et App.* **80**(1).

Wylie, F.R., Johnston, P.J.M. and Forster, B.A. (1993) *Decline of Casuarina and Eucalyptus in the Mary River Catchment*, Queensland Forest Service Forest Research Institute Research Paper No. 17, 34 pp.

Yamazaki, S. (1987) Serious damage to mahogany by the shoot borer *Hypsipyla grandella* Zeller (Lepidoptera: Pyralidae). *Tropical Forestry* **8**, 26–34.

Yamazaki, S., Taketani, A., Fujita, K. *et al.* (1990) Ecology of *Hypsipyla grandella* and its seasonal changes in population density in Peruvian Amazon forest. *Japan Agricultural Research Quarterly* **24**(2), 149–155.

THE IMPACTS OF CLIMATE CHANGE AND POLLUTION ON FOREST PESTS

13

Maureen Docherty, David T. Salt and Jarmo K. Holopainen

13.1 INTRODUCTION

The quality of air in the environment we inhabit has been changing for over two centuries, since the industrial revolution. Industrial processes have caused increases in airborne pollutants, among them SO_2 and NO_x, leading to acid rain, ozone formation and increased nitrogen deposition, alongside increased ultraviolet-B (UV-B) exposure. Increased combustion of fossil fuels has also produced rising levels of atmospheric CO_2 likely to exceed 600 μl/l in the next century (Houghton *et al.*, 1992), contributing to predicted rises in global temperatures and changes in climate. Recent predictions are that a rise of up to 2°C will occur over the next century (Houghton *et al.*, 1992). Changes in land use have also had important impacts on the environment and can exacerbate the effects of these changes (Vitousek, 1994). The aim of this chapter is to review the effects of pollution and climate change on the performance of insect herbivores, particularly forest pest species. We have tried, where possible, to report studies on tree-feeding insects. However, where there is a paucity of such studied interactions, we have selected other studies as a guide to possible effects.

The literature on the effects of air pollutants on insects has been extensively reviewed, most recently by Brown (1995) and Whittaker (1994) and also by Riemer and Whittaker, 1989 (concentrating on field observations), McNeill and Whittaker, 1990 (on tree-feeding aphids), Whittaker and Warrington, 1990 (concentrating on causal mechanisms) and by others (e.g. Alstad *et al.*, 1982; Hain, 1987; Hughes, 1988; Heliövaara and Väisänen, 1990a (on pine-feeding insects). The literature on the effects of elevated CO_2 on insects has been covered by only one review (Watt *et al.*, 1995) though wider reviews have been undertaken (e.g. Bazzazz, 1990). Effects of increased UV-B on herbivores are represented by only three published studies (Hatcher and Paul, 1994), whereas aspects of insect behaviour and performance due to changing temperature have been examined by several authors (in Harrington and Stork, 1995).

The data collected on effects of pollution and climate change on insects arise from both field observations near pollution sources and a number of experimental systems (Brown, 1995). Although a few field manipulation experiments using O_3 and SO_2 have been undertaken on trees (e.g. McLeod, 1995) and free air CO_2 enrichment (FACE) experiments are under way (Ellsworth *et al.*, 1995), problems of scale and providing realistic treatments combining pollutants with other climate change treatments pose difficulties. Thus most studies on trees and associated herbivores have been conducted in chambers under natural or artificial light, some partially open to the environment and generally using single pollutant treatments.

While concentrations of CO_2 and UV-B exposure rise globally, local concentrations of oxides of nitrogen and sulphur in western Europe are declin-

---

Forests and Insects. Edited by A.D. Watt, N.E. Stork and M.D. Hunter. Published in 1997 by Chapman & Hall, London. ISBN 0 412 79110 2.

ing, though they are still rising in industrializing areas of the world (Brown, 1995). Trees are likely to experience changes in pollutant concentrations and climate during their lifetime. As trees are major regulators of global carbon flux dynamics (Eamus and Jarvis, 1989), their response and adaptation to these conditions will be important for interactions within the biosphere. Changes in growth, phenology, plant chemistry and ultimately genetic shifts will have important consequences for insect growth, population dynamics and phytophage interactions within forests.

13.2 CARBON DIOXIDE AND CLIMATE CHANGE

13.2.1 EFFECTS OF CARBON DIOXIDE AND CLIMATE CHANGE ON TREES

Experimental evidence is consistent with the theory that CO_2 and gaseous pollutants act via the plant in modifying food quality and not directly on the insects at the concentrations normally experienced (Braun and Fluckiger, 1989; Nicolas and Sillans, 1989; Houlden et al., 1990). By contrast, heavy metal pollutants are thought to act directly on insects (Alstad et al., 1982). Some guilds, e.g. leaf-miners (Koricheva and Haukioja, 1992, 1995) and root-feeding aphids (Salt and Whittaker, 1995), are not generally exposed to such high concentrations of gaseous pollutants as insects feeding at other sites on the trees, yet can still respond to plant-mediated changes.

Many studies, both short- and long-term, have examined the effects of elevated CO_2 on phenology, productivity, photosynthesis and plant chemistry of broadleaved and coniferous tree species commonly found in temperate woodlands (Ceulemans and Mousseau, 1994; Murray et al., 1994). Tree biomass generally increases under elevated CO_2 but its partitioning depends on the species and nutrient status (Norby et al., 1992; El Kohen et al., 1993; Mousseau, 1993). In general, trees growing in low nutrient forest soils have enhanced biomass partitioning to the roots in elevated CO_2, compared with those trees grown in soils of high nutrient availability. Extrapolation from work on individual tree species in elevated CO_2 indicates that woodland composition may gradually change due to the variability in responses, and hence modified competitive interactions in both temperate and tropical forests.

Bud dormancy in temperate forest trees is broken by an accumulation of chilling, followed by a period of higher temperatures (Murray et al., 1989; Cannell, 1990). In contrast, the cessation of growth in autumn is daylight dependent. Therefore any alteration in air temperature would be most important in spring. In a study of 15 tree species, Murray et al. (1989) determined that budburst in lowland British sites should not be markedly affected by increased temperatures but that upland sites should experience earlier budburst. This change in the phenology has important implications for insect–plant interactions (Dewar and Watt, 1992). Unlike trees of temperate forests, the phenology of tropical trees is influenced by precipitation, through changes in soil moisture and plant water status (Reich and Borchert, 1984; Reich, 1995). Hence climatic changes affecting water balance can influence phenology.

Recently effects of elevated CO_2 have been found to cause increased leaf surface temperatures (Idso et al., 1993) which occur alongside changes in food quality and have important consequences for those insects living close to the leaf surface (Brooks, 1995). The localized increase in temperature where insects feed and develop would occur as a result of CO_2 modifying stomatal functioning, independently of global temperature rises. Until recently these effects have been largely ignored, though the importance of reduced evaporative cooling as a result of increased stomatal closure was recognized on ozone fumigated *Asclepias* plants, by Bolsinger et al. (1991), who suggested that it may lead to faster development of monarch butterfly larvae.

It is generally accepted that food quality for chewing herbivores will decline at elevated levels of atmospheric CO_2 (Watt et al., 1995); effects for sucking insects are still under investigation (Docherty, unpublished data). Reductions in concentrations of foliar nitrogen in elevated CO_2 are well established (Watt et al., 1995). However, the impact of secondary plant chemistry, particularly phenolic compounds, has only recently received attention (Lincoln, 1993; Lincoln et al., 1993; Lindroth et al., 1995; Docherty et al., 1996). Fast and slow growing tree species allocate different amounts of resources into defence compounds (Price, 1991) but defensive chemicals are also influenced by nutrients and water supply (Lindroth et al., 1993b). In elevated CO_2 extra fixed carbon may be allocated to the production of carbon-based defence compounds or stored as starch. It is generally considered that investment in defensive compounds is lowest in plant species adapted to high nutrient conditions and that these will show greater plasticity in the production of such chemicals as atmospheric CO_2 levels increase (Coley et al., 1985; Lindroth et al., 1993b). The concentration of foliar water can also decrease in elevated CO_2, which can affect the digestibility of plant tissue (Watt et al., 1995).

13.2.2 EFFECTS OF ELEVATED CO_2 ON INSECTS

Most tree-feeding insects investigated in recent studies have responded to feeding on elevated CO_2-grown foliage by a reduction in growth rate (Figure 13.1), in line with the decrease in available plant nitrogen (Figure 13.2) (Lindroth et al., 1993b, 1995; Roth and Lindroth, 1994; Williams et al., 1994). Although food consumption (Figure 13.3) often increased, final weights attained were lower (Roth and Lindroth, 1994; Lindroth et al., 1993b), indicating that compensation for lower food quality was incomplete, which may in turn affect fecundity (Honek, 1993). Measurement of nutritional indices indicated that digestive and metabolic efficiencies decreased on average in elevated CO_2, possibly as a result of the increased concentrations of secondary compounds. These studies have largely been conducted on chewing insects on detached leaves but recently leaf-mining and sap-sucking insects on growing trees have been tested (Docherty et al., 1996, 1997), the former indicating no within-generation changes in performance except increased consumption by female larvae. A recent investigation into effects of elevated CO_2 on herbivory by a generalist leaf-chewer, Spodoptera, in a mixed tropical community including Ficus and Cecropia led to suggestions that nutrient limitation may be responsible for a lack of effects on herbivore food quality and thence larval performance (Arnone et al., 1995). At the population level, the abundance of four out of six insects arising from natural infestations was significantly lower in elevated CO_2 on non-tree hosts in greenhouse and open-topped chamber experiments (Butler, 1985; Butler et al., 1986; Tripp et al., 1992). Insect guilds involved included sap-suckers and leaf-chewers. Experiments using the leaf beetle Gastrophysa viridula on the herb Rumex obtusifolius indicate that it may take several generations before the effects of elevated CO_2 on insect performance are realized (Brooks, 1995). The evidence so far, from small-scale experiments, indicates that elevated CO_2, in the absence of global warming, may cause slower growth rates, lower fecundity and compensatory feeding by herbivorous insects. However, until between-generation performance is also considered, with respect to elevated CO_2, and then put into the context of the dynamics of the species in the field, firm predictions of the fate of herbivorous insects in an environment with increasing atmospheric CO_2 cannot be made.

13.2.3 EFFECTS OF ELEVATED TEMPERATURE ON INSECTS

Optimal development and reproduction of herbivorous insects in Britain occur between 20 and 25°C (Harrington et al., 1995). An increase in average temperature should hasten development and the rate of reproduction of aphids in countries like Britain, in which higher temperatures are rarely sustained. Parthenogenetic development of tree-feeding aphids is unlikely to be extended by higher autumn temperatures as tree dormancy uses photoperiod as its cue (Murray et al., 1989). Fecundity of the population of grey pine aphid, Schizolachnus pineti, in central Finland has a peaked response curve with maximum numbers of offspring associated with a 6°C temperature increase over current mean daily temperature in June, while the developmental period of nymphs from birth to reproduction decreased linearly with the increase in mean

232 *Impacts of climate change and pollution*

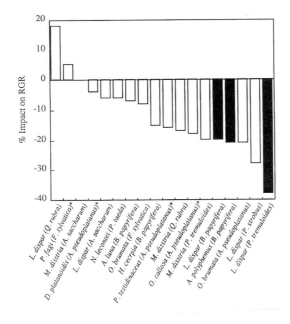

Figure 13.1 Impact of elevated CO_2 on the relative growth rate of insect herbivores. Each column represents the percentage increase or decrease in MRGR of insects reared on elevated CO_2-grown foliage. Significant differences due to the impact of elevated CO_2 ($P <0.05$) shown by ■; * denotes non-chewing insects. (Sources: Lindroth *et al.*, 1993b, 1995; Roth and Lindroth, 1994; Williams *et al.*, 1994; Docherty, unpublished data.)

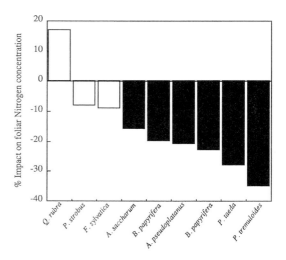

Figure 13.2 Impact of elevated CO_2 on the nitrogen concentration of plant foliage. (Details as for Figure 13.1.)

daily temperature (Hoiopainen and Kainulainen, unpublished). However, the effect of climate warming on performance of herbivorous insects is not a simple function of temperature increase, because negative effects of plant allelochemicals on insect performance may decrease at higher temperatures (Yang and Stamp, 1995).

An increase in average temperatures may be accompanied by a decrease in the incidence of frost (Bennetts, 1995). Overwintering eggs requiring a combination of chilling and thermal units before hatching may, like certain trees (Murray *et al.*, 1989), have an altered phenology at elevated temperatures. The timing of egg hatch may be delayed due to reduced accumulation of chilling, resulting in enhanced mortality for those insect–host phenologies that become uncoupled. Simulation modelling by Dewar and Watt (1992) demonstrate this projected phenological asynchrony. However, development of surviving nymphs in spring, enhanced by higher temperatures, may allow them

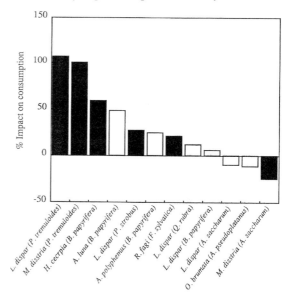

Figure 13.3 Impact of elevated CO_2 on food consumption by insect herbivores. (Details as for Figure 13.1.)

to 'escape' from predation, thus compensating to some extent for earlier mortality (Price et al., 1980). Disruption of the synchrony between insects and their hosts would be detrimental for many insects, especially univoltine spring feeders (Landsberg and Smith, 1992). Landsberg and Smith suggest that those insects which use photoperiod as their post-diapause activity cue will suffer under climatic warming, but determination of specific cues is difficult. There is much scope for further research in this area. Additional broods of multivoltine species and even bivoltine species (in southern ranges) may be predicted with climate warming, due to enhanced development.

At a given daylength, the production of sexually reproducing aphids may be delayed at higher temperatures (Dixon, 1973) though the success of such a strategy would be determined by the performance of the host plant. The production of sexual morphs by the fundatrix of the Arctic aphid, *Acyrthosiphon svalbardicum*, feeding on *Dryas octopetala* guarantees egg production in the short summer season. The production of a small number of parthenogenetic individuals by the fundatrix gives this species the flexibility to increase reproduction under favourable conditions (Strathdee et al., 1993). This strategy may become more important as the climate changes. At increased average temperatures, time required to initiate flight activity for oviposition or food collection by insects will be reduced (Robert, 1987) which may increase pest potential if flight activity limits the success of the species. Changing weather patterns, including increased rainfall, may contribute to greater stochastic variation in population dynamics of insect herbivores (Harcourt, 1966; Dixon, 1973). Predicted drought stress in some regions will also influence population dynamics (Porter, 1995).

A 2°C rise in temperature may well be within the physiological and phenological tolerance of many species except possibly those individuals at the southern limit of their range. More extreme environmental changes may promote species extinction, distribution changes and/or adaptation *in situ*. The slower migration rates of vegetation are thought to constrain migration rates of insects (Davis, 1986) but the fact that pest insects are capable of invasion over long distances to favourable habitats (Lienhold et al., 1992) suggests that insect species with distribution currently limited by climate (Tauber and Tauber, 1986) may be able to exploit further forest resources in the future. Changes in distributions caused by increased temperatures may cause unpredictable uncoupling of community interactions and formation of new ones (Davis et al., 1995). Stone and Sunnocks (1993) highlight the loss of genetic variability of the cynipid gallwasp, *Andricus quercuscalicis*, from its core range in Hungary, through Europe to Britain. It is unclear how reduced variability will affect the ability of such insects to adapt to climate change (Lawton, 1995).

13.3 EFFECTS OF INCREASED UV-B EXPOSURE ON HOST PLANTS

Damage to the stratospheric ozone layer by emissions of ozone-depleting chemicals has been shown to cause increased exposure to UV-B radiation (280–320 nm) of between 5 and 10% per decade at northern temperate latitudes (Blumthaler and Ambach, 1990). Elevated UV-B radiation causes reduced photosynthesis, resulting in reduced leaf area, biomass and plant height (Tevini, 1993). Indeed loblolly pine has demonstrated some of the most clear-cut effects (Sullivan and Teramura, 1989). Increases in secondary compounds in exposed tissue, particularly phenolics, have been recorded which can afford greater UV-B protection for plants (Caldwell and Flint, 1994). McCloud and Berenbaum (1994) observed that foliar concentrations of furanocumarins, which are toxic to generalist insect herbivores, increase in the tropical tree *Citrus jamnhri* when the plants are grown under enhanced UV-B. Experiments on crop plants have indicated that increased UV-B levels induce higher concentrations of secondary compounds in the plant material but, as Hatcher and Paul (1994) suggest, an increase in secondary metabolite concentrations does not necessarily mean reduced growth for insect herbivores on those plants. Changes in leaf structure produced by increased UV-B, such as increased leaf thickness (Bornman and Vogelmann, 1991) and epicuticular waxes (Berenbaum, 1988), may also influence insect feeding.

13.3.1. UV-B EFFECTS ON INSECTS

Studies on elevated UV-B effects on herbivores are in their infancy with few published studies (Yazawa *et al.*, 1992; McCloud and Berenbaum, 1994). At this stage direct effects of increased fluxes on insects and effects mediated by changes in plant structure cannot be discounted (Hatcher and Paul, 1994), though studies to date have concentrated on changes in food quality. Hatcher and Paul (1994) showed that larvae of the noctuid moth *Autographa gamma* grew slightly better on UV-B exposed pea plants (*Pisum sativum*), but consumption was reduced whilst the efficiency of conversion of ingested food increased. Hatcher and Paul (1994) also found increased foliar concentrations of nitrogen and total phenolics in treated plants. McCloud and Berenbaum (1994) found that young larvae of *Triplochiton ni* developed more slowly on *C. jambhri* plants exposed to increased UV-B fluxes. On artificial diets containing the same levels of furanocumarins in the plant material, this same developmental pattern was observed.

13.4 GASEOUS AIR POLLUTANTS: O_3, SO_2 AND NO_X AND ACID PRECIPITATION

13.4.1 EFFECTS OF AIR POLLUTION ON HOST PLANTS

Tree damage was attributed to pollution in the nineteenth century. Ozone and the oxides of nitrogen and sulphur have been considered responsible for the tree mortality and poor performance in parts of Europe and North America, known as forest decline, which became apparent in the 1980s (Cramer, 1951; Krause, 1989). The formation of these major pollutants is discussed in more detail in the recent review by Brown (1995). Often, insects demonstrate effects of pollutants before visible symptoms become apparent in the plants themselves (e.g. Jones and Coleman, 1988; Hiltbrunner and Flückiger, 1992). Indeed Last and Watling (1991) regard the responses of their herbivores as the most distinct of the responses that plants show to gaseous pollutants.

Gaseous pollutants cause a stress response in exposed plants (White, 1984; Larsson and Bjorkman, 1993) which can be demonstrated by the breakdown of proteins in the plant and disruption to amino acid metabolism leading to changes in both the balance and the quantity of free amino acids in the plant (Bolsinger and Flückiger, 1984, 1989; Dohmen *et al.*, 1984). In addition, phloem feeders, which are often concentrated at sinks of amino acids in the plant, are likely to be influenced by changes in partitioning of biomass in the plant, e.g. between root and shoot, as a result of pollutant exposure (Kasana and Mansfield, 1986; Whittaker, 1994).

Effects of pollutants, particularly ozone, on volatile chemicals emitted by the plants have received little consideration in the literature but pollutants may modify the chemical cues used by insects in locating host plants (Bolsinger *et al.*, 1992; Salt and Whittaker, 1995). Although changes in secondary compounds due to SO_2 exposure have been detected in an experimental study on Scots pine (*Pinus sylvestris*) and Norway spruce (*Picea abies*), responses depended on seed origin and the dosage applied (Kainulainen *et al.*, 1995). Kainulainen *et al.* (1993) concluded that increased pest potential of aphids is likely to be caused by improved food quality rather than reductions in secondary compounds in mainly SO_2 polluted conditions. Jordan *et al.* (1991) found that ozone treatment led to increased concentrations of tannins in loblolly pine (*Pinus taeda*) foliage. An interesting observation made by Lin *et al.* (1990) on soybean (*Glycine max*) was that ozone fumigation (25–100 nl/l for 5 hours over 2–4 days) may inhibit the induction of defensive chemicals stimulated by herbivore damage to protect the plant against further herbivore attack. Although gaseous pollution generally occurs as a mixture of chemicals rather than as single pollutants from emission sources, most experimental studies have concentrated on effects of single pollutants on trees and insects. This makes it difficult to predict the effects of mixtures and to interpret field observations. There is evidence that effects of mixtures of pollutants (NO_2

and SO_2) can produce more than additive reductions in growth of broadleaved and coniferous trees when compared with the treatments applied singly (Freer-Smith, 1984; Freer-Smith and Mansfield, 1987).

13.4.2 POLLUTANT EFFECTS ON INSECTS

Both sulphur dioxide and oxides of nitrogen have consistently produced enhanced performance of insect herbivores in controlled experiments at concentrations of the pollutants routinely encountered in the field, both on individuals (Dohmen et al., 1984; Warrington, 1987; Bolsinger and Flückiger, 1989) and populations (Warrington and Whittaker, 1990; Whittaker, 1994). This response has been shown with shoot-phloem feeders and leaf-chewers used in experiments and reflects the pattern of responses shown by field observations (Figure 13.5, later) (Riemer and Whittaker, 1989). Short-term experiments conducted over the summer months on aphids of broadleaved trees are the only exceptions to this pattern for herbivores in chamber experiments (McNeill and Whittaker, 1990). In a study on transplanted Scots pine and Norway spruce trees along a gradient of mixed pollutants (mostly SO_2), Holopainen et al. (1993) found that although *Cinara pilicornis* showed increased performance with exposure, *Sitobion pineti* failed to show a significant response.

Many studies of gaseous pollutants on aphids have measured relative growth rates (RGR) as an indicator of performance (van Emden, 1969). This parameter is useful in that enhanced growth rates of aphids occur alongside increased reproduction (Dixon et al., 1982). Warrington and Whittaker (1990) demonstrated that in elevated SO_2 (25 nl/l) increased populations of the aphid *Elatobium abietinum* were reached on Sitka spruce (*Picea sitchensis*). This was the result predicted from increased RGR of the aphids on fumigated trees (McNeill et al., 1987). Larger populations and consequently greater damage by phloem-feeding insects accords with the evidence from descriptive studies near polluted sites (Riemer and Whittaker, 1989). Warrington (1987) demonstrated that the relationship of RGR to concentration of SO_2 is linear up to about 100nl/l for the pea aphid and that only at concentrations above 300 nl/l is RGR greater in unfumigated controls than in polluted air. This suggests that any exposure to SO_2 encountered in the field affects aphid performance (Whittaker and Warrington, 1990). The pea aphid also showed a 19% increase in the rate of production of nymphs compared with aphids on plants subjected to filtered air (Warrington et al., 1987). *C. pilicornis* nymphs on Norway spruce showed a bell-shaped response to SO_2 between 0 and 150 nl/l. During fumigation, first and third instar nymphs had their highest RGR at 100 nl/l, while in a post-fumigation experiment RGR peaked at a concentration of 50 nl/l (Holopainen et al., 1995). This experiment with SO_2 also confirmed observations that nymphal growth rate of *C. pilicornis* on spruce is more responsive to SO_2 than *Cinara pinea* or *S. pineti* on Scots pine. A model created by Kidd (1991) to predict changes in population size resulting from RGR increases in *C. pinea* brought about by acid deposition suggested that only a 10% increase in RGR through the season would lead to doubled populations of aphids. Experimental evidence for effects of SO_2 and NO_2 mixtures on population growth of *E. abietinum* suggests that the effects of mixtures were much greater than additive effects of equivalent concentrations of pollutants supplied singly (Whittaker, 1994). Interestingly, root aphids, unlike shoot aphids, have shown poorer performance when feeding on trees exposed to gaseous pollutants compared with relatively unpolluted trees. The spruce-feeding root aphid *Pachypappa vesicalis*, on Sitka spruce (*P. sitchensis*) subjected to 85 nl/l SO_2, showed reduced RGR on fumigated plants compared with plants in ambient air (Salt, 1993). A further experiment indicated the same response, whilst the needle-feeding green spruce aphid *Elatobium abietinum* showed the opposite response of increased RGR (Figure 13.4). Negative effects of SO_2 and O_3 treatments, singly and in combination, were found in the Liphook field fumigation experiment after three years fumigation on mixed species assemblages of root aphids feeding on Norway spruce (Salt and Whittaker, 1995). Changes in biomass partitioning in fumigated plants may be responsible for this effect on root feeders (Kasana and Mansfield, 1986; Whittaker, 1994).

Ozone in the troposphere is formed some distance away from pollutant sources by complex chemical reactions (Brown, 1995). Therefore field data are more difficult to evaluate for ozone as there are no point sources to allow observations along concentration gradients (Krause, 1989). Ozone pollution is likely to remain a major problem in countries where bright sunlight and vehicle emissions interact. The effects of ozone on herbivorous insects have been difficult to predict, especially for aphids. However, recently it has been suggested that the response to ozone is determined by the temperature at which the exposure occurs and the concentration and duration of the exposure (Brown, 1995; Jackson, 1995). Brown et al. (1993) reported a range of responses of C. pilicornis feeding on Sitka spruce in 100 nl/l O_3 over a range of exposure periods. When the ambient temperatures were considered during the experiments it was revealed that RGR was increased at maximum temperatures below 20°C whereas above that temperature it was reduced. The same species also had increased RGR near Basel in ambient air, which had diurnal fluctuating ozone concentrations as the main pollutant, compared with filtered air (Holopainen et al., 1994). Ozone reaches its highest concentrations during hot weather and thus global warming may be particularly important for interactions with this pollutant (Brown, 1995). The complexity of responses to ozone are illustrated by a leaf palatability experiment using the gypsy moth (Jeffords and Endress, 1984) where fumigation with 90 nl/l O_3 before exposure to insects reduced palatability, whereas higher concentrations increased it. The latter response was shown by the leaf beetle *Plagiodera versicolora* on *Populus* foliage exposed to high ozone concentrations (200 nl/l) (Jones and Coleman, 1988) but oviposition preference showed the opposite response. As these workers state, the result of ozone stress on plants subjected to herbivory depends on taking behavioural, growth and reproductive processes into account.

Although increased consumption has been reported in experiments with chewing insects in ozone-fumigated conditions (e.g. Colemann and Jones, 1988; Endress and Post, 1985), only some experiments report increased performance of the insects themselves (Figure 13.5) (Lee et al., 1988; Whittaker et al., 1990; Bolsinger et al., 1992). Lindroth et al. (1993a) demonstrated an 8% reduction in growth of the gypsy moth when fed on ozone-fumigated foliage. Field observations of attack by bark beetle (*Dendroctonus* spp.) on *Pinus ponderosa* indicated that ozone-damaged trees were less resistant to attack because of a reduced rate of resin flow and exudation (Cobb et al., 1968; Stark et al., 1968; Stark and Cobb, 1969). However, insect performance on these trees is often reduced (Hain, 1987). Effects of ozone on pheromones have been investigated by Arndt (1995), who demonstrated that ozone can affect behaviour of fruit flies by modifying the pheromones that cause the insects to aggregate.

Observational evidence of pest outbreaks of several insect species has indicated that effects of insects on trees are greater in the presence of gaseous pollutants than in unpolluted conditions (Riemer and Whittaker, 1989). In an experimental study Warrington et al. (1989) demonstrated that feeding punctures on sycamore leaves left by typhlocybine leafhoppers acted as routes on entry for SO_2 gas and caused further damage to the plants. In a study that sought to investigate effects of pollutants and *Elatobium* infestation on *P. sitchensis*, Warrington and Whittaker (1990) found

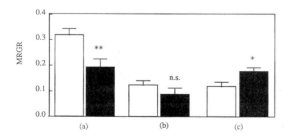

Figure 13.4 Mean relative growth rates of root and shoot aphids on Sitka spruce in cabinets in ambient (☐, 10 nl/l) and elevated (■, 85 nl/l) SO_2 concentrations. (a) Root aphid *Pachypappa vesicalis*, March 1991; (b) *P. vesicalis*, May 1992; (c) shoot aphid *Elatobium abietinum*, May 1992. (Source: Salt, 1993.)

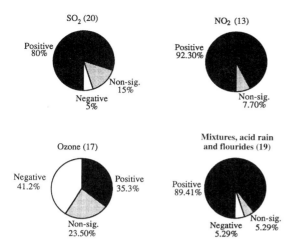

Figure 13.5 Effects of air pollutants on aphid performance on plants in pollutant fumigation and filtration experiments. Number of reports in parentheses. Charts mainly based on data from Whittaker (1994). Other sources include, for SO_2: Holpainen *et al.* (1995), Salt and Whittaker (1995); for NO_2: Holpainen *et al.* (1995); for ozone: Summers *et al.* (1994), Kainulainen *et al.* (1994), Holopainen *et al.* (1995), Salt and Whittaker (1995); for mixtures, acid rain and fluorides: Kidd (1990), Holopainen *et al.* (1994, 1995), Hautala and Holopainen (1995), Salt and Whittaker (1995).

over 20% more reduction in root dry weights and leader lengths when both aphids and 25 nl/l SO_2 were applied than would have been expected from the additive effects of these factors. In polluted conditions, increased populations of herbivores would be expected to produce more damage to trees than in clean air.

13.4.3 ACID PRECIPITATION AND NITROGEN DEPOSITION: EFFECTS ON INSECTS

Although SO_2 and oxides of nitrogen affect plants in the gaseous phase, wet deposition of these pollutants as acid rain has well known effects on forest ecosystems (Kauppi *et al.*, 1990). Simulated acid rain on the host plant (pH 3.0 and 4.0) by the fifth year of treatment reduced the susceptibility of the pine sawfly (*Neodiprion sertifer*) to polyhedrosis virus infection, which may enhance its outbreak frequency (Saikkonen and Neovonen, 1993).

Heliövaara *et al.* (1992) observed increased survival of overwintering egg clusters of *N. sertifer* after treatment with simulated acid (pH 2–6) precipitation containing a mixture of sulphuric and nitric acid. The eggs with and without needles gave similar results, indicating direct acid-induced effects. Numbers of some spiders and carabids that consume pupae of many folivorous insects were lower in forest soil plots treated with simulated acid rain (Heliövaara and Väisänen, 1993). However, acid precipitation (pH 4.0) did not affect growth of the sheetweb spider *Pityohyphantes phrygianus* on Norway spruce (Gunnarsson and Johnsson, 1989).

Field and laboratory studies have indicated that aphids, especially on conifers, benefit from acid rain. Kidd (1990) found increased nymphal and population growth rates of the aphids *S. pineti* and *Eulachnus agilis* and increased population growth rate of *Cinara pini* on Scots pine seedlings treated with acid mist (pH 2.99) compared with controls. Neuvonen *et al.* (1992) observed a slightly (but not significantly) higher density of *C. pinea* on pine seedlings treated with simulated acid rain. The heavy damage of the balsam woolly adelgid (*Adelges picea*) on *Abies fraseri* at higher elevations is ascribed to pollution mist (Hain, 1987). Neuvonen and Lindgren (1987) compared reproduction of the aphid *Euceraphis betulae* on silver birch treated with simulated acid rain (pH 3.5) or water. Four of eight separate experiments showed increased reproduction rate of aphids on trees treated with acid rain. On beech, acid precipitation produced the opposite response. Population growth of the beech aphid *Phyllaphis fagi* was inhibited by spraying acid mists (pH 3.6 and 2.6) compared with controls (Braun and Flückiger, 1989). Neuvonen *et al.* (1990) did not find consistent effects of simulated acid rain (sulphuric and nitrogen acid mixture) treatment on the quality of pine foliage as food for *N. sertifer* larvae. However, acid rain treatment (pH 4.0) reduced fecundity of the elm leaf beetle (*Xanthogalerucea luteola*) compared with that of beetles fed on untreated elm foliage (Heliövaara and Väisänen, 1993).

Nitrogen is emitted as oxides from fossil fuel combustion and as ammonia from agriculture, at rates exceeding 500 kg/ha per annum in parts of

Central Europe (van Dijk and Roelofs, 1988). Deposition of these nitrogenous compounds in soils can cause acidification but they can also act as fertilizers (van Dijk and Roelofs, 1988; Redinbaugh and Campbell, 1991). This results in accumulation of nitrogen in the form of free amino acids and deficiencies of Mg, K and P relative to nitrogen in plant tissue (van Dijk and Roelofs, 1988). The availability of nitrogen is often limiting for plant growth and herbivore performance. Waring and Cobb (1992) analysed 186 nitrogen fertilization studies which included measurements of arthropod herbivore performance. The majority (60%) of these studies showed enhanced performance on nitrogen fertilized plants compared with unfertilized plants. Improved performance was seen with all feeding guilds investigated (suckling, chewing and gall-forming mites and insects). Therefore where nitrogen availability to forest trees is limiting, increased deposition may result in greater pest potential of herbivores.

13.5 EFFECTS OF OTHER POLLUTANTS

Small amounts of fluoride, in common with CO_2 and SO_2, can be produced by volcanic activity (Hughes, 1988; Heliövaara and Väisänen, 1993) but fluoride mainly results from the operation of aluminium smelters and fertilizer factories. Fluoride accumulates in plants, and is one of the most phytotoxic of the common pollutants (Hughes, 1988). There are few studies on the effects of fluoride on forest insects and thus the mechanisms are poorly understood. The studies that exist may confuse effects due to fluoride with other pollutants emitted from the same or distant sources. Mitterbock and Fuhrer (1988) reported a clear positive correlation between nun moth (*Lymantria monacha*) mortality and fluoride concentration of up to 365 µg/g in spruce needles. Davies *et al.* (1992) did not find a relationship between fluoride content of Corsican pine (*Pinus nigra*) needles (up to 170 µg/g) and pupal weight of the large pine sawfly (*Diprion pini*) near a fluoride-emitting aluminium smelter. However, outbreaks of the saddleback looper (*Ectropus crepuscularia*) and spruce budworm (*Choristoneura orea*) appeared in the vicinity of an aluminium smelter in British Columbia on the exposed area, where the concentration of fluoride in the foliage of hemlock (*Tsuga* sp.) was increased (24–100 µg/g) (Hughes, 1988). Aphid growth on conifers is enhanced by fluoride exposure of approximately 23 µg/g (Thalenhorst, 1974; Villemant, 1981; Holopainen *et al.*, 1991) but RGR and reproduction of *Rhopalosiphum padi* on barley was unaffected by fluoride exposure of between 60 and 360 µg/g (Hautala and Holopainen, 1995). Fluoride exposure of crop plants can result in increased concentrations of free amino acids in foliage, but also enhanced foliar levels of indole alkaloids that inhibit aphid growth (Hautala and Holopainen, 1995). The balance between these opposite changes in plant quality may affect the final outcome of herbivore performance on fluoride-exposed plants.

There are several examples of the harmful effects of heavy metal contamination near industrial sources on forest soil animals (Heliövaara and Väisänen, 1993). Bioaccumulation of heavy metals is often found in folivores. Pupal size of the pine beauty moth, *Panolis flammea*, and the pine looper, *Bupalus piniarius*, was reduced when the pupae were reared on needles collected near factories that emit sulphur dioxide and heavy metals. The elemental composition of needles explained 24–53% of the variation of pupal weight and most of the explained variation was associated with concentration of heavy metals (Heliövaara *et al.*, 1989). When reared on needles collected from the same industrial area in southwestern Finland, larvae of the large pine sawfly (*D. pini*) developed at a slower rate and hatched later, and cocoon mortality was higher, compared with larvae reared on needles from unpolluted sites (Heliövaara and Väisänen, 1990b). The results from experiments on Scots pine suggest that heavy metal pollution has detrimental effects on folivorous insect performance. This is mainly caused by reduced quality of food, because parasitism rates are not reduced at forest sites affected by heavy metals (Heliövaara *et al.*, 1991; Koricheva, 1994).

13.6 POLLUTANT AND CLIMATE CHANGE EFFECTS ON NATURAL ENEMIES

McNeill and Whittaker (1990) drew attention to the difficulty of extrapolating results from fumigation experiments in chambers to field situations, where the tendency for insect populations to increase may be countered by density-dependent responses of natural enemies. It has been suggested that entomophagous insects may be more susceptible to pollutants than herbivores due to greater mobility and hence greater exposure to the pollutants (Koricheva et al., 1995). Heliövaara et al. (1991) reported that sawfly larval mortality due to parasitism was reduced nearest to a factory complex, whereas mortality due to pathogens was greater at that point. Nuclear polyhedrosis virus killed larvae of *Neodiprion sertifer* more frequently near the factory, suggesting that the pollutants (including dust and mineral particles) on the needles may indirectly promote the virus infection, increasing larval mortality (Heliövaara et al., 1991). Riemer and Whittaker (1989) drew attention to several studies reporting negative correlations between natural enemies and pollutant concentrations. However, the number of host predator/parasite interactions examined in polluted environments has been small (Koricheva et al., 1995). Studies on both ant predation and parasitism of *Eriocrania* leaf-miners (Koricheva, 1994; Koricheva et al., 1995) indicated no effect of the sulphur and heavy-metal pollution from a copper–nickel smelter on mortality due to natural enemies. Similar findings on parasitism of the pine resin gall moth (*Petrova resinella*) at the same site were reported by Heliövaara and Väisänen (1986). Braun and Flückiger (1985) assessed the impact of natural enemies of *Aphis pomi* on *Crataegus* in a study of pollution from a motorway. They suggested that aphid enemies were less active nearer the motorway. An open-air fumigation experiment on the grain aphid (*Sitobion avenae*) showed that numbers increased in SO_2 fumigation (21–57 nl/l) but numbers of specialist and generalist predators were generally not influenced by SO_2 pollution, though two staphylinids were negatively correlated with SO_2 concentrations and the increased densities of aphids in elevated SO_2 attracted greater numbers of one coccinellid (Aminu-Kano et al., 1991). The overall evidence is that, in the short term, insect parasites and predators may be unable to compensate for the increase in numbers of insect herbivores in polluted situations, and that negative effects on these natural enemies are only likely at concentrations of pollutants higher than those that cause the prey to increase.

13.7 CASE STUDY: GYPSY MOTH

Experiments involving different types of pollution and changes in environmental parameters have been conducted on relatively few species. One species that has been investigated is the gypsy moth, *Lymantria dispar* L., a pest of deciduous trees in Europe and Asia, which has become a serious pest in North America since its introduction in 1869 (Montgomery, 1990). The mechanisms causing population outbreaks are unknown, though food quality and weather have been implicated (Weseloh, 1990). Parasitoids are ineffectual in controlling this species in America but they have achieved some success in Europe (Mills, 1990).

It is thought that the high concentrations of foliar gallotannins in some species, e.g. red oak and chestnut oak (*Quercus prinus*), make these hosts more susceptible to large populations and persistent outbreaks of the pest. Outbreaks tend to be less common on hosts containing low levels of gallotannins in their foliage, with the notable exception of aspen, *Populus tremuloides* Mich., which is devoid of gallotannins and on which outbreak populations persist (Houston and Valentine, 1977). Therefore, the presence or absence of these chemicals may not drive the dynamics of this insect, but they do have a role to play in the interaction between some tree hosts and this pest. High levels of these chemicals confer resistance to the LdNPV virus, allowing gypsy moth populations to reach high densities before epizootics are thought to reduce the population. Lindroth et al. (1993b) demonstrated that average gallotannin concentrations of red oak decreased by approximately 14% in elevated CO_2 but that there was an enhancement of about 20% in maple foliage. There were no detectable changes in the concentration of foliar nitrogen in elevated CO_2 but leaf toughness increased and the concentration of starch in the leaves significantly increased by 1.4 times. Decreases in the average concentration of gallotannins may potentially increase gypsy moth susceptibility to virus, thus affecting gypsy moth performance, outbreak potential and patterns of defoliation, both spatially and temporally (Foster et al., 1992). The quality of two other hosts of gypsy moth – paper birch (*Betula papyrifera*) and white pine (*Pinus strobus*) – expressed as foliar nitrogen concentration is reported to decline in elevated CO_2, causing concomitant reductions in the performance of the larvae (Roth and Lindroth, 1994). Ozone pollution decreases the palatability of *Acer saccharum* and *Populus* foliage to gypsy moth, causing a reduction in fecundity of the moth (Jeffords and Endress, 1984). Whilst the mechanisms underlying changes in the populations of insects on these hosts are unknown, rising CO_2 occurring with ozone pollution seems more likely to decrease the potential of this species to outbreak than to increase it. Further fundamental information on the mechanisms involved in the ecology of this species is required before firm predictions can be made concerning the effects of climate change.

13.8 CONCLUSIONS

Over the last 20 years, much consideration has been given to the importance of food quality in insect/plant interactions, particularly the importance of the availability and quality of nitrogen or free amino acids in the food plant. Nitrogen is the most important nutritional chemical for insects (Mattson, 1980; Prestidge and McNeill, 1983; White, 1984). Effects of pollution-induced stress on the host plant, by increasing nitrogen availability, generally result in enhanced performance of insects, particularly aphids. Although changes have been recorded in several other aspects of plant chemistry as a result of pollutant exposure, likely physical changes such as increased leaf surface temperatures (Hughes, 1988; Bolsinger et al., 1991) have largely gone uninvestigated. Effects of ozone compared with other gaseous pollutants are complex and still require further elucidation (Brown, 1995). In comparison with general effects of gaseous pollutants, carbon dioxide often reduces food quality, particularly for chewing insects, by increasing the C/N ratio of plant material. In some cases (at least in the first generation) insects can compensate for this reduction in food quality by increased consumption without their growth being affected; however, recent evidence on multivoltine species over three generations (Brooks, 1995) suggests that negative effects on performance become apparent in later generations.

Both gaseous pollutants and increased CO_2 concentrations are likely to alter the amount of insect damage to trees. In addition to direct tree damage by pollutants, damage is often increased through larger populations of herbivores. The evidence from experimental fumigation systems and studies at polluted sites suggests that resultant increases in herbivore populations cannot be controlled by natural enemies. Greater consumption to compensate for poorer food quality has generally been reported for chewing insects on plant material grown in elevated CO_2, though at present there are few studies of the consequences at the population level. FACE (free air CO_2 enrichment) experiments should allow the extent of predation and parasitism to be assessed on herbivores that show slower development in elevated CO_2 and may therefore be more susceptible to mortality due to natural enemies. Although consumption by individuals may increase, populations may be adversely affected by rising CO_2 concentrations. It is not yet clear whether the increase in plant growth in elevated CO_2 will compensate for increased herbivore damage under those conditions (Salt et al., 1995).

Brown (1995) pointed out that it is likely that in the developed world pollution-related problems of SO_2 and oxides of nitrogen will diminish, whereas in the developing world they are likely to remain serious problems for some time to come. Exposure to pollutant gases since the industrial revolution has occurred against a background of rising global concentrations of CO_2. These changes are likely to produce greater contrasts in insect performance between sites subjected to SO_2 and NO_2 pollution and comparatively unpolluted sites by increasing the contrast in nitrogen availability, thereby increasing the likelihood of outbreaks in polluted areas.

To date, few studies have combined pollutant treatments and combinations involving realistic increases in temperature and CO_2 but one study (Heagle et al., 1994) investigated effects of combined O_3 and CO_2 on mites. Despite continued discussions of the scale of likely temperature rises, rising CO_2 concentrations are well established and facilities used for pollutant fumigation could also deliver combined pollution and elevated CO_2 treatments. There is a need for ecological studies that combine all the main factors in a future environment. These include enhanced O_3, CO_2, temperature and UV-B. The results in plant and insect responses may differ between single and multiple treatment experiments. The provision of these combinations applied factorially is extremely expensive, but possible with the combined efforts of multiple research groups.

Current knowledge of the effects on the range of factors modified by a changed environment makes it difficult to predict the consequences of changes in these factors acting in combination on insect/plant interactions. The pattern is well established in the case of oxides of sulphur and nitrogen but ozone is more problematic. Recent work has highlighted the changes in insect micro-environment that are likely to occur in elevated CO_2. These factors, particularly leaf temperature, may have been overlooked in earlier pollution studies and may help to explain some observed effects.

The amount and types of adaptation insects and their hosts show to a changed gaseous environment and changed climate are difficult to predict and investigate experimentally. For example, we can only speculate how selective pressure caused by natural enemies may change against insects whose development time may be lengthened by reduced food quality on the one hand, but shortened by rising ambient temperatures and increased leaf temperatures on the other.

REFERENCES

Alstad, D.N., Edmunds, G.F. and Weinstein, L.H. (1982) Effects of air pollutants on insect populations. *Annual Review of Entomology* **27**, 369–384.

Aminu-Kano, M., McNeill, S. and Hails, R.S. (1991) Pollutant, plant and pest interactions: the grain aphid *Sitobion avenae* (F.) *Agriculture, Ecosystems and Environment* **33**, 233–243.

Arndt, U. (1995) Air pollutants and pheromones – A problem. *Chemosphere* **30**, 1023–1031.

Arnone, J.A., Zaller, J.G., Zieger, C. and Zandt, H. (1995) Leaf quality and insect herbivory in model tropical plant communities after long term exposure to elevated atmospheric CO_2. *Oecologia* **104**, 72–78.

Bazzazz, F.A. (1990) Responses of natural ecosystems to the rising global CO_2 levels. *Annual Review of Ecology and Systematics* **21**, 167–196.

Bennetts, D.A. (1995) The Hadley Centre Transient Climate Experiment. In *Insects in a Changing Environment: 17th Symposium of the Royal Entomological Society, 7–10 September 1993, Rothamsted Experimental Station, Harpenden, England* (eds R. Harrington and N.E. Stork), pp. 50–58, Academic Press, London.

Berenbaum, M. (1988) Effects of electromagnetic radiation in insect–plant interactions. In *Plant Stress–Insect Interactions* (ed. E.A. Heinrichs), pp. 167–185, John Wiley and Sons, New York.

Blumthaler, M. and Ambach, W. (1990) Indication of increasing solar ultraviolet-B radiation flux in alpine regions. *Science* **248**, 206–208.

Bolsinger, M. and Flückiger, W. (1984) Effects of air pollution at a motorway on the infestation of *Viburnum opulus* by *Aphis fabae*. *European Journal of Forest Pathology* **14**, 256–260.

Bolsinger, M. and Flückiger, W. (1989) Ambient air pollution induces changes in amino acid pattern on phloem sap in host plants – relevance to aphid infestation. *Environmental Pollution* **56**, 209–235.

Bolsinger, M., Lier, M.E., Lansky, D.M. and Hughes, P.R. (1991) Influence of air-pollution on plant–herbivore interactions. Part 1: Biochemical changes in ornamental milkweed (*Asclepias curassavica* L.; Asclepiadaceae) induced by ozone. *Environmental Pollution* **72**, 69–83.

Bolsinger, M., Lier, M.E. and Hughes, P.R. (1992) Influence of ozone air pollution on plant–herbivore interactions. Part 2: Effects of ozone on feeding preference, growth and consumption rates of monarch butterflies (*Danaus plexippus*). *Environmental Pollution* **77**, 31–37.

Bornman, J.F. and Vogelmann, T.C. (1991) Effect of UV-B radiation on leaf optical properties measured with fibre optics. *Journal of Experimental Botany* **42**, 547–554.

Braun, S. and Flückiger, W. (1984) Increased population of the aphid *Aphis pomi* at a motorway: Part 1, Field evaluation. *Environmental Pollution* **33**, 107–120.

Braun, S. and Flückiger, W. (1989) Effect of ambient ozone and acid mist on aphid development. *Environmental Pollution* **56**, 177–187.

Brooks, G.L. (1995) The effect of elevated CO_2 on herbivorous insects. PhD thesis, Lancaster University.

Brown, V.C. (1995) Insect herbivores and gaseous air pollutants – current knowledge and predictions. In *Insects in a Changing Environment: 17th Symposium of the Royal Entomological Society, 7–10 September 1993, Rothamsted Experimental Station, Harpenden, England* (eds R. Harrington and N.E. Stork), pp. 219–249, Academic Press, London.

Brown, V.C., Ashmore, M.R. and McNeill, S. (1993) Experimental investigations of the effects of air pollutants on aphids on coniferous trees. *Forstwissenschaftliches Centralblatt* **112**, 128–132.

Butler, G.D. (1985) Populations of several insects on cotton in open-top carbon dioxide enrichment chambers. *The Southwestern Entomologist* **10**, 264–267.

Butler, G.D., Kimball, B.A. and Mauney, J.R. (1986) Populations of *Bemisia tabaci* (Genn.) (Homoptera: Aleyrodidae) on cotton grown in open-top field chambers enriched with CO_2. *Environmental Entomology* **15**, 61–63.

Caldwell, M.M. and Flint, S.D. (1994) Stratospheric ozone reduction, solar UV-B radiation and terrestrial ecosystems. *Climatic Change* **28**, 375–394.

Cannell, M.G.R. (1990) Modelling the phenology of trees. *Silva Genetica* **15**, 11–27.

Ceulemans, R. and Mousseau, M. (1994) Tansley Review No. 71: Effects of elevated atmospheric CO_2 on woody plants. *New Phytologist* **127**, 425–446.

Cobb, F.W., Wood, D.L., Stark, R.W. and Parmenter, J.R. (1968) IV. Theory on the relationship between oxidant injury and bark beetle infestation. *Hilgardia* **39**, 121–152.

Coleman, J.S. and Jones, G.C. (1988) Plant stress and insect performance: Eastern cottonwood, ozone and a leaf beetle. *Oecologia* **76**, 56–61.

Coley, P.D., Bryant, J.P. and Chapin, F.S. (1985) Resource availability and plant anti-herbivore defense. *Science* **230**, 895–899.

Corcuera, L.J. (1990) Plant chemicals and resistance of cereals to aphids. *Ambio* **19**, 365–367.

Cramer, H.H. (1951) Die geographischen Grundlagen des Massenwechsels von *Epiblema tedella*. *Forstwissenschaftliches Centralblatt* **70**, 42–53.

Davies, M.T., Davison, A.W. and Port, G.R. (1992) Fluoride-loading of larvae of pine sawfly from a polluted site. *Journal of Applied Ecology* **29**, 63–69.

Davis, A.J., Jenkinson, L.S., Lawton, J.H. *et al.* (1995) Global warming, population dynamics and community structure in a model insect assemblage. In *Insects in a Changing Environment: 17th Symposium of the Royal Entomological Society, 7–10 September 1993, Rothamsted Experimental Station, Harpenden, England* (eds R. Harrington and N.E. Stork), pp. 431–439, Academic Press, London.

Davis, M.B. (1986) Climatic instability, time lags, and community disequilibrium. In *Community Ecology* (eds J. Diamond and T.J. Case), pp. 269–284, Harper and Row, New York.

Dewar, R.C. and Watt, A.D. (1992) Predicted changes in the synchrony of larval emergence and budburst under climate warming. *Oecologia* **89**, 557–559.

Dixon, A.F.G. (1973) *Biology of Aphids*, Edward Arnold, London.

Dixon, A.F.G., Chambers, G.R.J. and Dharma, T.R. (1982) Factors affecting size in aphids with particular reference to the black bean aphid, *Aphis fabae*. *Entomologia experimentalis et Applicata* **32**, 123–128.

Docherty, M., Hurst, D.K., Holopainen, J.K. *et al.* (1996) Carbon dioxide-induced changes in beech foliage cause female beech weevil larvae to feed in a compensatory manner. *Global Change Biology* **2**, 335–341.

Docherty, M., Wade, F.A., Hurst, D.K. *et al.* (1997) The response of tree sap- and mesophyll-feeding herbivores to elevated CO_2. *Global Change Biology*, **3**, 51–59.

Dohmen, G.P., McNeill, S. and Bell, J.N.B. (1984) Air pollution increases *Aphis fabae* pest potential. *Nature* **307**, 52–53.

Eamus, D. and Jarvis, P.G. (1989) The direct effects of increase in global atmospheric CO_2 concentration on natural and commercial temperate trees and forests. *Advances in Ecological Research* **19**, 1–55.

El Kohen, A., Venet, L. and Mousseau, M. (1993) Growth and photosynthesis and two deciduous forest species at elevated carbon dioxide. *Functional Ecology* **7**, 480–486.

Ellsworth, D.S., Oren, R., Huang, C. *et al.* (1995) Leaf and canopy responses to elevated CO_2 in a pine forest under free air CO_2 enrichment. *Oecologia* **104**, 139–146.

Endress, A.G. and Post, S. (1985) Altered feeding preference of Mexican bean beetle *Epilachna varivestis* for ozonated soybean foliage. *Environmental Pollution* **39**, 9–16.

Foster, M.A., Shultz, J.C. and Hunter, M.D. (1992) Modeling gypsy moth–virus–leaf chemistry interactions: implications of plant quality for pest and pathogen dynamics. *Journal of Animal Ecology* **61**, 509–520.

Freer-Smith, P.H. (1984) The response of six broad-leaved trees during long term exposure to SO_2 and NO_2. *New Phytologist* **97**, 49–62.

Freer-Smith, P.H. and Mansfield, T.A.M. (1987) The combined effects of low temperature and $SO_2 + NO_2$ pollution on the new season's growth, development and gas exchange and water relations of *Picea sitchensis*. *New Phytologist* **106**, 237–250.

Gunnarsson, B. and Johnsson, J. (1989) Effect of simulated acid rain on growth rate in a spruce-living spider. *Environmental Pollution* **56**, 311–317.

Hain, F.P. (1987) Interactions of insects, trees and air pollutants. *Tree Physiology* **3**, 93–102.

Harcourt, D.G. (1966) Major factors in survival of the immature stages of *Pieris rapae* (L.) *Canadian Entomologist* **98**, 653–662.

Harrington, R. and Stork, N.E. (eds) (1995) *Insects in a Changing Environment: 17th Symposium of the Royal Entomological Society, 7–10 September 1993, Rothamsted Experimental Station, Harpenden, England*, Academic Press, London.

Harrington, R., Bale, J.S. and Tatchell, G.M. (1995) Aphids in a changing climate. In *Insects in a Changing Environment: 17th Symposium of the Royal Entomological Society, 7–10 September 1993, Rothamsted Experimental Station, Harpenden, England* (eds R. Harrington and N.E. Stork), pp. 126–155, Academic Press, London.

Hatcher, P.E., and Paul, N.D. (1994) The effect of elevated UV-B radiation on herbivory of pea by *Autographa gamma*. *Entomologia experimentalis et Applicata* **71**, 227–233.

Hautala, E.L. and Holopainen, J.K. (1995) Gramine and free amino acids as indicators of fluoride-induced stress in barley and its consequences to insect herbivory. *Ecotoxicology and Environmental Safety* **31**, 238–245.

Heagle, A.S., Brandenburg, R.L., Burns, J.C. and Miller, J.E. (1994) Ozone and carbon dioxide effects on spider mites in white clover and peanut. *Journal of Environmental Quality* **23**, 1168–1176.

Heliövaara, K. and Väisänen, R. (1986) Parasitization in *Petrova resinella* (Lepidoptera: Tortricidae) galls in relation to industrial air pollutants. *Silva Fennica* **20**, 233–236.

Heliövaara, K. and Väisänen, R. (1990a) Changes in population dynamics of pine insects induced by air pollution. In *Population Dynamics of Forest Insects* (eds A.D. Watt, S.R. Leather, M.D. Hunter and N.A.C. Kidd), pp. 209–218, Intercept, Andover.

Heliövaara, K. and Väisänen, R. (1990b) Prolonged development in *Diprion pini* (Hymenoptera, Diprionidae) reared on pollutant affected pines. *Scandanavian Journal of Forest Research* **5**, 127–131.

Heliövaara, K. and Väisänen, R. (1993) *Insects and Pollution*, CRC Press, Boca Raton, Florida.

Heliövaara, K., Väisänen, R. and Kemppi, E. (1989) Change of pupal size of *Panolis flammea* (Lepidoptera; Noctuidae) and *Bupalus piniarius* (Geometridae) in response to concentration of industrial pollutants in their food plant. *Oecologia* **79**, 179–183.

Heliövaara, K., Väisänen, R. and Varama, M. (1991) Larval mortality of pine sawflies (Hymenoptera: Diprionidae) in relation to pollution: a field experiment. *Entomophaga* **36**, 315–321.

Heliövaara, K., Väisänen, and and Varama, M. (1992) Acidic precipitation increases egg survival in *Neodiprion sertifer*. *Entomologia experimentalis et Applicata* **62**, 55–60.

Hiltbrunner, E. and Flückiger, W. (1992) Altered feeding preference of beech weevil, *Rhynchaenus fagi* L., for beech foliage under ambient air pollution. *Environmental Pollution* **75**, 333–336.

Holopainen, J.K., Kainulainen, E., Oksanen, J. et al. (1991) Effect of exposure to fluoride, nitrogen compounds and SO_2 on the numbers of Spruce shoot aphids on Norway spruce seedlings. *Oecologia* **86**, 51–56.

Holopainen, J.K., Mustaniemi, A., Kainulainen, P. et al. (1993) Conifer aphids in an air polluted environment. I. Aphid density, growth and accumulation of sulphur and nitrogen by Scots pine and Norway spruce seedlings. *Environmental Pollution* **80**, 185–191.

Holopainen, J.K., Braun, S. and Flückiger, W. (1994) The response of spruce shoot aphid *Cinara pilicornis* Hartig to ambient and filtered air at two elevations and pollution climates. *Environmental Pollution* **86**, 233–238.

Holopainen, J.K., Kainulainen, P. and Oksanen J. (1995) Effects of gaseous air pollutants on aphid performance on Scots pine and Norway spruce seedlings. *Water, Air, Soil Pollution* **85**, 1431–1436.

Honek, A. (1993) Intraspecific variation in body size and fecundity in insects: a general relationship. *Oikos* **66**, 483–492.

Houghton, J.T., Callander, B.A. and Varney, S.K. (1992) *Climate Change 1992. The Supplementary Report of the IPCC Scientific Association*, Cambridge University Press, Cambridge.

Houlden, G., McNeill, S., Aminu-Kano, M. and Bell, J.N.B. (1990) Air pollution and agricultural aphid pests. I: Fumigation experiments with SO_2 and NO_2. *Environmental Pollution* **67**, 305–314.

Houston, D.R. and Valentine, H.T. (1977) Comparing and predicting forest stand susceptibility to gypsy moth. *Canadian Journal of Forest Science* **7**, 447–461.

Hughes, P.R. (1988) Insect populations on host plants subjected to air pollution. In *Plant Stress–Insect Interactions* (ed. E.A. Heinrichs), pp. 249–319, John Wiley, Chichester.

Idso, S.B., Kimball, B.A., Akin, D.E. and Kridler, J. (1993) A general relationship between CO_2-induced reductions in stomatal conductance and concomitant increases in foliage temperature. *Environmental and Experimental Botany* **33**, 443–446.

Jackson, G.E. (1995) The effect of ozone, nitrogen dioxide and nitric oxide fumigation of cereals on the rose grain aphid *Metopolophium dirhodum*. *Agriculture, Ecosystems and Environment* **54**, 187–194.

Jeffords, M.R. and Endress, A.G. (1984) A possible role of ozone in tree defoliation by the gypsy moth (Lepidoptera: Lymantridae) *Environmental Entomology* **13**, 1249–1252.

Jones C.G. and Coleman, J.S. (1988) Plant stress and insect behaviour – cottonwood, ozone and the feeding and oviposition preference of a beetle. *Oecologia* **76**, 51–56.

Jordan, D.N., Green, T.H., Chappelka, A.H. *et al.* (1991) Response of total tannins and phenolics in loblolly pine foliage exposed to ozone and acid rain. *Journal of Chemical Ecology* **17**, 505–513.

Kainulainen, P., Satka, H., Holopainen, J.K. and Oksanen, J. (1993) Conifer aphids in an air-polluted environment. II Host plant quality. *Environmental Pollution* **80**, 193–200.

Kainulainen, P., Holopainen, J.K., Hyttinen, H. and Oksanen, J. (1994) Effects of ozone on the biochemistry and aphid infestation of Scots pine. *Phytochemistry* **35**, 39–42.

Kainulainen, P., Holopainen, J.K. and Oksanen, J. (1995) Effects of SO_2 on the concentrations of carbohydrates and secondary compounds in Scots pine (*Pinus sylvestris* L.) and Norway spruce (*Picea abies* (L.) Karst.) seedlings. *New Phytologist* **130**, 231–238.

Kasana, M.S. and Mansfield, T.A.M. (1986) Effects of air pollutants on the growth and functioning of roots. *Proceedings of the Indian Academy of Sciences (Plant Sciences)* **96**, 429–444.

Kauppi, P., Anttila, P. and Kenttämies, K. (1990) *Acidification in Finland*, Springer-Verlag, Berlin.

Kelliher, F.M., Kostner B.M.M., Hollinger, D.Y. *et al.* (1992) Evaporation, xylem sap flow and tree transpiration in a New Zealand broad-leaved forest. *Agriculture and Forest Meteorology* **62**, 53–73.

Kidd, N.A.C. (1990) The effects of simulated acid mist on the growth rates of conifer aphids and the implications for tree health. *Journal of Applied Entomology* **110**, 524–529.

Kidd, N.A.C. (1991) The implications of air pollution for conifer aphid population dynamics: a simulation analysis. *Journal of Applied Entomology* **111**, 166–171.

Koricheva, J. (1994) Can parasitoids explain density patterns of *Eriocrania* (Lepidoptera: Eriocraniidae) miners in a polluted area? *Acta Oecologia* **15**, 365–378.

Koricheva, J. and Haukioja, E. (1992) Effects of air pollution on host plant quality, individual performance and population density of *Eriocrania* miners (Lepidoptera: Eriocraniidae) *Environmental Entomology* **21**, 1386–1392.

Koricheva, J. and Haukioja, E. (1995) Variations in chemical composition of birch foliage under air pollution stress and their consequences for *Eriocrania* miners. *Environmental Pollution* **88**, 41–50.

Koricheva, J., Lappalainen, J. and Haukioja, E. (1995) Ant predation of *Eriocrania* miners in a polluted area. *Entomologia experimentalis et Applicata* **75**, 75–82.

Krause, G.H.M. (1989) Forest decline in central Europe: The unravelling of multiple causes. In *Toward a More Exact Ecology* (eds P.J. Grubb and J.B. Whittaker), pp. 377–399, Blackwell, Oxford.

Landsberg, J. and Stafford Smith, M. (1992) A functional scheme for predicting the outbreak potential of herbivorous insects under global atmospheric change. *Australian Journal of Botany* **40**, 565–577.

Larsson, S. and Bjorkman, C. (1993) Performance of chewing and phloem-feeding insects on stressed trees. *Scandanavian Journal of Forest Research* **8**, 550–559.

Larsson, S., Wiren, A., Lundgren, L. and Ericsson, T. (1986) Effects of light and nutrient stress on leaf phenolic chemistry in *Salix dasyclados* and susceptibility to *Calerucella lineola* (Coleoptera). *Oikos* **47**, 205–210.

Last, F.T. and Watling, R. (eds) (1991) *Acidic Deposition: Its Nature and Impacts. Proceedings of the International Symposium, Glasgow, Scotland, 16–21 September 1991*, Royal Society of Edinburgh, Edinburgh.

Lawton, J.H. (1995) The response of insects to environmental change. In *Insects in a Changing Environment: 17th Symposium of the Royal Entomological Society, 7–10 September 1993, Rothamsted Experimental Station, Harpenden, England* (eds R. Harrington and N.E. Stork), pp. 4–46, Academic Press, London.

Lee, E.H., Wu, Y., Barrows, E.M. and Mulchi, C.L. (1988) Air pollution, plants and insects: growth and feeding preferences of Mexican bean beetles on bean foliage stressed by SO_2 or O_3. *Environmental Pollution* **53**, 441–442.

Liebhold, A.M., Halverston, J.A. and Elmes, G.A. (1992) Gypsy moth invasion in North America – a quantitative analysis. *Journal of Biogeography* **19**, 513–520.

Lin, H., Kogan, M. and Endress, A.G. (1990) Influence of ozone on induced resistance in soybean to the Mexican bean beetle (Coleoptera: Coccinellidae). *Environmental Entomology* **19**, 854–858.

Lincoln, D.E. (1993) The influence of plant carbon dioxide and nutrient supply on susceptibility to insect herbivores. *Vegetatio* **104/105**, 273–280.

Lincoln, D.E., Fajer, E.D. and Johnson, R.H. (1993) Plant–insect herbivore interactions in elevated CO_2

environments. *Trends in Ecology and Evolution* **8**, 64–68.

Lindroth, R.L., Jung, S.M. and Fenker, A.M. (1993a) Detoxification activity in the gypsy moth: effects of CO_2 and NO_3 availability. *Journal of Chemical Ecology* **19**, 357–367.

Lindroth, R.L., Kinney, K.K. and Platz, C.L. (1993b) Responses of deciduous trees to elevated atmospheric CO_2: productivity, phytochemistry and insect performance. *Ecology* **74**, 763–777.

Lindroth, R.L., Arteel, G.E. and Kinney, K.K. (1995) Responses of three saturniid species to Paper Birch grown under enriched CO_2 atmospheres. *Functional Ecology* **9**, 306–311.

Mattson, W.J. (1980) Herbivory in relation to plant nitrogen content. *Annual Review of Ecology and Systematics* **11**, 119–161.

McCloud, E.S. and Berenbaum M.R. (1994) Stratospheric ozone depletion and plant–insect interactions: effects of UV-B radiation on foliage quality of *Citrus jambhiri* for *Trichoplusia ni*. *Journal of Chemical Ecology* **20**, 525–539.

McLeod, A. (1995) An open-air system for exposure of young forest trees to sulphur dioxide and ozone. *Plant, Cell and Environment* **18**, 215–225.

McNeill, S. and Whittaker, J.B. (1990) Air pollution and tree dwelling aphids. In *Population Dynamics of Forest Insects* (eds A.D. Watt, S.R. Leather, M.D. Hunter and N.A.C. Kidd), pp. 195–208, Intercept, Andover.

McNeill, S., Aminu-Kano, M., Houlden, G. *et al.* (1987) The interactions between air pollution and sucking insects. In *Acid Rain: Scientific and Technical Advances* (eds R. Perry, R.M. Harrisson, J.N.B. Bell and J.W. Lester), pp. 602–607, Selper, London.

Mills, N.J. (1990) Are parasitoids of significance in endemic populations of forest defoliators? Some experimental observations from gypsy moth, *Lymantria dispar* (Lepidoptera: Lymantriidae). In *Population Dynarnics of Forest Insects* (eds A.D. Watt, S.R. Leather, M.D. Hunter, and N.A.C. Kidd), pp. 183–192, Intercept, Andover.

Mitterbock, F. and Fuhrer, E. (1988) Wirkungen fluorbelasteter Fichtnnadeln auf Nonnenraupen, *Lymantria monacha* L. (Lep., Lymantriidae). *Journal of Applied Entomology* **105**, 19–27.

Montgomery, M.E. (1990) The role of site and insect variables in forecasting defoliation by the gypsy moth. In *Population Dynamics of Forest Insects* (eds A.D. Watt, S.R. Leather, M.D. Hunter and N.A.C. Kidd), pp. 73–84, Intercept, Andover.

Mousseau, M. (1993) Effects of elevated CO_2 on growth, photosynthesis and respiration of sweet chestnut (*Castanea sativa* Mill.) *Vegetatio* **104/105**, 413–419.

Murray, M.B., Cannell, M.G.R. and Smith, R.I. (1989) Date of budburst of fifteen tree species in Britain following climatic warming. *Journal of Applied Entomology* **26**, 693–700.

Murray, M.B., Smith, R.I., Leith, I.D. *et al.* (1994) The effect of elevated CO_2, nutrition and climatic warming on bud phenology in Sitka spruce (*Picea sitchensis* (Bong.) Carr.) and their impact on frost tolerance. *Tree Physiology* **14**, 691–706.

Neuvonen, S. and Lindgren, M. (1987) The effect of simulated acid rain on perfomance of the aphid *Euceraphis betulae* (Koch) on Silver birch. *Oceologia* **74**, 77–80.

Neuvonen, S., Saikkonen, K. and Suomela, J. (1990) Effect of simulated acid rain on the growth performance of the European pine sawfly (*Neodiprion sertifer*). *Scandinavian Journal of Forest Research* **5**, 541–550.

Neuvonen, S., Routio, I. and Haukioja, E. (1992) Combined effects of simulated acid rain and aphid infestation on the growth of Scots pine (*Pinus sylvestris*) seedlings. *Annales Botanici Fennici* **29**, 101–106.

Nicolas, G. and Sillans, D. (1989) Immediate and latent effects on pasture plants and cornmunities. *Annual Review of Entomology* **34**, 97–116.

Norby, R.J., Gunderson, C.A., Wullschleger, S.D. *et al.* (1992) Productivity and compensatory responses of yellow-poplar trees in elevated CO_2. *Nature* **357**, 322–324.

Porter, J. (1995) The effects of climate change on the agricultural environment for crop insect pests with particular reference to the European corn borer and grain maize. In *Insects in a Changing Environment: 17th Symposium of the Royal Entomological Society, 7–10 September 1993, Rothamsted Experimental Station, Harpenden, England* (eds R. Harrington and N.E. Stork), pp. 93–123, Academic Press, London.

Prestidge, R.A. and McNeill, S. (1983) The role of nitrogen in the ecology of grassland Auchenorrhyncha. In *Nitrogen as an Ecological Factor: British Ecological Society Symposium 22* (eds J.A. Lee, S. McNeill and I.H. Rorison), pp. 257–287, Blackwell Scientific Publications, Oxford.

Price, P.W. (1991) The plant vigor hypothesis and herbivore attack. *Oikos* **62**, 244–251.

Price, P.W., Bouton, C.E., Grass, P. *et al.* (1980) Interactions among three trophic levels: influence of plants on interactions between insect herbivores and natural enemies. *Annual Review of Ecology and Systematics* **11**, 41–65.

Redinbaugh, M.G. and Campbell, W.H. (1991) Higher plant response to environmental nitrate. *Physiologia Plantarum* **82**, 640–650.

Reich, P.B. (1987) Quantifying plant responses to ozone: a unifying theory. *Tree Physiology* **3**, 63–91.

Reich, P.B. (1995) Phenology of tropical forests: patterns, causes, and consequences. *Canadian Journal of Botany* **73**, 164–174.

Reich, P.B. and Borchert, R. (1984) Water stress and tree phenology in a tropical dry forest in the lowlands of Costa Rica. *Journal of Ecology* **72**, 61–74.

Riemer, J. and Whittaker, J.B. (1989) Air pollution and insect herbivores: observed interactions and possible mechanisms. In *Insect Plant Interactions*, Vol. 1 (ed. E.A. Bernays), pp. 73–105, CRC Press, Boca Raton.

Robert, Y. (1987) Dispersion and migration. In *Aphids, their Biology, Natural Enemies and Control* (eds A.K. Minks and P. Harrewijn), pp. 299–313, Elsevier, Amsterdam.

Roth, S.K. and Lindroth, R.L. (1994) Effects of elevated carbon dioxide-mediated changes in paper birch and white pine chemistry on gypsy moth performance. *Oecologia* **98**, 133–138.

Saikkonen, K.T., and Neuvonen, S. (1993) Effects of larval age and prolonged simulated acid rain on the susceptibility of European pine sawfly to virus infection. *Oecologia* **95**, 134–139.

Salt, D.T. (1993) Effects of gaseous air pollution on root-feeding aphids of spruce. PhD thesis, Lancaster University.

Salt, D.T. and Whittaker, J.B. (1995) Populations of root-feeding aphids in the Liphook forest fumigation experiment. *Plant Cell and Environment* **18**, 321–325.

Salt, D.T., Brooks, G.L. and Whittaker, J.B. (1995) Elevated carbon dioxide affects leaf-miner performance and plant growth in docks (*Rumex* spp.) *Global Change Biology* **1**, 153–156.

Smith, W.H. (1990) *Air pollution and Forests: Interactions Between Air Contaminants and Forest Ecosystems*, Springer-Verlag, New York.

Stark, R.W. and Cobb, F.W. (1969) Smog injury, root diseases and bark beetle damage in ponderosa pine. *Californian Agriculturalist* **23**, 13–15.

Stark, R.W., Miller, P.R., Cobb, F.W. *et al.* (1968) Incidence of bark beetle infestation in injured trees. *Hilgardia* **39**, 121–126.

Stone, G.N. and Sunnocks, P. (1993) Genetic consequences of an invasion through a patchy environment – the cynipid gallwasp *Adricus quercuscalicis* (Hymenoptera: Cynipidae) *Molecular Ecology* **2**, 251–268.

Strathdee, A.T., Bale, J.S., Block, W.C. *et al.* (1993) Extreme adaptive life-cycle in a high arctic aphid, *Acyrthosiphon svalbardicum*. *Ecological Entomology* **18**, 254–258.

Sullivan, J.H. and Teramura, A.H. (1989) The effects of ultraviolet-B radiation on loblolly pine. I. Growth, photosynthesis and pigment production in greenhouse-grown seedlings. *Physiologia Plantarum* **77**, 202–207.

Summers, C.G., Retzlaff, W.A. and Stephenson, S. (1994) The effect of ozone on the mean relative growth rate of *Diuraphis noxia* (Mordvilko) (Homoptera: Aphididae) *Journal of Agricultural Entomology* **11**, 181–187.

Tauber, C.A. and Tauber, M.J. (1986) Ecophysiological responses in life-history evolution – Evidence for the importance in a geographically widespread insect species complex. *Canadian Journal of Zoology* **64**, 875–884.

Tevini, M. (1993) Effects of enhanced UV-B radiation on terrestrial plants. In *UV-B Radiation and Ozone Depletion: Effects on Humans, Animals, Plants, Microorganisms, and Materials* (ed. M. Tevini), pp. 125–253, Lewis Publishers, Boca Raton.

Thalenhorst, W. (1974) Untersuchungen über den Einfluss fluorhaltigen Abgase auf die Disposition der Fichte für den Befall durch die Gallenlaus *Sacchiphantes abietis* (L.) *Zeitschrift für Pflanzenkrankheiten und Pflanzenschudz* **81**, 717–727.

Tripp, K.E., Kroen, W.K., Peet, M.M. and Willits, D.H. (1992) Fewer whiteflies found on CO_2-enriched greenhouse tomatoes with high C:N ratios. *Hortscience* **27**, 1079–1080.

Van Dijk, H.F.G. and Roelofs, J.G.M. (1988) Effects of excessive ammonium deposition on the nutritional status and condition of pine needles. *Physiologia Plantarum* **73**, 494–501.

Van Emden, H.F. (1969) Plant resistance to *Myzus persicae* induced by a plant regulator and measured by aphid relative growth rate. *Entomologia experimentalis et Applicata* **12**, 125–131.

Villemant, C. (1981) Influence de la pollution atmosphérique sur les populations d'aphides du pin sylvestre en forêt de Roumare (Seine-Maritime) *Environmental Pollution Ser. A* **24**, 245–262.

Vitousek, P.M. (1994) Beyond global warming: ecology and global change. *Ecology* **75**, 1861–1876.

Waring, G.L. and Cobb, N.S. (1992) The impact of plant stress on herbivore population dynamics. In *Insect–Plant Interactions*, Vol. 4 (ed. E.A. Bernays), pp. 167–226, CRC Press, Boca Raton.

Warrington, S. (1987) Relationship between sulphur dioxide dose and growth of the pea aphid *Acyrthosiphon pisum* on peas. *Environmental Pollution* **43**, 155–162.

Warrington, S. and Whittaker, J.B. (1990) Interactions between Sitka spruce, the green spruce aphid, sulphur dioxide pollution and drought. *Environmental Pollution* **65**, 363–370.

Warrington, S., Mansfield, T.A. and Whittaker, J.B. (1987) Effect of SO_2 on the reproduction of pea aphids *Acyrthosiphon pisum* and the impact of SO_2 and aphids on the growth and yield of peas. *Environmental Pollution* **48**, 285–294.

Warrington, S., Cottam, D.A. and Whittaker, J.B. (1989) Effects of insect damage on photosynthesis, transpiration and SO_2 uptake by sycamore. *Oecologia* **80**, 136–139.

Watt, A.D., Whittaker, J.B., Docherty, M. *et al.* (1995) The impact of elevated atmospheric CO_2 on insect herbivores. In *Insects in a Changing Environment: 17th Symposium of the Royal Entomological Society, 7–10 September 1993, Rothamsted Experimental Station, Harpenden, England* (eds R. Harrington and N.E. Stork), pp. 198–217, Academic Press, London.

Weseloh, R.M. (1990) Gypsy moth predators: an example of generalist and specialist natural enemies. In *Population Dynamics of Forest Insects* (eds A.D. Watt, S.R Leather, M.D. Hunter and N.A.C. Kidd), pp. 233–244, Intercept, Andover.

White, T.C.R. (1984) The abundance of invertebrate herbivores in relation to the availability of nitrogen in stressed food plants. *Oecologia* **63**, 90–105.

Whittaker, J.B. (1994) Interactions between insects and air pollutants. In *Plant Responses to the Gaseous Environment. Molecular, Metabolic and Physiological Aspects* (eds R.G. Alscher and A.R. Wellburn), pp. 365–384, Chapman & Hall, London.

Whittaker, J.B. and Warrington, S. (1990) Effects of atmospheric pollutants on interactions between insects and their food plants. In *Pests, Pathogens and Plant Communities* (eds J.J. Burdon and S.R. Leather), pp. 97–110, Blackwell Scientific Publications, Oxford.

Whittaker, J.B., Kristiansen, L., Mikkelsen, T. and Moore, R. (1990) Responses to ozone of insects feeding on a crop and a weed species. *Environmental Pollution* **62**, 89–101.

Williams, R.S., Lincoln, D.E. and Thomas, R.B. (1994) Loblolly pine grown under elevated carbon dioxide affects early instar pine sawfly performance. *Oecologia* **98**, 64–71.

Wolfenden, J. and Mansfield, T.A.M. (1991) Physiological disturbances in plants caused by air pollutants. *Proceedings of the Royal Society of Edinburgh* **97B**, 117–138.

Yang, Y.L. and Stamp, N.E. (1995) Simultaneous effects of night-time temperature and an allelochemical on performance of an insect herbivore. *Oecologia* **104**, 225–233.

Yazawa, M., Shimizu, T. and Hirao, T. (1992) Feeding response of the silk worm, *Bombyx mori*, to UV irradiation of mulberry leaves. *Journal of Chemical Ecology* **18**, 561–569.

PART FIVE
INSECT DIVERSITY

14

PATTERNS OF USE OF LARGE MOTH CATERPILLARS (LEPIDOPTERA: SATURNIIDAE AND SPHINGIDAE) BY ICHNEUMONID PARASITOIDS (HYMENOPTERA) IN COSTA RICAN DRY FOREST

Daniel H. Janzen and Ian D. Gauld

14.1 INTRODUCTION

This chapter is a brief eco-taxonomic exploration of a small and taxonomically defined portion of the results that are accumulating during descriptive mapping of a caterpillar fauna (Lepidoptera) on its natural host plants, and simultaneous mapping of the parasitoid fauna (various families of Hymenoptera, Tachinidae, fungi, viruses and bacteria) on this caterpillar fauna. The chapter considers the interactions between some monophyletic lineages of the Campopleginae and Ophioninae, two parasitoid subfamilies of Ichneumonidae, and their host caterpillars in two families of Lepidoptera, the Saturniidae and Sphingidae.

The study site is all of the 1200 km² of conserved Neotropical lowland dry forest, the Guanacaste Conservation Area (GCA) in northwestern Costa Rica (10°50'N, 85°38'W), and its adjacent wet forest habitats (Janzen, 1986c, 1988a,c,e, 1992a, 1993b). This description of the GCA is deliberately designed to document the caterpillar–host and parasitoid–host interaction as a whole that is composed of many small parts, rather than to test specific ecological hypotheses. Description will reveal pattern and suggest process that can be further explored (e.g. Bernays and Janzen, 1988; Janzen, 1987a,b, 1993b; Gauld, 1987; Gaston and Gaul, 1993; Gauld et al., 1992; Gauld and Gaston, 1994; Gauld and Janzen, 1994; Memmot et al., 1994), and provide the natural history base for a very wide variety of users (e.g. Janzen, 1992b, 1994a,b; Reid et al., 1993). This action is explicitly supported by the taxasphere (Janzen, 1992c, 1993a) as an act of biodiversity development through inventory and biodiversity use (Gámez, 1991a,b; Janzen, 1994b; Janzen and Hallwachs, 1994).

The study was initiated in 1978 by D.H. Janzen and W. Hallwachs with the help of a large team of paraecologists and taxonomists and will continue into the indefinite future, with the rearing records currently being stored in an event-based Filemaker Pro 3.0 public domain database on the WWW server at http://janzen.sas.vpenn.edu/index.html/. The rearing voucher specimens are primarily being deposited at the Instituto Nacional de

---

Forests and Insects. Edited by A.D. Watt, N.E. Stork and M.D. Hunter. Published in 1997 by Chapman & Hall, London. ISBN 0 412 79110 2.

Biodiversidad (INBio). This description touches on the taxonomy and ecology of thousands of species of Lepidoptera and their parasitoids, since the study site contains most of the major habitats that are found in Costa Rica. At the time of writing (1995), more than 54 000 wild-caught caterpillars and 'trap caterpillars', placed out in the forest for later recovery, have been reared. All parasitoids that have emerged have been collected and documented in the same database. Simultaneously, most of the adult macrolepidopteran fauna and much of the microlepidopteran fauna of the study have been collected and are being taxonomically processed at INBio and elsewhere.

Since 1984, I.D. Gauld has focused on discriminating Costa Rica's approximately 2000 species of ichneumonid wasps (e.g. Gauld, 1988b, 1991, 1995; Ward and Gauld, 1987). This has provided the taxonomic foundation to be able not only to begin documenting what species of wasps are using what species of caterpillars in the GCA, but also to reconstruct their phylogeny in the context of their use of hosts.

The combination of these efforts has yielded an unusually thorough picture of the caterpillars and ichneumonids present in a single area, the larval food plants, and the caterpillars that are actually parasitized (and are not parasitized) by the ichneumonids. The patterns addressed in this chapter are only those where we feel there is sufficient information to reach preliminary conclusions and generate hypotheses that have a solid taxonomic and natural history foundation. This is why we focus this eco-taxonomic exploration only on the larvae of two families of moths, the Sphingidae and the Saturniidae, and on the only two subfamilies of Ichneumonidae that contain evolutionary lineages that parasitize both of these families (in the study area), the Ophioninae and the Campopleginae. Other groups of ichneumonids, with isolated species in Costa Rican dry forests, that attack either saturniids (e.g. an undescribed species of *Podogaster* (Anomaloninae) which attacks *Hylesia*: Gauld, 1995) or sphingids (such as the ichneumonine genus *Tricyphus*: Ward and Gauld, 1987) are not discussed here.

14.2 SITE AND HABITAT

The GCA study site is a conserved wildland stretching from sea level to 2000 m elevation and ranging from dry forest, with only six months of rainfall, to ever-wet rainforest and cloud forest (Janzen, 1988a, 1992a, 1993b). When the study began in 1978 it was focused on a 50 km^2 dry forest portion central to the GCA at 200–300 m elevation on the Pacific coastal plain. This area is today known as Sector Santa Rosa, and was formerly the eastern portion of Santa Rosa National Park. Nearly all of the records reported here are from this part of the GCA.

Sector Santa Rosa is a mosaic of tropical dry forest fragments and regenerating forest of different ages (10–400 years old) occupying land previously and incompletely cleared for pasture, rice fields and timber harvest, and frequently subject to anthropogenic fires (Janzen, 1986d). This dry forest has a six-month dry season (December to early May) during which it receives, on average, less than 1% of its total 900–2600 m of annual rainfall. In the wet season there are two peaks of rainfall: one in late May/June, when on average about one-third of the annual precipitation occurs, and a more substantial peak in September/October, when about half of the annual rainfall is received (Janzen, 1987a, 1993b). Much of the secondary successional forest is highly deciduous during the dry season (the younger the age of the succession, the more deciduous) and most new foliage is produced at the beginning of the wet season.

14.3 OVERALL LEPIDOPTERA AND PARASITOID PHENOLOGY

Virtually all species of medium-sized to large moths in this forest have a larval generation during the first two months of the rainy season (e.g. Janzen, 1987b, 1988c,d, 1993b). From late May until early July a very large array of different species of caterpillars are feeding on the leaves of trees, saplings, shrubs, herbs and vines, presenting any parasitoids present with a wide variety of what appear to be potential hosts. During the remainder of the rainy season some species of Lepidoptera have one or two subsequent larval generations in substantially reduced numbers (Janzen, 1993b). During the dry season there are almost no exophytic folivorous caterpillars present in the habitat.

Concomitant with this upsurge in caterpillar numbers and diversity in the first two months of the rainy season there is a major change in the composition of the ichneumonid fauna, as represented by adults appearing in Malaise traps and light traps. In the late dry season the active trappable adult ichneumonid fauna is extremely depauperate and sparse, and largely comprises a few idiobiont species – species that stop their host's development upon oviposition (Askew and Shaw, 1986). These wasps, such as *Neotheronia* and *Camera* spp., attack Lepidoptera pupae and prepupae (and their parasitoids) in cocoons and litter. With the onset of the wet season, the diversity and abundance of ichneumonids in Malaise trap samples increases dramatically (Figure 14.1). This is mainly due to the appearance of large numbers of koinobiont ichneumonids (e.g. Figure 28 in Gauld, 1991), that is, wasps that allow the host to continue to develop for a time following parasitization (Askew and Shaw, 1986). Most koinobionts parasitize exophytic or weakly concealed (i.e. non-leaf-mining) caterpillars. These include, for example, the saturniid parasitoid *Enicospilus lebophagus* (Figure 14.3) and the sphingid parasitoids *Thyreodon atriventris* and *Cryptophion inaequalipes* (Figures 14.2, 14.4). Most of the September to December peak in ichneumonid species numbers is due to the increasing abundance of idiobiont 'pupal' parasitoids, such as *Neotheronia tacubaya* (Figure 14.5), which is a hyperparasitoid attacking tachinid larvae, or pupae in their puparia, inside the pupae of sphingids and saturniids in the caterpillar's pupation chamber (Gauld, 1991).

14.4 CATERPILLARS AND WASPS

We consider the interactions between some monophyletic lineages of the Campopleginae and Ophioninae (two subfamilies of Ichneumonidae) and their host caterpillars in two families of Lepidoptera (the Saturniidae and Sphingidae).

14.4.1 CATERPILLARS

The Santa Rosa forest saturniid fauna comprises 30 breeding species in the subfamilies Arsenurinae, Ceratocampinae, Hemileucinae and Saturniinae (Janzen, 1984a, 1985b, 1986a). These saturniid caterpillars feed on about 50 families and 155 species of woody plants (detailed records can be found by searching the project database at http://janzen.sas.vpenn.edu/index.html), ranging from extremely polyphagous species, like *Hylesia lineata* and *Automeris zugana*, to monophages such as *Ptiloscola dargei* and *Copaxa moinieri* (Janzen, 1984b). The sphingid fauna is somewhat larger and less polyphagous (the most polyphagous being *Eryinnis ello*, selectively using various species of latex-bearing plants, and *Manduca rustica*). It comprises 64 regularly breeding species in the subfamilies Sphinginae and Macroglossinae. The 59 species of sphingid caterpillars that have been located feed on 23 families and 125 species of woody and (a few species of) herbaceous plants (Janzen, 1984a, 1985b, 1986a). All 89 species of these large caterpillars are microgeographically available to the ichneumonids discussed here, and during the third to fifth weeks of the rainy season all are present in the Santa Rosa dry forest.

14.4.2 ICHNEUMONIDS

The Santa Rosa ichneumonid fauna is far less well known than the moth fauna, but the Ophioninae compresses about 76 species collected as adults in Malaise traps and at lights, and reared, whilst the Campopleginae comprises 60 species collected in Malaise traps and reared. These two subfamilies are phylogenetically quite closely related (Gauld, 1985) and comprise part of a complex of ichneumonid subfamilies – the ophioniformes. Like other members of the ophioniformes, both subfamilies are solitary koinobiont endoparasitoids of the larvae of holometabolous insects. Although in extratropical regions some species of campoplegines and very few ophionines attack the larvae of Symphyta, Coleoptera or Neuroptera (Townes, 1970, 1971), in Costa Rica they have only been reared from Lepidoptera. Despite very intensive

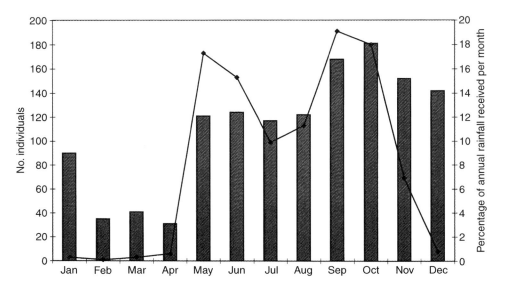

Figure 14.1 Abundance of Ichneumonidae in a Townes-style Malaise trap in the understorey of 70-year-old secondary dry forest non-riparian succession (Bosque San Emilio) in Santa Rosa during 1986. Solid line shows percentage of mean annual rainfall received per month.

collecting with Malaise traps throughout the country (Hanson and Gauld, 1995) none of the campoplegine genera (such as *Olesicampe*, *Bathyplectes* and *Rhimphoctona*) that are known to focus on non-lepidopterous hosts have been found in Costa Rica.

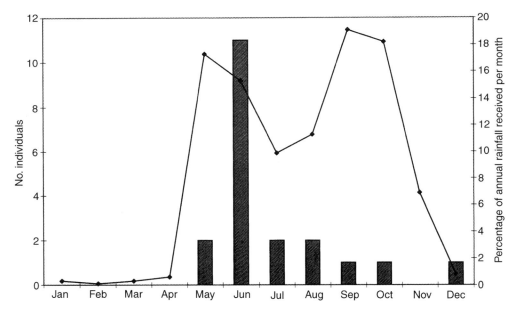

Figure 14.2 Abundance of *Thyreodon atriventris* (Ichneumonidae: Ophioninae) in a Malaise trap (see Figure 14.1) in Santa Rosa during 1986. Solid line shows percentage of mean annual rainfall received per month.

Caterpillars and wasps 255

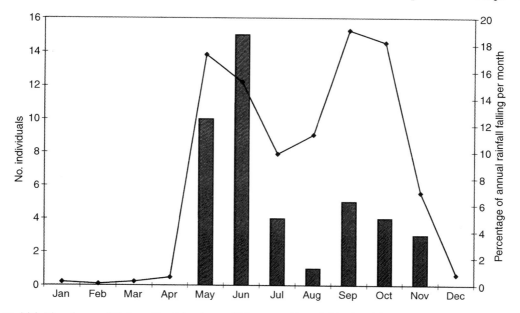

Figure 14.3 Abundance of *Enicospilus lebophagus* (Ichneumonidae: Ophioninae) in a Malaise trap (see Figure 14.1) in Santa Rosa during 1986. Solid line shows percentage of mean annual rainfall received per month.

All ophionines oviposit in caterpillars (rather than in eggs or pupae), but the wasp larva does not kill its host until after the caterpillar has completed feeding and sought out a pupation site. At this time the wasp larva rapidly consumes the caterpillar, (usually) exits from the caterpillar cuticle and spins

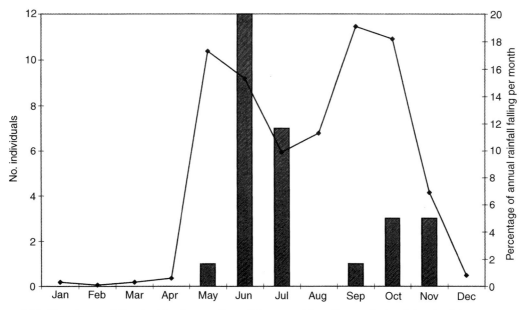

Figure 14.4 Abundance of *Cryptophion inaequalipes* (Ichneumonidae: Campopleginae) in a Malaise trap (see Figure 14.1) in Santa Rosa during 1986. Solid line shows percentage of mean annual rainfall received per month.

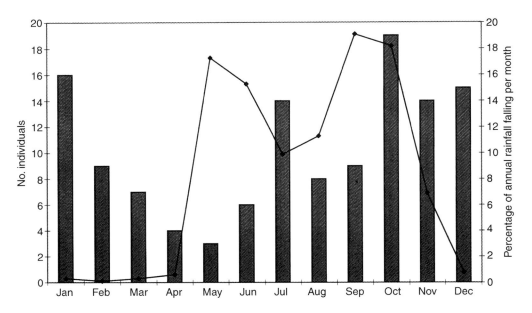

Figure 14.5 Abundance of *Neotheronia tacubaya* (Ichneumonidae: Pimplinae) in a Malaise trap (see Figure 14.1) in Santa Rosa during 1986. Solid line shows percentage of mean annual rainfall received per month.

a cocoon in the larval pupation site. Many of the structurally less derived campoplegine genera, such as *Diadegma*, have a very similar biology, but in the tropics a fairly large proportion of the campoplegine fauna comprises members of the *Hyposoter* genus-complex. The species in the *Hyposoter* genus-complex generally kill and consume the interior of their host whilst the caterpillar is still a pre-ultimate instar larva feeding on its foodplant. These campoplegines, which are not in a pupation retreat, construct black-and-white mottled cocoons, resembling bird droppings, that are often suspended from the vegetation by a long thread (e.g. *Microcharops*), or partially concealed under the dead shrivelled cadaver of the host larva (e.g. *Cryptophion*). In Santa Rosa, Ophioninae are amongst the largest of Ichneumonidae, while Campopleginae are medium-sized.

Members of three ophionine genera in the dry forests of Sector Santa Rosa of the GCA, *Thyreodon*, *Rhynchophion* and *Enicospilus*, attack either sphingid or saturniid larvae but not both. The members of the first two genera only attack sphingids or saturniids, whilst *Enicospilus* is more distributed in a very oligophagous (rarely) to quite monophagous (usually) manner across Notodontidae, Noctuidae, Lasiocampidae, Arctiidae and Saturniidae. All the saturniid parasitoids in *Enicospilus* belong to one monophyletic subgroup, the *E. americanus* species-group, though not all members of this subgroup are restricted to attacking saturniids (Gauld, 1988a). The non-use of Sphingidae by neotropical *Enicospilus* is striking, given that sphingid caterpillars are of the large size used by most *Enicospilus* and that sphingid caterpillars of many species are common in the habitats occupied by hundreds of species of *Enicospilus*. Members of only one campoplegine genus, *Cryptophion*, attack both sphingids and saturniids, though even in this case a given species does one or the other.

Before discussing the host relationships of these ichneumonids in Santa Rosa dry forest, it is necessary to outline briefly their phylogenetic interrelationships as based on their comparative morphology.

14.5 SPHINGID- AND SATURNIID-EATING OPHIONINAE

The Ophioninae comprises five monophyletic lineages, the *Ophion*, *Sicophion*, *Eremotylus*, *Thyreodon* and *Enicospilus* genus-groups (Gauld, 1985). The last two of these are the most derived in the subfamily and are sister-groups comprising the terminal clade. It is within some lineages of this clade that the sphingid and saturniid parasitoids occur.

14.5.1 *THYREODON* GENUS-GROUP

Rhynchophion and three species-groups of *Thyreodon* occur in the Santa Rosa dry forests (Porter, 1986; Gauld, 1988b). *Thyreodon* and *Rhynchophion* are closely related genera that comprise part of the *Thyreodon* genus-group with three other Old World genera, *Barytatocephalus*, *Euryophion* and *Dictyonotus*. Gauld (1985) undertook a cladistic analysis of this group and hypothesized that *Barytatocephalus* is the most plesiomorphic genus, whilst *Euryophion* is the sister lineage to *Dictyonotus* + *Rhynchophion* + *Thyreodon*, and *Dictyonotus* + *Rhynchophion* is the sister lineage of *Thyreodon* (Figure 14.6). *Dictyonotus* and *Rhynchophion* are both very small generally, but *Thyreodon* is much larger with at least 40 distinct taxa, most of which occur in the New World tropics. *Thyreodon* comprises three distinct species-groups (Cushman, 1947; Porter, 1984), one of which, the *atriventris* species-group, is sometimes treated as a separate genus, *Athyreodon* (e.g. Porter, 1989). The *atriventris* species-group is the sister-lineage of the *laticinctus* + *atricolor* species-groups (Figure 14.6). Despite these groupings, the species-level phylogenetic interrelationships of *Thyreodon* have yet to be resolved.

14.5.2 *ENICOSPILUS* GENUS-GROUP

In the New World only a single evolutionary lineage in this group, the *Enicospilus americanus* complex, is known to attack saturniid caterpillars. This group is a derived and clearly monophyletic assemblage characterized by the loss of alar sclerites and having a modified blunt gonolacinia, with a hook-like subterminal appendage (Gauld, 1988a). Within the group, phylogenetic patterns are not clearly resolved but several very close sister-species pairs are discernible, such as *E. texanus*/*E. cushmani* (Gauld, 1988a) and *E. robertoi*/*E. scuintlei* (Gauld, 1988b). *E. glabratus*, which has a slightly more evenly tapered gonolacinia and a complete occipital carina, may represent a sister-lineage to all other species.

14.6 SPHINGID- AND SATURNIID-EATING CAMPOPLEGINAE

The classification of the subfamily Campopleginae is still in extreme disarray, and many Neotropical groups or species are not clearly assignable to any described genus. Virtually nothing is known of the phylogenetic interrelationships of the genera, but one more or less distinctive group, the *Hyposoter* genus-group, is apparent. This comprises rather small, stout campoplegines with very short, almost straight ovipositors. Many have a ventral row of close hairs present on the ventral side of the hind basitarsus. In the New World tropics, species in only one genus in the *Hyposoter* genus-group, *Cryptophion*, use the larvae of Sphingidae and Saturniidae. Six of the eight species of *Cryptophion* occur in Santa Rosa dry forest, and all six have been reared there. A hypothesis of the phylogenetic interrelationships of these six species was presented by Gauld and Janzen (1994) and their cladogram is shown here in Figure 14.7.

14.7 HOST RELATIONSHIPS OF THE ICHNEUMONIDS

14.7.1 *THYREODON* GENUS-GROUP

Worldwide, all species in this genus-group are parasitoids of moths of the superfamily Bombycoidea (*sensu* Scoble, 1992). All reared specimens of *Dictyonotus*, *Rhynchophion* and the overwhelming majority of *Thyreodon* species, including all known members of the *atriventris* and *laticinctus* species-groups, have come from

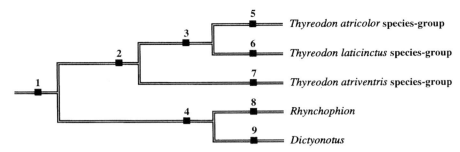

Figure 14.6 Presumed phylogenetic interrelationships of species-groups of *Thyreodon*, *Rhynchophion* and *Dictyonotus*. Derived characters supporting this hypothesized phylogeny: (1) centrally pointed clypeal margin; short stout flagellum; extremely elongate propodeal spiracle; absence of umbo on tergite II; small ocelli; loss of glabrous area in discosubmarginal cell; (2) presence of a pronotal crest; reduction in size of metapleuron; (3) massively enlarged propodeum, reaching to level of end of hind coxa; (4) loss of notauli; transverse impression on mesopleuron; (5) enlarged head with broad genae; (6) coarse, rugose notauli; (7) secondarily enlarged ocelli; swollen second segment of maxillary palp; (8) elongate, anthophilus mouthparts; reduced lower mandibular tooth; (9) enlarged metanotal protuberance; complete posterior transverse carina on mesosternum. (More details of characters and analysis in Gauld, 1985.)

the larvae of Sphingidae. For example, in Asia *Dictyonotus purpurascens* has been reared from the sphingine sphingid *Ampelophaga rubiginosa* and an unidentified species of *Smerinthus* (Gauld and Mitchell, 1981) and in North America *Thyreodon atricolor* is known to attack the sphingine sphingids *Paonias exaecatus*, *Lapara coniferarum* and *L. bombycoides* (Carlson, 1979, and specimens in USNM). In the wet forests of Costa Rica, *Thyreodon laticinctus* is known to parasitize the macroglossine *Xylophanes anubis* (Gauld, 1988b).

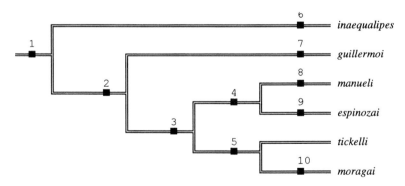

Figure 14.7 Presumed phylogenetic interrelationships of species of *Cryptophion*. Derived characters supporting this hypothesized phylogeny: (1) hindleg long in comparison to forewing; propleural flange present; vein $2m$-cu opposite or slightly distal to $3rs$-m; tergite II depressed; propodeum short, centrally excavate; (2) hind tarsal claw only pectinate basally; clypeal ridge present; propleural flange enlarged; propodeal denticle present; mandibular flange truncated; (3) scutellar carinae weak; hind tibia strongly compressed; (4) scutellum densely pubescent; distal apex of hind femur yellow; (5) mesoscutum smooth and impunctate medially; occipital carina produced into a flange; (6) scutellum pyramidal; mid coxa yellow; (7) hind tarsus black; (8) scape yellow ventrally; (9) posterior transverse carina of mesosternum centrally emarginate; (10) interocellar distance shorter than 1.8 times ocellar diamater; flagellum with more than 53 segments; hind coxa red. (More details of characters and analysis in Gauld and Janzen, 1994.)

All rearings of the species in the *Thyreodon* genus-group from Santa Rosa dry forests have been from sphingine and macroglossine sphingid caterpillars and from ceratocampine saturniid caterpillars (Table 14.1). All but two species ever collected in 36 Malaise trap years in the Santa Rosa dry forest have been reared. The two species not reared (*Thyreodon rivinae* and *T. erythrocera*) have only very rarely been collected, and in the most open dry forest habitats (seven specimens in about 10 000 Ichneumonidae collected in the survey), and are more generally found in very open habitats (Porter, 1980).

In the dry forests it is very striking that these large wasps are using such a tiny portion of the entire sphingid and saturniid caterpillars that are microgeographically and chronologically available. First, the overall rate of parasitism is clearly

Table 14.1 The caterpillar hosts of species of *Thyreodon* and *Rhynchophion* (Ichneumonidae) recorded in the dry forests of Sector Santa Rosa, Guanacaste Conservation Area, northwestern Costa Rica[1]

Parasitoid	Host	Subfamily	Number caterpillars reared	Percentage attacked[4]
Rhynchophion flammipennis	*Manduca lefeburei*	Sphinginae	113	2
Rhynchophion sp. 1 (93-SRNP-1755)	*Manduca corallina*	Sphinginae	50	4
	Manduca barnesi	Sphinginae	81	6
	Manduca rustica	Sphinginae	168	1
	Manduca muscosa	Sphinginae	52	2
Thyreodon atriventris	*Pachylia ficus*	Macroglossinae	267	11
	Pachylia syces	Macroglossinae	17	1
	Pachylioides resumens[2]	Macroglossinae	173	6
	Pachygonidia drucei	Macroglossinae	12	8
Thyreodon maculipennis	*Perigonia ilus*[3]	Macroglossinae	355[3]	10[3]
	Perigonia lusca[3]	Macroglossinae		
Thyreodon new species	*Xylophanes turbata*	Macroglossinae	397	4
Thyreodon apricus	*Erinnyis lassauxii*	Macroglossinae	9	1
Thyreodon santarosae	*Syssphinx molina*	Ceratocampanae	159	8
	Syssphinx colla	Ceratocampinae	41	1
	Othorene purpurascens	Ceratocampinae	44	11
	Othorene verana	Ceratocampinae	27	1
	Ptiloscola dargei	Ceratocampinae	38	8
Total			2003	6%

[1] For the purposes of this analysis, wasps in these genera that died in their cocoons or as larvae newly emerged are assumed to be of the species of wasp in these genera invariably reared from that host, since the larvae and cocoons within these genera cannot be reliably distinguished at the level of species, and to postulate otherwise would be to postulate host-parasitoid relationships that have not been encountered. If only 'reared to adult wasp' were recorded in this table, the host-specificity pattern would be the same but the sample sizes reduced to about half.

[2] This genus has been treated as a synonym of *Pachylia* by various authors, but we choose to follow Hodges (1971) and consider them to be separate but very closely related. The larva of *Pachygonidia* has the same life form as *Pachylia* and *Pachylioides*, but is not closely related to them.

[3] It has not been possible to discriminate these two species of caterpillars as larvae (and indeed, they are together generally called "*Perigonia lusca*"; Janzen, 1984a). The wasp appears to attack them indiscriminately on all of their host plants. The percentage attacked has been pooled across all larvae.

[4] This can only be a minimal figure, since both tachinids and other ichneumonids kill the caterpillar (which could have a *Rhynchophion* or *Thyreodon* larva in it) before it reaches the prepupal stage. Furthermore, the act of collecting the caterpillar removes it from being exposed to host-searching wasps, and the parasitism rates of wild-caught caterpillars will be less than those of caterpillars that are fully exposed to parasitoids until the pupal stage.

hovering between 1% and 10% (6% for records presented in Table 14.1 as a whole). Second, of the 89 species of sphingids and saturniids whose larvae have been reared in the Santa Rosa dry forest, only 18 (20%) are being used by *Rhynchophion* and *Thyreodon*. Put another way, a sample of slightly more than 12 000 wild-caught sphingid and saturniid caterpillars of 71 species have generated no *Thyreodon* genus-group wasps, even though this genus-group specializes on Sphingidae. A few more species may be found to be parasitized by these wasps with larger sample sizes (e.g. when the hosts of *Thyreodon rivinae* and *T. erythrocera* have been located – if, that is, they attack dry forest large moths), but the pattern of very low use will remain.

14.7.2 *ENICOSPILUS AMERICANUS* SPECIES-GROUP

All members of this species-group are parasitoids of the caterpillars of large moths, including species of Lasiocampidae, Arctiidae and Lymantriidae, but the majority are restricted to parasitizing saturniids. There are no authenticated records of any species in this complex developing on sphingid caterpillars. In North America the two widespread species, *E. americanus* and *E. texanus*, parasitize a wide range of saturniid species (Gauld, 1988a), the former specializing on saturniines (except for a short series of specimens which is questionably conspecific and was reared from a hemileucine) and the latter primarily on hemileucines (except for one record from *Agapema galbina* that requires repetition to be assured it was not an ecological accident) (Table 14.2). The members of this species-group that attack saturniids in Santa Rosa dry forest, however, are much more monophagous (Table 14.2). This tropical/extra-tropical contrast is fully consistent with the overall behaviour of *Enicospilus* on all of its different dry forest hosts (Table 14.3).

It is very striking that these large wasps are using such a small portion of the entire sphingid (none) and saturniid caterpillars that are microgeographically and chronologically available. First, just as was the case with the *Thyreodon* genus-group, the overall rate of parasitism on saturniids is clearly hovering between 1% and 10%, except for *Enicospilus lebophagus*, which averages 33% (Table 14.3) and can be as high as 90% in certain circumstances, such as when the caterpillars are placed out in the forest at artificially high density (DHJ, unpublished field experiment; and see Olson, 1995). Second, of the 30 species of saturniids that breed in Santa Rosa dry forest, only four (13.3%) are being used by *Enicospilus* despite the existence of nearly 70 species of *Enicospilus* in this forest (Gauld, 1988b). Put another way, a sample of slightly more than 4000 wild-caught saturniid caterpillars of 26 species have generated no *Enicospilus* wasps. A few more species may be found to be parasitized by these wasps after larger samples of the scarce species of saturniids have been obtained, but the pattern of very low use of species and numbers of caterpillars will remain. In summary, the host relationships are strikingly specific and it is clear that there are many hosts 'unused' by these large wasps, whatever the ecological and evolutionary causes.

14.7.3 GENUS *CRYPTOPHION*

All members of this genus are parasitoids of Sphingidae, or exceptionally of Saturniidae (Table 14.4). The structurally more primitive species, *Cryptophion inaequalipes* and *C. guillermoi*, are monophagous, attacking macroglossines feeding on rubiaceous understorey treelets and shrubs. In one of the more derived sister-species pairs, each attacks several sphingid host species; *C. espinoza* specializes in macroglossine sphingids in sunnier earlier stages of succession, while *C. manueli* specializes in the more rough-skinned sphingine sphingids in shadier understorey of later stages of succession. The two species comprising the other derived sister-species pair are each monophagous, one on a macroglossine sphingid and the other on a ceratocampine saturniid in tree crowns, both living in sunnier habitats and very similar in general body morphology (Gauld and Janzen, 1994).

14.8 DISCUSSION

14.8.1 PROPORTION OF FAUNA ATTACKED

Perhaps the most striking feature about Tables 14.1, 14.2 and 14.4 is that although these evolu-

Table 14.2 The known host of Nearctic American and Costa Rican saturniid-parasitizing ichneumonids in the *Enicospilus americanus* species-complex (Ichneumonidae: Ophioninae); only authenticated specimen-based records have been included.

Parasitoid	Host	Subfamily
Enicospilus americanus[1]	*Actias luna*	Saturniinae
	Antheraea polyphemus	Saturniinae
	Automeris io[3]	Hemileucinae
	Callosamia promethea	Saturniinae
	C. securifera	Saturniinae
	Hyalophora cecropia	Saturniinae
	H. euryalis	Saturniinae
	Rothschildia orizaba[4]	Saturniinae
	Samia cynthia	Saturniinae
Enicospilus texanus[1]	*Agapema galbina*[5]	Saturniinae
	Hemileuca juno	Hemileucinae
	Hemileuca magnifica	Hemileucinae
	Hemileuca maia	Hemileucinae
	Hemileuca oliviae	Hemileucinae
	Hemileuca peigleri	Hemileucinae
	Hemileuca tricolor	Hemileucinae
Enicospilus bozai[2]	*Copaxa moinieri*	Saturniinae
Enicospilus lebophagus[2]	*Rothschildia lebeau*	Saturniinae
Enicospilus robertoi[2]	*Hylesia lineata*	Hemileucinae
Enicospilus tenuigena[2]	*Automeris tridens*	Hemileucinae
Enicospilus ugaldei[2]	*Automeris tridens*	Hemileucinae

[1] Nearctic American
[2] Costa Rican
[3] There is some doubt that the short series reared from this host is conspecific with *E. americanus*: it may represent a new species (Gauld, 1988a).
[4] This record was obtained from Mexico in the extreme southern end of the North American distribution of *E. americanus*.
[5] Only a single instance of this host being used has been recorded; *A. galbina* has a more open cocoon than other saturniines.

tionary lineages comprise sphingid and saturniid oligophages they are only developing on a very small proportion of the sphingid and saturniid caterpillars microgeographically available as hosts in Santa Rosa. The 18 dry forest ophionine and campoplegine species for which we have data successfully parasitize only 34 (38%) of the 89 species of breeding sphingids and saturniids. Intensive Malaise trap and light trap collecting in the area has revealed only six additional species of the *Enicospilus americanus* complex (*E. aktites*, *E. alvaroi*, *E. chiriquensis*, *E. enigmus*, *E. gamezi* and *E. scuintlei*). *Enicospilus gamezi* has been reared once from a very hairy Apatelodidae, and *E. scuintlei* from hairy Lasiocampidae (Table 14.3), so these two probably do not use Saturniidae. Trapping has produced two other species of *Thyreodon* (*T. erythrocera* and *T. rivinae*), and no additional species of either *Rhynchophion* or *Cryptophion*. On a pro rata basis, assuming a similar low degree of polyphagy for the species we have not reared, the entire array of these parasitoids in Santa Rosa is likely to be using only about 40% of the potential host species present, and most of those at an extremely low frequency. Many of the sphingid larvae present do not seem to be attacked at all by any *Thyreodon* species, even though they may be attacked by other species of *Thyreodon* or

Table 14.3 The host records of *Enicospilus* species (Ichneumonidae: Ophioninae) in Santa Rosa dry forest

Parasitoid	Host	Host family	Number caterpillars reared	Percentage attacked
Enicospilus bozai	*Copaxa moinieri*	Saturniidae	226	6.2
Enicospilus colini	*Hapigiodes sigifredamdrini*	Notodontidae	97	4.1
Enicospilus dispilus	*Melipotis fasciolaris*	Noctuidae	22	4.5
	Malocampa hibrida	Notodontidae	18	11.1
	Dicentria rustica	Notodontidae	262	1.1
	Schizura rivalis	Notodontidae	455	1.1
Enicospilus echeverri	94-SRNP-5470	Noctuidae	1	100.0
Enicospilus glabratus	*Ecpantheria icasia*	Arctiidae	124	2.4
	Ecpantheria suffusa	Arctiidae	106	6.6
Enicospilus lebophagus	*Rothschildia lebeau*	Saturniidae	321	33.6
Enicospilus liesneri	*Eulepidotis rectimango*	Noctuidae	26	19.2
Enicospilus luisi	*Elymiotis* 2 spp. on Malphighiaceae	Notodontidae	126	6.4
Enicospilus maritzai	*Thysania zenobia*	Noctuidae	7	14.3
Enicospilus monticola	*Gonodonta bidens*	Noctuidae	8	37.5
	Gonodonta clotilda	Noctuidae	14	28.6
	Gonodonta incurva	Noctuidae	39	9.7
	Gonodonta pyrgo	Noctuidae	119	10.8
Enicospilus pescadori	82-SRNP-344a	Noctuidae	36	2.8
Enicospilus randalli	*Euclystis guerini*	Noctuidae	20	5.0
Enicospilus robertoi	*Hylesia lineata*	Saturniidae	1454	0.6
Enicospilus scuintlei	*Euglyphis melancholica*	Lasiocampidae	15	20.0
Enicospilus simoni	*Lirimiris lignitecta*	Notodontidae	76	9.2
Enicospilus tenuigena	*Automeris tridens*	Saturniidae	382	1.6
Enicospilus ugaldei	*Automeris tridens*	Saturniidae	382	7.0

Cryptophion in other parts of their range. For example, *Xylophanes anubus* is extremely common in Santa Rosa's dry forest feeding on *Psychotria horizontalis*, *P. pubescens* and *P. nervosa*. It is found as apparently a breeding population throughout the remainder of lowland Costa Rica, feeding also on these *Psychotria* and others. In Costa Rican wet forest habitats it is attacked by *Thyreodon laticinctus* (Gauld, 1988b) and by *Cryptophion* sp. 1. The former is a very distinctive large black-and-yellow ichneumonid found throughout Costa Rican rainforests, from low to intermediate elevations (but not dry forest) and as close as 10 km to the east (and upwind) of Santa Rosa dry forest. However, 124 rearing events from Santa Rosa have generated no hymenopterous parasitoids of *Xylophanes anubus*. That is to say, this is a host that is demonstrably suitable for use by *Thyreodon* and *Cryptophion* but is not used by either in Santa Rosa, both because the species there do not develop in it, and because its parasitoids have not invaded Santa Rosa.

The dry forest caterpillar fauna appears to be used by fewer ichneumonid species than is a nearby mid-elevation wet forest caterpillar fauna. Ichneumonid species richness in Costa Rica shows the same mid-altitudinal bulge displayed by many other insects (Janzen, 1973; Janzen *et al.*, 1976), with greatest species richness being encountered between 1000 and 1500 m (Gauld, Ugalde and Hanson, unpublished). It is noticeable that an intensive study of pimpliform ichneumonids in the Guanacaste Conservation Area has shown that whilst 67% of the species present in the dry forests

Table 14.4 The caterpillar host of species of *Cryptophion* (Ichneumonidae) recorded in the dry forests of Sector Santa Rosa, Guanacaste Conservation Area, northwestern Costa Rica[1].

Parasitoid	Host	Subfamily	Number caterpillars reared	Percentage attacked
Cryptophion inaequalipes	Xylophanes turbata	Macroglossinae	397	4
Cryptophion guillermoi	Xylophanes tyndarus	Macroglossinae	16	13
Cryptophion tickelli	Eumorpha satellita	Macroglossinae	255	1
Cryptophion moragai	Syssphinx molina	Ceratocampinae	159	7
Cryptophion espinozai	Aellopos titan	Macroglossinae	469	1
	Erinnyis crameri	Macroglossinae	163	1
	Erinnyis lassauxii	Macroglossinae	9	10
	Erinnyis ello	Macroglossinae	376	0.3
	Hemeroplanes triptolemus	Macroglossinae	11	27
	Pachylia ficus	Macroglossinae	267	3
Cryptophion manueli	Enyo ocypete	Macroglossinae	1095	0.1
	Manduca florestan	Sphinginae	235	4
	Manduca lefeburei	Sphinginae	113	1
	Manduca dilucida	Sphinginae	445	0.2
	Protambulyx strigilis	Sphinginae	72	7
	Adhemarius gannascus	Sphinginae	65	3
Total			4147	3.3%

[1] For the purposes of this analysis, wasps in these genera that died in their cocoons or as larvae newly emerged are assumed to be of the species of wasp in these genera invariably reared from that host, since the larvae and cocoons within these genera cannot be reliably distinguished at the level of species, and to postulate otherwise would be to postulate host–parasitoid relationships that have not been encountered. If only 'rearing to adult wasp' were recorded in this table, the host-specificity pattern would be the same but the sample sizes reduced to about half.

[2] Since *Cryptophion* species kill the host in the early instars, late instar caterpillars cannot generate *Cryptophion*. Late instar caterpillars are easier to find than are early instar caterpillars, and at least 70% of each species' sample is made up of wild-caught caterpillars too large to produce *Cryptophion*. This means that the actual percentage killed in nature by *Cryptophion* could be as high as three times that recorded here.

of Santa Rosa also occur in the more humid forests at 1000 m elevation to the east on Volcán Cacao, only 38% of the ichneumonid species present at the humid forest site also occur in the dry forests 10–20 km to the west of Santa Rosa (Gauld, 1991). Despite the huge number of species and biomass of caterpillars present in the Santa Rosa dry forest during the first half of the rainy season, there is something clearly inhospitable about the site for many species of Ichneumonidae. However, this picture is complicated by the high species richness of the set of parasitoids of sphingids and saturniids in the Santa Rosa dry forest. Despite the large biomass and species richness of 'unused' caterpillars in Santa Rosa, this dry forest apparently has many more species of all three groups than does the humid intermediate elevation site on Volcán Cacao.

We propose a relatively simple scenario to reconstruct how the low use of the dry forest sphingid and saturniid caterpillars by these ichneumonids has come about – a scenario that can be examined through hypothesis generation and testing with further field work. To begin, we note that there is nothing special about the dry forest habitats in Santa Rosa, being part of a Neotropical habitat that was very widespread in pre-Columbian times but is today almost entirely converted to pastures and crops (e.g. Janzen, 1986a, 1988b, 1992d; Bullock *et al.*, 1995).

There is no reason to postulate that the Santa Rosa dry forest itself has been the site of evolution of any of the species present, be they caterpillars or parasitoids. All of these species, and their parasitoid relationships, evolved somewhere (else) as small populations isolated by this or that accident of colonization, climatic change or biotic interaction. Then these species expanded their ranges, again through the perfectly ordinary processes that occur as populations colonize new land masses, climates change and biotic interactions change. Each caterpillar arrived in Santa Rosa through this range expansion, encountered conditions for survival (at whatever density), and constituted yet one more food source for parasitoids. Equally, each parasitoid arrived at Santa Rosa as the result of the same processes. As with caterpillars encountering 'their' (i.e. at least one) foodplant, and non-eradicating climate and carnivore regimes, the parasitoids encountered caterpillars that stimulated whatever host-selection behaviour they came with, and they possessed appropriate mechanisms for circumventing the caterpillar's immunodefensive system. These caterpillars were sufficiently abundant to sustain a population. The parasitoids persisted. In short, the array is constructed through a process of ecological fitting (Janzen, 1985a, 1986a) rather than on-site evolutionary interaction.

The fact that various amounts and kinds of other potential hosts and friendly conditions are also present is probably in great part irrelevant. *Xylophanes anubus* means as much to *Cryptophion moragai* as does a stone. With respect to species of sphingid and saturniid parasitoids of the types discussed here, we view the number found in the Santa Rosa dry forest as a serendipitous outcome of evolutionary events at distant points, range changes over time, and whether an 'adequate for survival' set of hosts, climate and predator regimes is present for each species as it arrives. Whether any one of the present species has competitively or by disease blocked the arrival of other non-present species is moot and unknowable (but unlikely, considering the small number of species of these sorts of wasps present in the Neotropics, their low frequency of host attack, and the low potential for interspecific interactions). In other words, one could probably stack at least another 20 species of *Thyreodon*, *Enicospilus* and *Cryptophion* on to just the saturniids of Santa Rosa, to say nothing of the sphingids, without having any significant effect on those already present.

We could evoke any number of complex second- and third-order interactions that could conceivably be occurring between these large wasps and the other organisms that eat the species of sphingids and saturniids that they are not (currently) using. A more parsimonious hypothesis is that the Santa Rosa dry forest has a large number of sphingids and saturniid species that are not used by the wasps under discussion here because those wasps present were simply never evolutionarily programmed to seek or recognize them. Also, they are not used by other (additional) species in these groups of wasps because those other species have not arrived. They have not yet evolved or have not yet got there. Or, if they did arrive, they did not encounter adequate conditions to survive as a population. These 'conditions' are of all sorts, but probably have little or nothing to do with the other species of wasps in these groups that are present in this dry forest. That is to say, the patterns disclosed to date do not display traits suggesting that they are the result of interspecific competition within these groups, or on-site evolution that was rich in interspecific interactions within these groups.

Whenever a herbivore colonizes a new habitat it may encounter a parasitoid that recognizes it as a host, and there is considerable evidence that native parasitoids add such new immigrants to the array of hosts they attack (Cornell and Hawkins, 1993). In North America the recent immigrant saturniid *Samia cynthia* is attacked by the native saturniid parasitoid *Enicospilus americanus*, showing that these wasps are capable of using a newly arrived host. It is therefore particularly striking that in the tens of thousands of years that the suite of Central American dry forest herbivores and parasitoids have coexisted, more sphingids and saturniids have not been colonized as hosts by the sphingid and saturniid parasitoids. We have already suggested elsewhere (Gauld *et al.*, 1992; Gauld and Gaston, 1994) a chemical mechanism that might constrain tropical endoparasitoids to a narrower range of hosts than their extra-tropical relatives, and thus prevent easy host recruitment by sympatric parasitoids in tropical forest ecosystems.

14.8.2 DEGREE OF POLYPHAGY OF EACH SPECIES OF PARASITOID

As is implicit above, all of the species of *Enicospilus*, *Thyreodon*, *Rhynchophion* and *Cryptophion* in Santa Rosa that attack sphingids and saturniids use one or a very few species of hosts. There are no 'generalists' that attack many species of a given life form, or many life forms, of caterpillars. In particular, each of the species of the *Enicospilus americanus* species-group that attacks saturniids is restricted to a single host in Santa Rosa, even though closely related species in North America attack a large number of species (Table 14.2). This difference is reinforced by the presence of closely related species pairs of caterpillars, one of which is used by the wasp and the other not. For example, *Rothschildia lebeau* is heavily parasitized by *Enicospilus lebophagus*. However, *Rothschildia erycina*, feeding on the same host plant as (and sometimes alongside) *R. lebeau*, is never successfully parasitized by this wasp. A second example is *Automeris tridens*, used (solely) by both *Enicospilus ugaldei* and *E. tenuigena*. This is a common caterpillar in the habitat along with *Automeris zugana* on the same foodplants, yet *A. zugana* is not used by any hymenopterous parasitoid.

Four of the six species of *Cryptophion* in Santa Rosa are restricted to a single host species (Table 14.4). The use of a small group of different sphingids (out of the many species present) by each member of a species pair – *Cryptophion espinozai* and *C. manueli* – is both taxonomically and ecologically restricted. The former species uses only a small part of the macroglossine sphingids present, and does so in the deeply shaded understorey of late secondary succession. The latter species uses a small portion of the sphingine caterpillars present, and does so in the more sunny habitats of earlier secondary succession.

14.8.3 EVOLUTIONARY PATTERNS OF HOST UTILIZATION

In each of two separate evolutionary lineages, *Cryptophion* and *Thyreodon*, there is a species that has moved from the original state of parasitizing Sphingidae to parasitize ceratocampine saturniids (Figures 14.8, 14.9). It may be significant that ceratocampine caterpillars are the most 'sphingid-like' – in caterpillar anatomy and larval natural history – of the common Neotropical Saturniidae. Perhaps even more indicative, *Syssphinx molina* is one of the saturniid hosts in both cases. (Owing to our not having had the opportunity to sample many early instar *Syssphinx colla*, *Othorene* or *Ptiloscola*, the possibility remains open that they are also used by *Cryptophion moragai*.)

It is easy to envisage that the ancestors of each of these wasp species found itself in an evolution-prone circumstance whereby ceratocampine caterpillars were present and its sphingid host relatively absent, or that the former was in some sense 'better' as a host. The outcome has been the generation of two lineages of wasps, each of which is now perhaps 'set up' to radiate evolutionarily among the large number of caterpillar species and life forms in the Saturniidae (and as such generate new 'genera' or other clades). It could be said that the ceratocampine saturniids are now more likely to acquire, ecologically or evolutionarily, species of ichneumonid parasitoids than before, given that there are now at least two ichneumonid lineages that have made the change to these saturniids.

Although two lineages have switched from Sphingidae to Saturniidae, we have no reverse examples. No species of *Enicospilus*, with its hundreds of Neotropical species attacking medium-sized to large exophytic caterpillars, has moved on to Sphingidae. In the case of the *E. americanus* species-group, the host switching has been from Saturniidae to Lasiocampidae, and therefore remains within the Bombycoidea. This has happened in at least two lineages. In North America *E. texanus* is a parasitoid of various species of hemileucine saturniids (Table 14.2) and its close relative *E. cushmani* parasitizes *Malacosoma* spp. (Gauld, 1988a). In the Santa Rosa dry forest, *E. robertoi* parasitizes *Hylesia lineata* (Table 14.2) and its sibling species *E. scuintlei* parasitizes *Euglyphis melancholia* (Table 14.3; Gauld, 1988b). Perhaps it is significant that, in both of these switches, the new host group is a hairy life form and spins a cocoon, just as is the case with the 'apparent' host lineage.

266 *Large moth caterpillars in Costa Rican dry forest*

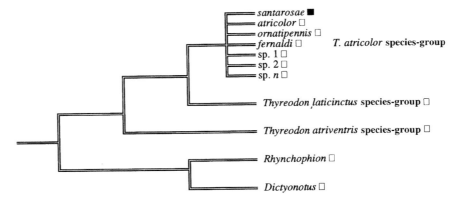

Figure 14.8 Patterns of host utilization superimposed on cladogram of presumed phylogenetic interrelationships of species-groups of *Thyreodon*, *Rhynchophion* and *Dictyonotus*, with the *atricolor* species-group shown diagramatically as an unresolved polychotomy; *T. santarosae*, a parasitoid of ceratocampine saturniids, arises from this polychotomy. □, sphingid parasitoids; ■, ceratocampine saturniid parasitoids.

Among the two large genera that attack sphingids there is no apparent evolutionary concordance in pattern of caterpillar use. For example, *Cryptophion espinozai* attacks both *Erinnyis lassauxii* and *Pachylia ficus*, but each of these two sphingids is attacked by a *Thyreodon* from a different species-group – *Erinnyis lassauxii* by *Thyreodon apricus* of the *Thyreodon laticinctus* species-group, and *Pachylia ficus* by *Thyreodon atriventris* of the *Thyreodon atriventris* species-group.

14.8.4 ENVIRONMENTAL INFLUENCE ON HOST USE

Environmental conditions limit a parasitoid's ability to use hosts, just as with all other organisms. A good example is that mentioned above of *Thyreodon laticinctus* being broadly distributed throughout Costa Rica rainforest but not occurring in dry forest (specimen records in INBio, more than 60 Malaise trap years of collecting in Costa Rica: Hanson and Gould, 1995; and 36 Malaise trap years in Santa Rosa dry forest), yet feeding on *Xylophanes anubus*, whose caterpillar is common in both forest types. While we have no idea what habitat trait actually excludes *Thyreodon laticinctus* from Costa Rican dry forest, a reasonable hypothesis is that it either avoids the sunny, hot, windy (etc.) habitats represented by dry forest (its normal microhabitat is cool, deeply shady rainforest understorey), or invades them gradually during each rainy season and then is eliminated by the dry

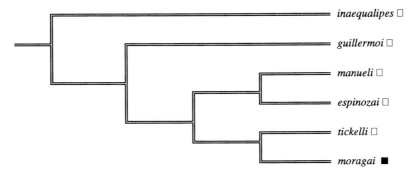

Figure 14.9 Patterns of host utilization superimposed on cladogram of presumed phylogenetic interrelationships of species of *Cryptophion*. □, sphingid parasitoids; ■, ceratocampine saturniid parasitoids.

season, or both. The dry season could eliminate this wasp either through direct physical impact, or through the absence of caterpillars for about six months – the Santa Rosa species of *Thyreodon* pass this caterpillar-free period as dormant immatures in cocoons in the litter or below ground, or, in some cases, as free-flying adults. It is possible that *Thyreodon laticinctus* cannot tolerate the dry weather as an adult or that it does not have the capacity for dormancy in a resistant cocoon.

The two Santa Rosa ichneumonids that have shifted from Sphingidae to Saturniidae are distinctive in another manner that is very relevant to future potential evolutionary radiations. These two wasps oviposit in caterpillars high in the crowns of large trees (*Syssphinx* spp. and *Othorene* spp.) or the vines in them (*Ptiloscola dargei*). Quite notably, 36 Malaise trap years of sampling in the Santa Rosa dry forests (evenly distributed among 20-year-old, 70-year-old and primary forest understorey) rich in these wasps (Gauld and Janzen, unpublished) have not yielded a single specimen of these two distinctive wasps (but have yielded specimens of all the other wasps discussed here). The strong implication is that the adults of these two wasps spend their entire foraging lives well above the ground (*Thyreodon santarosae* cocoons are in the litter, while *Cryptophion moragai* cocoons directly on the leaves of the host plant in the canopy), while the others forage in the space of about 0–4 m above the ground, which is where their host caterpillars occur. This shift in foraging microhabitat should also increase the chance that one or both of these wasps will be the root of a radiation on to other Saturniidae, since a very large number of saturniid larvae occur in the tree canopies and many species of dry forest Saturniidae are entirely or largely found only in the canopy (and see Janzen, 1985b). An alternative hypothesis would be that these two wasps made their evolutionary move on to Saturniidae through 'competitive displacement' upward from the habitat occupied by their sphingid-parasitizing ancestors. There is, quite frankly, no apparent way to distinguish between being 'forced out' of the understorey habitat by competition, or 'attracted out' of the understorey habitat by a new food source, or 'moved out' by an evolution-prone circumstance (e.g. small isolated population) whereby the only food available was in the canopy, with subsequent evolutionary adaptation to that microhabitat and its caterpillars. Of course, all three actions could well have been in play in a given circumstance.

A key part of the environment is the foodplant of the caterpillar, since volatile plant chemicals are known to attract parasitoids (Read *et al.*, 1970; Whitman, 1988; Vinson, 1991; Turlings and Tumlinson, 1992). We do not generally have a large enough sample size of wasp records to conduct an analysis of the impact of foodplant species on caterpillar use by wasps, but one example suggests potential in this area. *Xylophanes turbata* larvae are commonly and exclusively found on two species of rubiaceous understorey/edge plants – the treelet *Hamelia patens* and the small bush *Psychotria microdon*. Both of these plants grow in similar areas, frequently intermixed in the study site. To date 278 caterpillars of *X. turbata* have been collected from *H. patens* and 115 from *P. microdon*, but all 16 *Cryptophion inaequalipes* exited from larvae feeding on *H. patens*. On the other hand, *Thyreodon* n. sp. attacked 3% of each of the samples from *Hamelia* and *Psychotria*.

14.9 IN CLOSING

The large and growing caterpillar and parasitoid database for the Guanacaste Conservation Area is beginning to allow a picture not only of what parasitoid wasps use what caterpillars but, equally revealing, of what caterpillars they do not use. Many species of microgeographically available caterpillars (defined as living within a few metres of each other) are not used by the ichneumonids that demonstrably can use their confamilials. The overall percentage of caterpillar attack is also very low. These two observations suggest the hypothesis that this habitat is far from containing as many species of wasps as it could contain if more species were available (to arrive, geographically or evolutionarily). It also suggests a specific application of a general 'ecological fitting' hypothesis (Janzen, 1985a, 1986a): the host specificity of the wasps is not driven by on-site evolutionary response to intense interspecific competition, but rather by the wasps entering the habitat with their host selection traits established in previous evolutionary circumstances, and simply applying these traits. This application in contemporary ecological circumstances results in percentage attack ranging from very high (e.g. *Enicospilus lebophagus* killing as many as a third of the caterpillars of *Rothschildia lebeau*) to the very common case of fewer than 10% of the acceptable species of caterpillars being used by a given wasp species. All the 'zero' cases, where an incoming wasp population does not survive (e.g. when *Thyreodon laticinctus* flies out into the dry forest during the rainy season), for all practical purposes cannot be observed even though it is evident that such cases must occur.

The student of plant–herbivore–parasitoid interactions such as these needs to be aware of a major perturbation in this real-world case study. Not only did all of the players evolve elsewhere under unquantifiably different conditions, immigrate to this site, and form an array as best they could (e.g. 'ecological fitting', *sensu* Janzen, 1985a, 1986a), but also the ecological circumstances today bear a highly variable degree (usually low) of resemblance to the circumstances that these organisms encountered in pre-human and pre-Columbian times (e.g. 'no park is an island': Janzen, 1983, 1986b,d, 1988b,e, 1990). Somewhat ironically, the anthropogenically and arbitrarily fragmented habitats encountered today in the Guanacaste Conservation Area (e.g. Janzen, 1986d, 1988a, 1992d) – and on other tropical landscapes under extensive agriculture – are likely to be more similar to the circumstances of evolution of these species, in some kind of ecologically insular circumstance, than they are to the large expanses of interdigitated macrohabitats visualized as existing in pre-human, and especially pre-Columbian, times. The later circumstances, occupied by large populations, each subject to a wide variety of conflicting directions of selection and varied habitat circumstances, were not likely places to encounter total-population-applied directed selection focused on a small population – the kinds of circumstances likely to generate new species. It is an unfortunate conservation reality that the quickest way to generate new species would be to fragment a large widespread population into a few small units, each unit being large enough to survive but each to survive in quite a different microhabitat. This is precisely the fate of much Neotropical biodiversity.

The pragmatic aspects of this study cause us to reflect on host–parasitoid records as commonly accumulated in collections and on summer field trips. The occasional record acquired here and there begins to give an impression of host use by parasitoids. Then a study such as this begins and the very large number of rearing records gives a more solid impression. However, when this larger pool of records is subjected to a level of scrutiny infeasible with casually obtained records, it is again found to be wanting. Large sample sizes across many habitats consolidate patterns (e.g. *Thyreodon santarosae* is clearly a saturniid parasite despite the obvious radiation of the *Thyreodon* genus-group on to Sphingidae). But they also generate 'unique' records that require yet larger samples or more direct sampling to clarify their meaning. For example, the single specimen of *Thyreodon santarosae* reared from *Othorene verana* is orange instead of black and seems to differ ever so slightly in morphology. Does this then represent the beginning of further *Thyreodon* radiation on to Saturniidae, or is it just an 'aberrant' specimen? The two *Thyreodon* cocoons obtained from *Syssphinx quadrilineata* in an open field 10 m from oak forest, whose spinners died of disease and thus remain unknown, could be from a *Thyreodon santarosae* that descended to these caterpillars at ground level, or could be one of the two species of *Thyreodon* not yet reared in the Santa Rosa dry forest.

Finally, we note that this brief report, extracted from an ever-growing computerized event-based database, differs from our centuries of traditional taxonomic and ecological publications. The taxonomist traditionally lists host records (usually from the museum collections at hand and as extracted from previous taxonomic publications) in taxonomic monographs and species descriptions. The only context is taxonomic and geographical. The ecologist traditionally poses a question about interspecific and/or intraspecific interactions focused on a specific habitat or specific taxon. Here, the positive statements are many, but 'absence' statements are impoverished. With database-derived syntheses such as this one, it is now possible (as noted earlier) to suggest what focal species do not do, as well as what they do do.

When we state that four species of the *Enicospilus americanus* group are monophagous parasitoids of saturniid caterpillars in the Santa Rosa dry forest (in contrast to their polyphagous siblings in North America), the negative part of that statement cost far more than did the positive part. Furthermore, while this brief publication is a snapshot of the data in that database as at the end of 1994, the database will have evolved well beyond this point by the time this chapter is first read by the public. And the database will continue to evolve and fill. Additionally, the actual base for the statements appearing here is precisely recoverable. In the past it was not, owing to the vagaries of specimen and notebook loss, constraints of labelling material, memory failure, taxonomic changes, etc. But computerized databases contain their own kinds of changes. They are continually added to, modified through new identifications (and corrections of old identifications), and extend across far longer time horizons than is normal for person-based observations. Our science is changing, as well as the organisms and their habitats.

ACKNOWLEDGEMENTS

This study and its database have been supported by US-AID, US NSF, US NIH, the government of Costa Rica in general and the Ministry of Natural Resources (MIRENEM), the administration of the Guanacaste Conservation Area, Costa Rica's National Biodiversity Institute (INBio), the University of Pennsylvania, Universidad de Costa Rica and the Natural History Museum (London). We wish specifically to thank the Costa Rican team of caterpillar collectors, caterpillar rearers and parasitoid rearers who gathered the bulk of the rearing records summarized here: Winnie Hallwachs, Guillermo Pereira, Lucia Ríos, Manuel Pereira, Elieth Cantillano, Osvaldo Espinosa, Ruth Franco, Harry Ramirez, Gloria Sihezar, Roberto Espinosa and Roster Moraga. The caterpillar collecting has also been very strongly supported by a wide variety of other researchers, biodiversity prospectors, administrators and visitors to the Guanacaste Conservation Area, and especially Eric Olson, Felipe Chavarria, Sandy Salas, Vanessa Nielsen, Adrian Guadamuz, Mariano Pereira, Camilo Camargo, Jon Sullivan, Roger Blanco, Isabel Salas, Hazel Mora, Heiner Araya and Daniel Perez. Pam Mitchell has provided invaluable help by sorting the large Malaise trap collections. This overall study is totally dependent on the myriad of taxonomists who have contributed through identifying the specimens in the caterpillar rearing database. In this particular case, I.D. Gauld identified all of the ichneumonids while D.H. Janzen identified the larvae, with these names in turn being based on confirmations of adult sphingid and saturniid specimens from Santa Rosa by J.-M. Cadiou and C. Lemaire.

REFERENCES

Askew, R.R. and Shaw, M.R. (1986) Parasitoid communities: their size, structure and development. In *Insect Parasitoids* (eds J. Waage and D. Greathead), pp. 225–264, Academic Press, London.

Bernays, E.A. and Janzen, D.H. (1988) Saturniid and sphingid caterpillars: two ways to eat leaves. *Ecology* 69, 1153–1160.

Bullock, S.H., Mooney, H.A. and Medina, E. (1995) *Seasonally Dry Tropical Forests*, Cambridge University Press, Cambridge, 450 pp.

Carlson, R.W. (1979) Ichneumonidae. In *Catalog of Hymenoptera in America North of Mexico, 1* (eds K.V. Krombein, P.D. Hurd Jr, D.R. Smith and B.D. Burks), Smithsonian Institution Press, Washington DC.

Cornell, H.V. and Hawkins, B.A. (1993) Accumulation of native parasitoid species on introduced herbivores: a comparison of hosts as natives and hosts as invaders. *American Naturalist* 141, 847–865.

Cushman, R.A. (1947) A generic revision of Ichneumonflies of the tribe Ophionini. *Proceedings of the US National Museum* 96, 417–482.

Gámez, R. (1991a) El Instituto Nacional de Biodiversidad de Costa Rica: poniendo la biodiversidad a trabajar sostenidamente para la sociedad. *Biodiversity* 7, 86–88.

Gámez, R. (1991b) Biodiversity conservation through facilitation of its sustainable use: Costa Rica's National Biodiversity Institute. *Trends in Ecology and Evolution* 6, 377–378.

Gaston, K.J. and Gauld, I.D. (1993) How many species of pimplines (Hymenoptera: Ichneumonidae) are there in Costa Rica? *Journal of Tropical Ecology* **9**, 491–499.

Gauld, I.D. (1985) The phylogeny, classification and evolution of parasitic wasps of the subfamily Ophioninae (Ichneumonidae). *Bulletin of the British Museum (Natural History) Entomology* **51**, 61–185.

Gauld, I.D. (1987) Some factors affecting the composition of tropical ichneumonid faunas. *Biological Journal of the Linnean Society* **30**, 299–312.

Gauld, I.D. (1988a) The species of the *Enicospilus americanus* complex (Hymenoptera: Ichneumonidae) in eastern North America. *Systematic Entomology* **13**, 31–53.

Gauld, I.D. (1988b) A survey of the Ophioninae (Hymenoptera: Ichneumonidae) of tropical Mesoamerica with special reference to the fauna of Costa Rica. *Bulletin of the British Museum (Natural History) Entomology* **57**, 1–309.

Gauld, I.D. (1991) The Ichneumonidae of Costa Rica, 1. *Memoirs of the American Entomological Institute* **47**, 1–589.

Gauld, I.D. (1995) The Ichneumonidae of Costa Rica, 2. *Memoirs of the American Entomological Institute* (in press).

Gauld, I.D. and Gaston, K.J. (1994) The taste of enemy-free space: parasitoids and nasty hosts. In *Parasitoid Community Ecology* (eds B.A. Hawkins and W. Sheehan), pp. 279–299, Oxford University Press, Oxford.

Gauld, I.D. and Janzen, D.H. (1994) The classification, evolution and biology of the Costa Rican species of *Cryptophion* (Hymenoptera: Ichneumonidae). *Zoological Journal of the Linnean Society* **110**, 297–324.

Gauld, I.D. and Mitchell, P.A. (1981) *The Taxonomy, Distribution and Host Preferences of Indo-Papuan Parasitic Wasps of the Subfamily Ophioninae*, CAB, Slough, 611 pp.

Gauld, I.D., Gaston, K.J. and Janzen, D.H. (1992) Plant allelochemicals, tritrophic interactions and the anomalous diversity of tropical parasitoids: the 'nasty' host hypothesis. *Oikos* **65**, 353–357.

Hanson, P.E. and Gauld, I.D. (eds) (1995) *The Hymenoptera of Costa Rica*, Oxford University Press, Oxford, 893 pp.

Hodges, R.W. (1971) *The Moths of America North of Mexico. (21) Sphingoidea*, E.W. Classey, London, 158 pp.

Janzen, D.H. (1973) Sweep samples of tropical foliage insects: effects of seasons, vegetation types, elevation, time of day, and insularity. *Ecology* **54**, 687–708.

Janzen, D.H. (1983) No park is an island: increase in interference from outside as park size decreases. *Oikos* **41**, 402–410.

Janzen, D.H. (1984a) Two ways to be a tropical big moth: Santa Rosa saturniids and sphingids. *Oxford Surveys in Evolutionary Biology* **1**, 85–140.

Janzen, D.H. (1984b) Natural history of *Hylesia lineata* (Saturniidae: Hemileucinae) in Santa Rosa National Park, Costa Rica. *Journal of the Kansas Entomological Society* **57**, 490–514.

Janzen, D.H. (1985a) On ecological fitting. *Oikos* **45**, 308–310.

Janzen, D.H. (1985b) A host plant is more than its chemistry. *Illinois Natural History Bulletin* **33**, 141–174.

Janzen, D.H. (1986a) Biogeography of an unexceptional place: what determines the saturniid and sphingid moth fauna of Santa Rosa National Park, Costa Rica, and what does it mean to conservation biology? *Brenesia* **25/26**, 51–87.

Janzen, D.H. (1986b) The future of tropical ecology. *Annual Review of Ecology and Systematics* **17**, 305–324.

Janzen, D.H. (1986c) *Guanacaste National Park: Tropical Ecological and Cultural Restoration*, Editorial Universidad Estatal a Distancia, San José, Costa Rica, 103 pp.

Janzen, D.H. (1986d) The eternal external threat. In *Conservation Biology: the Science of Scarcity and Diversity* (ed. M.E. Soulé), pp. 286–303, Sinauer Associates, Sunderland, Mass.

Janzen, D.H. (1987a) How moths pass the dry season in a Costa Rican dry forest. *Insect Science and its Application* **8**, 489–500.

Janzen, D.H. (1987b) When, and when not to leave. *Oikos* **49**, 241–243.

Janzen, D.H. (1988a) Guanacaste National Park: tropical ecological and biocultural restoration. In *Rehabilitating Damaged Ecosystems*, Vol. ii (ed. J.J. Cairns), pp. 143–192, CRC Press, Boca Raton, Florida.

Janzen, D.H. (1988b) Management of habitat fragments in a tropical dry forest. *Annals of the Missouri Botanic Garden* **75**, 105–116.

Janzen, D.H. (1988c) Ecological characterization of a Costa Rican dry forest caterpillar fauna. *Biotropica* **20**, 120–135.

Janzen, D.H. (1988d) The migrant moths of Guanacaste. *Orion Nature Quarterly* **7**, 38–41.

Janzen, D.H. (1988e) Tropical dry forests; the most endangered major tropical ecosystem. In *Biodiversity* (ed. E.O. Wilson), pp. 130–137, National Academy Press, Washington, DC.

Janzen, D.H. (1990) An abandoned field is not a tree fall gap. *Vida Silvestre Neotropical* **2**, 64–67.

Janzen, D.H. (1992a) A dry tropical forest ecosystem restored. *Earth Summit Times, New York Daily News* **28** (April 1, 1992), 1.

Janzen, D.H. (1992b) A south–north perspective on science in the management, use, and economic development of biodiversity. In *Conservation of Biodiversity for Sustainable Development* (eds O.T. Sandlund, K. Hindar and A.H.D. Brown), pp. 27–52, Scandinavian University Press, Oslo.

Janzen, D.H. (1992c) What does tropical society want from the taxonomist? In *Hymenoptera and Biodiversity* (eds J. LaSalle and I.D. Gauld), pp. 295–307, CAB, Wallingford.

Janzen, D.H. (1992d) The neotropics. A broad look at prospects for restoration in Central and South America raises some basic questions about methods, about goals, and about the restorationist's role in evolution. *Restoration and Management Notes* **10**, 8–13.

Janzen, D.H. (1993a) Taxonomy: universal and essential infrastructure for development and management of tropical wildland biodiversity. In *Proceedings of the Norway/UNEP Expert Conference on Biodiversity, Trondheim, Norway* (eds O.T. Sandlund and P.J. Schei), pp. 100–113, NINA, Trondheim, Norway.

Janzen, D.H. (1993b) Caterpillar seasonality in a Costa Rican dry forest. In *Caterpillars. Ecological and Evolutionary Constraints on Foraging* (eds N.E. Stamp and T.M. Casey), pp. 448–477, Chapman & Hall, New York.

Janzen, D.H. (1994a) Wildland biodiversity management in the tropics: where we are now and where are we going? *Vida Silvestre Neotropical* **3**, 3–15.

Janzen, D.H. (1994b) Priorities in tropical biology. *Trends in Ecology and Evolution* **9**, 365–367.

Janzen, D.H. and Hallwachs, W. (1994) *All Taxa Biodiversity Inventory (ATBI) of Terrestrial Systems. A generic protocol for preparing wildland biodiversity for non-damaging use.* Report of a NSF Workshop, 16–18 April 1993, Philadelphia, Pennsylvania, 132 pp.

Janzen, D.H., Ataroff, M., Fariñas, M. *et al.* (1976) Changes in the arthropod community along an elevational transect in the Venezuelan Andes. *Biotropica* **8**, 193–203.

Memmott, J., Godfray, H.C.J. and Gauld, I.D. (1994) The structure of a tropical host–parasitoid community. *Journal of Animal Ecology* **63**, 521–540.

Olson, E.J. (1995) Seasonality of growth rate and mortality of a polyphagous caterpillar in a lowland moist tropical forest. Unpublished PhD Thesis, University of Pennsylvania, Philadelphia.

Porter, C.C. (1980) A new *Thyreodon* Brullé (Hymenoptera: Ichneumonidae) from south Texas. *Florida Entomologist* **63**, 242–246.

Porter, C.C. (1984) *Laticinctus* group of *Thyreodon* in the northern Neotropics (Hymenoptera: Ichneumonidae). *Wasmann Journal of Biology* **42**, 40–71.

Porter, C.C. (1986) A new arboricolous *Thyreodon* from Costa Rica (Hymenoptera Ichneumonidae: Ophioninae). *Psyche* **93**, 133–139.

Porter, C.C. (1989) A new Floridan *Athyreodon* Ashmead (Hymenoptera: Ichneumonidae), with comments on related species in the northern Neotropics. *Florida Entomologist* **72**, 294–304.

Read, D.P., Feeny, P.P. and Root, R.B. (1970) Habitat selection by the aphid parasite *Diaeretiella rapae* (Hymenoptera: Braconidae) and hyperparasite *Charips brassicae* (Hymenoptera: Cynipidae). *Canadian Entomologist* **102**, 1567–1578.

Reid, W.V., Laird, S.A., Gámez, R. *et al.* (eds) (1993) *Biodiversity Prospecting*, World Resources Institute, Washington, DC, 341 pp.

Scoble, M.J. (1992) *The Lepidoptera*, Oxford University Press, Oxford, 404 pp.

Townes, H. (1970) Genera of Ichneumonidae 3. *Memoirs of the American Entomological Institute* **13**, 1–307.

Townes, H. (1970) Genera of Ichneumonidae 3. *Memoirs of the American Entomological Institute* **17**, 1–372.

Turlings, T.C.J. and Tumlinson, J.H. (1992) Systemic release of chemical signals by herbivore-injured corn. *Proceedings of the National Academy of Science of the United States of America* **89**, 8399–8402.

Vinson, S.B. (1991) Chemical signals used by parasitoids. *Redia*, Appendice, **74**(3), 15–42.

Ward, S. and Gauld, I. (1987) The callajoppine parasitoids of sphingids in Central America (Hymenoptera: Ichneumonidae). *Systematic Entomology* **12**, 503–508.

Whitman, D.W. (1988) Plant natural products as parasitoid cuing agents. In *Biologically Active Natural Products Potential Use in Agriculture* (ed. H.G. Cuttler), American Chemical Society, Washington, DC, 496 pp.

# IMPACT OF FOREST LOSS AND REGENERATION ON INSECT ABUNDANCE AND DIVERSITY

Allan D. Watt, Nigel E. Stork, Paul Eggleton, Diane Srivastava, Barry Bolton, Torben B. Larsen, Martin J.D. Brendell and David E. Bignell

15.1 INTRODUCTION

Tropical forests are exceptionally rich in biodiversity and, therefore, the current rate of tropical forest clearance is certain to have a profound impact on global biodiversity. However, the rate at which species are becoming extinct is unclear (Whitmore and Sayer, 1992; UNEP, 1995), and the degree to which tropical forest management can minimize the loss of biodiversity when forests are exploited is poorly understood (Blockhus *et al.*, 1992). Several recent studies have tried to measure the impact of forest clearance and the effect of different forest management practices on insect abundance and diversity. This paper reviews the results of these studies. In particular, we review the results of recent studies of the effects of forest clearance and different silvicultural practices on a number of arthropod taxa in Cameroon.

15.2 DIVERSITY OF ARTHROPODS IN TROPICAL FORESTS

Studies of the canopy of tropical forest trees have shown that it is extremely rich in arthropod species. In the first study of its kind, Erwin and Scott (1980) used knockdown insecticides to collect and sort over 1200 beetle species from a single tree species in Panama. Few studies have attempted to quantify fully the species richness of canopy arthropod communities, but Stork (1987, 1991), using the same techniques, recorded more than 4000 species of arthropods from fewer than 25 000 individuals collected from the canopy of only 10 trees (five species) in lowland rain forest in Borneo. Biodiversity in tropical forests is not, however, limited to tree canopies. A study of the lowland rainforest of Seram, Indonesia, showed that 24% of the total number of arthropods were to be found in the canopy, and 70% in the soil and leaf litter (Stork, 1988; Stork and Brendell, 1990). As yet, no studies have measured the species richness of all arthropod groups in all forest habitat strata, but Hammond (1990), using a wide range of sampling methods, recorded 4500 Coleoptera species in a 500 ha lowland forest study area in Sulawesi, and estimated that the actual number of beetle species in the area was over 6000. Further detailed analysis of the stratum preference for each of these species (Hammond *et al.*, 1996) would indicate that at least three times as many of these beetle species are ground specialists as canopy specialists. The arthropod fauna of the soil, therefore, would appear to be as rich or indeed much richer than that of the canopy.

A few of the studies of tropical biodiversity have been used to produce global species richness

Forests and Insects. Edited by A.D. Watt, N.E. Stork and M.D. Hunter. Published in 1997 by Chapman & Hall, London. ISBN 0 412 79110 2.

estimates of 10 to 30 million. Current estimates for the number of species on Earth are around 10–15 million, but there is considerably uncertainty over the numbers of undescribed species in tropical forests and elsewhere (Stork, 1996; Hammond, 1995). Whatever the true level of global species richness, tropical forests, despite covering only 8% of the land surface, probably contain 50–80% of the world's species (Myers, 1986, 1988, 1990; Stork, 1988; WCMC, 1992; Bibby et al., 1992; Davis et al., 1994a,b, 1995). Myers (1988), for example, suggests that some 13% of all plant species are found in 10 'hot spots' around the world and that deforestation and land use change in these areas threatens the survival of these species. Thus deforestation poses a serious threat to global biodiversity.

15.3 DEFORESTATION AND REFORESTATION

The average annual rate of tropical deforestation stands at about 1% overall (Whitmore and Sayer, 1992; WCMC, 1992). The highest regional rate of deforestation is in West Africa: 2.2% overall (and 4.1% in closed forest). Within the region, 2.7% and 5.2% of the total forest area was lost annually in the 1980s in Nigeria and Côte d'Ivoire, and 5% and 6.5% respectively in closed forest areas of these countries (WRI, 1992). Deforestation rates in Cameroon are lower: about 0.4–0.6% per annum. Although the above estimates are based on data collected in the 1980s, recent figures for tropical log production show that the pressure on tropical forests for timber remains high (Johnson, 1995). Log production in Cameroon, in particular, has recently increased dramatically (Johnson, 1995).

Reforestation projects are small in scale in comparison with deforestation. For example, during the 1980s, when the rate of deforestation in West Africa was estimated to have been about 12 000 km^2 annually (FAO, 1991), only 360 km^2 of plantation forests were established each year (Lawson, 1994). This is equivalent to just 3% of the area deforested. Moreover, these plantations have had a poor record of maintenance and survival, and were usually established in areas of existing forest (Lawson, 1994).

15.4 IMPACT OF FOREST LOSS ON BIODIVERSITY

Various attempts have been made to estimate the impact of deforestation on biodiversity. These attempts have been based on (1) extrapolations from known relationships between species richness and habitat area and (2) estimates of actual and projected deforestation. This approach has led to predictions of global species extinction rates of 1–10% per decade (Reid, 1992; Stork, 1991). Such predictions, however, take no account of the ability of many species to survive in the habitats created after forest clearance (Lugo, 1988). Forest clearance varies greatly in severity, from selective logging, to partial clearance prior to tree planting, to complete clearance prior to the establishment of a new crop. The new habitats created include agricultural crops, plantations of exotic tree species and plantations of indigenous species. Different levels of forest clearance and the establishment of various habitats are likely to have different impacts on insect diversity, and several studies have been established to quantify the impact of several forms of forest clearance on a range of insect taxa. This chapter presents research on this area, concentrating on recent work in Cameroon.

15.5 RESPONSE OF INSECTS TO FOREST CLEARANCE

Published research (other than in Cameroon) on the impact of tropical forest management practices on arthropods is currently limited to studies on moths, dung and carrion beetles and leaf-litter ants. Holloway et al. (1992) sampled macrolepidoptera in 14 sites in Danum, Sabah and Gunung Mulu, Sarawak. Diversity, measured as alpha diversity, was greatest in undisturbed forest sites and lowest in *Acacia mangium* and *Pinus caribaea* plantation sites. Similarly, the diversity of macrolepidoptera in 20 sites in Sulawesi was greater in forest than in disturbed, open or cultivated habitats. However, Chey (1994) obtained somewhat different results from a study of moths in plantations and natural forests in Sabah. Sampling was carried out in plantations of *Acacia mangium*, *Eucalyptus deglupta*, *Gmelina arborea*, *Albizia falcataria*, *Pinus carib-*

aea and logged secondary forest. A total of 1642 moth species was recorded, from 675 species in the *Gmelina arborea* plantation to 872 species in the *Eucalyptus deglupta* plantation and 1048 species in the secondary forest. Alpha diversity was similar in the secondary forest and the *Eucalyptus deglupta* plantation, greater than that found in the plantations of the other species. Chey (1994) concluded that plantations of exotic tree species are not as species-poor as had been thought. Other similar studies have also found high arthropod diversity in exotic plantations. For example, Gadagkar *et al.* (1990) found that the highest and lowest diversity they obtained through sampling insects, using a range of methods in 12 different types of forest, was in *Eucalyptus* and teak plantations, respectively, with natural forest plots being between these two extremes.

Other studies have concentrated on the fauna associated with the soil and leaf litter. Belshaw and Bolton (1993) surveyed the leaf-litter ant fauna of 34 sites in Ghana. These sites comprised 14 in primary forest, 10 in secondary forest and 10 in cocoa. A total of 176 ant species was recorded, but there was no difference between the number of species found in the three site types – on average 26–28 species in each site. They concluded that forest clearance and the establishment of cocoa farms in Ghana had little or no effect on either the species richness or the species composition of the leaf-litter ant fauna (Belshaw and Bolton, 1993). Similarly, the diversity of dung and carrion beetles has been found to be unaffected by forest clearance and the establishment of plantations (Hanski and Krikken, 1991).

Thus, published information on arthropod diversity tends to suggest that the impact of forest clearance is small, at least when plantations of timber trees and cocoa are established. However, given the importance of this subject, there have been very few studies, only a limited number of arthropod taxa have been examined, and no studies of the impact of forest clearance on canopy-dwelling arthropods have been carried out. One of the main aims of the projects based in Cameroon described below, therefore, was to focus on arthropod taxa that had not been studied in relation to forest clearance, e.g. termites, and to include a canopy sampling component in the research.

15.6 INSECTS AND FOREST MANAGEMENT IN CAMEROON

15.6.1 FOREST MANAGEMENT

The Mbalmayo Forest Reserve, southern Cameroon, is the site of a number of projects designed to demonstrate the potential of sustainable forestry with native timber trees, utlizing a range of silvicultural techniques from line-planting (following limited clearance of vegetation by hand) to complete clearance by chainsaw and bulldozer (Lawson, 1994). The Reserve also contains large areas of old growth secondary forest and parts where only light, selective logging has ever taken place, and which therefore have a largely primary character. The Reserve thus affords the opportunity to sample insects from sites along a gradient of disturbance. In additon, experimental components of some previous projects within the reserve have investigated the effects of different silvicultural methods on the physical and chemical properties of the soil, endomycorrhizal fungi and tree physiology, and in consequence a large body of background biological and pedological information is available (Ngeh, 1989; Eamus *et al.*, 1990; Leakey, 1991; Mason *et al.*, 1992). The sampling described in this chapter formed part of the Forest Management and Regeneration Project – FMRP, 1990–1996, a bilateral Government of Cameroon (Office National de Développement des Forêts) and UK (Overseas Development Administration) project – or was an integral part of one of the TIGER Cameroon Projects (Terrestrial Initiative in Global Environmental Research, Natural Environment Research Council, UK). Full details of these projects are given by Eggleton and Bignell (1995), Eggleton *et al.* (1995, 1996), Watt *et al.* (1995), and Chapter 8 of this volume. A summary of the forest treatments sampled is given below.

There are several approaches to tropical forest management, including: sustainable management of natural forest; enrichment planting and other techniques where areas of forest are partially cleared before planting; taungya (see below) and other agroforestry methods; and the establishment of plantations after complete clearance. In 1987–88, two sets of plots 3 km apart were established at Bilik and Ebogo within the Mbalmayo Forest Reserve, under the following treatments.

- **Complete clearance**, where all large trees were felled by chainsaw; then these, the smaller trees and all remaining vegetation were cleared from the plot by bulldozer. The resulting effect was that all above-ground plant material was removed and there was considerable soil compaction due to the bulldozer.
- **Partial mechanical clearance**, where a bulldozer was used to remove most of the undergrowth and approximately 50% of the large trees, resulting in a reduction of about 50% in the canopy cover. The use of the bulldozer also resulted in some soil compaction.
- **Partial manual clearance**, where ground vegetation and some small trees were removed by machete, and some other small trees were felled by chainsaw. This technique resulted in minimal soil compaction.
- **Uncleared forest control**, where no trees were cut or removed, there was no ground disturbance by machinery, no trees were planted, and very little ground vegetation was cleared for narrow access lines.

In 1991, a more extensive series of plots was established in the Mbalmayo Forest Reserve. These plots, all near Eboufek village, included the following treatments (Lawson, 1994).

- **Complete clearance**, as in Bilik and Ebogo described above except that felled trees were pushed into windrows every 40–50 m (i.e. two per 1 ha plot).
- **Line planting**, where widely spaced lines of existing vegetation (including trees) were cleared.
- **Taungya**, an agroforestry technique where, after clearing and burning of existing vegetation, crops such as cassava, maize, yams and peanuts were grown by villagers for the first 2–4 years of the rotation.
- **Uncleared forest control**, as for Bilik and Ebogo.

An additional non-forest site was also used:

- **Weeded fallow**, completely cleared of trees and other vegetation in 1990 (except that some of the felled trees were sawn up, spread and left to decompose naturally to provide nutrient input to the soil) and weeded periodically to prevent tree regeneration (part of the International Institute of Tropical Agriculture experimental farm in the Mbalmayo Forest Reserve).

In all, three uncleared forest plots were sampled in the Mbalmayo Forest Reserve. While it is known that logging has taken place in the Reserve several times, there are no records on where and when this was. However, the evidence – in particular the presence/absence of logging trails and the size of trees in each plot – strongly suggests that the Bilik plot had not experienced disturbance for some considerable time. This plot has therefore been defined as 'near primary', and the Eboufek plot (which contains an old logging trail that was widened during the course of the study) as 'old secondary' (Eggleton *et al.*, 1995). The Ebogo plot may also be classified as 'near primary'. All plantation plots sampled were planted with *Terminalia ivorensis*. This species is found in Cameroon but it does not occur in the wild in the Mbalmayo area, though the related species *Terminalia superba* does occur within the Reserve.

15.6.2 ARTHROPOD SAMPLING IN THE MBALMAYO FOREST RESERVE

A full range of plot treatments were, therefore, available to assess the impact on arthropod diversity of forest clearance and the establishment of plantations in different ways. These plots are listed in Table 15.1, together with the arthropod groups that were sampled and subsequently sorted to species. It should be noted that the apparent incompleteness of the data is largely due to the fact that the groups studied occupy different micro-habitats and forest strata. While the overall aim of those working on different groups was the same – to quantify the impact of forest clearance – it was clearly possible to sample the butterflies and soil-dwelling groups in more habitats than the arboreal groups.

Eggleton, Bignell and co-workers have been studying termites in the Mbalmayo Forest Reserve since 1992 (Eggleton et al., 1995, 1996; Eggleton and Bignell, 1995; Chapter 8 of this volume). Sampling was extremely labour intensive, taking over 30 man-days per plot. Transects (100m × 2 m) were run across each plot to assess species richness, while a stratified quantitative sampling programme in dead wood, soil and termite mounds/nests was undertaken to assess abundance and biomass. The transects sampled:

- all mounds, purse nests, and runways;
- logs, branches, loose soil under logs and soil around tree buttresses, soil pits.

Termites were classified as far as possible into four functional groups as follows:

- **soil feeders** – species apparently feeding on mineral soil and living in the soil profile, surface litter and/or epigeal mounds;
- **soil/wood interface feeders** – termites found only or mainly within soil under logs, within soil plastered to the surface of rotten logs or within highly decayed wood;
- **wood feeders** – termites feeding on wood and excavating galleries in woody litter, including species with arboreal nests and others with subterranean nests with cultivated fungus gardens;
- **litter foragers** – species that forage for leaf litter and small woody litter.

Further details of termite sampling are given in Eggleton et al. (1995, 1996) and Eggleton and Bignell (1995).

Tree canopy sampling was carried out in the Reserve on four occasions between 1991 and 1993. The aim of this work was to assess the overall abundance of different arthropod groups, and to measure the diversity of ants and beetles. On each occasion, approximately 15 T. ivorensis trees in each 1 ha plantation plot were either fogged or sprayed with permethrin, an insecticide with a rapid knockdown effect on insects. Clearly, although insect sampling was centred on (previously planted) T. ivorensis trees, the samples reflected the abundance and diversity of insects on both this species and the other trees whose canopies overlapped with the T. ivorensis canopies. In 1993, the fogging programme was extended to include the uncleared forest control plots in Ebogo and Bilik in order to provide natural forest controls. Fogging in the uncleared forest plots centred on wild T. superba trees of unknown age. On each occasion 10–40 (generally 25) collecting trays, each 1 m², were used.

In addition to the monitoring of ants in the canopy of T. ivorensis, a survey of leaf-litter-dwelling ants was carried out in November 1993 at various sites within the Mbalmayo Forest Reserve. Ten samples 1 m² were collected in an approximately 50 m transect across the centre of each of

Table 15.1 Sampling programmes for different insect groups in the Mbalmayo Forest Reserve, Cameroon (E, Ebogo; B, Bilik; K, Eboufek; F, farm fallow): note that the complete clearance plot at Eboufek was cleared shortly before the insect sampling, whereas the other complete clearance plots were cleared (and replanted) several years before sampling

| Insect group | Farm fallow | Plantation clearance treatment | | | Uncleared forest |
		Complete	Partial manual	Partial mechanical	
Termites	F	K	E		K, B
Leaf-litter ants		K, E	E	E	K, E, B
Canopy ants		E, B	E, B	E, B	
Canopy beetles		E, B	E, B	E, B	E, B
Butterflies	F	K, E	E	E	K, E, B

the plots and across the rows of planted trees such that half the samples were from tree rows and half were from between tree rows. The leaf litter and top few millimetres of soil were scraped up and placed in a bag. Each sample was sieved with a coarse 1 cm mesh sieve and the residue placed in mesh bags. These were suspended in 'Winkler bags' to allow the soil and leaf litter residues to dry out slowly, causing the insects to drop down into a pot of alcohol at the bottom of the Winkler bag.

During February–March 1994, assessments of the impact of silvicultural practice on butterfly abundance and diversity were carried out. A total of 434 specimens was collected on eight plots by: (1) one of the authors (DS), netting individuals for a standard amount of time per plot (42% of all specimens); (2) four locally hired butterfly collectors, netting individuals for a standard amount of time per plot, and rotated among plots to avoid collector bias (45% of specimens); and (3) using fruit-baited traps in each plot (12% of specimens). Abundance was estimated as the mean number of butterflies seen during several (10–16) standard five-minute walks through each plot.

The termites, ants and butterflies were identified or sorted to morphospecies in the entomology department of the Natural History Museum, London. Five per cent of the butterflies, 30% of the termites and 40% of the ants were previously undescribed. No attempt was made to identify the beetles from the canopy and they were sorted to morphospecies.

15.6.3 OVERALL ABUNDANCE AND DIVERSITY

Figure 15.1a shows the mean level of abundance of each group as assessed by different sampling methods. Although sampling of each group was carried out in different ways and the results are not directly comparable, they demonstrate the overall differences in the abundance of different arthropod groups. Termites were the most abundant group sampled, on average over 3000/m². From the canopy, a mean total of 196 arthropods/m² was obtained from *T. ivorensis* plantation plots in the Mbalmayo Forest Reserve by insecticide knockdown sampling. Ants were the commonest arthropod group in the canopy, making up 63% of the overall total, with a mean of 136 ants/m². Ants were also abundant in the leaf litter, on average 52/m². No other group of canopy arthropods approached the level of abundance of ants, but among the other groups, adult Diptera (10% of the arthropods found in the canopy), Hymenoptera other than ants (4%), Homoptera (4%), Thysanoptera (4%), adult Coleoptera (3%) and larval Diptera (3%) were most abundant. These seven groups made up 90% of the total overall.

The overall species richness of each group of insects is shown in Figure 15.1b. With 114 species, the Mbalmayo Forests Reserve is the richest known site for termites in the world (Eggleton *et al.*, 1995). A total of 96 ant species has been

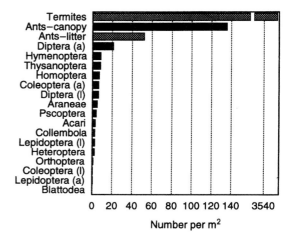

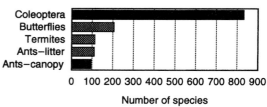

Figure 15.1 (a) Mean abundance and (b) number of Coleoptera, butterfly and ant species recorded in the Mbalmayo Forest Reserve, Southern Cameroon (■, canopy samples; ▨, soil and other samples). Only the number of Coleoptera and canopy ant species are directly comparable, but the number of termite species was obtained from a similar sampling intensity; the numbers of butterfly and litter ant species were derived from a lower level of sampling than the other groups.

recorded in the canopy and, despite a much reduced amount of sampling, 111 ant species in the leaf litter. The canopy beetle samples, derived from roughly the same amount of sampling as the canopy ants, showed this group to be the most diverse examined: 833 species were recorded. Sampling of the butterflies resulted in 206 species, including a number of species sampled in a preliminary survey in 1993.

The species richness of each group was assessed by using one of two non-parametric techniques which can also be used to calculate standard errors on the estimates, either rarefaction (Simberloff, 1972; Krebs, 1989) or the jack-knife technique (Heltshe and Forrester, 1983a,b; Krebs, 1989; Colwell and Coddington, 1994). Further papers on individual taxa will consider different approaches to estimating species richness, other measures of diversity and comparisons of the species composition found under different levels of forest disturbance.

15.6.4 TERMITES

Sampling was carried out in a farm fallow plot, two plantation plots – one recently completely cleared (prior to the establishment of a plantation plot) and the other a partial manual clearance plantation – an old secondary forest plot and a near primary forest plot (Table 15.1) (Eggleton et al., 1995). Species richness was found to be highest in the partial manual clearance plantation and old secondary forest plots (Figure 15.2). The number of species in these plots was slightly, and non-significantly (as estimated by the first-order jack-knife), greater than the number recorded in the near primary forest plot. However, many fewer species were found in the most disturbed plots – the completely cleared plot and the farm fallow plot. Termite abundance was found to be considerably greater in the forest and partial manual clearance plantation plots than in the completely cleared plot and the farm fallow plot (Figure 15.2) (Eggleton et al., 1996).

Amongst different functional groups, soil-feeding termites were most affected by disturbance, and wood- and litter-feeding species least affected by disturbance (Eggleton et al., 1996). Although the numbers of soil-feeding species were similarly

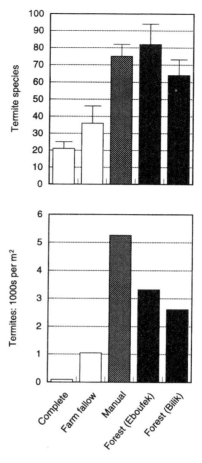

Figure 15.2 (a) Termite species richness (estimated by first-order jack-knife with 95% confidence limits) and (b) mean abundance in five plots in the Mbalmayo Forest Reserve, Southern Cameroon, 1992. ■, forest plots; ▨, partial clearance plantation plots; □, other plots.

high in the three forest/forest-plantation plots, very few species were found in the completely cleared and the farm fallow plots. The highest abundance of wood feeders was found in the old secondary and manual clearance plantation plots, with fewer, generally, in the near primary plot and very few in the cleared plots.

Thus, forest clearance has a profound impact on termite diversity and abundance; complete forest clearance reduces the number of termite species by approximately 50%. However, slight disturbance, such as the establishment of a plantation of a native tree species, does not have a detrimental effect on termites. As decomposers, termites require dead plant material and do not appear to be critically limited by the number and type of tree species in an area (Eggleton et al., 1996). Indeed, there is some evidence from the data that termites benefit from some forest disturbance, the heterogeneous habitat in an old secondary forest or a partially cleared plantation with a rich supply of dead wood providing a diversity of micro-habitats for a wide range of termite species. Eggleton et al. (1995, 1996) suggest that habitats, such as 'enrichment' plantations with lines of forest cleared prior to tree planting, contain the equivalent of natural tree-fall gaps in primary forest.

15.6.5 LEAF-LITTER ANTS

Leaf-litter sampling was carried out in two complete clearance plots, the partial manual and partial mechanical clearance plots in Ebogo, and all three uncleared forest plots (Table 15.1). The number of ant species recorded was greatest in the two partial clearance plantation plots and the Eboufek uncleared forest plot (Figure 15.3). The latter plot was the most disturbed of the 'uncleared' forest plots, and was the 'old secondary' plot where the greatest number of termite species were recorded. Ant species richness was lower in the complete clearance plots and, surprisingly, in the least disturbed (uncleared) forest plots (in Ebogo and Bilik). Ant abundance roughly paralled ant species richness except that ant abundance was notably lower in the recently established Eboufek complete clearance plot, which contained small trees, than the Ebogo complete clearance plot, which contained 6-year-old *T. ivorensis* (Figure 15.3). This difference in ant abundance probably relates to the difference in leaf litter between the two plots.

Thus, complete forest clearance and the establishment of plantations appears to have a negative impact on leaf-litter ant species diversity, which can be seen best by comparing the complete clearance plots with the uncleared forest controls in Ebogo and Eboufek (as pairs of plots). However, slight disturbance, in the form of establishing par-

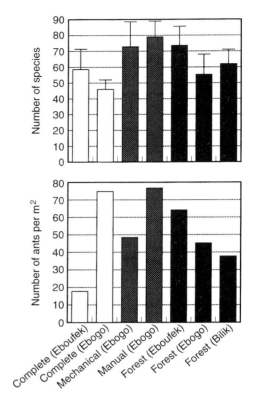

Figure 15.3 (a) Leaf-litter ant species richness (estimated by first-order jack-knife with 95% confidence limits) and (b) mean abundance in seven plots in the Mbalmayo Forest Reserve, Southern Cameroon, November 1993. (Bar shading as Figure 15.2).

tial clearance plantations, appears to lead to an increase in leaf-litter ant species diversity. The sampling programme (for litter ants) did not allow us to assess the impact of more severe forest disturbance, such as the farm fallow treatment, and the complete clearance plot sampled for both groups had changed somewhat between the termite (1992) and the ant (1993) sampling. In conclusion, however, the results for the leaf-litter ants are comparable to the results for termites: slight forest disturbance appears to lead to an increase in species richness. These results appear to differ somewhat from those from Ghana, where similar levels of leaf litter ant diversity were found in secondary forest, primary forest and cocoa plantations (Belshaw and Bolton, 1993). However, that study did not include the type of forest plantation studied here – enrichment plantations in partially cleared forest, where the highest levels of ant diversity were recorded.

15.6.6 BUTTERFLIES

Butterflies were sampled in the same plots as were used to sample leaf litter ants, plus the farm fallow plot (Table 15.1). Butterfly species richness was similar in all the forest plantation and forest plots except for the Eboufek (uncleared) forest plot. Significantly more species were recorded in this plot, the one which had experienced the most 'disturbance' of the forest plots, and significantly fewer species were recorded in the farm fallow plot (Figure 15.4). Butterfly abundance closely followed butterfly species richness (Figure 15.4). Thus butterfly diversity and abundance appear to be negatively affected by forest clearance and conversion to farm fallow, but remain high across most afforested plots, whether natural, partial clearance or complete clearance plantation.

15.6.7 CANOPY-DWELLING ANTS

Canopy-dwelling ants were sampled in the six plantation plots in Ebogo and Bilik. Ants were much more abundant in the partial manual clearance plot in Ebogo than in all the others (Figure 15.5). In this plot, 970 ants/m² were recorded, most of which were from a single species of *Technomyrmex*. Although ants were more abundant in the partial manual clearance plot than elsewhere in Ebogo, at Bilik ants were equally abundant across all plots.

At the Bilik site, canopy-dwelling ant species richness declined with forest disturbance i.e. more species were found in the partial manual clearance plot (Figure 15.5). At Ebogo, however, the larger numbers of individuals in the manual clearance plot were not accompanied by a larger number of species. Thus partial clearance appeared to have provided a suitable environment for one particular species at the Ebogo site but the presence of this dominant *Technomyrmex* species depressed the overall diversity of ants.

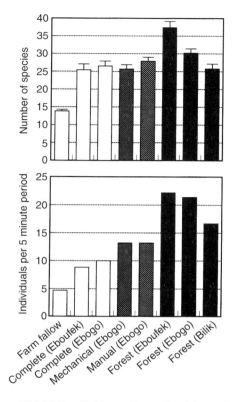

Figure 15.4 (a) Rarefied butterfly species richness (with standard deviations) and (b) mean abundance in eight plots in the Mbalmayo Forest Reserve, Southern Cameroon, February 1994. (Bar shading as Figure 15.2.)

15.6.8 CANOPY-DWELLING COLEOPTERA

Canopy-dwelling beetles, like the ants, were sorted to species in the six plantation plots in Ebogo and Bilik, but unlike the ants, species identification was also carried out in the uncleared forest plots at each of these sites (Table 15.1). A preliminary examination of the partially sorted data, rarefied to 100 individuals, shows a steady increase in species richness between complete clearance plots and uncleared forest plots: about 50 species in the complete clearance plantation plots, 80 species in the uncleared forest plots (Figure 15.6). Unlike other insect groups, there was no evidence that moderate levels of forest disturbance were beneficial to beetle species diversity. Comparable data on beetle abundance are only available in the plantation plots

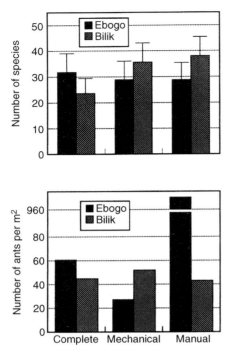

Figure 15.5 (a) Canopy-dwelling ant species richness (estimated by first-order jack-knife with 95% confidence limits) and (b) mean abundance in different plantation treatment plots in two sites in the Mbalmayo Forest Reserve, Southern Cameroon, 1992–1993.

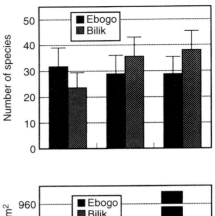

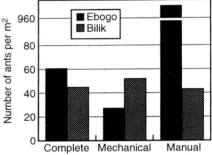

Figure 15.6 (a) Rarefied beetle species richness (with standard deviations) and (b) mean abundance in plots of three different silvicultural treatments and (above only) two uncleared forest plots in two sites in the Mbalmayo Forest Reserve, Southern Cameroon, 1992–1993.

where at one site (Bilik) no difference in abundance was observed, but at the other site (Ebogo) beetles were more abundant in the partial manual clearance plot than elsewhere (Figure 15.6).

15.6.9 OTHER ARTHROPOD GROUPS

We have no further data on the species richness of different arthropod groups in the Mbalmayo Forest Reserve, but the abundance of other canopy-dwelling groups was measured by insecticide knockdown sampling and was found to be affected by plantation treatment type. For example, spiders and Hymenoptera (other than ants) tended to be least abundant in complete clearance plots and most abundant in partial manual clearance plots (Watt et al., 1997).

15.7 DISCUSSION AND CONCLUSIONS

15.7.1 LIMITATIONS

The research in the Mbalmayo Forest Reserve described here has several limitations. First, the single hectare plots were probably too small. A particular concern is that the surrounding areas of uncleared forest influenced the abundance and diversity of insects in the research plots. Thus the results should be used with caution in predicting, for example, the impact of the establishment of large areas of complete clearance plantation on insect diversity.

A second problem is that the data discussed here were collected over a three-year period, usually during short sampling 'campaigns', with limited amount of replication of treatments, and on only five groups of arthropods. Longer-term sampling with more emphasis on seasonal variability in arthropod abundance and diversity is clearly needed. Third, the work reviewed here concentrated on forest clearance and the establishment of plantations of native tree species. Research in Sabah and elsewhere has produced valuable data on the impact of forest clearance and the establishment of plantations of *Acacia mangium*, *Pinus caribaea* and other exotic species (Holloway et al., 1992; Chey, 1994). However, comparisons of arthropod diversity in plantations of exotic and native tree species (as well as uncleared forest) in the same areas are urgently needed.

Finally, in these studies, we have only examined local, or alpha, diversity. It should be emphasized that these results infer nothing about regional, or gamma, diversity. Even if the conversion of near primary forest to a plantation does not result in reduced diversity at a particular site, if the species composition is changed from one with many regionally rare species to one with many common species it is clear that the likelihood of some species becoming regionally extinct increases. Similarly, loss of habitats with relatively low levels of adult insect diversity (e.g. butterflies) may ultimately be catastrophic for populations if the larval stages are dependent on host plant species found only in those habitats.

Taking these various limitations together in relation to the research in Cameroon presented here, and summarized below, they probably mean that the results underestimate the impact of forest clearance on arthropod diversity.

15.7.2 FOREST DISTURBANCE AND THE FUNCTIONAL IMPORTANCE OF DIFFERENT ARTHROPOD GROUPS

Arthropods cannot be seen only in terms of abundance and diversity: they perform important functional roles within forests, and forest disturbance may disrupt this role. Chapter 8 describes the role of termites as mediators of carbon fluxes in tropical forests.

Arthropods are also important as pests of tropical forest plantations. Mahoganies, for example, apparently cannot be grown in plantations because of the impact of shoot-boring moths (*Hypsipyla* spp.) (Newton et al., 1993). This has, no doubt, been an important factor in the establishment of plantations of exotic tree species such as *Eucalyptus deglupta* and *Gmelina arborea*. Indeed, most 'reforestation' projects in West Africa and other tropical areas have been with exotic species (G.L. Lawson, personal communication).

Forest plantations have encountered pest problems for many reasons (Chapter 12). These problems may be ameliorated by adopting methods other than establishing plantations after complete clearance. *Zonocerus variegatus* (Orthoptera), for example, is an abundant species in areas cleared of forest (Chapman et al., 1986). It can be found in large numbers on trees in young plantations established this way, but not in partial clearance plantations (Watt and Stork, unpublished).

15.7.3 CONCLUSIONS

The main purpose of this chapter has been to discuss the impact of forest clearance on insect diversity. Despite some of the concerns expressed above, we can make the following conclusions.

1. Although only measured for two groups – termites and butterflies – the process of forest clearance and conversion to farm fallow has a large negative impact on diversity: a reduction of approximately 50% in species richness was recorded in the farm fallow plot compared with the forest plots in the Mbalmayo Forest Reserve.
2. Similarly, the process of forest clearance and conversion to forest plantation prepared by complete clearance of vegetation before replanting had a negative impact on the diversity of all groups studied: a reduction in species richness of approximately 15% in butterflies and leaf-litter ants, but 40–70% in arboreal beetles and termites. Although for the reasons outlined above most of these estimates will err on the side of underestimation, it should be noted that the estimate of species richness for termites in complete clearance plantations comes solely from a

recently cleared area, and that the figure for reduction in the species richness of this group is likely to be an overestimate.

3. In contrast, conversion from forest to partial clearance plantation had different effects on different insect groups. The species richness of termites and leaf-litter ants increased after partial manual clearance and replanting, probably because of the increased amount and variety of dead wood in these plantations. In contrast, the species richness of both butterflies and canopy-dwelling beetles was found to be greater overall in forest plots than in the partial manual clearance plots. Unfortunately, at present we only have data for diversity of one group of canopy-dwelling insects in uncleared forest plots in the Mbalmayo Forest Reserve: the beetles. This shows that this group is more diverse in uncleared forest than in all other habitats surveyed. However, this situation may change as partial clearance plantations mature, and their epiphytic floras and associated faunas develop.

4. Finally, it is clear that plantation forest management has a marked impact on insect species diversity. All insect groups assessed – termites and leaf-litter ants, butterflies, and two canopy-dwelling groups, beetles and ants – were found to be more diverse in partial manual clearance plantation plots than in plantation plots established after complete clearance.

Thus, in conclusion: research in the Mbalmayo Forest Reserve has confirmed that forest clearance has a profound negative impact on insect diversity. Apart from uncleared forest, plantations of native tree species in previously forested areas are rich in insect species. For some insect groups, relatively minor levels of disturbance (from uncleared forest to partial clearance plantation) may lead to increases in species richness, though the potentially detrimental impact of even slight forest disturbance on species composition is poorly understood. Plantation forestry can, therefore, help to conserve insect diversity in areas where tropical moist forests are the natural vegetation, but the way in which plantations are established is critically important: partial clearance plantations of native trees are much better, in terms of species richness, than complete clearance plantations.

ACKNOWLEDGEMENTS

The research in Cameroon was financed by the Overseas Development Administration (ODA), and by the Natural Environment Research Council through its Terrestrial Initiative in Global Environment Reseach Programme (Award no. GST/02/625). This work would have been impossible without the support and encouragement of David Bignell, Gerry Lawson, John Lawton, Roger Leakey, Paulinus Ngeh, Andy Roby and Zac Tchoundjeu, and technical help from, in particular, Colin McBeath, Marcel Mboglen, Tommy Brown, Jacqueline Fanguem, Janvier Tchoupa, Julius Tipa, Melanie West, Eileen Wright, Emma Lindsay and Jonathan Mason. Jack-knife analysis on the data was done with the program *EstiMateS* (R. K. Colwell, University of Connecticut, unpublished).

REFERENCES

Belshaw, R. and Bolton, B. (1993) The effect of forest disturbance on the leaf-litter ant fauna in Ghana. *Biodiversity and Conservation* 2, 656-666.

Bibby, C.J., Crosby, M.J., Heath, M.F. *et al.* (1992) *Putting Biodiversity on the Map: Global Priorities for Conservation*, ICBP, Cambridge.

Blockhus, J.M., Dillenbeck, M., Sayer, J.A. and Wegge, P. (1992) *Conserving Biological Diversity in Managed Tropical Forests*, IUCN, Gland.

Chapman, R.F., Page, W.W. and McCaffery, A.R. (1986) Bionomics of the variegated grasshopper (*Zonocerus variegatus*) in West and Central Africa. *Annual Review of Entomology* 31, 479–505.

Chey, V.K. (1994) Comparison of biodiversity between plantation and natural forests in Sabah using moths as indicators. PhD thesis, University of Oxford.

Colwell, R.K. and Coddington, J.A. (1994) Estimating terrestrial biodiversity through extrapolation. *Philosophical Transactions of the Royal Society of London Series B – Biological Sciences* **345**, 101–118.

Davis, S.D., Heywood, V.H. and Hamilton, A.C. (1994a) *Centers of Plant Diversity: A Strategy for their Conservation, Volume 1, Europe, Africa, South West Asia and the Middle East*, WWF and IUCN, Cambridge.

Davis, S.D., Heywood, V.H. and Hamilton, A.C. (1994b) *Centers of Plant Diversity: A Strategy for their Conservation. Volume 2. Asia, Australia and the Pacific*, WWF and IUCN, Cambridge.

Davis, S.D., Heywood, V.H. and Hamilton, A.C. (1995) *Centers of Plant Diversity: A Strategy for their Conservation. Volume 3. The Americas*, WWF and IUCN, Cambridge.

Eamus, D., Lawson, G.J., Leakey, R.R.B. and Mason, P.A. (1990) Enrichment planting in the Cameroon moist deciduous forest: microclimatic and physiological effects. Proceedings of the 19th INFRO World Congress, Montreal, 258–270.

Eggleton, P. and Bignell, D.E. (1995) Monitoring the response of tropical insects to changes in the environment: troubles with termites. In *Insects in a Changing Environment* (eds R. Harrington and N.E. Stork), pp. 474-497, Academic Press, London.

Eggleton, P., Bignell, D.E., Sands, W.A. *et al.* (1995) The species richness of termites (Isoptera) under differing levels of forest disturbance in the Mbalmayo Forest Reserve, southern Cameroon. *Journal of Tropical Ecology* **11**, 1–14.

Eggleton, P., Bignell, D.E., Sands, W.A. *et al.* (1996) The diversity, abundance and biomass of termites under differing levels of disturbance in the Mbalmayo Forest Reserve, southern Cameroon. *Philosophical Transactions of the Royal Society of London Series B – Biological Sciences* (in press).

Erwin, T.L. (1983) Beetles and other insects of tropical forest canopies at Manaus, Brazil, sampled by insecticidal fogging. In *Tropical Rain Forest: Ecology and Management* (eds S.L. Sutton, T.C. Whitmore and A.C. Chadwick), pp. 59–75, Blackwell Scientific Publications, Oxford.

Erwin, T.L. (1991) How many species are there – revisited. *Conservation Biology* **5**, 330–333.

Erwin, T.L. and Scott, J.C. (1980) Seasonal and size patterns, trophic structure, and richness of Coleoptera in the tropical arboreal ecosystem: the fauna of the tree *Luehea seemannii* Triana and Planch in the canal zone of Panama. *Coleopterists Bulletin* **34**, 305–322.

FAO (1991) *Second Interim Report on the State of Tropical Forests by Forest Resources Assessment 1990 Project*, Tenth World Forestry Congress, Paris.

Gadagkar, R., Chandrashekara, K. and Nair, P. (1990) Insect species diversity in the tropics: sampling methods and a case study. *Journal of the Bombay Natural History Society* **87**, 337–353.

Hammond, P.M. (1990) Insect abundance and diversity in the Dumoga-Bone National Park, North Sulawesi, with special reference to the beetle fauna of lowland rainforest in the Toraut region. In *Insects and the Rain Forests of South East Asia (Wallacea)* (eds W.J. Knight and J.D. Holloway), pp. 197–254, Royal Entomological Society, London.

Hammond, P.M. (1995) The correct magnitude of biodiversity. In *Global Biodiversity Assessment* (ed. UNEP), pp. 113–138, Cambridge University Press, Cambridge.

Hammond, P.M., Stork, N.E. and Brendell, M.J.D. (1997)Tree-crown beetles in context: a comparison of canopy and other ecotone assemblages in a [?] tropical forest in Sulawesi. In *Canopy Arthropods* (eds N.E. Stork, J. Adis and R.K. Didham), p.184–223, Chapman & Hall, London.

Hanski, I. and Krikken, J. (1991) Dung beetles in tropical forests in South-east Asia. In *Dung Beetle Ecology* (eds I. Hanski and Y. Cambefort), pp. 179–197, Princeton University Press.

Heltshe, J.F. and Forrester, N.E. (1983a) Estimating diversity using quadrat sampling. *Biometrics* **39**, 1073–1076.

Heltshe, J.F. and Forrester, N.E. (1983b) Estimating species richness using the jackknife procedure. *Biometrics* **39**, 1–11.

Holloway, J.D., Kirk-Spriggs, A.H. and Chey, V.K. (1992) The response of some rain forest insect groups to logging and conversion to plantation. In *Tropical Rain Forest: Disturbance and Recovery (Phil. Trans. Royal Society B 335)* (eds A.G. Marshall and M.D. Swaine), pp. 425–436, Alden Press, Royal Society, Oxford.

Johnson, S. (1995) Production and trade of tropical logs. *Tropical Forest Update* **5**(1), 21–23.

Krebs, C.J. (1989) *Ecological Methodology*, Harper Collins, New York.

Lawson, G.L. (1994) Indigenous trees in West African forest plantations: the need for domestication by clonal techniques. In *Tropical Trees: the Potential for Domestication and the Rebuilding of Forest Resources* (eds R.R.B. Leakey and A.C. Newton), pp. 112–123, HMSO, London.

Leakey, R.R.B. (1991) Clonal forestry: towards a strategy. Some guidelines based on experience with tropical trees. In *Tree Breeding and Improvement* (ed. J.E. Jackson), pp. 27–42, Royal Forestry Society, Tring.

Lugo, A.E. (1988) Estimating reductions in the diversity of tropical forest species. In *Biodiversity* (ed. E.O. Wilson), pp. 58–70, National Academy Press, Washington, DC.

Mason, P.A., Musoko, M.O. and Last, F.T. (1992) Short-term changes in vesicular–arbuscular mycorrhizal spore populations in *Terminalia* plantations in Cameroon. In *Mycorrhizas in Ecosystems* (eds D.J. Read, D.H. Lewis, A.H. Fitter and I.J. Alexander), pp. 261–267, CAB International, Wallingford.

Myers, N. (1986) Tropical deforestation and a mega-extinction spasm. In *Conservation Biology: the Science of Scarcity and Diversity* (ed. M.E. Soule), pp. 394–409, Sinauer, Sunderland, USA.

Myers, N. (1988) Threatened biotas: 'hot spots' in tropical forests. *The Environmentalist* **8**, 187–208.

Myers, N. (1990) The biodiversity challenge: expanded hot-spots analysis. *The Environmentalist* **19**, 243–256.

Newton, A.C., Baker, P., Ramnarine, S. et al. (1993) The mahogany shoot borer – prospects for control. *Forest Ecology and Management* **57**, 301–328.

Ngeh, P. (1989) Effects of land clearing methods on a tropical forest ecosystem and the growth of *Terminalia ivorensis* (A. Chev.). PhD thesis, University of Edinburgh.

Reid, W.V. (1995) How many species will there be? In *Tropical Deforestation and Species Extinction* (eds T.C. Whitmore and J.A. Sayer), pp. 55–73, Chapman & Hall, London.

Simberloff, D.S. (1972) Properties of the rarefaction diversity measurement. *American Naturalist* **106**, 414–418.

Stork, N.E. (1987) Guild structure of arthropods from Bornean rain-forest trees. *Ecological Entomology* **12**, 69–80.

Stork, N.E. (1988) Insect diversity – facts, fiction and speculation. *Biological Journal of the Linnean Society* **35**, 321–337.

Stork, N.E. (1991) The composition of the arthropod fauna of Bornean lowland rain-forest trees. *Journal of Tropical Ecology* **7**, 161–180.

Stork, N.E. (1997) Measuring global biodiversity and its decline. In *Biodiversity*, National Academy Press, Washington, DC.

Stork, N.E. and Brendell, M.J.D. (1990) Variation in the insect fauna of Sulawesi trees with season, altitude and forest type. In *Insects and the Rain Forests of South East Asia (Wallacea)* (eds W.J. Knight and J.D. Holloway), pp. 173–190, Royal Entomological Society, London.

UNEP (1995) *Global Biodiversity Assessment*, Cambridge University Press, Cambridge.

Watt, A.D., Stork, N.E., McBeath, C. et al. (1995) *Ecology of Insects in Cameroon Plantation Forests*. Final report to UK Overseas Development Administration (ODA)/GOC Office National de Developpement des Forêts (ONADEF): Forest Management and Regeneration Project, Mbalmayo, Cameroon.

Watt, A.D., Stork, N.E., McBeath, C. and Lawson, (1997) Biodiversity II. Understanding and protecting our biological resources (eds M.L. Reaka-Kindle, D.E. Wilson and E.O. Wilson) pp. 41–68. *J. Applied Ecology* (in press).

WCMC (World Conservation Monitoring Centre) (1992) *Global Biodiversity: Status of the Earth's Living Resources*, Chapman & Hall, London.

Whitmore, T.C. and Sayer, J.A. (1992) *Tropical Deforestation and Species Extinction*, Chapman & Hall, London.

WRI (World Resources Institute) (1992) *World Resources 1992–93*, Oxford University Press, Oxford.

BEETLE ABUNDANCE AND DIVERSITY IN A BOREAL MIXED-WOOD FOREST

John R. Spence, David W. Langor, H.E. James Hammond and Gregory R. Pohl

16.1 INSECTS AND THE BOREAL MIXED-WOOD

Much of the western half of Canada, from northeastern British Columbia to southeastern Manitoba (Figure 16.1), is covered by a vast, relatively undeveloped wilderness of boreal mixed-wood forest. Within Alberta, this forest type covers more than 290 000 km² (Rowe, 1972) as a mosaic of stands that vary in age, dimension and dispersion (Samoil, 1988; Peterson and Peterson, 1992). The canopy of this forest type is dominated largely by trembling aspen (*Populus tremuloides* Michaux.) and white spruce (*Picea glauca* [Moench] Voss) (Thorpe, 1992), with stands of black spruce (*Picea mariana* [Mill.] B.S.P.) developing on wetter or north-facing slopes. Balsam poplar (*Populus balsamifera* L.), paper birch (*Betula papyrifera* Marsh.), and balsam fir (*Abies balsamea* [L.] Miller) are more minor canopy elements. Deciduous and coniferous canopies can be a fine-grained mix or can occur in a patchwork of stands dominated by either evergreen or broadleaf trees (Stelfox, 1995). Relatively pure stands of aspen are common as climax communities in southern parts of the western boreal mixed-wood (Thorpe, 1992; Stelfox, 1995), but black spruce becomes the predominant climax tree even in upland settings further north (Van Cleve *et al.*, 1991).

These mixed-wood forests are a major and extensive component of nearctic vegetation. Their palearctic counterpart is rare because *Populus tremula* L., a species very similar to trembling aspen (Graham *et al.*, 1963), has been largely eliminated in northern Europe during directed regeneration of forests after logging (Siitonen and Martikainen, 1994). Further east on the Russian taiga, where there has been natural succession following wildfire, *Betula* species are numerically more important than aspen in mixed-wood sequences leading to re-establishment of Norway spruce (*Picea abies* [L.] Karst.) (Syrjänen *et al.*, 1994). In contrast, trembling aspen is the main pioneer species following wildfire (Wein and MacLean, 1983; Bonan and Shugart, 1989), giving way in time to conifers at many sites throughout western Canada. Fire and other natural disturbances constantly set the clock back, contributing to a landscape dominated by deciduous trees (Van Cleve *et al.*, 1991; LaRoi, 1992).

The boreal mixed-wood harbours a diverse but poorly studied northern fauna of arthropods. In a recent and extensive review, Danks and Foottit (1989) estimated that *c.* 22 000 species of insects occur in Canada's boreal forests. They further showed that the orders Diptera and Hymenoptera appear to be disproportionately represented in northern nearctic faunas and that overall diversity of Coleoptera falls with increasing latitude. More than half of the insects that occur in the boreal zone do not have strictly boreal distributions but appear to be forest habitat generalists, also abundant in other life zones. This observation contributes to the

---

Forests and Insects. Edited by A.D. Watt, N.E. Stork and M.D. Hunter. Published in 1997 by Chapman & Hall, London. ISBN 0 412 79110 2.

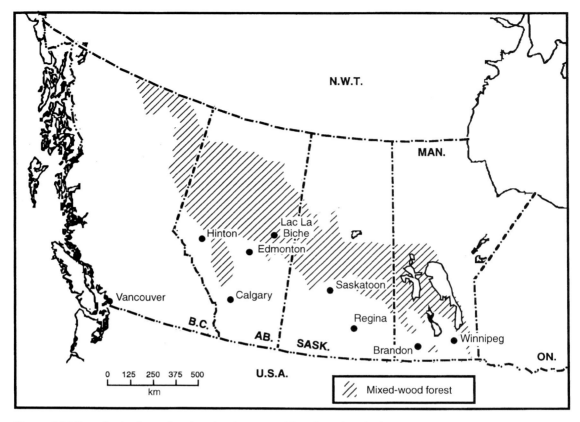

Figure 16.1 Boreal mixed-wood region showing general location of study site near Lac La Biche, Alberta.

evidence that southern glacial refugia are the most important sources of the current boreal fauna (Danks and Foottit, 1989). It appears that only about 8% of all insects found in the Canadian boreal forest have holarctic distributions, despite the substantial effect of connections like Beringia on development of the modern arctic fauna.

Nonetheless, some insects are characteristic of and largely restricted to the boreal zone. In contrast to the general rule, many biting flies have holarctic distributions restricted to northern forest and subarctic habitats (Danks and Foottit, 1989). These species have received considerable attention because of their obvious impact on human activity in the north (e.g. Wood *et al.*, 1979; Danks, 1981). Although biting flies parasitize forest vertebrates and are a likely factor in their population dynamics, the abundance of mosquitoes, blackflies and tabanids in the boreal region depends mainly on aquatic habitats.

Among actual forest dwellers, phytophagous species may be most restricted to boreal regions through host plant relationships. For example, the geographical range of the spruce budworm (*Choristoneura fumiferana* [Clemens]) mirrors that of its principal host, white spruce. The population dynamics of this economically important species has been well studied (reviewed by Royama, 1992) and, furthermore, it and other phytophagous species are known to exert strong influences on development of forested landscapes throughout the boreal life zone (Holling, 1992). *Choristoneura fumiferana* shows locally adaptive features (Volney, 1985; Weber, 1994), suggesting that insect life styles can be fine-tuned to environmental variation that is characteristic of the boreal zone. However, such detailed biological information is available for only a small fraction of insect species of the boreal forest. We know little more than names for described species that are not recognized as significant pests, and an estimated 45% of insect species inhabiting the boreal region remain unknown to science (Danks, 1979; Danks and Foottit, 1989). In short, our present understanding of boreal forest insect assemblages, their variation in space and time and their ecological significance is rudimentary at best.

More information about insects of little or unrecognized economic importance and their broad ecological roles is available about taiga forests of northern Europe. Increased understanding of linkages between insects and lichens or fungi promises to be valuable for a more complete picture of forest processes involving insects and in developing forest management strategies. For example, Pettersson et al. (1995) have shown that regenerated coniferous forests in Sweden provide significantly fewer of the canopy-dwelling invertebrates important as food for passerine bird assemblages than do natural forests. Reductions in Scandinavian forest passerines may be attributable to stand-level changes in invertebrate production brought about by several decades of harvest and the associated loss of lichen species essential for certain faunal elements. Nilsson et al. (1995) showed that the lichen, *Lobaria pulmonaria* (L.) Hoff., was a good predictor for the presence of red-listed lichen species in Sweden but was only weakly correlated with the presence of red-listed beetles using dead wood as habitat. Kaila et al. (1994) demonstrated strong associations between the presence of saproxylic beetles in unharvested forests and the infection of birch by specific species of polypore fungi. Siitonen (1994) collected more than 200 species of saproxylic species from old spruce forests in Finland and emphasized the importance of this ecological assemblage in relation to the conservation of biodiversity in northern forests subjected to harvest (see also Siitonen and Martikainen, 1994). Increasingly, it appears that temporal continuity of forests is a significant issue for development of some arthropod assemblages, even in forest types strongly affected at various locations on the landscape by periodic disturbance.

Forest disturbances such as harvesting and fire alter various physical, chemical and biological factors in the litter and soil (Simms, 1976; Kubin and Kemppainen, 1991; Cortina and Vallejo, 1994) and coarse woody debris (Muona and Rutanen, 1994). This, in turn, modifies these sites as habitats for invertebrates. A number of studies in forest ecosystems have shown dramatic effects of logging on invertebrate communities inhabiting soil, litter and coarse woody debris (e.g. Huhta, 1971; Vlug and Borden, 1973; Lenski, 1982; Heliövaara and Väisänen, 1984; Speight, 1989; McIver et al., 1992; Niemalä et al., 1993a,b, 1994; Langor et al., 1993; Siitonen and Martikainen, 1994) but there are fewer studies of the effects of wildfire (Holliday, 1991, 1992).

Some argue that use of harvest regimes resembling the spatial pattern of wildfire and other disturbances on forested landscapes will ensure conservation of biodiversity and simultaneously permit extraction of economically desirable volumes of forest fibre (Hunter, 1993; Bunnell, 1995). Others point out that fire and harvest are not ecologically equivalent, especially with respect to effects on invertebrate faunas (Zackrisson, 1977; Muona and Rutanen, 1994). Thus, biotic components critical to ecosystem function may be at risk in landscapes harvested without consideration of detailed ecological process, no matter what pattern of cutting is adopted. Hence, there is a need to determine whether succession from burned and logged sites generates forest structure suitable as habitat for the normal mixed-wood biota, especially those species characteristic of older stands. If, for example, stand-level processes like those documented by Pettersson et al. (1995) lead to reductions in abundance of food for larger wildlife, population sizes of valued forest vertebrates will surely decrease no matter what pattern of harvest is imposed on the landscape.

16.2 OLD-GROWTH STANDS IN THE MIXED-WOOD

In Alberta, the aspen–mixed-wood forest is now subject to major economic development (Peterson and Peterson, 1992; Stelfox, 1995). Despite our rudimentary understanding of the boreal mixed-wood ecosystem and how it will respond to intensive development, a steadily increasing demand for forest fibre inexorably pushes the configuration of the mixed-wood landscape on a course of dramatic change (Pratt and Urquart, 1994; Navratil and Chapman, 1991). Harvesting in boreal mixed-wood forests in the past concentrated largely on pockets of white spruce. Currently, however, the majority of fibre harvested in Alberta is aspen, which is the main resource base for a growing pulp and lumber industry (Peterson and Peterson, 1992). Thus, logging has now emerged as a primary disturbance in these forests. Continued development of the aspen resource in Alberta will transform much of the mixed-wood landscape into a mosaic of stands in various stages of harvest rotation. There will be little primeval forest left over large areas of this landscape in a generation or two. Such development is guided by the interplay between two basic sets of values (Kuusipalo and Kangas, 1994; Stelfox, 1995). On the one hand, use of forest resources for fibre is seen as vital from a political and socio-economic perspective. On the other hand, forestry practices must be sensitive to non-fibre components, increasingly defined in relation to the myriad of other forest organisms that maintain ecosystem integrity and the welfare of which is thought to reflect resource sustainability.

The significance of 'old-growth' forests for biodiversity is at the centre of discussion about the impact of northern forestry practices (Berg *et al.*, 1994; Hanski and Hammond, 1995), though what actually constitutes an old-growth forest is subject to disagreement (Barnes, 1989; Maser, 1990; Duchesne, 1994). However, it is clear that stand age distributions will be markedly truncated in a landscape dominated by forestry activity. In general, only preserves and relatively unproductive or inaccessible stands are left unharvested once a land base is brought into rotation. Because biodiversity hot spots for various taxa are not spatially congruent (Prendergast *et al.*, 1993; Nilsson *et al.*, 1995) it is questionable whether any particular selection of such remnants will maintain a full complement of the biota. Given this uncertainty, it is difficult to predict the implications of large-scale forest harvesting on old-growth stands and their constituent biota.

In response to concerns about ecological integrity of the mixed-wood ecosystem in the wake of extensive harvesting, we have launched a programme of research about insect assemblages of these forests. Limited resources and availability of taxonomic knowledge restrict our initial efforts to study of groups that are relatively well known. Through a combination of community ecology and faunistics (e.g. Niemalä *et al.*, 1993 a,b; Langor *et al.*, 1993; Spence *et al.*, 1996), we have studied beetles of litter and coarse woody debris to answer specific questions about arthropod conservation. We began our work with this ecological group of beetles because the necessary reference collections and other taxonomic resources were readily available, because some of these beetles seem to have potential as indicators of environmental change in general (Freitag *et al.*, 1973; Pearson and Cassola, 1992; Niemalä *et al.* 1993b) and because recent evidence suggests that species associated with fungi and coarse woody debris, in particular, may be threatened by forestry practices presently employed in northern forests (Siitonen and Martikainen, 1994; Kaila *et al.*, 1994; Nilsson *et al.*, 1995). Our long-term goals are to incorporate information about arthropods into development of forest management plans. In this chapter we focus on the following two questions.

1. Are there insect species that depend on old-growth, northern *Populus*-dominated mixed-wood forests?
2. If these exist, what should be done to ensure that such species retain viable populations in the face of the extensive logging activity that is planned?

16.3 BEETLE ASSEMBLAGES

16.3.1 STUDY DESIGN

We studied a chronosequence of even-aged, mixed-wood stands located in northeastern Alberta in a 40 km^2 area near the town of Lac La Biche (Figure 16.1). The stands varied in age from 8 to > 120 years, and were dominated by *Populus* (> 75%

of stems), but also contained willows (*Salix* spp.), paper birch and white spruce. All stands except those in the youngest age class had developed naturally in the wake of wildfire. The two youngest stands had developed after clearcut harvest; no young fire-origin stands were available for study.

We grouped the stands into four age classes based on estimated time since last disturbance, as determined from the ages of the dominant canopy trees:

1. The '**old**' stands harboured trees > 120 years old and were characterized by relatively large trees with diameters of 29–42 cm (at breast height). The canopy was 22–27 m above the ground, and contained many gaps. Large-diameter dead and decaying tree stems were common either as standing snags and/or fallen logs.
2. Stands characterized as '**mature**' had been undisturbed for *c.* 60 years and had reached rotation age. These stands had a slightly lower (18–19 m) but more continuous canopy with fewer gaps. Trees had diameters of 15–16 cm. Mature stands had less coarse woody debris and both logs and snags were of smaller diameter than in the old stands.
3. '**Young**' forest stands, *c.* 40 years of age, were coming into the final self-thinning stage described by Graham *et al.* (1963). Trees were smaller (11–13 cm in diameter) but stem density was much greater than in older forests.
4. The '**new**' stands had been harvested 8 years previously and had regenerated naturally to dense thickets of young hardwoods. One of the new stands had been scarified and planted with lodgepole pine, *Pinus contorta* Douglas var. *latifolia* Engelm., but the surviving young conifers were minor components in a thicket of natural deciduous growth.

Our samples included large numbers of individuals of two litter-dwelling groups, the Carabidae (*c.* 15 000) and the Staphylinidae (*c.* 6100), and somewhat smaller samples of many species from a large number of families that use rotting wood as habitat (*c.* 2900). Carabid beetles are widely employed in conservation studies (Eyre and Rushton, 1989; Eyre *et al.*, 1989; Stork, 1990; Niemalä, 1993 a,b) because they are species-rich and because excellent guides to their taxonomy are generally available (e.g. Lindroth, 1961–1969). Staphylinids are equally diverse in litter environments of northern forests. They are relatively well known taxonomically (e.g. Smetana, 1971, 1982; Campbell, 1973; Moore and Legner, 1979) with the exception of the Aleocharinae, and also have been employed as indicators in conservation studies (Buse and Good, 1993). Together, carabids and staphylinids represent a range of trophic roles in litter habitats, including predation, herbivory (seed-feeding) and fungivory. Conservation issues surrounding insects inhabiting coarse woody debris, especially saproxylic beetles, are receiving increased attention in boreal forests (Speight, 1989; Martin, 1989; Warren and Key, 1991; Kaila *et al.*, 1994; Siitonen and Martikainen, 1994).

Carabids and staphylinids were collected in pitfall traps (Spence and Niemalä, 1994) set out in trap lines of six traps each. Intertrap distances were 50–60 m, and no trap was set closer than 50 m to an edge, ensuring that catches would be independent samples (Digweed *et al.*, 1995). Beetles were trapped continuously into ethylene glycol for the entire snow-free seasons (May–October) of 1992 and 1993 with the following number of lines in stands of each age class: six in old stands, six in mature stands, two in young stands and two in new stands.

Work with beetle species associated with coarse woody material was limited to mature and old stands. Two methods of collection were employed to sample this community. First, in 1993 and 1994, beetles were reared from dead wood collected from the forest and held in emergence cages for one year at ambient outdoor temperature. Wood chosen for rearing represented the range of natural variation in amount of decay and diameter of snags and logs in the stands. Second, in 1994 a total of 72 transparent window traps (modified from Kaila, 1993) were attached to snags in two old and two mature stands. Each window trap consisted of a cloth funnel with a small plastic collecting bag (whirl-pak™) attached to the bottom and a plexiglass window (20 × 30 cm, 15 mm thick) set perpendicularly to the bark and dividing the trap in half at the top. A thin welded wire tube was attached to the rim of the funnel and suspended around the bag to protect it from animal damage. Two traps were placed on each snag, one at c. 1.3 m and the second at c. 2.6 m above the ground. These sampled all insects coming to dead and decaying wood, not just the saproxylic species (Siitonen, 1994), but allowed sampling of more trees than did rearings.

16.3.2 PATTERNS OF DIVERSITY AND ABUNDANCE

Data from traps were used to compare relative abundances of species among age classes of stands. As sampling effort was not identical in all stands, catches were standardized to 1000 trap-days (number of traps × number of trapping days). To compare species richness among stands, the data were further adjusted to minimize bias due to sampling errors and uneven catches. Rarefaction (Simberloff, 1978; software from Krebs, 1989) was used to estimate the number of species in a random subsample of the entire sample. The resulting value can also be interpreted as a measure of diversity because the method takes into account both species richness and abundance. Standardized abundances were compared among age classes and between years using two-way analysis of variance. Relationships among beetle assemblages from different stand age classes were depicted using cluster analysis of Bray–Curtis percentage similarity values, with group averaging as the weighting procedure (software by Ludwig and Reynolds, 1988).

(a) Beetles of the leaf litter

Overall, 44 species of carabids were collected, 10–17 from each stand age class. Species diversity, as estimated by rarefaction in a sample of 200 individuals, was greater in 1993 than in 1992 for all age classes (Figure 16.2a). Predicted diversity was considerably greater in newly cut areas than in the older age classes, reflecting primarily colonization by open-habitat specialists in the genera *Amara*, *Bembidion* and *Harpalus*. However, there was no obvious trend toward greater diversity in old-growth stands. Conversely, carabid abundance varied significantly with stand age ($F = 21.3$; d.f. = 3, 24; $P < 0.0001$) and was lowest in the newly cut stands (Figure 16.2b). The pattern of means and variances suggests that carabids were somewhat less abundant in old-growth than in either young or mature stands. Overall carabid abundance was significantly higher in 1993 ($F = 23.8$; d.f. = 1, 24; $P < 0.0001$), but the pattern of abundance among age classes was similar between years (F-test for interaction was not significant). Cluster analyses of Bray–Curtis percentage similarity for the carabid assemblages were nearly identical in both years, in that the three oldest age classes clustered together at 65–70% similarity (Figure 16.3). The fauna of newly cut stands was most dissimilar from the three older age classes.

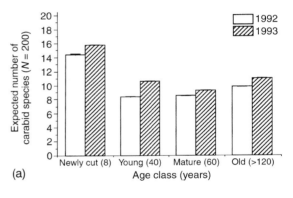

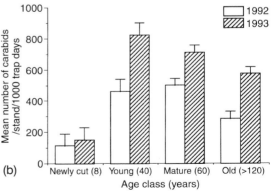

Figure 16.2 Characteristics of carabid assemblages collected by pitfall trapping in four age classes: (a) diversity (± 1 SD) as reflected by a rarefaction analysis; (b) least-square means for activity-abundance (± 1 SE) over snow-free periods of two years.

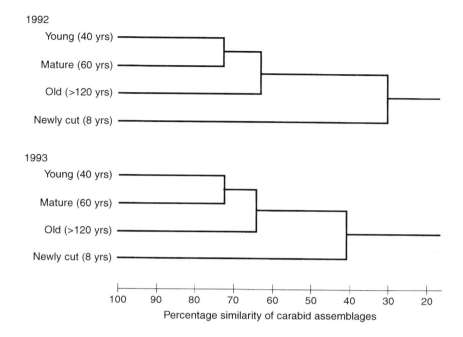

Figure 16.3 Cluster analysis of carabids caught by pitfall trapping in four stand age classes for 1992 and 1993.

Overall, 86 species of staphylinids were collected. The Aleocharinae were not identified to species and were treated as a single taxon in the analyses. The patterns of diversity for rove beetles (Figure 16.4a), as represented by rarefaction estimates based on a sub-sample of 100 individuals, is rather different from that of carabids. There is a trend toward greater diversity in the old-growth forest in 1993, and there was significant variation in diversity within newly cut stands between years. Patterns of staphylinid abundance did not differ significantly between years ($F = 1.40$; d.f. = 1, 24; $P = 0.25$) and so data are pooled in Figure 16.4b. Stand age had a significant effect on staphylinid abundance ($F = 7.69$; d.f. = 3, 24; $P < 0.001$) and, as for carabids, staphylinids were least abundant in newly cut stands. Unlike the patterns shown for ground beetles, however, staphylinds did not appear to be less abundant in old-growth compared with young and mature stands. The cluster analyses (Figure 16.5) showed a pattern similar to the carabid data, in that the three oldest age classes clustered together at just under 70% similarity, though the clustering pattern was not identical in both years.

(b) Beetles of coarse woody debris

Because the assemblage of beetles living in dead and decaying wood is more speciose than the litter-dwelling assemblage, we focus mainly on the comparison of old-growth and mature stands. Species diversity, as estimated by rarefaction in a subsample of 400 individuals, did not differ between old and mature stands, based on data from rearings (Figure 16.6a). However, rarefaction estimates for window trap samples (subsample of 1300 individuals) suggests a richer fauna in mature stands than in old stands (Figure 16.6b). Bray–Curtis measures of faunal similarity from the two age classes were more similar when based on data from window traps (58% similarity) than from rearings (45%).

Comparisons of beetle abundance are restricted to window trap data (Figure 16.7). These data suggest that the general trophic structure of this community does not vary significantly with stand age

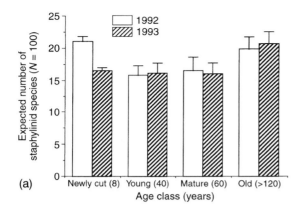

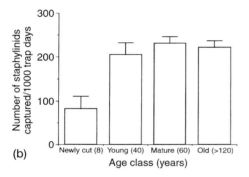

Figure 16.4 Characteristics of staphylinid assemblages collected by pitfall trapping in four age classes: (a) diversity (± 1 SD) as reflected by a rarefaction analysis; (b) least-square means for activity–abundance (± 1 SE) pooled over the snow-free periods of 1992–1993.

(P values ranged from 0.17 to 0.75 for individual trophic groups), and so data are pooled in Figure 16.7. Fungivores and predators clearly predominate in this community.

16.3.3 OLD-GROWTH DEPENDENCY

For the purposes of this study, we define old-growth as having existed undisturbed for longer than the projected rotation times for the land base under consideration. Given that 50–70-year rotations are to be used in the boreal mixed-wood (Stelfox, 1995), only the oldest age class clearly represents old-growth forest. However, under the shortest envisioned rotations even mature stands will be cut.

Although mainly concerned with conserving whole communities representative of particular forest types, we must consider the species involved to determine how dependent a fauna is on a particular habitat. Therefore, we compared the abundance of each beetle species in old-growth with that observed in stands of pre-rotational age to assess its dependency on old-growth habitat. The analysis was done separately for carabids and staphylinids from pitfall traps and for all beetles of coarse woody material captured with window traps. Only species comprising > 0.2% of the total sample were included in the analysis because data for rare species are too sparse to estimate accurate ratios. Pitfall data about staphylinids and carabids

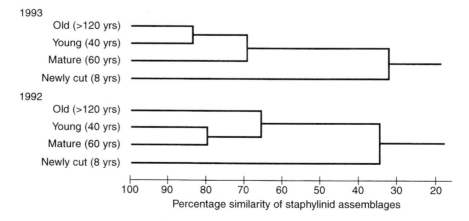

Figure 16.5 Cluster analysis of staphylinids caught by pitfall trapping in four stand age classes for 1992 and 1993.

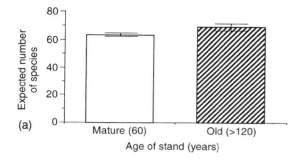

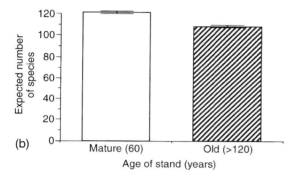

Figure 16.6 Diversity of beetles using coarse wood debris as determined by rarefaction analyses of data from (a) rearings and (b) window trap samples in four stand age classes (error bars represent ± 1 standard deviation).

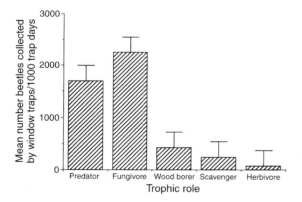

Figure 16.7 Mean number of beetles associated with coarse woody debris (± 1 SE) captured in window traps at Lac La Biche, 1994. Means for each trophic group are based on pooled estimates from two old and two mature stands.

from the newly cut-overs were not used because we have no such data for the window trap samples. Also, some beetle species, mainly characteristic of old-growth, exist in declining residual populations on clearcuts before disappearing altogether from young regenerating forests (Niemalä et al., 1993a; Spence et al., 1996).

We calculated an old-growth dependency index by comparing the mean standardized catch (MSC) from post-rotational age stands with the sum of the mean catches in post- and pre-rotational age stands. For window trap samples this amounted to (MSC in old stands)/(MSC in old stands + MSC in mature stands). We calculated the index under two possible scenarios for carabids and staphylinids: a short rotation of 50 years (data from mature stands in both numerator and denominator) and a long rotation of 70 years (data from mature stands in denominator only). A species was designated as old-growth dependent if > 80% of the mean standardized catch of that species occurred in old-growth stands. Any criterion for this designation is somewhat arbitrary. The criterion of 80% allows for the possibility that some emigrants from old-growth establish unstable populations in younger forests, and seemed to delineate a natural division in the outcome of our analyses. The old-growth index is indirectly proportional to the ability of young forest stands to support a particular species. For example, after harvest of a virgin forest landscape, we could expect that population sizes of an old-growth dependent species would not exceed 20% of former values throughout forest regeneration, even with unrestricted colonization.

Within the litter fauna, eight species of Staphylinidae (c. 30% of common species) but only one of Carabidae (c. 11% of common species) met our criterion for old-growth dependency (Figure 16.8). All but *Quedius velox* did so under both short and long rotation periods. Data for the carabid, *P. riparius*, are difficult to interpret as all specimens were captured in one trap on one date during 1993. For beetles in coarse woody debris, the relative importance of old-growth for particular species was assessed by comparing window-trap catches between the old and mature stands. Because these taxa were not sampled in young

296 *Beetle abundance/diversity in boreal mixed-wood forest*

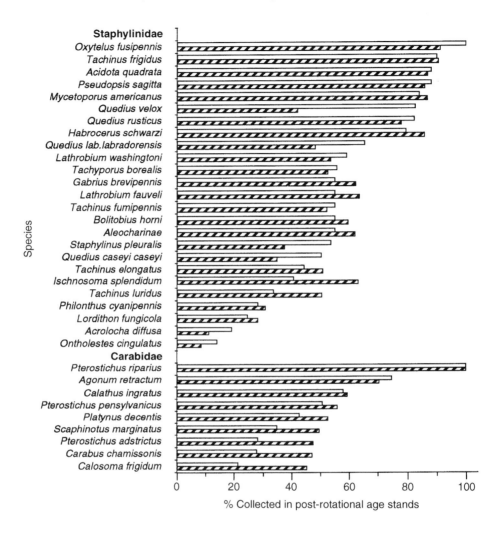

Figure 16.8 Old-growth dependence in carabid and staphylinid beetles captured by pitfall trapping. (See text for explanation of short and long rotations and percentage collected in post-rotational age stands.) Analysis based on pooled data from 1992 and 1993. ☐, short rotation; ▨, long rotation.

stands only the long-rotation scenario can be evaluated (Figure 16.9). Fourteen species, or about 20% of the species included in the assemblage, met the criterion for old-growth dependence.

16.4 BIODIVERSITY CONSIDERATIONS

The simple question about whether old-growth forests have distinctive beetle assemblages does not have a simple answer. In our study, the extent of old-growth dependence varied among higher taxa and between litter and coarse woody debris habitats. We found no compelling evidence for old-growth specialists among Carabidae in the mixed-wood forest, but in montane lodgepole pine forests only a few hundred kilometres to the west a rather large component of such species has been identified (Niemalä *et al.*, 1993 a,b; Spence *et al.*, 1996). In contrast, we found clear evidence in this study for old-growth dependency at Lac La Biche among

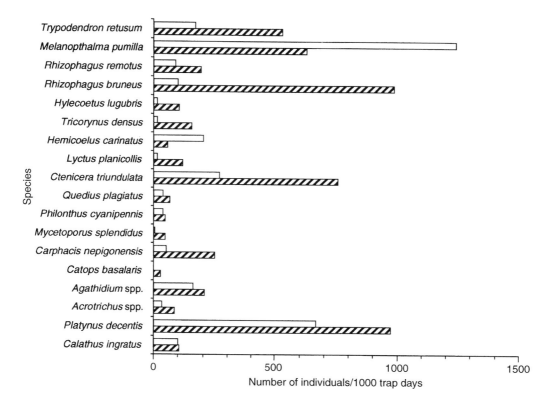

Figure 16.9 Capture of saproxylic beetles in nature and old growth forests by window traps in 1994. □, mature forest; ▨, old-growth forest.

staphylinid species using litter habitats and the beetle assemblage living in or on coarse woody debris.

Old-growth forests provide habitats for arthropods that may be rare or altogether absent in younger forests. Thus, logging plans should consider future landscape patterns of interstand distance and connectivity in relation to dispersal power of invertebrate and some lower vertebrate taxa (Harris, 1984). Because climax old-growth communities are more stable than those of earlier successional stages, many epigeal arthropods specializing in old-growth have diminished powers of interstand movement. Among carabid beetles, for example, reduced dispersal power is associated with wing-reduction (den Boer, 1970, 1990) and larger body size (Spence et al., 1996). Although such species may persist in suitably large old-growth fragments, their recolonization of widely separated patches of forest that have recently acquired the characteristic of old-growth may be problematic (Nield, 1990). In contrast to epigeal species occupying litter habitats that are relatively continuous within stands, species of naturally patchy habitats such as coarse woody material have been selected for strong colonization ability (Chandler and Peck, 1992; Ås 1993). These species may be more successful in colonizing stands when they become suitable as old-growth habitats.

It might be argued that worry about logging planned for the boreal mixed-wood is misplaced because the biota of this forest type is well adapted to periodic natural disturbance caused by wildfire. Fires have long created a shifting spatial mosaic of forest (Bonan and Shugart, 1989). Holliday (1991, 1992), for example, argued that carabid assemblages of burned stands recover so as to be virtually indistinguishable from those of unburned stands in 11 years. Although our findings are similar for carabids, data for epigeal staphylinids do not show this pattern. Three species of old-growth dependent staphylinids, *Oxytelus fuscipennis*, *Acidota quadrata* and *Mycetoporus americanus*, were common in new cuts but altogether absent from the 40-year-old stands. Thus, some epigeal arthropods may be slow to colonize regenerated stands even when embedded in a matrix of mainly older stands.

Although harvest is not a 'natural process', there is considerable impetus to adopt a 'natural disturbance' model to guide harvest patterns (e.g. Hunter, 1993; Bunnell, 1995). Essentially, this is taken to mean that patterns of forest harvesting should approximate patterns of wildfire on the landscape. This strategy should protect vertebrate faunas adapted to fire cycles, but may be insufficient to initiate the full range of natural processes that develop habitats of some old-growth arthropods. Unfortunately, less attention has been given to how fire intensity and other natural processes – e.g. windthrow (Veblen *et al.*, 1989), insects and disease (Mattson and Addy, 1975; Geiszler *et al.*, 1980; Veblen *et al.*, 1991) affect development of stand structure. Such more subtle effects may have serious implications for maintenance of biodiversity, especially of invertebrates. In short, fires lead to stands of different physical structure from those characteristic of a landscape dominated by forestry practices (e.g. Barnes, 1989) and such structure may be crucial for invertebrates (Haila *et al.*, 1994). Therefore, we argue that relationships between stand structure and biodiversity deserve explicit study and that understanding the faunal implications of alternative cutting and reforestation plans is a necessary feature of sustainable forest development.

ACKNOWLEDGMENTS

We thank Alberta-Pacific Forest Industries Inc. and Diashowa-Marubeni International Ltd. for their cooperation and interest in this study. R. Barrington-Leigh, N. Berg, H. Cárcamo, T. Clarke, S. Francis, R. Lucas, M. Maximchuck, S. Rasmussen, D. Raven, P. Rodriguez, P. Shipley, T. Spanton, K. Sytsma and D. Williams have assisted ably and enthusiastically with field and/or laboratory aspects of the work. We thank R. Anderson, G. Ball, E. Fuller, D. Pollack, D. Shpeley, Y. Bousquet, D. Bright, J. Campbell, A. Davies, L. LeSage, S. Peck, A. Smetana and Q. Wheeler for encouragement and cheerful assistance with identifications. The paper was improved by constructive criticism from N. Stork. Financial needs of the project were supported by the Alberta Forest Development Research Trust Fund, the Alberta Environmental Research Trust Fund, the Canada–Alberta Partnership Agreement in Forestry, Diashowa-Marubeni International Ltd, Forestry Canada, and the National Science and Engineering Research Council of Canada.

REFERENCES

Ås, S. (1993) Are habitat islands islands? Woodliving beetles (Coleoptera) in deciduous forest fragments in boreal forest. *Ecography* **16**, 219–228.

Barnes, B.V. (1989) Old-growth forests of the northern lake states: a landscape ecosystem perspective. *Natural Areas Journal* **9**, 45–57.

Berg, Å., Ehnström, B., Gustafsson, L. *et al.* (1994) Threatened plant, animal and fungus species in Swedish forests: distributions and habitat associations. *Conservation Biology* **8**, 718–731.

Bonan, G.B. and Shugart, H.H. Jr (1989) Environmental factors and ecological processes in boreal forests. *Annual Review of Ecology and Systematics* **20**, 1–28.

Bunnell, F. (1995) Forest-dwelling vertebrate faunas and natural fire regimes in British Columbia: patterns and implications for conservation. *Conservation Biology* **9**, 636–644.

Buse, A. and Good, J.E.G. (1993) The effects of conifer forest design and management on abundance and diversity of rove beetles (Coleoptera: Staphylinidae): implications for conservation. *Biological Conservation* **64**, 67–76.

Campbell, J.M. (1973) *A Revision of the Genus* Tachinus *(Coleoptera: Staphylinidae) of North and Central America*, Memoirs of the Entomological Society of Canada No. 90, 137 pp.

Chandler, D.S. (1987) Species richness and abundance of Pselaphidae (Coleoptera) in old-growth and 40-

year-old forests in New Hampshire. *Canadian Journal of Zoology* **65**, 608–615.

Chandler, D.S. (1991) Comparison of some slime-mold and fungus feeding beetles (Coleoptera: Eucinetoidea, Cucujoidea) in an old-growth and 40-year-old forest in New Hampshire. *Coleopterists Bulletin* **45**, 239–256.

Chandler, D.S. and Peck, S.B. (1992) Diversity and seasonality of leiodid beetles (Coleoptera: Leiodidae) in an old-growth and a 40-year-old forest in New Hampshire. *Environmental Entomology* **21**, 1283–1293.

Cortina, J. and Vallejo, V.R. (1994) Effects of clearfelling on forest floor accumulation and litter decomposition in a radiata pine plantation. *Forest Ecology and Management* **70**, 299–310.

Danks, H.V. (ed.) (1979) *Canada and its Insect Fauna*, Memoirs of the Entomological Society of Canada No. 108, 573 pp.

Danks, H.V. (1981) *Arctic Arthropods. A review of systematics and ecology with particular reference to the North American fauna*, Entomological Society of Canada, Ottawa, Ontario, 608 pp.

Danks, H.V. and Foottit, R.G. (1989) Insects of the boreal zone of Canada. *Canadian Entomologist* **121**, 625–690.

den Boer, P.J. (1970) On the significance of dispersal power for populations of carabid beetles (Coleoptera, Carabidae). *Oecologia (Berlin)* **4**, 1–28.

den Boer, P.J. (1990) The survival value of dispersal in terrestrial arthropods. *Biological Conservation* **54**, 175–192.

Digweed, S.C., Currie, C.R., Cárcamo, H.A. and Spence, J.R. (1995) Digging out the 'digging-in effect' of pitfall traps: influences of depletion and disturbance on catches of ground beetles (Coleoptera: Carabidae). *Pedobiologia* **39**, 561-576

Duchesne, L. C. (1994. Defining Canada's old-growth forests – problems and solutions. *Forestry Chronicle* **70**, 739–744.

Eyre, M.D. and Rushton, S.P. (1989) Quantification of conservation criteria using invertebrates. *Journal of Applied Ecology* **26**, 159–171.

Eyre, M.D., Luff, M.L., Rushton, S.P. and Topping C.J. (1989) Ground beetles and weevils (Carabidae and Curculionoidea) as indicators of grassland management practices. *Journal of Applied Entomology* **107**, 508–517.

Freitag, R., Hastings, L., Mercer, W.R. and Smith, A. (1973) Ground beetle populations near a kraft mill. *Canadian Entomologist* **105**, 299–310.

Geiszler, D.R., Gara, R.I., Driver, C.H. et al. (1980) Fires, fungi and beetle influences on a lodgepole pine ecosystem of south-central Oregon. *Oecologia (Berlin)* **46**, 239–243.

Graham, S., Harrison, R.P. Jr and Westall, C.E. Jr (1963) *Aspens: Phoenix Trees of the Great Lakes Region*, University of Michigan Press, Ann Arbor, 272 pp.

Haila, Y., Hanski, I.K., Niemalä, J. et al. (1994) Forestry and boreal fauna: matching management with natural forest dynamics. *Annales Zoologici Fennici* **31**, 187–202.

Hanski, I. and Hammond, P. (1995) Biodiversity in boreal forests. *Trends in Ecology and Evolution* **10**, 5–6.

Harris, L.D. (1984) *The Fragmented Forest: Island Biogeography Theory and the Preservation of Biotic Diversity*, University of Chicago Press, Chicago, 211 pp.

Heliövaara, K. and Väisäinen, R. (1984) Effects of modern forestry on northwestern European forest invertebrates: a synthesis. *Acta Forestalia Fennica* **189**, 1–32.

Holliday, N.J. (1991) Species responses of carabid beetles (Coleoptera: Carabidae) during post-fire regeneration of boreal forest. *Canadian Entomologist* **123**, 1369–1389.

Holliday, N.J. (1992) The carabid fauna (Coleoptera: Carabidae) during postfire regeneration of boreal forest: properties and dynamics of species assemblages. *Canadian Journal of Zoology* **70**, 440–452.

Holling, C.S. (1992) The role of forest insects in structuring the boreal landscape. In *A Systems Analysis of the Boreal Forest* (eds H.H. Shugart Jr, R. Leemans and G.B. Bonan), pp. 170–191, Cambridge University Press, Cambridge.

Huhta, V. (1971) Succession in the spider communities of the forest floor after clear-cutting and prescribed burning. *Annales Zoologici Fennici* **8**, 483–542.

Hunter, M.L. Jr (1993) Natural fire regimes as spatial models for managing boreal forests. *Biological Conservation* **65**, 115–120.

Kaila, L. (1993) A new method for collecting quantitative samples of insects associated with decaying wood or wood fungi. *Entomologica Fennica* **4**, 21–23.

Kaila, L., Martikainen, P., Punttila, P. and Yakovlev, E. (1994) Saproxylic beetles (Coleoptera) on dead birch trunks decayed by different polypore species. *Annales Zoologici Fennici* **31**, 97–107.

Krebs, C.J. (1989) *Ecological Methodology*, Harper & Row, New York, 654 pp.

Kubin, E. and Kemppainen, L. (1991) Effect of clearcutting of boreal spruce forest on air and soil temperature conditions. *Acta Forestalia Fennica* **225**, 1–42.

Kuusipalo, J. and Kangas, J. (1994) Managing biodiversity in a forestry environment. *Conservation Biology* **8**, 450–460.

Langor, D.W., Spence, J.R., Niemalä, J. and Cárcamo, H. (1993) Insect biodiversity studies in the boreal forests of Alberta, Canada. *Metsäntutkimuslaitoksen Tiedonantoja* **42**, 25–31.

LaRoi, G. (1992) Classification and ordination of southern boreal forests from the Hondo-Slave Lake area of central Alberta. *Canadian Journal of Botany* **70**, 614–628.

Lenski, R.E. (1982) The impact of forest cutting on the diversity of ground beetles (Coleoptera: Carabidae) in the southern Appalachians. *Ecological Entomology* **7**, 385–390.

Lindroth, C.H. (1961–1969) The ground-beetles of Canada and Alaska. *Opuscula Entomologica* (Suppl.) **20**, **24**, **29**, **33** and **34**, 1192 pp.

Ludwig, J.A. and Reynolds, J.F. (1988) *Statistical Ecology*, John Wiley & Sons, New York, 337 pp.

Martin, O. (1989. Smaeldere (Coleoptera, Elateridae) fra gammel lövskov i Danmark. *Entomologiske Meddelelser* **57**, 1–107.

Maser, C. (1990) *The Redesigned Forest*, Stoddart, Toronto, 224 pp.

Mattson, W.J. and Addy, N.D. (1975) Phytophagous insects as regulators of forest primary production. *Science* **190**, 515–522.

McIver, J.D., Parsons, G.L. and Moldenke, A.R. (1992) Litter spider succession after clear-cutting in a western coniferous forest. *Canadian Journal of Forest Research* **22**, 984–992.

Moore, I. and Legner, E.F. (1979) *An Illustrated Guide to the Genera of the Staphylinidae of America North of Mexico, Exclusive of the Aleocharinae*, Publication 4093, University of California, Riverside, 332 pp.

Muona, J. Rutanen, I. (1994) The short-term impact of fire on the beetle fauna in a boreal coniferous forest. *Annales Zoologici Fennici* **31**, 109–122.

Navratil, S. and Chapman, P.B. (1991) *Aspen Management for the 21st Century*, Forestry Canada, Northwest Region, Edmonton, Alberta, Cat. No. Fo42-165/199E, 174 pp.

Nield, C.E. (1990) Is it possible to age woodlands on the basis of their carabid beetle diversity? *The Entomologist* **109**, 136–145.

Niemalä, J., Spence, J., Langor, D. *et al.* (1993a) Logging and boreal ground-beetle assemblages on two continents: implications for conservation. In *Perspectives in Insect Conservation* (eds K.J. Gaston, T.R. New and M.J. Samways), pp. 29–50, Intercept, Andover.

Niemalä, J., Langor, D. and Spence, J.R. (1993b) Effects of clear-cut harvesting on boreal ground-beetle assemblages (Coleoptera: Carabidae) in western Canada. *Conservation Biology* **7**, 551–561.

Nilsson, S.G., Arup, U., Baranowski, R. and Ekman, S. (1995) Tree-dependent lichens and beetles as indicators in conservation forests. *Conservation Biology* **9**, 1208–1215.

Pearson, D.L. and Cassola, F. (1992) World-wide species richness patterns of tiger beetles (Coleoptera: Cicindelidae): indicator taxon for biodiversity and conservation studies. *Conservation Biology* **6**, 376–391.

Peterson, E.B. and Peterson, N.M. (1992) *Ecology, Management, and Use of Aspen and Balsam Poplar in the Prairie Provinces*, Spectial Report 1, Forestry Canada, Northern Forestry Centre, Edmonton, Canada, 252 pp.

Petterson, R.B., Ball, J.P., Renhorn, K.-E. *et al.* (1995) Invertebrate communities in boreal forest canopies as influenced by forestry and lichens with implications for passerine birds. *Biological Conservation* **74**, 57–63.

Pratt, L. and Urquart, I. (1994 *The Last Great Forest*, NeWest Press, Edmonton, Canada, 222 pp.

Prendergast, J.R., Quinn, R.M., Lawton, J.H. *et al.* (1993) Rare species, the coincidence of diversity hotspots and conservation strategies. *Nature* **365**, 335–337.

Rowe, J.S. (1972) *Forest Regions of Canada*, Canadian Forest Service Publications, Ottawa, 172 pp.

Royama, T. (1992) *Analytical Population Dynamics*, Chapman & Hall, London, 371 pp.

Samoil, J. (ed.) (1988) *Management and Utilization of Northern Mixed-woods*, Inf. Rep. NOR-X-926, Northern Forestry Centre, Edmonton, Alberta, 163 pp.

Siitonen, J. (1994) Decaying wood and saproxylic Coleoptera in two old spruce forests: a comparison based on two sampling methods. *Annales Zoologici Fennici* **31**, 89–96.

Siitonen, J. and Martikainen, P. (1994) Occurrence of rare and threatened insects living on decaying *Populus tremula*: a comparison between Finnish and Russian Karelia. *Scandinavian Journal of Forest Research* **9**, 185–191.

Simberloff, D.S. (1978) Use of rarefaction and related methods in ecology. In *Biological Data in Water Pollution Assessment: Quantitative and Statistical Analysis* (eds K.L. Dickson, J. Garins Jr and R.J. Livingston), pp. 150–165, American Society for Testing and Materials, STP 652.

Simms, H.P. (1976) The effect of prescribed burning on some physical soil properties of jack pine sites in southeastern Manitoba. *Canadian Journal of Forest Research* **6**, 58–68.

Smetana, A. (1971) *A Revision of the Tribe Quediini of North America North of Mexico (Coleoptera: Staphylinidae)*, Memoirs of the Entomological Society of Canada No. 79, 303 pp.

Smetana, A. (1982) *A Revision of the Subfamily Xantholininae of North America North of Mexico (Coleoptera: Staphylinidae)*, Memoirs of the Entomological Society of Canada No. 120, 389 pp.

Speight, M.C.D. (1989) *Saproxylic Organisms and their Conservation*, Council of Europe, Strasborg, 82 pp.

Spence, J.R. and Niemalä, J. (1994) Sampling carabid assemblages with pitfall traps: the method and the madness. *Canadian Entomologist* **126**, 881–894.

Spence, J.R., Langor, D.W., Niemalä, J. *et al.* (1996) Northern forestry and carabids: the case for concern about about old growth species. *Annales Zoologici Fennici* **33**, 173–184.

Stelfox, J.B. (ed.) (1995) *Relationships between Stand Age, Stand Structure and Biodiversity in Aspen Mixed-wood Forests in Alberta*, jointly published by the Alberta Environmental Centre (AECV95-R1), Vegreville, AB, and Canadian Forest Service (Project No. 0001A), Edmonton, AB, 308 pp.

Stork, N.E. (ed.) (1990) *The Role of Ground Beetles in Ecological and Environmental Studies*, Intercept, Andover, 424 pp.

Syrjänen, K., Kalliola, R., Puolasmaa A. and Mattsson, J. (1994) Landscape structure and forest dynamics in subcontinental Russian European taiga. *Annales Zoologici Fennici* **31**, 19–34.

Thorpe, P.T. (1992) Patterns of diversity in the boreal forest. In *The Ecology and Silviculture of Mixed-species Forests* (ed. M.J. Kelty), pp. 65–79, Kluwer Academic Publ., Dordrecht, The Netherlands.

Van Cleve, K., Chapin, F.S. III, Durness, C.T. and Viereck, L.A. (1991) Element cycling in taiga forests: state-factor control. *Bioscience* **41**, 78–88.

Veblen, T.T., Hadley, K.S., Reid, M.S. and Rebertus, A.J. (1989) Blowdown and stand development in a Colorado subalpine forest. *Canadian Journal of Forest Research* **19**, 1218–1225.

Veblen, T.T., Hadley, K.S., Reid, M.S. and Rebertus, A.J. (1991) The response of subalpine forests to spruce beetle outbreak in Colorado. *Ecology* **72**, 213–231.

Vlug, H. and Borden, J.H. (1973) Soil Acari and Collembola populations affected by logging and slash burning in a coastal British Columbia coniferous forest. *Environmental Entomology* **2**, 1019–1023.

Volney, W.J.A. (1985) Comparative population biologies of North American spruce budworms. In *Recent Advances in Spruce Budworms Research. Proceedings of CANUSA Spruce Budworm Research Symposium, Bangor, ME* (eds C.J. Saunders, R.W. Stark, E.J. Mullins and J. Murphy), pp. 71–84, Canadian Forest Service, Ottawa.

Warren, M.S. and Key, R.S. (1991) Woodlands: past, present and potential for insects. In *The Conservation of Insects and Their Habitats, 15th Symposium of the Royal Entomological Society of London, 14–15 September, 1989* (eds N.M. Collins and J.A. Thomas), pp. 155–211, Academic Press, Toronto.

Weber, J. (1994) Latitude, physiological time, and the spruce budworm. MSc thesis, Department of Entomology, University of Alberta, Edmonton, Canada, 84 pp.

Wein, R.W. and MacLean, D.A. (1983) An overview of fire in northern ecosystems. In *The Role of Fire in Northern Circumpolar Ecosystems* (eds. R.W. Wein and D.A. MacLean), pp. 1–18, Wiley, New York.

Wheeler, Q.D. (1995) Systematics and biodiversity. *Bioscience Supplement, Science and Biodiversity Policy*, S21–S28.

Wilson, E.O. (1992) *The Diversity of Life*, W.W. Norton and Company, New York, 424 pp.

Wood, D.M., Dang, P.T. and Ellis, R.A. (1979) The mosquitoes of Canada (Diptera: Culicidae). *The Insects and Arachnids of Canada, Part 6*, Agriculture Canada Publication No. 1686, 390 pp.

Zackrisson, O. (1977) Influence of forest fires on the North Swedish boreal forest. *Oikos* **29**, 22–32.

AN OVERVIEW OF INVERTEBRATE RESPONSES TO FOREST FRAGMENTATION

17

Raphael K. Didham

17.1 INTRODUCTION: FOREST FRAGMENTATION IN OVERVIEW

During the last three decades extensive data have been compiled (e.g. Burgess and Sharpe, 1981; Harris, 1984; Turner *et al.*, 1990; Whitmore and Sayer, 1992; Skole and Tucker, 1993; Zipperer, 1993; Vitousek, 1994) to show that the earth's forest ecosystems have been greatly impacted by fragmentation and isolation. But exactly what impact forest fragmentation has on animal assemblages, particularly invertebrates, is poorly documented empirically, and theoretically not well understood.

Island biogeographic theory (MacArthur and Wilson, 1967; Simberloff, 1974; Diamond, 1975; Diamond and May, 1976) fuelled the assumption that the effects of forest fragmentation are driven largely by fragment area and isolation. However, this model fails to recognize that the penetration of adverse biotic and abiotic edge effects from outside the forest fragment increases the proportion of edge habitat with decreasing area. Many other factors vary with area as well. The principal processes shaping biotic communities and abiotic conditions within forest fragments (Laurance and Yensen, 1991; Saunders *et al.*, 1991; Taylor *et al.*, 1993; Andrén, 1994; Murcia, 1995) are:

- area effects;
- edge effects;
- the shape of the fragment;
- the degree of spatio-temporal isolation;
- the degree of habitat connectivity in the landscape.

These factors apply generally to all communities in all landscapes, though local responses will be taxon- and site-specific and will be tempered by history, scale and chance.

The best synthesis of fragmentation processes in a single model is Laurance's core-area model (Laurance, 1991; Laurance and Yensen, 1991) which incorporates an area component, an edge penetration distance and a shape index that is independent of fragment area. However, the core-area model does not include any measure of the degree of spatio-temporal isolation or forest connectivity (for fragments that are joined by narrow corridors and thus not strictly isolated). Combining these simple concepts, based on the core-area model, allows the development of a five variable-attribute state which defines the physical characteristics of a forest fragment as habitat for forest organisms (Figure 17.1). Naturally, different organisms will be more affected by some fragment attributes than others. For example, where metapopulation dynamics are important, the degree of fragment isolation may be a key factor in the long-term survival of a population (Hanski *et al.*, 1995). For a widely dispersing species, fragment shape may determine the likelihood of fragment colonization (or recolonization) (e.g. Game, 1980). In contrast, an organism that has a large area requirement will be most affected by the size of the core-area. Potentially, a far more detailed characterization of the physical characteristics of different forest fragments and the surrounding matrix (e.g. Stouffer

Forests and Insects. Edited by A.D. Watt, N.E. Stork and M.D. Hunter. Published in 1997 by Chapman & Hall, London. ISBN 0 412 79110 2.

and Bierregaard, 1995) may lead to better ecological and conservation management decisions (Laurance *et al.*, 1997).

This chapter makes distinctions between the confounding influences of fragment area and edge effects by using the simple protocol of comparing invertebrate samples collected at known distances from the forest edge along edge-to-interior gradients in isolated versus non-isolated forests. As an experimental system the leaf-litter invertebrate fauna of an experimentally fragmented tropical forest landscape in Central Amazonia was sampled. The edge-gradient samples are used to highlight how a greater mechanistic understanding of fragmentation processes can be achieved by controlling for two very important variables – area and edge effects – while holding the shape index, degree of connectivity, time since isolation and distance of isolation relatively constant. The layout of forest fragments and sampling constraints meant that scale-dependent isolation processes other than area and distance from forest edge were not explicitly controlled (Doak *et al.*, 1992), but they were not markedly variable at relevant ecological scales. These latter factors almost certainly work synergistically and may be difficult to discriminate, even in properly replicated factorial designs. Explicit conservation implications are drawn with respect to edge penetration distance and habitat area requirements for invertebrates.

17.2 INVERTEBRATE RESPONSES TO FRAGMENTATION

17.2.1 INVERTEBRATES, ISLANDS AND AREA

Whether area is the subject of interest, or merely a factor to control for in the analysis of biogeographic patterns, it is one of the most important variables to take into account in the study of forest fragmentation. Fragmentation, by definition, results in a reduction in habitat area and a concomitant decrease in living space for plants and animals, with an almost inevitable reduction in species richness in remaining habitat fragments. Invertebrate species–area relationships are well documented; for example, Dean and Bond (1990) and others (Goldstein, 1975; Vepsäläinen and Pisarski, 1982; Boomsma *et al.*, 1987) found that ant diversity on islands decreased significantly with decreasing island area. Similarly, Toft and Schoener (1983) found that the species richness of orb-web spiders decreased significantly on small islands in the Bahamas. Jaenike (1978) found that population densities of *Drosophila* spp. may also decrease markedly (and non-linearly) with island area; temporal fluctuations in abundance thus make small island populations particularly vulnerable to extinction (Jaenike, 1978). Particularly clear examples of the importance of island area as a factor determining invertebrate species richness are provided by the island defaunation–recolonization experiments of Simberloff and Wilson (1970) and Rey and Strong (1983). The invertebrate fauna of mangrove islands (Simberloff and Wilson, 1970) and *Spartina* salt marsh islands (Rey and Strong, 1983) of varying area (and hence differing species richness) returned to their respective pre-treatment species richness levels within one year of total defaunation.

Unlike true islands, larger habitat islands do not always support richer faunas because of the nature

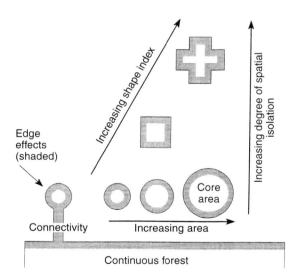

Figure 17.1 Representation of five principal fragmentation processes operating on invertebrates in forest fragments. (1) Area effects, measured as either total fragment area or interior core area (i.e. area unaffected by proximity to forest edge); (2) edge effects; (3) shape of fragment; (4) degree of spatio-temporal isolation (temporal isolation not shown); (5) degree of habitat connectivity in landscape.

of the surrounding landscape matrix (e.g. Janzen, 1983; Webb *et al.*, 1984; Bauer, 1989; Brown and Hutchings, 1997), producing negative species–area relationships in some cases. Further examples are provided below, and this situation may be more common than previously suspected.

There are many cases where factors other than area come into effect in both true island and habitat fragment systems. Habitat availability and suitability are important for the maintenance of invertebrate populations in small fragments (e.g. trapdoor spiders in Australia: Main, 1987), a generality highlighted by Zimmerman and Bierregaard (1986). Population dynamics, too, give different insights into the effects of area on invertebrate communities. Bach (1988a) found that three closely related species of chrysomelid beetles feeding on the same host plant had different population densities in habitat patches of different sizes: *Diabrotica virgifera* was most abundant on plants in large host plant patches, *D. undecimpunctata howardi* was most abundant on intermediate-sized patches and first generation *Acalymma vittatum* adults were most abundant on plants in small patches. Some form of spatio-temporal interaction between species seems apparent, as indicated in a study by Sowig (1989) where long-tongued bumblebees were forced to inhabit small habitat patches as a result of the high abundance of nectar-robbing, short-tongued bees in large patches. For highly dispersive invertebrates, distance to the nearest source population may be a better determinant of population size than habitat area, habitat heterogeneity or plant species richness (e.g. the Bay Checkerspot Butterfly, *Euphydryas editha bayensis*: Launer and Murphy, 1994). For these species, small patches have higher emigration rates and act as net exporters of migrants, while large patches are net importers (McCauley, 1991).

17.2.2 INVERTEBRATE ABUNDANCE AND SPECIES RICHNESS IN FRAGMENTED FOREST ECOSYSTEMS

Forest fragments are complex habitat islands. As in non-forest systems, responses to forest fragmentation are varied and contrasting (Robinson *et al.*, 1992). Some butterfly assemblages show a significant decrease in species richness in small forest fragments (Shreeve and Mason, 1980; Rodrigues *et al.*, 1993; Daily and Ehrlich, 1995; see also Thomas *et al.*, 1992), whereas others show the reverse trend due to invasion of species from outside the fragments (Bierregaard *et al.*, 1992; Brown and Hutchings, 1997). Pollinating insects in general are greatly affected by forest fragmentation. Aizen and Feinsinger (1994a) found that the abundance and species richness of native flower pollinators in Argentina was significantly lower in small forest fragments (< 1 ha) than in either large fragments (> 2.2 ha) or continuous forest. In the small fragments there was a complementary increase in the abundance of a generalist exotic pollinator, *Apis mellifera*. Powell and Powell (1987) found similar patterns for euglossine bees in forest fragments, but this may have been a transient response to disturbance as differences between small and large fragments largely disappeared after five years (Becker *et al.*, 1991). Although not studying a forest habitat, Jennersten (1988) found a decrease in the abundance and diversity of insects visiting flowers in fragmented versus continuous meadows.

Termites show a significant and positive species–area relationship in forest fragments (Souza, 1989; Souza and Brown, 1994). Small forest fragments had lower termite species richness, more rare species and more desiccation-resistant litter-feeding species, in contrast to the predominance of soft-bodied soil-feeders in continuous forest. Habitat resource partitioning was also more uneven in small forest fragments (Souza and Brown, 1994). Ants are another invertebrate group affected by deforestation (e.g. Verhaagh, 1991), though no good evidence exists for a strong decline in ant species richness in small forest fragments. Army ants disappeared from some small (1 and 10 ha) fragments in Central Amazonia (Lovejoy *et al.*, 1986; Harper, 1989), as did the leaf-cutter ant *Atta cephalotes* (Vasconcelos, 1988). A second, more common species of leaf-cutter ant, *A. s. sexdens*, showed no change in mean colony density in small forest fragments (Vasconcelos, 1988).

A handful of further examples show invertebrate responses to forest fragmentation to be equally context-sensitive. Ås (1993) found no species–area effect operating on beetle diversity in large (> 120 ha) and small (< 20 ha) patches of deciduous forest within a matrix of coniferous forest. Martins (1989) showed that the abundance and species richness of *Drosophila* increased in small (1 and 10 ha) forest fragments due to an influx of disturbed-habitat species. Finally, Margules *et al.* (1994) reiterated the conclusion of Robinson *et al.* (1992) that communities, species and populations respond differently, and at different scales, to forest fragmentation; they found no change in the abundance of the scorpion *Cercophonius squama* with decreasing fragment area, whereas amphipods (Tallitridae) were found to decrease markedly in abundance, particularly in small fragments.

17.2.3 A CASE STUDY FROM CENTRAL AMAZONIA

At the Biological Dynamics of Forest Fragments Project (BDFFP) 80 km north of Manaus, Central Amazonia, Brazil (Bierregaard *et al.*, 1992), I studied the effects of forest fragmentation on leaf-litter invertebrates during the January to May rainy season 1994. Leaf-litter invertebrate samples were collected using the Winkler method (Besuchet *et al.*, 1987; Didham, 1997). Identical edge-to-interior transects were sampled at seven distances from the forest edge (0, 13, 26, 52, 105, 210 and 420 m) in two isolated 100 ha fragments, two non-isolated continuous forest edges and at an identical series of distances at two sites deep within continuous forest (> 10 km from the nearest edge). Sites were independent and at each sampling point twenty leaf-litter samples of 1 m² were collected by rapidly scraping the leaf litter down to the compact soil layer and then sieving to remove large debris. The fine litter material containing invertebrates was placed in Winkler sacks and hung for three days to extract invertebrates. Additionally, the same number of samples were collected at 105 m into two 10 ha fragments and at 52 m into two 1 ha fragments, giving a total of 46 sampling sites and 920 m² of leaf litter sampled. For further details of the sampling protocol see Didham (1997).

Site similarity analysis was performed on a preliminary sample of 295 species of Staphylinidae, Carabidae and Scarabaeidae beetles; total abundance 2111 individuals. Percentage similarity was calculated using the Normalized Expected Species Shared (NESS) index (Grassle and Smith, 1976; Wolda, 1983), at $m = 50$.

There is evidence for a negative effect of area on the density and species richness of beetles and ants in forest fragments of 1, 10 and 100 ha (Figures 17.2 and 17.3, respectively). However, while beetle species richness is highest in 1 ha fragments, the species composition similarity to continuous forest is lowest, indicating the influx of numerous disturbed-area species (Figure 17.4). At the level of sampling conducted, the average similarity of beetle species composition amongst the 14 continuous forest sites is 43%. Beetle species composition in 1 ha forest fragments is, on average, only 7% similar to continuous forest (Figure 17.4).

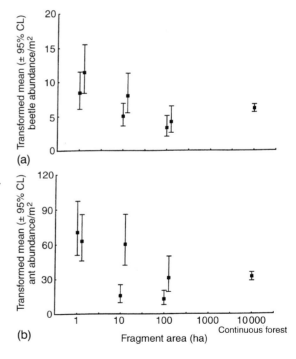

Figure 17.2 Relationship between invertebrate density (abundance/m²) and fragment area for (a) beetles and (b) ants sampled from leaf litter at centres of 1, 10 and 100 ha forest fragments and in continuous forest in Central Amazonia. Forest fragments, $n = 20$ m²; continuous forest, $n = 280$ m². Means are back-transformed.

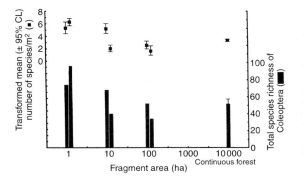

Figure 17.3 Relationship between species richness and area for leaf-litter beetles in forest fragments of 1, 10 and 100 ha and in continuous forest in Central Amazonia. Species richness per m² is mean of $n = 20$ m² for forest fragments and $n = 280$ m² for continuous forest. Means are back-transformed. Total species richness in continuous forest is mean ($\pm$ 95% CL) of 14 sites (each site, $n = 20$ m²).

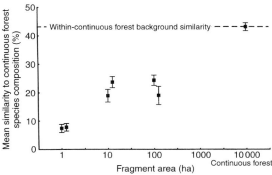

Figure 17.4 Percentage similarity of leaf-litter beetle species composition between continuous forest and forest fragments of 1, 10 and 100 ha in Central Amazonia. Figures are based on analysis of 295 species of Staphylinidae, Carabidae and Scarabaeidae beetles; total abundance 2111 individuals. Percentage similarity is mean ($\pm$ 1 SE) Normalized Expected Species Shared (NESS) index for each fragment site compared with each of 14 continuous forest sites. Within-continuous forest similarity is mean ($\pm$ 1 SE) similarity of species composition between all pairwise combinations of 14 continuous forest sites.

17.2.4 FUNCTIONAL RESPONSES TO FRAGMENTATION

Didham *et al.* (1996) consider a functional approach to the study of invertebrates in fragmented forests to be a useful one in analysing whole ecosystem dynamics and the importance of invertebrates in forest fragments. The functional role of an insect refers to its role in ecosystem processes, such as decomposition and nutrient cycling. At least four major functional groups can be recognized: pollinators, herbivores, predators (including parasitoids) and decomposers. Each group carries out essential ecosystem services without which forests would be profoundly altered, and members of each are known to be adversely affected by forest fragmentation. The benefit of applying a functional framework to such findings is that changes in ecosystem-level rate processes give a measure of the importance of changes in the abundance and diversity of invertebrates in forest fragments. This entails measuring both empirical species-abundance patterns and associated rate functions (e.g. decomposition). This has rarely been attempted in the same system.

A reduction in insect pollinator abundance and species richness may result in decreased flower pollination, lower seed set and fruit production and hence in lower reproductive success and (potentially) reduced genetic variability of progeny in small forest fragments (Chan, 1980; Spears, 1987; Jennersten, 1988; Aizen and Feinsinger, 1994b). Altered insect seed-predator activity may have impacts on plant reproductive success and plant community dynamics (Janzen, 1975; Louda, 1982; Burkey, 1993). Other forms of insect herbivory have not been studied in fragmented forest systems, although Nielsen and Ejlersen (1977) found that some herbivore species preferentially fed on leaves at the edge of a 3 ha pure beech forest fragment, resulting in greater levels of herbivory than in the forest interior, while other species fed more on leaves in the interior of the fragment.

Natural enemies (predators and parasitoids), living at relatively low population densities, are thought to be more susceptible to habitat fragmentation than their prey (Kruess and Tscharntke, 1994). In a non-forest system, Kareiva (1987) found that the aphid *Uroleucon nigrotuberculatum* (Hemiptera: Aphididae) was released from predation pressure on small, isolated habitat fragments because fragmentation interfered with the non-random searching behaviour of the predator (*Coccinella septempunctata*, Coleoptera: Coccinellidae). Similarly, Kruess and Tscharntke (1994) found that the rate of parasitism of herbivores was lower on small, isolated habitat fragments. In a study of citrus tree invertebrates by DeBach *et al.* (1976), the removal of just two parasitoid species resulted in the defoliation and death of the trees by herbivores (LaSalle and Gauld, 1991). Münster-Swendsen (1980) found that the rate of parasitism of *Epinotia tedella* (Lepidoptera: Tortricidae) by *Pimplopterus dubius* (Hymenoptera: Ichneumonidae) increased at the forest edge, while parasitism by *Apanteles tedellae* (Hymenoptera: Braconidae) on the same host decreased at the forest edge. Thus, host–parasitoid interactions varied markedly with forest fragmentation.

Only Klein (1989) has measured a change in the rate of decomposition with a decrease in the diversity of decomposer organisms following forest fragmentation. A two-fold decrease in species richness and a seven-fold decrease in average population densities of continuous-forest dung beetles (Scarabaeinae) resulted in a dramatic decrease in the rate of dung decomposition in 1 ha and 10 ha forest fragments (Klein, 1989; Didham *et al.*, 1996). Loss of biodiversity is widely assumed to affect ecosystem processes such as decomposition, with accumulating empirical support (Springett, 1976; Klein, 1989; Hobbs, 1992; Naeem *et al.*, 1994, 1995).

Clearly, fragmentation-induced changes in the abundance and diversity of invertebrates can result in pervasive changes to ecosystem functioning. Such effects have almost always been attributed to fragment isolation and reduction in area, and ignore the concomitant decrease in distance from fragment edge to centre with decreasing fragment size.

17.3 EDGE EFFECTS: A BETTER MECHANISTIC UNDERSTANDING

The penetration of external landscape influences into forest fragments (e.g. Murcia, 1995; Pickett and Cadenasso, 1995) include hotter, drier and windier conditions at the edge than the forest interior, with a higher light intensity and modified plant composition and habitat structure (Ranney *et al.*, 1981; Kapos, 1989; Williams-Linera, 1990a,b, 1993; Chen *et al.*, 1992, 1995; Kapos *et al.*, 1993, 1997; Matlack, 1993; Fraver, 1994; Malcolm, 1994; Young and Mitchell, 1994; Camargo and Kapos, 1995). The penetration of physical changes in forest structure and microclimate extends for up to 80 m into the edge, and sometimes considerably further (Laurance *et al.*, 1997). At forest edges, an increase in the abundance and diversity of insects was documented early this century (Leopold, 1933) and at one stage the subdivision of forests to create edge habitat was considered to be a useful management tool for promoting species diversity (Mattiske, 1987)! An increase in insect abundance and diversity at the forest edge is almost certainly due to the invasion of generalist species from disturbed habitats outside the forest fragment. Many of these species may be tree-fall gap specialists and can be extremely common at edges.

17.3.1 EDGE EFFECTS IN REVIEW

In Central Amazonia, Fowler *et al.* (1993) found that total invertebrate abundance and biomass were significantly higher at the edge of a 10 ha forest fragment than near the fragment centre (160 m from the edge). Both biomass and abundance were also more variable at the forest edge under varying climatic conditions. Fowler *et al.* (1993) suggested that the average body size of invertebrates was greater at the forest edge, consistent with a possible detrimental effect of edge microclimate on small-bodied animals. In a more comprehensive study at the same site, Malcolm (1997) found that both overstorey and understorey insect biomass and abundance increased markedly toward the forest edge, but that there was a proportional shift in insect biomass to the ground layer near the edge, correlated with a reduction in canopy height and an increase in low secondary vegetation (Malcolm, 1997). Similarly, Winnett-Murray (1986, cited in Malcolm, 1997) found elevated insect biomass at forest edges in Costa Rica, and Helle and Muona (1985) found elevated insect abundance at forest edges in Finland.

Other invertebrate community studies have shown an increase in overall invertebrate species richness, as well as abundance, at forest edges (e.g. Báldi and Kisbenedek, 1994). Most diversity analyses, however, have concentrated on individual invertebrate orders or families. Results at this level of resolution are taxon-specific, at times showing no edge pattern and at other times a decrease in abundance or species richness at the forest edge. For example, Gunnarson (1988) found that there was no significant difference in overall spider abundance between spruce forest edge sites and sites 50 m into the forest patch. However, when spiders were classified into small (< 2.5 mm) and large (> 2.5 mm) individuals, trees at the forest edge supported a significantly greater abundance of large spiders, while the reverse was true for small spiders. At the guild level, there were significantly fewer orb-web spider species at the forest edge (Gunnarson, 1988). Ozanne et al. (1997) found similar negative edge responses for canopy invertebrates in spruce forests: most invertebrate taxa had significantly lower abundances at the forest edge. Species richness of spiders (notably woodland specialists) was 57% lower at the forest edge and the edge fauna included more generalist predators and fewer specialist predators than the forest interior (Ozanne et al., 1997). In general, however, such results are the exception to the rule of higher invertebrate diversity at the forest edge (Duelli et al., 1990; Buse and Good, 1993; Báldi and Kisbenedek, 1994).

In a comprehensive study of the three-dimensional nature of forest edges, Toda (1992) analysed the microdistributional abundance patterns of a *Drosophila* community at temperate forest edges in Japan. In addition to strong vertical stratification in species-abundance patterns, Toda found a significant increase in the abundance and diversity of *Drosophila* at the forest edge, due to an influx of grassland species and other invasive species associated with human disturbance. The edge-zone *Drosophila* community had more in common with the canopy fauna than the understorey fauna (Toda, 1992), supporting the conclusions of Nielsen and Ejlersen (1977), Münster-Swendsen (1980) and Malcolm (1997) that invertebrate communities near the ground at the forest edge have more in common with invertebrate communities of the canopy of interior habitats than they have with the ground layer of interior habitats.

The species richness of Staphylinidae (Buse and Good, 1993), Carabidae (Duelli et al., 1990) and an entire beetle assemblage (Báldi and Kisbenedek, 1994) were found to be significantly higher at the forest edge. Numerous studies have investigated other aspects of invertebrate (particularly beetle) responses to landscape boundaries, including edge permeability, dispersal and gene flow (Mader, 1984; Stamps et al., 1987; Liebherr, 1988; Burel, 1989; den Boer, 1990; Dempster, 1991).

At the species level, the wood-boring beetle, *Megacyllene robiniae* (Cerambycidae), was found to be more abundant at forest edges in the USA, resulting in increased attack rates on black locust trees (*Robinia pseudoacacia*) located at forest edges and in the open (McCann and Harman, 1990). The latter example is similar to studies showing the importance of edge effects and forest fragmentation in increasing forest susceptibility to Lepidoptera outbreaks (Bellinger et al., 1989; Roland, 1993; Chapter 7, this volume). Forest edges also support a greater diversity of enchytraeid worms (Heck et al., 1989) and a greater abundance of some mite species, including chiggers, *Eutrombicula alfreddugesi* (Acari: Trombiculidae), an important pest of humans (Clopton and Gold, 1993).

There are also many non-forest examples of increased invertebrate diversity at habitat boundaries. Often such studies are more convincing than descriptive forest examples because they are conducted in well controlled experimental set-ups such as crop-patches or in low-diversity shrub communities. Notable amongst these is the heathland invertebrate study of Webb and co-workers (Hopkins and Webb, 1984; Webb et al., 1984; Webb and Hopkins, 1984; Webb, 1989). An increase in total beetle species richness in small heathland fragments was found to be the result of high beetle species richness at edges and the increase in such edge habitat in fragments (Webb and Hopkins, 1984). The observed edge effect was due to an invasion of beetle species from non-heathland areas. Webb *et al.* (1984) showed that heathland edges surrounded by structurally complex vegetation (e.g. forest, with many insect species) had an increased insect species richness, while edges adjacent to less structurally complex vegetation showed few changes in insect species richness. However, true heathland specialists, with poor powers of dispersal, were rare or absent in small fragments (Hopkins and Webb, 1984). Bauer (1989) found an almost identical response for beetle communities on limestone habitat islands in England, and Webb (1976) found a similar increase in invertebrate diversity at the edge of a *Spartina* salt marsh, perhaps attributable to increased plant diversity.

Cusson *et al.* (1990) found that the tuber flea beetle (*Epitrix tuberis*, Chrysomelidae) was more abundant at the edge of potato (*Solanum tuberosum*) patches. Cappuccino and Root (1992) highlighted the significance of host patch edges in the colonization and development of *Corythucha marmorata* (Hemiptera: Tingidae) on goldenrod (*Solidago altissima*): adult *C. marmorata* were more abundant at habitat edges, females laid a greater number of eggs at edges than at interior patch sites and egg development on edge plants was significantly faster. Behavioural studies of herbivorous chrysomelid beetles on Cucurbitaceae plants (Bach, 1988b) showed that a significantly lower proportion of beetles left host-plant patches with an edge border composed of non-host plants, suggesting that such a border acted as a reflective boundary, retaining insects within the patch.

Drawing these examples and arguments together, invertebrate abundance and diversity, like other edge variables, may (1) be positively or negatively correlated with distance from the forest edge, (2) genuinely show no response to edge effects, or (3) be undetectable due to an interaction between two or more confounding variables that cancel one another out (Murcia, 1995). Despite dependence on site and taxon, it appears that a common response in invertebrate communities is for there to be an increase in the abundance and diversity of invertebrates at the forest edge. This is the result of an *in situ* increase in the abundance of many disturbed-habitat species and an influx of species from other human-modified habitats, rather than an increased abundance of interior-forest species. On the contrary, many interior-forest invertebrates are edge-avoiders, as opposed to edge-insensitive or edge-seeking species.

17.3.2 EDGE EFFECTS ON LEAF-LITTER INSECTS IN CENTRAL AMAZONIA

In Central Amazonia, the increase in density of leaf-litter beetles and ants with decreasing forest fragment area can be most parsimoniously explained by a monotonic edge effect model (Figure 17.5). This simple edge function is typical of those fitted to all manner of edge-responsive variables. There is a growing awareness, however, that edge functions are not necessarily as simple as this and may be both additive, where multiple edge effects interact (Malcolm, 1994), and complex (Camargo and Kapos, 1995; Murcia, 1995; Didham, 1997; Laurance *et al.*, 1997). Analysis of edge-to-interior gradient transects in Central Amazonia show this to be the case: edge functions are variable, but show distinct bimodal fluctuations in the density of beetles and ants (Figure 17.6) and in the species richness of beetles (Figure 17.7) with increasing distance from the forest edge. The bimodal pattern is consistent across different sites and different taxa. Mechanistic explanations for the bimodal pattern are not easily definable in terms of simple correlations with environmental variables (Didham, 1997). For beetles, at least, the bimodal density pattern is derived from two separate, overlapping faunas – one at the forest edge (dominated by Scolytidae, Pselaphidae and Scydmaenidae) and the other in the interior (dominated by Staphylinidae, Carabidae and Scarabaeidae). The mid-distance density peak appears to be an 'ecotone-type' interface between the edge zone and the interior zone (Didham, 1997).

17.4 DISCRIMINATING FRAGMENT AREA AND EDGE EFFECTS

While both area effects and edge effects are undoubtedly of importance in determining inverte-

NESS species similarity between beetle assemblages in continuous forest and those in forest fragments of differing area and distance from forest edge (Figure 17.8) indicates that there are both strong edge effects and strong area effects influencing beetle species composition in forest fragments in Central Amazonia. If one considers the slope of the faunal similarity surface at any given distance from the forest edge (Figure 17.8), there is a marked decline in faunal similarity with decreasing fragment area; that is, area affects beetle species composition independently of edge effects. Equally, distance from forest edge affects beetle species composition independently of fragment area. This is a unique result amongst fragmentation studies. It is interesting to note that edge and area effects seem to contribute almost equally to changes in beetle species composition in small forest fragments (Figure 17.8). In studies that only sample the centres of different sized forest fragments as a 'total' fragmentation effect (section 17.2.3), the relative influences of edge and area are clearly confounded (Figure 17.4).

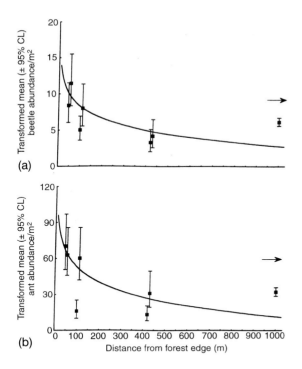

Figure 17.5 Mean densities of (a) beetles and (b) ants at centres of 1, 10 and 100 ha forest fragments and in continuous forest in Central Amazonia, plotted with respect to distance from forest edge at which samples were taken. Simple monotonic edge effect model explains observed increase in invertebrate abundance in small forest fragments. For clarity, continuous forest values are plotted here at 1000 m from edge (sample collected at 10 000 m from edge). For forest fragments, $n = 20$ m^2; for continuous forest, $n = 280$ m^2. Means are back-transformed.

brate community structure in forest fragments, the relative importance of the two factors has not been investigated before: many 'forest fragmentation' studies have confounded different fragmentation processes and used 'area' as the surrogate explanation, while many 'edge effect' studies have ignored area and other variables. This section provides a preliminary analysis that distinguishes between distance effects and fragment area effects for the leaf-litter beetle assemblage discussed in section 17.2.3 – 295 species of Staphylinidae, Carabidae and Scarabaeidae, total abundance 2111 individuals.

Two-Way Indicator Species (Twinspan: Hill, 1979) multivariate analysis of the preliminary beetle assemblage described above, excluding 134 'singleton' species, indicates a strong dichotomy in species composition between isolated and non-isolated sites (Figure 17.9). Within the non-isolated site grouping, continuous forest control sites separate strongly from continuous forest edge sites. Within the isolated forest fragment grouping, there is an indistinct ordering of edge to interior sites, but no apparent separation based on fragment area when fragments are between 1 and 100 ha (Figure 17.9). Thus, isolation of fragments, irrespective of fragment area, seems to be the main determinant of beetle species composition. This is an indication that even forest fragments of 100 ha are too small to preserve an intact continuous forest beetle fauna. Indicator species for non-isolated sites (Figure 17.9) support this conclusion, as all six species decline markedly in abundance from continuous forest sites to forest fragments. Notably, for the data set analysed, these six species are the dominant (most abundant) species in continuous forest, representing 30% of total beetle abundance, yet all

312 *Overview of invertebrates and forest fragmentation*

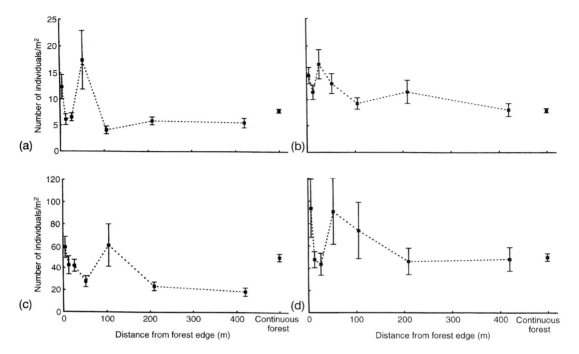

Figure 17.6 Edge-to-interior gradient transects for leaf-litter invertebrates in Central Amazonia (mean ± 1 SE abundance/m^2). (a) Beetle density at 100 ha site 1; (b) beetle density at continuous forest edge transect 1; (c) ant density at 100 ha site 2; (d) ant density at 100 ha site 1; showing the complex, bimodal response curve evident across different sites and different taxa. (Reproduced from Didham, 1997.)

are rare or locally extinct in forest fragments of 1–100 ha (e.g. Figures 17.10 and 17.11). While both edge and area effects are important in the decline in abundance of these species, the relative importance of the two factors varies markedly between different species. For example, at one extreme *Coproporus* sp. 0002 (Staphylinidae: Tachyporinae) is predominantly affected by edge effects (Figure 17.10), while *Tachys* sp. 0268 and sp. 0269 (Carabidae: Bembidiini) are apparently entirely unaffected by edge effects, yet are dramatically affected by a reduction in fragment area (Figure 17.11), such that they may never be encountered in forest fragments smaller than 1000 ha. The minimum forest fragment size necessary to support these species is unknown, as is the mechanistic basis for their local demise.

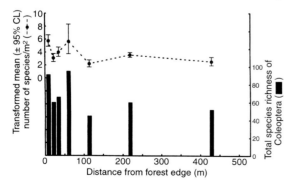

Figure 17.7 Example of a bimodal edge effect pattern for leaf-litter beetle species richness at 100 ha site 1 in Central Amazonia.

17.5 CONSERVATION IMPLICATIONS

Discriminating edge, area and other effects of forest fragmentation provides a greater degree of predictive power in conservation planning. Guidelines for reserve preservation may be more clearly outlined in terms of the individual components of

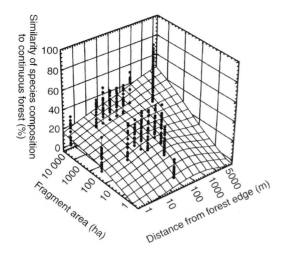

Figure 17.8 Three-dimensional least-squares surface showing observed trend in percentage similarity of beetle species composition between continuous forest and forest fragments of differing area and distance from forest edge. Continuous forest is enumerated as 10 000 ha and 5000 m from the edge. Percentage similarity is Normalized Expected Species Shared (NESS) index for each fragment site compared with each of 14 continuous forest sites. Within-continuous forest species similarity is similarity of species composition between all pairwise combinations of the 14 continuous forest sites. Data points are not independent, hence the fit is illustrative only and the P-value is not meaningful. Data are based on analysis of 295 species of Staphylinidae, Carabidae and Scarabaeidae beetles; total abundance 2111 individuals. With area held constant, effect of distance from edge on beetle species composition can be determined independently of fragment area; equally, with distance from edge held constant, effect of fragment area can be judged independently of edge effects.

fragmentation. As a simple example, in existing fragmented landscapes in which remaining forested area is fixed, the significance of management decisions to buffer edge effects may be more accurately assessed given knowledge of how edge effects operate both independently and in conjunction with area. However, the cost of management strategies must be weighed against the perceived benefits. Extrapolation from data presented here (e.g. Figure 17.8) suggests that if management plans, such as planting a buffer strip along the forest edge, reduce edge penetration distance by even 30–50 m, species similarity with continuous forest can double. In other words, a significantly greater percentage of the original fauna is preserved. Very small (1 ha) fragments will benefit proportionately more from the amelioration of edge effects (provided that the fauna has not been irretrievably lost, in which case even simple transplants of leaf litter, soil or captured biota may prove cost-effective).

Invasion of forest edges by disturbed-habitat species and the hyper-abundance of edge communities is mixed fortune for forest fragments. Edge invertebrates may displace some interior forest species by competition or predation, but equally, a high abundance and biomass of invertebrates provides an increased food supply for many vertebrate and invertebrate predators (Malcolm, 1997). This may be significant in the hyperabundance of many insectivorous mammal and bird species (particularly near the ground) at edges and in small forest fragments (Malcolm, 1988, 1997; Laurance et al., 1997; Terborgh, 1997).

Abundance and species richness patterns for beetles and ants in Central Amazonia indicate an edge penetration distance of approximately 100 m (Didham, 1997), depending on site and taxon. However, similarity of beetle species composition between forest fragments and continuous forest does not approach continuous forest background levels for at least 200–400 m into the forest, when area effects are factored out. However, over-rash use of the latter figure should be guarded against, as the data-set has not been fully analysed. A 200 m edge penetration distance for modification of invertebrate species composition in Central Amazonia is a firm but conservative estimate.

Area exerts a considerable influence on invertebrate population density, species richness and species composition. What is unexpected is that an isolated area as large as 100 ha is faunistically distinct from undisturbed continuous forest due to the local extinction of some dominant species, and an unknown number of rare species. Extrapolating beyond the range of fragment sizes sampled to determine the minimum area needed to maintain an intact terrestrial invertebrate assemblage is fraught with difficulties, but such an area is probably

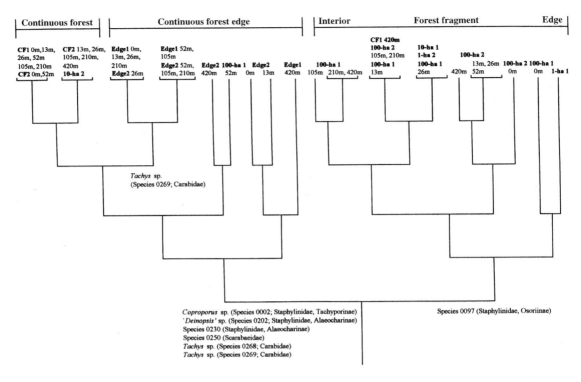

Figure 17.9 Two-Way Indicator Species (TWINSPAN) analysis of leaf-litter beetle species composition in Central Amazonia, showing strong dichotomy between non-isolated continuous forest sites and isolated fragment sites. CF = continuous forest. Site groupings based on analysis of 161 species of Staphylinidae, carabidae and Scarabaeidae beetles; total abundance 1977 individuals (singleton species excluded). Indicator species characteristic of principal site groupings are indicated on dendrogram. (Reproduced from Didham, 1997.)

greater than 100 ha and possibly 500–1000 ha. It is important to remember that all fragmentation processes depend on scale, site and taxon (Doak *et al.*, 1992). This should not detract from efforts to provide better estimates of edge effects and minimum area requirements, but advises caution. From another perspective, such caution may be conservative given that the observed changes are essentially first-order responses to fragmentation (changes in presence and abundance). Secondary and functional changes (for example, in decomposition rates) noted in other studies also modify community structure and whole ecosystem processes (Didham *et al.*, 1996). The significance of these factors should not be underestimated, for it may well be that they apply over far wider areas than simple first-order effects (Myers, 1987), and may take much longer to reveal themselves.

17.6 CONCLUSIONS

Scale is of crucial importance in the study of forest fragmentation (Murphy, 1989; Lord and Norton, 1990; Doak *et al.*, 1992). Scale modifies our perception of pattern and process and influences management evaluation. Scale is influential at all levels: scales of time and space (Wiens and Milne, 1989; Kotliar and Wiens, 1990; Danielson, 1991; Borcard *et al.*, 1992; Dutilleul, 1993; Pickett and Cadenasso, 1995), scales of fragmentation (Zipperer, 1993), scales of sampling (Norton *et al.*, 1995) and scales of area, edge, isolation, shape and

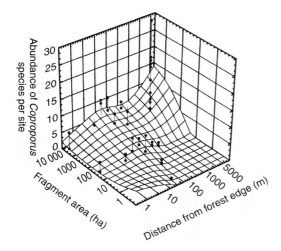

Figure 17.10 Three-dimensional least-squares fitted surface for population abundances of common continuous forest beetle, *Coproporus* sp. 0002 (Staphylinidae: Tachyporinae) in fragmented forests. Abundance is total abundance per site (20 m^2). Mean body length 2.62 mm; $N = 283$. *Coproporus* sp. population abundances are affected by distance from forest edge, and to a lesser degree by fragment area.

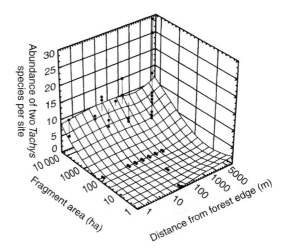

Figure 17.11 Three-dimensional least-squares fitted surface for population abundances of common continuous forest beetles, *Tachys* sp. 0268 and *Tachys* sp. 0269 (Carabidae: Bembidiini) in fragmented forests. Abundance is total abundance per site (20 m^2). Sp. 0268: mean body length 1.96mm; $N = 170$. Sp. 0269: mean body length 1.16mm; $N = 79$. Population abundances of *Tachys* spp. are apparently unaffected by edge effects, but decline dramatically (to local extinction) with decreasing fragment area.

connectivity (e.g. Lord and Norton, 1990; Laurance and Yensen, 1991; Doak *et al.*, 1992). This chapter has considered, in particular, the discrimination of the different intrinsic processes of forest fragmentation: area effects, edge effects, fragment shape, degree of spatio-temporal isolation and degree of habitat connectivity. These five principal processes have not been coherently factored into analyses of forest fragmentation, leading to poorly defined explanations for the mechanisms at work in fragmented ecosystems. When these factors are resolved, clearer causative inference can be made, though no studies have taken the next step toward proving clear mechanistic links between fragmentation processes and biotic abundance or diversity.

Studies of edge effects should be wary of assuming that such effects change monotonically with distance from the forest edge. As stated by Murcia (1995), edge effects are considerably more complex than previously thought (e.g. Malcolm, 1994; Camargo and Kapos, 1995) and unusual variations in edge response may not be merely noise in the data (or in the environment). Edge gradients in invertebrate abundance and species richness may even be bimodal. Core-area models and other analyses need to take this into consideration.

The proximate, mechanistic causes of population decline and extinction in habitat fragments are still largely conjecture. Perhaps the only way to determine these mechanisms is by experimental manipulation and the study of population dynamics (e.g. Chapter 7). Straw and Ludlow (1994) champion such small-scale dynamics studies, but their real applicability in a complex world may be questioned. There is a trade-off between their importance in resolving the central processes operating in specific instances and the limited general application of the results. The practical benefits of enhanced understanding from small-scale dynamics studies outweigh the investment in time, money and effort. The challenge to all fragmentation researchers is to take a clearer, more focused approach to the science.

ACKNOWLEDGEMENTS

I thank John H. Lawton, William F. Laurance and Nigel E. Stork for comments on an earlier draft of the manuscript. Mandy D. Tocher, Stuart J. Hine and various field technicians at the BDFFP provided assistance in the field. Peter M. Hammond gave invaluable taxonomic advice and checked the species sorting. Thanks to Claude Gascon for logistical support in Brazil. Funding was provided by the Commonwealth Scholarship Commission and the British Council (UK), the Natural History Museum (London), the Smithsonian Institution (Washington, DC), the Instituto Nacional de Pesquisas da Amazônia (Manaus, Brazil), the NERC Centre for Population Biology (Silwood Park, UK) and a University of Canterbury Doctoral Scholarship award (New Zealand). This is publication Number 174 in the BDFFP Technical Series.

REFERENCES

Aizen, M.A. and Feinsinger, P. (1994a) Habitat fragmentation, native insect pollinators, and feral honey bees in Argentine 'Chaco Serrano'. *Ecol. Appl.* **4**, 378–392.

Aizen, M.A. and Feinsinger, P. (1994b) Forest fragmentation, pollination, and plant reproduction in Chaco dry forest, Argentina. *Ecology* **75**, 330–351.

Andrén, H. (1994) Effects of habitat fragmentation on birds and mammals in landscapes with different proportions of suitable habitat: a review. *Oikos* **71**, 355–366.

Ås, S. (1993) Are habitat islands islands? Woodliving beetles (Coleoptera) in deciduous forest fragments in boreal forest. *Ecography* **16**, 219–228.

Bach, C.E. (1988a) Effects of host plant patch size on herbivore density: patterns. *Ecology* **69**, 1090–1102.

Bach, C.E. (1988b) Effects of host plant patch size on herbivore density: underlying mechanisms. *Ecology* **69**, 1103–1117.

Báldi, A. and Kisbenedek, T. (1994) Comparative analysis of edge effects on bird and beetle communities. *Acta Zool. Hung.* **40**, 1–14.

Bauer, L.J. (1989) Moorland beetle communities on limestone 'habitat islands'. I. Isolation, invasion and local species diversity in carabids and staphylinids. *J. Anim. Ecol.* **58**, 1077–1098.

Becker, P., Moure, J.S. and Peralta, F.J.A. (1991) More about Euglossine bees in Amazonian forest fragments. *Biotropica* **23**, 586–591.

Bellinger, R.G., Ravlin, F.W. and McManus, M.L. (1989) Forest edge effects and their influence on Gypsy Moth (Lepidoptera: Lymantriidae) egg mass distribution. *Environ. Ent.* **18**, 840–843.

Besuchet, C., Burckhardt, D.H. and Löbl, I. (1987) The 'Winkler/Moczarski' eclector as an efficient extractor for fungus and litter Coleoptera. *Coleopts Bull.* **41**, 392–394.

Bierregaard, R.O., Lovejoy, T.E., Kapos, V. *et al.* (1992) The biological dynamics of tropical forest fragments. A prospective comparison of fragments and continuous forest. *Bioscience* **42**, 859–866.

Boomsma, J.J., Mabelis, A.A., Verbeeck, M.G.M. and Los, E.C. (1987) Insular biogeography and distribution ecology of ants on the Frisian Islands. *J. Biogeog.* **14**, 21–37.

Borcard, D., Legendre, P. and Drapeau, P. (1992) Partialling out the spatial component of ecological variation. *Ecology* **73**, 1045–1055.

Brown, K.S. Jr. and Hutchings, R.W. (1997) Disturbance, fragmentation, and the dynamics of diversity in Amazonian forest butterflies. In *Tropical Forest Remnants: Ecology, Management and Conservation of Fragmented Communities* (eds W.F. Laurance and R.O. Bierregaard Jr), in press, University of Chicago Press, Chicago.

Burel, F. (1989) Landscape structure effects on carabid beetles spatial patterns in western France. *Landscape Ecol.* **2**, 215–226.

Burgess, R.L. and Sharpe, D.M. (eds) (1981) *Forest Island Dynamics in Man-Dominated Landscapes*, Ecological Studies Vol. 41, Springer-Verlag, New York.

Burkey, T.V. (1993) Edge effects in seed and egg predation at two neotropical rainforest sites. *Biol. Cons.* **66**, 139–143.

Buse, A. and Good, J.E.G. (1993) The effects of conifer forest design and management on abundance and diversity of rove beetles (Coleoptera: Staphylinidae): implications for conservation. *Biol. Cons.* **64**, 67–76.

Camargo, J.L. and Kapos, V. (1995) Complex edge effects on soil moisture and microclimate in central Amazonian forest. *J. Trop. Ecol.* **11**, 205–221.

Cappuccino, N. and Root, R.B. (1992) The significance of host patch edges to the colonization and development of *Corythucha marmorata* (Hemiptera: Tingidae). *Ecol. Ent.* **17**, 109–113.

Chan, H.T. (1980) Reproductive biology of some Malaysian dipterocarps. II. Fruiting biology and seedling studies. *Malay. Forester* **43**, 438–451.

Chen, J., Franklin, J.F. and Spies, T.A. (1992) Vegetation responses to edge environments in old-growth douglas-fir forests. *Ecol. Appl.* **2**, 387–396.

Chen, J., Franklin, J.F. and Spies, T.A. (1995) Growing-season microclimatic gradients from clearcut edges into old-growth douglas-fir forests. *Ecol. Appl.* **5**, 74–86.

Clopton, R.E. and Gold, R.E. (1993) Distribution and seasonal and diurnal activity patterns of *Eutrombicula alfreddugesi* (Acari: Trombiculidae) in a forest edge ecosystem. *J. Med. Ent.* **30**, 47–53.

Cusson, M., Vernon, R.S. and Roitberg, B.D. (1990) A sequential sampling plan for adult tuber flea beetles (*Epitrix tuberis* Gentner): dealing with 'edge effects'. *Can. Ent.* **122**, 537–546.

Daily, G.C. and Ehrlich, P.R. (1995) Preservation of biodiversity in small rainforest patches: rapid evaluations using butterfly trapping. *Biodivers. Conserv.* **4**, 35–55.

Danielson, B.J. (1991) Communities in a landscape: the influence of habitat heterogeneity on the interactions between species. *Am. Nat.* **138**, 1105–1120.

Dean, W.R.J. and Bond, W.J. (1990) Evidence for rapid faunal changes on islands in a man-made lake. I. Ants. *Oecologia* **83**, 388–391.

DeBach, P., Huffaker, C.B. and McPhee, A.W. (1976) Evaluation of impact of natural enemies. In *Theory and Practice of Biological Control* (eds C.B. Huffaker and P.S. Messenger), pp. 255–285, Academic Press, London.

Dempster, J.P. (1991) Fragmentation, isolation and mobility of insect populations. In *The Conservation of Insects and Their Habitats. The 15th Symposium of the Royal Entomological Society of London* (eds N.M. Collins and J.A. Thomas), pp. 143–153, Academic Press, London.

den Boer, P.J. (1990) The survival value of dispersal in terrestrial arthropods. *Biol. Cons.* **54**, 175–192.

Diamond, J.M. (1975) The island dilemma: lessons of modern biogeographic studies for the design of natural reserves. *Biol. Cons.* **7**, 129–146.

Diamond, J.M. and May, R.M. (1976) Island biogeography and the design of nature reserves. In *Theoretical Ecology. Principles and Application* (ed. R.M. May), pp. 163–186, Blackwell Scientific Publications, Oxford.

Didham, R.K. (1997) The influence of edge effects and forest fragmentation on leaf litter invertebrates in Central Amazonia. In *Tropical Forest Remnants: Ecology, Management and Conservation of Fragmented Communities* (eds W.F. Laurance and R.O. Bierregaard Jr), in press, University of Chicago Press, Chicago.

Didham, R.K., Ghazoul, J., Stork, N.E. and Davis, A. (1996) Insects in fragmented forests: a functional approach. *Trends Ecol. Evol.* **11**, 255–260.

Doak, D.F., Marino, P.C. and Kareiva, P.M. (1992) Spatial scale mediates the influence of habitat fragmentation on dispersal success: implications for conservation. *Theoret. Pop. Biol.* **41**, 315–336.

Duelli, P., Studer, M., Marchand, I. and Jakob, S. (1990) Population movements of arthropods between natural and cultivated areas. *Biol. Cons.* **54**, 193–207.

Dutilleul, P. (1993) Spatial heterogeneity and the design of ecological field experiments. *Ecology* **74**, 1646–1658.

Fowler, H.G., Silva, C.A. and Venticinque, E. (1993) Size, taxonomic and biomass distributions of flying insects in Central Amazonia: Forest edge vs understory. *Rev. Biol. Trop.* **41**, 755–760.

Fraver, S. (1994) Vegetation responses along edge-to-interior gradients in the mixed hardwood forests of the Roanoke River basin, North Carolina. *Cons. Biol.* **8**, 822–832.

Game, M. (1980) Best shape for nature reserves. *Nature* **287**, 630–632.

Goldstein, E.L. (1975) Island biogeography of ants. *Evolution* **29**, 750–762.

Grassle, J.F. and Smith, W. (1976) A similarity measure sensitive to the contribution of rare species and its use in investigation of variation in marine benthic communities. *Oecologia* **25**, 13–22.

Gunnarson, B. (1988) Spruce-living spiders and forest decline; the importance of needle-loss. *Biol. Cons.* **43**, 309–319.

Hanski, I., Pakkala, T., Kuussaari, M. and Lei, G. (1995) Metapopulation persistence of an endangered butterfly in a fragmented landscape. *Oikos* **72**, 21–28.

Harper, L.H. (1989) The persistence of ant-following birds in small Amazonian forest fragments. *Acta Amazonica* **19**, 249–263.

Harris, L.D. (ed.) (1984) *The Fragmented Forest*, University of Chicago Press, Chicago.

Heck, M., Nuss, D., Rurkowski, E. and Wegmann, G. (1989) Kleinringelwurmer (Enchytraeidae), Hornmilben (Oribatei) und bodenkundliche Parameter an einem Forstsaum an der Autobahn Avus in Berlin (West). *Verh. Ges. ™kol.* **18**, 397–401.

Helle, P. and Muona, J. (1985) Invertebrate numbers in edges between clear-fellings and mature forests in northern Finland. *Silva Fenn.* **19**, 281–294.

Hill, M.O. (1979) TWINSPAN – a FORTRAN program for arranging multivariate data in an ordered two-way table by classification of the individuals and attributes. Ecology and Systematics, Cornell University, Ithaca, New York 14850.

Hobbs, R.J. (1992) Is biodiversity important for ecosystem functioning? Implications for research and management. In *Biodiversity of Mediterranean Ecosystems in Australia* (ed. R.J. Hobbs), pp. 211–229, Surrey Beatty & Sons, Australia.

Hopkins, P.J. and Webb, N.R. (1984) The composition of the beetle and spider faunas on fragmented heathlands. *J. Appl. Ecol.* **21**, 935–946.

Jaenike, J. (1978) Effect of island area on *Drosophila* population densities. *Oecologia* **36**, 327–332.

Janzen, D.H. (1975) Behaviour of *Hymenaea courbaril* when its predispersal seed predator is absent. *Science* **189**, 145-146.

Janzen, D.H. (1983) No park is an island: increase in interference from outside as park size decreases. *Oikos* **41**, 402–410.

Jennersten, O. (1988) Pollination in *Dianthus deltoides* (Caryophyllaceae): effects of habitat fragmentation on visitation and seed set. *Cons. Biol.* **2**, 359–366.

Kapos, V. (1989) Effects of isolation on the water status of forest patches in the Brazilian Amazon. *J. Trop. Ecol.* **5**, 173–185.

Kapos, V., Ganade, G., Matsui, E. and Victoria, R.L. (1993) $\delta^{13}C$ as an indicator of edge effects in tropical rainforest reserves. *J. Ecol.* **81**, 425–432.

Kapos, V., Wandelli, E., Camargo, J.L. and Ganade, G. (1997) Edge-related changes in environment and plant responses due to forest fragmentation in Central Amazonia. In *Tropical Forest Remnants: Ecology, Management and Conservation of Fragmented Communities* (eds W.F. Laurance and R.O. Bierregaard Jr), in press, University of Chicago Press, Chicago.

Kareiva, P. (1987) Habitat fragmentation and the stability of predator–prey interactions. *Nature* **326**, 388–390.

Klein, B.C. (1989) Effects of forest fragmentation on dung and carrion beetle communities in central Amazonia. *Ecology* **70**, 1715–1725.

Kotliar, N.B. and Wiens, J.A. (1990) Multiple scales of patchiness and patch structure: a hierarchical framework for the study of heterogeneity. *Oikos* **59**, 253–260.

Kruess, A. and Tscharntke, T. (1994) Habitat fragmentation, species loss, and biological control. *Science* **264**, 1581–1584.

LaSalle, J. and Gauld, I.D. (1991) Parasitic hymenoptera and the biodiversity crisis. *Redia* **74**, 315–334

Launer, A.E. and Murphy, D.D. (1994) Umbrella species and the conservation of habitat fragments: a case of a threatened butterfly and a vanishing grassland ecosystem. *Biol. Cons.* **69**, 145–153.

Laurance, W.F. (1991) Edge effects in tropical forest fragments: application of a model for the design of nature reserves. *Biol. Cons.* **57**, 205–219.

Laurance, W.F. (1994) Rainforest fragmentation and the structure of small mammal communities in tropical Queensland. *Biol. Cons.* **69**, 23–32.

Laurance, W.F. (1997) Hyper-disturbed parks: edge effects and the ecology of isolated rainforest reserves in tropical Australia. In *Tropical Forest Remnants: Ecology, Management and Conservation of Fragmented Communities* (eds W.F. Laurance and R.O. Bierregaard Jr), University of Chicago Press, Chicago (in press).

Laurance, W.F. and Yensen, E. (1991) Predicting the impacts of edge effects in fragmented habitats. *Biol. Cons.* **55**, 77–92.

Laurance, W.F., Bierregaard, R.O. Jr, Gascon, C. et al. (1997) Tropical forest fragmentation: synthesis of a diverse and dynamic discipline. In *Tropical Forest Remnants: Ecology, Management and Conservation of Fragmented Communities* (eds W.F. Laurance and R.O. Bierregaard Jr), in press, University of Chicago Press, Chicago.

Leopold, A. (1933) *Game Management*, Charles Scribner Sons, New York.

Liebherr, J.K. (1988) Gene flow in ground beetles (Coleoptera: Carabidae) of differing habitat preference and flight-wing development. *Evolution* **42**, 129–137.

Lord, J.M. and Norton, D.A. (1990) Scale and the spatial concept of fragmentation. *Cons. Biol.* **4**, 197–202.

Louda, S.M. (1982) Distribution ecology: variation in plant recruitment over a gradient in relation to insect seed predation. *Ecol. Monogr.* **52**, 25–41

Lovejoy, T.E., Bierregaard, R.O., Rylands, A.B. et al. (1986) Edge and other effects of isolation on Amazon forest fragments. In *Conservation Biology. The Science of Scarcity and Diversity* (ed. M.E. Soulé), pp. 257–285, Sinauer Associates, Sunderland, Massachusetts.

MacArthur, R.H. and Wilson, E.O. (1967) *The Theory of Island Biogeography*, Princeton University Press, Princeton.

Mader, H.-J. (1984) Animal habitat isolation by roads and agricultural fields. *Biol. Cons.* **29**, 81–96.

Main, B.Y. (1987) Persistence of invertebrates in small areas: case studies of trapdoor spiders in Western Australia. In *Nature Conservation: The Role of Remnants of Native Vegetation* (eds D.A. Saunders, G.W. Arnold, A.A. Burbidge and A.J.M. Hopkins), pp. 29–39, Surrey Beatty & Sons in association with CSIRO and CALM, Australia.

Malcolm, J.R. (1988) Small mammal abundances in isolated and non-isolated primary forest reserves near Manaus, Brazil. *Acta Amazonica* **18**, 67–83.

Malcolm, J.R. (1994) Edge effects in central amazonian forest fragments. *Ecology* **75**, 2438–2445.

Malcolm, J.R. (1997) Insect biomass in Amazonian forest fragments. In *Canopy Arthropods* (eds N.E. Stork, J. Adis and R.K. Didham), 510–533, Chapman & Hall, London.

Margules, C.R., Milkovits, G.A. and Smith, G.T. (1994) Contrasting effects of habitat fragmentation on the scorpion *Cercophonius squama* and an amphipod. *Ecology* 75, 2033–2042.

Martins, M.B. (1989) Invasão de fragmentos florestais por espécies oportunistas de *Drosophila* (Diptera, Drosophilidae). *Acta Amazonica* 19, 265–271.

Matlack, G.R. (1993) Microenvironmental variation within and among forest edge sites in the eastern United States. *Biol. Cons.* 66, 185–194.

Mattiske, E.M. (1987) Creation of ecotones and management to control patch size. In *Nature Conservation: The Role of Remnants of Native Vegetation* (eds D.A. Saunders, G.W. Arnold, A.A. Burbidge and A.J.M. Hopkins), p. 383, Surrey Beatty & Sons in association with CSIRO and CALM, Australia.

McCann, J.M. and Harman, D.M. (1990) Influence of the intrastand position of Black Locust trees on attack rate of the Locust Borer (Coleoptera: Cerambycidae). *Ann. Ent. Soc. Am.* 83, 705–711.

McCauley, D.E. (1991) The effect of host plant patch size variation on the population structure of a specialist herbivore insect, *Tetraopes tetraophthalmus*. *Evolution* 45, 1675–1684.

Münster-Swendsen, M. (1980) The distribution in time and space of parasitism in *Epinotia tedella* (Cl.) (Lepidoptera: Tortricidae). *Ecol. Ent.* 5, 373–383.

Murcia, C. (1995) Edge effects in fragmented forests: implications for conservation. *Trends Ecol. Evol.* 10, 58–62.

Murphy, D.D. (1989) Conservation and confusion: wrong species, wrong scale, wrong conclusions. *Cons. Biol.* 3, 82–84.

Myers, N. (1987) The extinction spasm impending: synergisms at work. *Cons. Biol.* 1, 14–21.

Naeem, S., Thompson, L.J., Lawler, S.P. *et al.* (1994) Declining biodiversity can alter the performance of ecosystems. *Nature* 368, 734–737.

Naeem, S., Thompson, L.J., Lawler, S.P. *et al.* (1995) Empirical evidence that declining species diversity may alter the performance of terrestrial ecosystems. *Phil. Trans. R. Soc. Lond., B* 347, 249–262.

Nielsen, B.O. and Ejlersen, A. (1977) The distribution pattern of herbivory in a beech canopy. *Ecol. Ent.* 2, 293–299.

Norton, D.A., Hobbs, R.J. and Atkins, L. (1995) Fragmentation, disturbance, and plant distribution: mistletoes in woodland remnants in the Western Australian wheatbelt. *Cons. Biol.* 9, 426–439.

Ozanne, C.M., Hambler, C., Foggo, A. and Speight, M.R. (1997) The significance of edge effects in the management of forests for invertebrate biodiversity. In *Canopy Arthropods* (eds N.E. Stork, J. Adis and R.K. Didham), 534–550, Chapman & Hall, London.

Pickett, S.T.A. and Cadenasso, M.L. (1995) Landscape ecology: spatial heterogeneity in ecological systems. *Science* 269, 331–334.

Powell, A.H. and Powell, G.V.N. (1987) Population dynamics of male euglossine bees in amazonian forest fragments. *Biotropica* 19, 176–179.

Ranney, J.W., Bruner, M.C. and Levenson, J.B. (1981) The importance of edge in the structure and dynamics of forest islands. In *Forest Island Dynamics in Man-Dominated Landscapes*, Ecological Studies Vol. 41 (eds R.L. Burgess and D.M. Sharpe), pp. 67–95, Springer-Verlag, New York.

Rey, J.R. and Strong, D.R. (1983) Immigration and extinction of salt marsh arthropods on islands: an experimental study. *Oikos* 41, 396–401.

Robinson, G.R., Holt, R.D., Gaines, M.S. *et al.* (1992) Diverse and contrasting effects of habitat fragmentation. *Science* 257, 524–527.

Rodrigues, J.J.S., Brown, K.S. and Ruszczyk, A. (1993) Resources and conservation of Neotropical butterflies in urban forest fragments. *Biol. Cons.* 64, 3–9.

Roland, J. (1993) Large-scale forest fragmentation increases the duration of tent caterpillar outbreaks. *Oecologia* 93(1), 25–30.

Saunders, D.A., Hobbs, R.J. and Margules, C.R. (1991) Biological consequences of ecosystem fragmentation: a review. *Cons. Biol.* 5, 18–32.

Shreeve, T.G. and Mason, C.F. (1980) The number of butterfly species in woodland. *Oecologia* 45, 414–418.

Simberloff, D.S. (1974) Equilibrium theory of island biogeography and ecology. *Ann. Rev. Ecol. Syst.* 5, 161–182.

Simberloff, D.S. and Wilson, E.O. (1970) Experimental zoogeography of islands. A two-year record of colonization. *Ecology* 51, 934–937.

Skole, D. and Tucker, C. (1993) Tropical deforestation and habitat fragmentation in the Amazon: satellite data from 1978 to 1988. *Science* 260, 1905–1910.

Souza, O.F.F. (1989) Diversidade de Térmitas (Insecta: Isoptera) e sua relação com a fragmentação de ecossistemas na Amazônia Central. MSc Thesis, Viçosa, Minas Gerais, Brazil.

Souza, O.F.F. and Brown, V.K. (1994) Effects of habitat fragmentation on Amazonian termite communities. *J. Trop. Ecol.* 10, 197–206.

Sowig, P. (1989) Effects of flowering plant's patch size on species composition of pollinator communities, foraging strategies, and resource partitioning in bumblebees (Hymenoptera: Apidae). *Oecologia* 78, 550–558.

Spears, E.E. Jr (1987) Island and mainland pollination ecology of *Centrosema virginianum* and *Opuntia stricta*. *J. Ecol.* 75, 351–362

Springett, J.A. (1976) The effect of planting *Pinus pinaster* Ait. on populations of soil microarthropods and on litter decomposition at Gnangara, Western Australia. *Aust. J. Ecol.* **1**, 83–87.

Stamps, J.A., Buechner, M. and Krishnan, V.V. (1987) The effects of edge permeability and habitat geometry on emigration from patches of habitat. *Am. Nat.* **129**, 533–552.

Stouffer, P.C. and Bierregaard, R.O. (1995) Use of Amazonian forest fragments by understory insectivorous birds. *Ecology* **76**, 2429–2445.

Straw, N.A. and Ludlow, A.R. (1994) Small-scale dynamics and insect diversity on plants. *Oikos* **71**, 188–192.

Taylor, P.D., Fahrig, L., Henein, K. and Merriam, G. (1993) Connectivity is a vital element of landscape structure. *Oikos* **68**, 571–573.

Terborgh, J. (1997) Hyper-abundance as a response to fragmentation: transient states on the route to extinction. In *Tropical Forest Remnants: Ecology, Management and Conservation of Fragmented Communities* (eds W.F. Laurance and R.O. Bierregaard Jr), in press, University of Chicago Press, Chicago.

Thomas, C.D., Thomas, J.A. and Warren, M.S. (1992) Distributions of occupied and vacant butterfly habitats in fragmented landscapes. *Oecologia* **92**, 563–567.

Toda, M.J. (1992) Three-dimensional dispersion of drosophilid flies in a cool temperate forest of northern Japan. *Ecol. Res.* **7**, 283–295.

Toft, C.A. and Schoener, T.W. (1983) Abundance and diversity of orb spiders on 106 Bahamian islands: biogeography at an intermediate trophic level. *Oikos* **41**, 411–426.

Turner, B.L., Clark, W.C., Kates, R.W. *et al.* (eds) (1990) *The Earth as Transformed by Human Action*, Cambridge University Press, Cambridge, England.

Vasconcelos, H.L. (1988) Distribution of *Atta* (Hymenoptera – Formicidae) in 'Terra-firme' rain forest of central Amazonia: density, species composition and preliminary results on effects of forest fragmentation. *Acta Amazonica* **18**, 309–315.

Vepsäläinen, K. and Pisarski, B. (1982) Assembly of island ant communities. *Ann. Zool. Fenn.* **19**, 327–335.

Verhaagh, M. (1991) Clearing a tropical rain forest – effects on the ant fauna. In *Tropical Ecosystems: Systems Characteristics, Utilization Patterns, and Conservation Issues* (eds W. Erdelen, N. Ishwaran and P. Muller), pp. 59–68, Margalef Scientific Books, Weikersheim.

Vitousek, P.M. (1994) Beyond global warming: ecology and global change. *Ecology* **75(7)**, 1861–1876.

Webb, D.P. (1976) Edge effects on salt marsh arthropod community structures. *J. Georgia Ent. Soc.* **11**, 17–27.

Webb, N.R. (1989) Studies on the invertebrate fauna of fragmented heathland in Dorset, UK, and the implications for conservation. *Biol. Cons.* **47**, 153–165.

Webb, N.R. and Hopkins, P.J. (1984) Invertebrate diversity on fragmented *Calluna* heathland. *J. Appl. Ecol.* **21**, 921–933.

Webb, N.R., Clarke, R.T. and Nicholas, J.T. (1984) Invertebrate diversity on fragmented *Calluna*-heathland: effects of surrounding vegetation. *J. Biogeog.* **11**, 41–46.

Whitmore, T.C. and Sayer, J.A. (eds) (1992) *Tropical Deforestation and Species Extinction*, Chapman & Hall, London.

Wiens, J.A. and Milne, B.T. (1989) Scaling of 'landscape' in landscape ecology, or, landscape ecology from a beetle's perspective. *Landscape Ecol.* **3**, 87–96.

Wilcox, B.A. (1980) Insular ecology and conservation. In *Conservation Biology: An Evolutionary–Ecological Perspective* (eds M.E. Soulé and B.A. Wilcox), pp. 95–117, Sinauer Associates, Sunderland, Massachusetts.

Williams-Linera, G. (1990a) Origin and early development of forest edge vegetation in Panama. *Biotropica* **22**, 235–241.

Williams-Linera, G. (1990b) Vegetation structure and environmental conditions of forest edges in Panama. *J. Ecol.* **78**, 356–373.

Williams-Linera, G. (1993) Vegetación de bordes de un bosque nublado en el Parque Ecológico Clavijero, Xalapa, Veracruz, México. *Rev. Biol. Trop.* **41**, 443–453.

Winnett-Murray, K. (1986) Variation in the behavior and food supply of four Neotropical wrens. Unpublished PhD thesis, University of Florida, USA.

Wolda, H. (1983) Diversity, diversity indices and tropical cockroaches. *Oecologia* **58**, 290–298.

Young, A. and Mitchell, N. (1994) Microclimate and vegetation edge effects in a fragmented podocarp–broadleaf forest in New Zealand. *Biol. Cons.* **67**, 63–72.

Zimmerman, B.L. and Bierregaard, R.O. (1986) Relevance of the equilibrium theory of island biogeography and species–area relations to conservation with a case from Amazonia. *J. Biogeog.* **13**, 133–143.

Zipperer, W.C. (1993) Deforestation patterns and their effects on forest patches. *Landscape Ecol.* **8**, 177–184.

IMPACT OF FOREST AND WOODLAND STRUCTURE ON INSECT ABUNDANCE AND DIVERSITY

P. Dennis

18.1 INTRODUCTION

In recent centuries, there has been widespread deforestation and replacement of native forests by those composed of exotic species. This resulted in a cover of numerous remnants of semi-natural woodland which exist within a predominantly agricultural context. There has been an attempt to quantify the viability of insect populations in these fragmented woodlands in recent studies (Usher *et al.*, 1993; Dennis *et al.*, 1995). In Britain, there has been a recent modest expansion of lowland woodlands composed of native tree species, a consequence of agricultural reforms (Rodwell and Patterson, 1994). The effect of these changes in woodland cover on insect populations in existing fragmented woodlands are discussed in relation to woodland composition, geometry and juxtaposition with other woodlands. Woodland insect species have the potential to benefit from an extension of the existing woodland cover with new, native woodlands, but only if the longer-term development of the new woods can provide appropriate microhabitats within woodlands which are situated where there is a reasonable probability of colonization success for individual insect populations.

18.2 PREHISTORIC AND HISTORICAL DETERMINANTS OF WOODLAND INSECT ASSEMBLAGES

The extent of insect assemblages in contemporary northwestern European woodlands and forests can be related to four phases of biology. First, the evolutionary process of speciation was driven by increased complexity of woodland biotopes through the Mesozoic, particularly from the mid-Cretaceous when the Angiosperms first dominated terrestrial ecosystems (Strong *et al.*, 1984). Second, colonization by phytophagous insects of the newly expanded forest after the last ice age depended on the apparency of the different tree species to insects that dispersed northwards. Third, geographical fragmentation (Harris, 1984) of forest ecosystems by humans over 4000 years in northwest Europe has been a major determinant of woodland insect species assemblages. Geographical fragmentation has been defined for areas where forest clearance has occurred recently (Western Australia) as the point where the land covered by woodland falls below *c.* 20% and woodlands become significantly dispersed and isolated (R. Hobbs, personal communication). Finally, structural fragmentation (Lord and Norton, 1990)

Forests and Insects. Edited by A.D. Watt, N.E. Stork and M.D. Hunter. Published in 1997 by Chapman & Hall, London. ISBN 0 412 79110 2.

has occurred where human management over the last 500 years has reduced the habitat heterogeneity within woodlands (Warren and Key, 1991).

18.2.1 EVOLUTIONARY AND FUNCTIONAL ADAPTATIONS

Insect diversity in forests expanded in response to the progressive structural and chemical complexity that developed in forest biotopes from the mid-Cretaceous period (Strong *et al.*, 1984). The species richness of arboreal insects was determined by the cumulative ancient abundance of the different trees (Southwood, 1961), geographical range (Strong, 1974; Levin, 1976; Lawton and Schroder, 1977), the length of time that the tree has been available for colonization (Birks, 1980), and its taxonomic isolation (Lawton and Schroder, 1977), successional status (Cates and Orians, 1975), growth form (Maiorana, 1978; Scriber, 1978) and structural complexity (Price, 1977; Lawton and Price, 1979). Diversity in the guild structure also developed through this period and seven guilds have been identified in tree communities today: phytophages, epiphyte fauna, scavengers, predators, parasitoids, ants and tourists (Moran and Southwood, 1982). The proportions of species in the phytophage, epiphyte and predator guilds were consistent, both between trees of different species and between trees of the same species that are geographically separated. The proportional uniformity in the numbers of species within guilds both between and within ecological communities in general led to controversy because it was difficult to distinguish whether woodland insect communities were patterned structures of species, populations responding independently to environmental gradients, or independent random collections of species (Heatwole and Levin, 1972; Simberloff, 1976; Simberloff and Wilson, 1978).

18.2.2 HISTORICAL WOODLAND MANAGEMENT

Warren and Key (1991) produced an excellent account of the historical management of woodlands and the mechanisms that have affected the number of insect species in British woodlands. In summary, major woodland clearance had occurred by 2500 BP (perhaps 50% of cover) after initial human impacts from 6500 to 4000 BP (Warren and Key, 1991). Woodland clearance had a major effect on woodland insect populations and sub-fossil evidence suggests there were numerous extinctions (Warren and Key, 1991; Thomas and Morris, 1994). By 1086, the land covered by semi-natural woodland was *c*. 15%. Much of the area of woodland was managed by coppicing and continued to maintain habitats for the early and mid-successional insect species, but few trees were left for saproxylic insects to exploit. However, the practice of pollarding in pasture woodland maintained habitats for the saproxylic species which were denied resources in coppiced woodland.

By 1800, Britain still contained a rich mosaic of ancient, semi-natural woodland that supported insects associated with all of the woodland seral phases. Estimates suggested that 28 500 invertebrate species may have been supported by the woodland biotope (Hambler and Speight, 1995). Of these, 1500 were Coleoptera, of which 20–25% were deadwood specialists (Hanski and Hammond, 1995) and 58 were butterflies (Lepidoptera) (Hambler and Speight, 1995). Sixty per cent (*c*. 1338 species) of moths (Lepidoptera) were also associated with woodland habitats (Waring, 1989).

18.3 WOODLAND FRAGMENTATION AND HETEROGENEITY

Fragmentation was a feature of forests before humans became a predominant influence in Europe. Semi-permanent open spaces resulted from the dynamic interaction of treefall gaps provided by old-aged trees or windthrow events, beaver activity created wet meadows and grazers prolonged the process of regeneration in the clearings. Permanent open spaces in the woodland cover were maintained along river valleys, lakes, wetlands, cliffs and areas of poor soil (Warren and Key, 1991).

18.3.1 RECENT TRENDS IN WOODLAND MANAGEMENT

Towards the industrial revolution human influence grew and total forest cover declined. This geographical fragmentation exceeded natural limits and woodlands became geographically isolated (Peterken, 1993). Threatened woodland insects within fragmented patterns of woodlands evolved isolated races with reduced powers of dispersal (Dempster, 1991, Warren, 1987). Today, semi-natural woodland occupies 2% of land cover, but for 200 years there was also a progressive abandonment of the coppicing and pollarding of native tree species as the main management in favour of high forest, in which exotic replaced native tree species. Coppicing had previously replicated the natural balance of tree cover and open space within the fragments of woodland, though insects adapted to old growth declined due to a reduction in the number of old-growth trees and volume of deadwood.

The structural fragmentation caused a reduction in the habitat heterogeneity and loss of continuity of woodland cover across most of the woodland ecosystem, which reduced the habitats available to the early and late successional stages. Many woodland insect species required nectar sources close to the breeding habitat, which occurred only in glades and on woodland edges (Harding and Rose, 1986). A wide variety of invertebrates depended entirely on the rot holes and fungal growths of old growth in woodlands. Major losses of insect species therefore occurred during that period, most intensely in the last 50 years (Warren and Key, 1991). For example, saproxylic species represent 40% of Britain's extinct, endangered or vulnerable invertebrates (Thomas and Morris, 1994).

18.3.2 DISTRIBUTION OF SPECIES ACROSS FRAGMENTED WOODLANDS

Geographical isolation can reduce species richness because some species are poor dispersers or are restricted to specific habitat types. For example, *Hypebaeus flavipes* (Coleoptera: Melyridae) and related species occur in a few ancient woodland trees where there are features such as rot holes, heart rot, hollows, wood mould, snags and thick or loose bark (Warren and Key, 1991). Movement of the heath fritillary, *Mellicta athalia* (Rott.) (Lepidoptera: Nymphalidae), is restricted to around 100 m, which effectively prevents colonization of suitable, newly cut woodland habitats beyond 600 m from the source population (Warren, 1987; Warren and Key, 1991). The small pearl-bordered fritillary, *Boloria selene selene* ([D. & S.]) (Lepidoptera: Nymphalidae), and the black hairstreak, *Strymonidia pruni* (Linn.) (Lepidoptera: Lycaenidae), are similarly restricted in their dispersal ability (Thomas, 1989).

Different management approaches applied in Russia and Finland to a previously contiguous tract of boreal forest demonstrated the role of old-aged tree species, such as aspen. The more intensive forest management imposed in Finland removed aspen from spruce/pine forests to a point where too few patches remained to support the 33% of associated wood beetle species. Populations of these species remained viable on the Russian side because enough patches of aspen were tolerated in the forest management practice there (Hanski and Hammond, 1995). Isolation due to habitat fragmentation can lead to extinction of a species because of chance processes (MacArthur and Wilson, 1967; Hanski, 1989; Quinn and Hastings, 1987) or genetic changes and shifts in habitat quality (Dempster, 1991). The situation becomes complex when considering the interaction of subpopulations distributed amongst systems of woodland fragments (for example, butterflies in forests of Costa Rica: Thomas, 1991).

18.3.3 CONCEPTUAL MODELS OF WOODLAND INSECT DISTRIBUTIONS

Woodlands today are represented by discrete fragments of varied size and composition. One of the current aims of ecology has been to understand the fate of wildlife populations within such fragmented habitats (Simberloff, 1988). Several conceptual approaches have been applied to the study of insect populations in systems of fragmented woodlands.

(a) Island biogeography theory

The application of island biogeography theory to terrestrial islands such as fragmented woodlands has led to conflicts over appropriate sizes for nature reserves in conservation biology (Burkey, 1989). Controversy developed through conflicts between the results from studies that extrapolated the species–area relationship to favour a single, large fragment of habitat (Diamond, 1975; Wilson and Willis, 1975; Terborgh, 1976; Fahrig and Merriam, 1985) and empirical surveys of species numbers that found more species on several small islands (Simberloff and Abele, 1976, 1982; Abele and Connor, 1979; Gilpin and Diamond, 1980; Higgs and Usher, 1980: Higgs, 1981; Jarvinen, 1982; Margules et al., 1982). The appropriate long-term strategy for the maintenance of species was resolved by a consideration of the variation in the probability of extinction with habitat area and initial population size (Simberloff and Abele, 1982; Shaffer and Samson, 1985; Wilcox and Murphy, 1985).

The 'several small' approach maximized the species richness around the time of habitat isolation, perhaps through habitat differences due to the geographical spread of small fragments or equal extinction probabilities and communication between fragments (Burkey, 1989). Cole (1981) argued that higher species numbers could only be packed into a set of smaller fragments if exchange with a mainland source pool occurred (Cole, 1981). In the long term, minimization of extinction after isolation (large fragments) was more effective than maximization of species richness (small fragments) because isolation created greater extinction probabilities in smaller fragments (Burkey, 1989; Quinn and Hastings, 1987).

The value of the equilibrium theory of island biogeography (MacArthur and Wilson, 1967) was immense but the application of the species–area relationship was of limited value for the prediction of the distribution of species across fragmented terrestrial habitats (Haila, 1986). Many studies have used inappropriate data of species and area and fitted regression models devoid of ecological meaning (Connor and McCoy, 1979). Studies of different wildlife groups in which woodlands were treated as islands have proved either contradictory or inconclusive (Simberloff, 1988).

One problem has been dealing with the scale of interaction of different organisms with the apparent habitat pattern (Haila, 1990). The 'island' could be the individual host plant to a stenophagous insect (Dempster, 1991), a mosaic composed of several plants to a polyphagous insect, or a monocultured plantation to a mobile stenophagous species (Claridge and Evans, 1990). Many woodlands are not clearly definable as distinct insular units of habitat. Division of woodland by rivers, roads and railways may represent a complete or partial barrier to organisms active in the field layer and justify a division of the habitat into two 'islands'. If we considered organisms active in the canopy layer, a single 'island' definition would be appropriate if the leaf canopy interlocked above the barriers to epigeal organisms. The connection of the wood to other habitats that emulate woodland (such as hedgerows) and the context amongst other habitats raises a challenge to current definitions of 'island'. Many species are constrained by the resources available throughout a network of woodlands (Dennis et al., 1994; Waring, 1989). Further, many field-layer plant and arthropod species are common to habitats within and between woodlands (Usher et al., 1992, 1993).

(b) Metapopulation dynamics

Metapopulation dynamics complements island biogeography by focusing on populations of individual species and may provide some mechanisms underlying the equilibrium theory of island biogeography. Metapopulation dynamics provides a concept whereby local populations are changing (from both demographic and genetic points of view) but the metapopulation is stable (Olivieri et al., 1990). The concept depends on the free dispersal of species in a landscape made up of several fragments of the same habitat. Therefore, chance processes leading to extinction in one fragment would be counteracted by immigration of individuals from other extant subpopulations. The survival of the metapopulation assumes a limited correlation between the risks of extinction in each of the subpopulations, a key factor to the validity of the theory (Hanski and Gilpin, 1991). Extinction can be caused by four processes: demographic stochasticity, chronic environmental variation, environmental catastrophes and genetic effects (Shaffer, 1981).

Metapopulation persistence time increases exponentially with the number of island populations (Nisbet and Gurney, 1982). The models have been applied to different wildlife populations in fragmented habitat systems (Thomas and Harrison, 1992; Thomas and Jones, 1993; Verboom and Van Apeldoorn, 1990; Verboom et al., 1991; Nee and May, 1992). There are problems in the application of these models because they assume a closed system of habitats of equal quality, but empirical evidence contradicts that assumption. Therefore, the models cannot predict the appropriate number, size and spatial arrangement of habitats to maintain the viability of a specific metapopulation, though they can be used for general guidance (Harrison et al., 1988, 1994; Hanski, 1994a,b).

(c) Spatial ecological models

Source/sink models may be more appropriate than metapopulation models because the fitness of populations in different habitats can be quantified as birth, immigration, death and emigration rates (BIDE models; Pulliam, 1988) and can allow for patches of different quality (Watkinson and Sutherland, 1995). Some sink habitats may receive high densities of a species during periods of emigration from source habitats, which would contradict the assumption that population density relates to habitat quality (Van Horne, 1983). The value of spatial ecology has been to extend beyond a consideration of individual types of terrestrial habitat patches and to consider the context of these patches in a mosaic of mixed habitats.

Landscape ecology has included spatial ecological concepts and is ecology practised within a geographical context but with a consideration of human impacts. It has been applied to analyse the implications of changes to large-scale habitat patterns on population, community and ecosystem ecology (Naveh and Lieberman, 1984; Forman and Godron, 1986). There is no theoretical framework in landscape ecology but the concepts and terminology define spatial characteristics of landscapes which have proved useful in the study of population dynamics and species assemblages in heterogeneous habitats (Forman and Godron, 1986; McDonnell and Picket, 1988; Taylor et al., 1993). Landscapes are broken into constituent habitat elements and this has been used as the basis for analysing the responses of animal populations to habitat patches (Hanski, 1994b; Shure and Philips, 1991), mosaic structure (Baguette, 1993), habitat connectivity, context and contrast (Taylor et al., 1993), edge effects (Laurance and Yensen, 1991), ecotones (Gourov, 1994), boundary flows (Frampton et al., 1995) and corridors (Hobbs, 1992). The scale at which different organisms utilize their habitats defines the spatial scale of the dynamic processes that link their populations to the spatial patterns of the environment (Wiens and Milne, 1989a,b; Crist and Wiens, 1994; With and Crist, 1995; Crist et al., 1992; Wiens et al., 1992, 1993). The quality and context of habitat patches in the mosaic of habitats has consequences for population movement (Duelli et al., 1990; Wood and Samways, 1991) and the measurement of isolation of populations which provides realism for source/sink models compared with the models of island biogeography and metapopulation dynamics, where these parameters were assumed unimportant.

The spatial processes that influence insect populations in fragmented landscapes must be identified firstly by exploring spatial patterns (Liebhold et al., 1993). Data collected on different habitat parameters across appropriate spatial scales can be analysed with multiple regression to identify factors that significantly affect the distribution of a target species (van der Zee et al., 1992; Verboom and Van Apeldoorn, 1990; Van Apeldoorn et al., 1992; Roland, 1993).

18.4 ARBOREAL INSECTS AND FRAGMENTED WOODLAND – A CASE STUDY FROM SOUTHERN SCOTLAND

In circumstances where management of woodlands is required to realize nature conservation objectives (Harris and Harris, 1991; Peterken, 1993) we need to understand the processes which determine the distribution and abundance of wildlife. Guidelines based on sound scientific knowledge are required to improve management practices for wildlife conservation in existing and new lowland woodlands, promoted by incentives such as the UK's Farm

Woodland Premium Scheme (Rodwell and Patterson, 1994). Aspects of woodland structure and pattern are outlined from studies of phytophagous insects on broadleaved trees of established woodlands in Midlothian, Scotland, and from studies on other taxa in other woodland systems.

18.4.1 STUDY AREA AND METHODS

The study area comprised woodland cover in an area of Midlothian south of Edinburgh (NGR NT 252649), extending south through mixed agricultural land to Penicuik and rough-grazing land south to West Linton (NGR NT 169562). The area was bordered to the west by the Pentland Hills and to the east by the valley of the Esk. The woodlands comprised old plantation woods, networks of shelterbelts and small insular woods composed of derelict shelterbelt or individual semi-natural stands of trees (Dennis et al., 1995). All these woods included beech, *Fagus sylvatica* L. (Fagales: Fagaceae), but other species varied from mixtures with only *Pinus sylvestris* L. (Coniferae: Pinaceae), *Acer pseudoplatanus* L. (Sapindales: Aceraceae) and *Crataegus monogyna* Jacq. (Rosales: Rosaceae) in simple shelterbelts to semi-natural parts of woodland plantations dominated by *Quercus robur* L. (Fagales: Fagaceae), *Alnus glutinosa* (L.) (Fagales: Betulaceae) and *Betula pendula* Roth. (Fagales: Betulaceae) with a canopy comprising a wide mixture of native species.

The study investigated whether the abundance of insect species which mine and gall (endophage) and feed externally (ectophage) on leaves varied between the woodland types present in the study area (Dennis et al., 1995). Emphasis was placed on the relationship between phytophagous insects of *F. sylvatica* (Table 18.1) and characteristics of the woodland habitat to simplify the study, but it was recognized that some of these species were polyphagous. Three *F. sylvatica* were sampled in 15 elements of the woodland system and nine hedgerows. General ectophage diversity was sampled by chemical knockdown in 1990 and physical knockdown in 1991, in mid-summer and endophage species were sampled by leaf collection, from the canopies, in late summer. The essential resource for this phytophagous species assemblage comprised the three-dimensional extent of the leaf canopies of *F. sylvatica* in woodlands, tree lines and hedgerows between the woods. Measurements of woodland quality and spatial pattern reflected the context of the leaf resource of the arthropods at several spatial scales (Table 18.2). The structure and position of each tree and adjacent vegetation cover were recorded by direct measurements. The extent of woodland cover and the pattern of woodlands in the landscape were measured from maps and aerial photographs. These variables allowed meaningful interpretation and potential application of the results, a problem encountered in previous studies where woods were simply treated as islands (Simberloff, 1988).

Measurements of the leaf litter that remained under each tree after winter and the draughtiness of each woodland were made because it was believed these factors could affect endophage species richness and abundance. These measurements helped to interpret mechanisms from the pattern of arthropod species distributions detected by multiple regression analyses which could operate at different spatial scales (tree, wood and woodland system) to affect the distribution and abundance of arthropod species. Next, we tested whether the patterns of abundance of arboreal arthropods in general related to those of the endophage species (Dennis et al., 1994). In a further study, leaves of mature *F. sylvatica*, *Q. robur* and *B. pendula* were collected from the litter layer in winter to measure the phytophagous species richness in representative woodland types (Dennis et al., 1995). These data were used to explore whether the factors that accounted for the distribution of phytophagous arthropods on *F. sylvatica* could be used to estimate the size of these arthropod guilds on other tree species.

18.4.2 RESULTS AND DISCUSSION

(a) Woodland habitat heterogeneity and spatial pattern

The distributions of the phytophagous species varied amongst the available woodlands and it was possible to fit a species–area regression to these

Table 18.1 Leaf-dwelling arthropod species identified on leaves of *Fagus sylvatica* collected from woodlands in southern Scotland (from Bevan, 1987; Darlington, 1975; Emmet, 1976; Emmet *et al.*, 1985; Nielsen and Ejlersen, 1977; Pullin, 1985)

Arthropod family	Species	Evidence
Lepidoptera: Nepticulidae (micromoths)	*Stigmella tityrella*	Serpentine mine
	Stigmella hemargyrella	Serpentine mine
Lepidoptera: Gracillariidae (micromoths)	*Phyllonorycter messaniella*	Blister mine
	Phyllonorycter maestingella	Blister mine
Diptera: Cecidomyiidae (gall midges)	*Hartigiola annulipes*	Pouchgall
Coleoptera: Curculionoidea (weevils)	*Rhynchaenus fagi*	Beech leaf-mining weevil
	Phyllobius argentatus	Adult feeding holes
Homoptera: Callaphididae (aphids)	*Phyllaphis fagi*	Wooly aphids
Homoptera: Cicadellidae (leafhoppers)	*Fagocyba cruenta*	Beech leafhopper
Acarina: Eriophyidae (gall mites)	*Eriophyes nervisequus*	Filzgalls
	Eriophyes stenopis typicus	Rollgalls
	Eriophyes macrorhynchus ferrugineus	Pouchgalls

data (Figure 18.1). However, species richness of phytophages was greater in small fragments interconnected by a matrix of tree lines and hedgerows than insular fragments without such a network (Figure 18.2a; $P < 0.001$). Closer examination showed that there was also greater abundance of individual phytophage species in small fragments that were part of the matrix of tree lines and hedgerows, e.g. *Hartigiola annulipes* Hartig. (Diptera: Cecidomyiidae) (Figure 18.2b; initial galls, $P < 0.05$; successful galls, $P < 0.001$). Gall midge and micromoth species (Table 18.1) were

Table 18.2 Variables of different spatial scales which describe the context of the leaf canopy of *Fagus sylvatica* that could affect the distribution and abundance of arboreal arthropods throughout lowland woodlands

Woodland structure	Woodland situation	Tree structure	Tree situation
Number of tree species	Distance to the nearest woodland	Age	Abundance of tree species in the wood
Range of age classes	Distance to nearest woodland in direction of prevailing wind (SW)	Diameter at breast height	Distance to other trees of the same species
Distance between trees		Mean life size	
Complexity of below-canopy flora (species richness, structural dimensions)	Surrounding land use	Timing of budburst and senescence	
Edge structure	Altitude		Distance to the woodland edge
Presence of a network of hedges or tree lines between the woodlands	Level of stress (exposure to weather, soil type)		
Area of woodland			
Woodland cover per unit land area			
Woodland cover connected by hedgerows or treelines to target canopy			
Porosity of woodland (draughtiness)			

absent or in low abundance on *F. sylvatica* in small, insular woods or remnants of neglected shelterbelt systems.

Several variables were related to the distribution of endophage species richness (Table 18.3) and abundance (Table 18.4) in the multiple regression. Phytophage species number related to the draughtiness, litter density, height and species richness of the low vegetation of woodlands, and the extent of a surrounding network of tree lines and hedgerows (Table 18.3). The draughtiness of the woodland edge accounted for c. 67–82% of the between-woodland variation in the combined abundance of endophages (Table 18.4ai,bi). The density of leaf litter, date of budburst, and tree and field-layer plant species richness accounted for c. 58% of the variation in the combined abundance of endophages between trees (Table 18.4aii,bii). The distribution of the species of the endophage guild sampled on *F. sylvatica* covaried with the distribution of endophage species of similar life histories represented on *Quercus robur* and *Betula pendula* across the same woodland system (Dennis et al., 1995).

Assemblages of the general arthropods collected by search and knockdown showed a similar relationship with the tree and woodland variables to that shown for the phytophagous species (Table 18.5a). However, the number of Araneae and

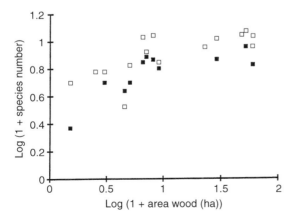

Figure 18.1 Species–woodland area relationships of phytophages on *Fagus sylvatica*. ■, 1990, -area regression, $b = 0.21$, $r^2 = 0.44$. □, 1991, -area, $b = 0.21$, $r^2 = 0.52$. All significant, $P < 0.01$. (Dennis, 1991.)

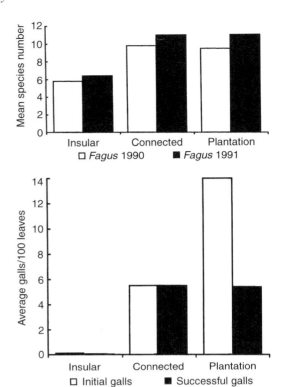

Figure 18.2 (a) Species richness of phytophagous insects in subjective woodland categories on *Fagus sylvatica*, leaves collected from leaf canopies in 1990 and 1991 ($F = 27.98$; $P < 0.001$) (b) Galls of *Hartigiola annulipes* on *F. sylvatica* in different woodland fragments in southern Scotland: initial galls (sum of squares of rank totals = 21 949, $P < 0.05$); successful galls (SOS rank totals = 7736, $P < 0.001$). (Dennis, 1991.)

Coleoptera species was higher at the edge of the woodlands but not where woodlands were draughty, which suggested that they responded to the architectural diversity of plants at the edge of less draughty woodlands (Table 18.5b). These results were consistent with studies of the habitat requirements of individual arthropod species, which can have very precise microhabitat requirements. Adults of many butterfly and moth species required the nectar sources provided by flowering plants and scrub that grew at the woodland edge and in woodland rides and clearings (Fuller and

Table 18.3 Stepwise variable selection and multiple regression of phytophage species number sampled on *F. sylvatica* in (a) 1990 and (b) 1991, compared with variables that characterized woodland structure and situation in southern Scotland

Variables	Coefficients (b, P < 0.05)
(a) $F_{2,33} = 21.64$, $P < 0.001$, $R^2 = 0.54$	
Draughtiness	−0.695
Litter density plus height of ground vegetation*	0.0001
(b) $F_{3,41} = 27.52$, $P < 0.001$, $R^2 = 0.64$	
Hedgerow/tree line connections	0.611
Height ground vegetation	0.095
Litter density	0.0001

*Variables with significant interaction that were combined for the analysis.

Warren, 1995; Greatorex-Davies *et al.*, 1993). Likewise, the dispersing adults of saproxylic Coleoptera species required nectar sources close to the deadwood of tree stumps in which the larvae had fed for many years (Harding and Rose, 1986).

Microclimate can also play a role. For example, the foodplant of the caterpillar of *Hamaeris lucina*, the Duke of Burgundy butterfly, was very common but the precise requirements of the butterfly for the conditions of shade, plant size and surrounding vegetation were not widely available (Sparks *et al.*, 1994). Some species of Staphylinidae (rove beetles) favoured a closed canopy where a few woodland specialist species were active in the forest litter, though abundance and diversity was greater overall, in the open habitats associated with

Table 18.4 Woodland and tree scale variables that affect the distribution and abundance of endophagous micromoth and gall midge species of *F. sylvatica* in (a) 1990 and (b) 1991. Dependent variable: combined abundance, log (1+n), of five endophagous insect species

Variables	Coefficients (b, P <0.05)	Mean square
(a) (i) Woodland variables		
$F_{1,11} = 26.32$ $P < 0.001$, $R^2 = 0.82$		
Draughtiness (SW)	−0.121	2.298
Draughtiness (NE)	−0.147	0.530
(a) (ii) Tree variables		
$F_{1,35} = 14.57$, $P < 0.001$, $R^2 = 0.58$		
Tree species number	0.089	3.373
Relative budburst	0.005	1.797
Litter density	0.002	1.337
(b) (i) Woodland variables		
$F_{1,14} = 26.21$, $P < 0.001$, $R^2 = 0.67$		
Draughtiness (NE)	−0.204	2.117
(b) (ii) Tree variables		
$F_{1,44} = 18.22$, $P < 0.001$, $R^2 = 0.57$		
Tree species number	0.072	1.904
Total plant species number	−0.011	1.180
Litter density	0.003	3.735

Table 18.5 Woodland and tree scale variables that affect the distribution and abundance of ectophagous insect species on *F. sylvatica* in southern Scotland. (a) Dependent variable: ordination on the abundance/distribution of twelve ectophagous insect species in 1991 (Correspondence Analysis, axis 2); (b) number of Araneae and Coleoptera species

Variables	Coefficients (b. $P < 0.05$)	Mean square
(a) (i) Woodland variables $F_{1,14} = 8.54, P < 0.01, R^2 = 0.68$		
Area (log)	1.362	8.638
Draughtiness (SW)	–0.268	1.523
(a) (ii) Tree variables $F_{1,44} = 59.32, P < 0.001, R2 = 0.73$		
Tree species number	0.072	1.904
Total plant species number	–0.011	1.180
Litter density	0.003	3.735
(b) (i) Woodland variables $F_{3,32} = 10.23, P < 0.001, R^2 = 0.46$		
Distance from edge	–0.004	–
Draughtiness	–0.040	–
Height of ground vegetation	0.010	–

forestry (Buse and Good, 1993). Likewise, phytophagous Coleoptera and Heteroptera species declined with increased shade in forest plantations, because most species fed on scrub and low vegetation present around breaks of the forest canopy (Greatorex-Davies *et al.*, 1994).

(b) Edge effects

The role of the woodland edges in providing habitats for numerous arthropod species has been recognized, and this included some rare species (Thomas, 1989). The increase in general arthropod species and abundance at forest edges was identified as an edge effect and interpreted as a positive aspect of woodland and forest structure (Hambler and Speight, 1995). However, increased clearing of boreal forest for agriculture and forestry in Canada exacerbates outbreaks of defoliation of *Populus tremuloides* Michx. by *Malacosoma disstria* Hbn. (Lepidoptera), the forest tent caterpillar. These herbivores accumulate at the edge of forest fragments where natural enemies are absent, by virtue of their movement patterns (Roland, 1993). In a fragmented woodland landscape, edge habitat is not a threatened aspect of the woodland ecosystem and too much emphasis has been placed on the role of the edge 'ecotone'. Many of the arthropod species could also utilize non-woodland habitats in the landscape.

Although fewer in number, the woodland interior species, particularly those that live in the tree canopy, are more specific to woodland and must be considered under greater threat in fragmented woodlands. One of the effects of woodland fragmentation is an increase in the edge exposed to other habitats and environmental influences (Brown, 1991). Edge woodland is warmer, drier and more illuminated and can contain non-woodland plant species (Saunders *et al.*, 1991). The concept of edge effects is well recorded in the literature (Laurance and Yensen, 1991; Brown, 1991; Stamps *et al.*, 1987) and it must be appreciated that the woodland edge is a dynamic entity and must be considered at different spatial scales to define its extent (Murcia, 1995). In the Midlothian study, the 'edge effect' was defined as the interaction between the weather, the structure of the wood and the spatial context of the wood in the woodland system. These factors have an impact on the habitat quality of the tree canopies for arboreal arthropods at different spatial scales and there was previous evidence for edge effects that affect canopy insects on the outer leaves of tree canopies (Nielsen and Ejlersen, 1977) and the outer trees of woodlands (Cappuccino and Root, 1992).

The pattern analyses suggested that wind penetration of woodlands could have direct consequences on the number of canopy species and their abundances in a given woodland fragment (section 18.4.2a). However, several interacting variables were implicated in this effect, such as measurements of the ground vegetation, distance from the woodland edge, litter density, draughtiness and hedgerow connections. To investigate potential mechanisms, further analyses were carried out on the distribution of endophagous species in tree canopies. The species number of phytophages (as listed in Table 18.1) was significantly inversely correlated with the density of leaves under individual trees (Figure 18.3b). In autumn and winter, winds resulted in the redistribution of the woodland litter layer, with local losses and gains. It was tested whether the structural measurements of the woodlands that were related to phytophage distribution and abundance determined the extent of wind penetration into woods.

Woodlands were further classified by association analysis of phytophage species presence or absence on *F. sylvatica* in each woodland (Dennis et al., 1995). Three classes of woodland were represented in sequence of declining species number by the presence of *Hartigiola annulipes*, *Phyllonorycter maestingella* Muller (Lepidoptera: Gracillariidae) and *Stigmella tityrella* Stainton (Lepidoptera: Nepticulidae) (Group 1), presence of *P. maestingella* or *S. tityrella* (Group 2), or the absence of each of these species (Group 3). The distribution of leaves of *F. sylvatica* and the height of the low vegetation was measured at 10 random distances along a southwest to northeast transect through each woodland. Draughtiness scores were calculated for each woodland as the proportion of space to vegetation, horizontally between tree canopies and vertically between the tree canopy and ground vegetation, for southwest and northeast woodland edges. These three measurements were

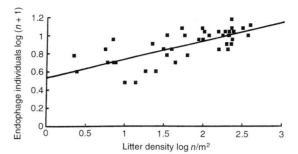

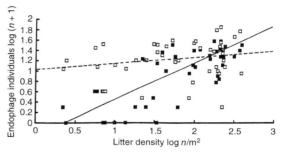

Figure 18.3 (a) Number of phytophagous species in relation to below-canopy leaf-litter density ($r = 0.72$, $P < 0.001$, $r^2 = 0.50$). (b) Response of individual species to litter density according to vulnerability of life history strategy. Broken line, *R. fagi* abundance: $r = 0.22$, not significant, $r^2 = 0.02$. Solid line, abundance of *H. annulipes*, *Phyllonorycter* spp. and *Stigmella* spp. combined: $r = 0.77$, $P < 0.001$, $r^2 = 0.58$. (Dennis, unpublished.)

compared across woodland groups using one-way ANOVA (Figure 18.4). Overall, woods of Group 3 (phytophagous species-poor) related to woods with greater draughtiness and a depressed litter density (Figure 18.4). The height of the low vegetation was not significantly different between these woodland classes. This resulted from high variance in vegetation height in plantation woodland, which may have reflected the localized depression of vegetation growth due to shade under the canopies of mature trees.

The different life histories of the canopy phytophagous species of *F. sylvatica* may have determined the extent of the impact of winter loss of leaf litter. Micromoth species that overwintered in the leaf within blister mines [*Phyllonorycter messaniella* Zeller and *P. maestingella* Muller (Lepidoptera: Gracillariidae)] or within galleries

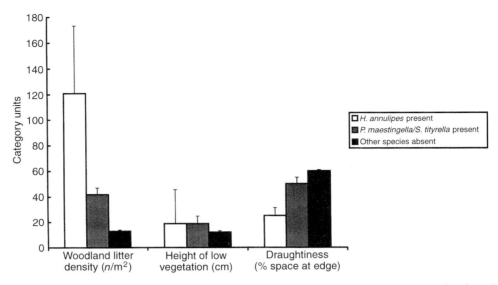

Figure 18.4 Characteristics of Midlothian woodlands of three classes based on association analysis of species presence/absence data (Dennis et al., 1995). Species unique to each group are indicated in key in sequence of declining species number. Analysis of variance on habitat measurements: litter density, $F_{2,79} = 10.15$, $P < 0.001$; height of low vegetation, $F_{2,79} = 2.20$, not significant; draughtiness at edge, $F_{2,23} = 23.54$, $P < 0.001$.

[(*Stigmella tityrella* Stainton and *S. hemargyrella* Kollar (Lepidoptera: Nepticulidae)] would be blown from the wood in the litter (Dennis et al., 1995). Indeed, the initial numbers of mines reflect the absence or low numbers of adult *P. maestingella* and *S. tityrella* available to oviposit in woods of Groups 2 and 3 in spring (Figure 18.5). Larvae of the gall midge *H. annulipes* which overwintered in leaf galls that fell to the ground before leaf fall (Darlington, 1975) may not be redistributed by leaf-litter loss by wind, but the larvae in fallen galls may have required the insulation of the litter layer through the winter. There was an absence of initial galls in woods of Groups 2 and 3 (Figure 18.5) that reflected the low densities of leaf litter (Figure 18.4). Finally, larvae of the leaf-mining weevil *Rhynchaenus fagi* L. (Coleoptera: Curculionidae) developed in leaves early in summer during a single generation and adults emerged on the low vegetation to locate a suitable overwintering site before leaf fall (Bale, 1984). There was no difference in the number of initiated mines of *R. fagi* across these woodland groups (Figure 18.5).

The data on phytophage abundance were divided according to vulnerability and compared with litter density under the sampled trees (Figure 18.3b). Weevil populations in winter avoided the problems of a winter redistribution of leaves in draughty woodlands whilst the vulnerable groups were positively correlated with litter density (Figure 18.3b). Survivorship of the endophage species through the summer of 1991 also varied across the woodland groups and suggested that exposure of the tree canopies to weather may reduce the suitability of leaves as habitat during that period (Figure 18.6). The extent of physical damage to leaves of the sampled tree was significantly different between the three woodland groups ($F_{2,23} = 3.77$, $P < 0.05$).

(c) Woodland connectivity

An index of tree line and hedgerow network connectivity also partially described the patterns of the phytophage guild (Table 18.3b). Closer examination of the population in the hedgerows explained the possible mechanism. Although woods in the area with *F. sylvatica*/*Crataegus monogyna* hedgerows were small, we found similar densities of the endophage (within-leaf feeding) species in

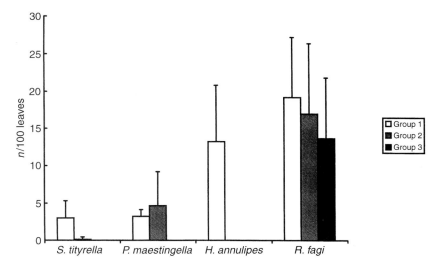

Figure 18.5 Density of mines or galls initiated on *F. sylvatica* in 1991 by four endophage species in woodlands classified by association analysis on species presence or absence. ANOVA between groups: *S. tityrella*, $F_{2,23} = 4.64$, $P < 0.05$; *P. maestingella*, $F_{2,23} = 7.48$, $P < 0.01$; *H. annulipes*, $F_{2,23} = 6.87$, $P < 0.01$; *R. fagi*, $F_{2,23} = 1.35$, not significant.

hedgerow *F. sylvatica* as on canopy *F. sylvatica* (Table 18.6). Moreover, there was an effect of woodland connectivity on hedgerow populations. Fewer species occurred along hedges with no connection to a woodland fragment (Figure 18.7). It was concluded that the population size could be influenced by the contiguous area of *F. sylvatica* canopy available to dispersing adults of the species, which would achieve a greater population. Hedgerows provide a low, closed canopy of leaves that emulates a woodland microclimate.

A further aspect of connectivity is the mutual shelter that a network of hedgerows/shelterbelts or closely spaced woods provided within the wood-

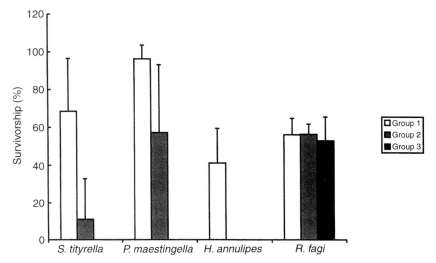

Figure 18.6 Survivorship of mines and galls of four endophage species in 1991 in woodlands classified by association analysis on species presence or absence. ANOVA on arcsine transformed data between groups: *S. tityrella*, $F_{2,23} = 8.66$, $P < 0.01$; *P. maestingella*, $F_{2,23} = 10.92$, $P < 0.001$; *H. annulipes*, $F_{2,23} = 12.29$, $P < 0.001$; *R. fagi*, $F_{2,23} = 0.46$, not significant.

Table 18.6 Comparison of the abundance of arboreal insect species between beech in woodlands and hedges in (a) 1990 and (b) 1991

Characteristic species	Mean mines/galls per 100 leaves	
	Woodlands	Hedges
(a) *1990*		
Rhynchaenus fagi	36.5	14.7*
Rhynchaenus – Chalcidoidea[a]	0.2	0.2
Phyllonorycter maestingella	6.3	10.0
Phyllonorycter – Chalcidoidea[a]	0.1	1.7
Stigmella tityrella	0.9	3.3
Hartigiola annulipes	0.5	0.7
(b) *1991*		
Rhynchaenus fagi	21.5	9.1*
Phyllonorycter messaniella	4.9	1.9*
Phyllonorycter maestingella	2.6	5.2*
Phyllonorycter – Chalcidoidea[a]	0.1	0.6
Stigmella tityrella	1.7	1.2
Hartigiola annulipes	9.1	11.8
Eriophyes nervisequus	25.5	3.4*
Eriophyes stenopsis typicus	31.4	18.7*
Eriophyes macrorhynchus ferrugineus	10.8	1.4*

* indicates significant difference between means derived from a *t*-test ($P < 0.05$).
[a] Chalcidoidea are parasitoid wasps associated with the indicated host.

land system. The wind velocity at a given woodland edge was related to the larger-scale influence of habitat geometry and the juxtaposition of habitat fragments (Heisler and Dewalle, 1988; Green *et al.*, 1991). Particularly for smaller fragments, the habitat quality of canopy leaves for populations of endophages was improved by the shelter effect (Figure 18.2). The reduced wind speeds within such a network of woods, shelterbelts and hedgerows could also increase immigration rates through an accumulation of dispersing insects (Pasek, 1988). Immigrants that reached exposed woods would find no favourable sites on host trees for feeding or oviposition.

18.5 CONCLUSIONS

The main factors that limit phytophages on *Fagus* (draughtiness, number of species and depth of field layer, density of leaf litter and connectivity through hedgerows and lines of trees) would seem to affect phytophages of similar life history strategy on

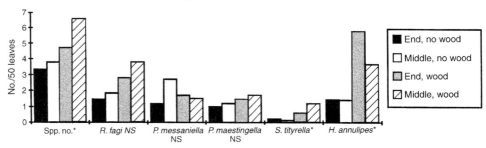

Figure 18.7 Mean number of species and abundance of individual endophage species in four hedgerow categories. Observed distributions of these species were tested against expected distributions which assumed individuals were randomly allocated to each sample (*G*-test: * = $P < 0.05$, NS = not significant). (Dennis, 1993.)

other tree species within the same woods (Dennis *et al.*, 1995). A closely spaced network of small woods, in particular in a matrix of hedgerows or shelterbelts, could generate mutual shelter so that under varied and seasonal weather conditions some sites will always avoid severe wind-mediated edge effects, and summer canopy or winter litter stages of arthropod populations will be maintained in at least a fraction of the available woods. Connor *et al.* (1994) found that when trees were isolated, and only colonized from the leaf litter, local population extinction occurred. Certainly, adults of the micromoths are poor fliers and usually fly at about 1 m above the ground (Woiwod and Stewart, 1990). This lack of flight power may restrict them to sheltered patches of woodland and limit dispersal between woods and reduce species viability (Dempster, 1991).

The corollary of the leaf-litter argument should also be considered. Leaf loss from one woodland may contribute to the leaf litter of other woods. It was assumed that endophagous larvae could survive this passive dispersal and that adult endophages could contribute to the population of a remote woodland site (an example of the sink populations referred to in section 18.3.3c), though they may not survive in an open habitat. In a proportion of the leaves, adult endophages might emerge and fly to other woodlands.

The mechanisms inferred from the results are testable as hypotheses. The first states that phytophagous species are excluded from woodland habitats by the combination of a vulnerable life history strategy, a deficiency in the structure and situation of the wood and an extrinsic environmental impact. The second states that the functional link between a phytophagous species' life history and the situation of its host tree may be emulated by several species adapted to different host tree species. This means that the pattern expressed for phytophagous arthropod species of known life history on a single tree species may indicate the extent of the same guild associated with other tree species of a woodland system. These topics may represent profitable areas of future research.

The results have suggested that the small woods (*c.* 1–5 ha) being established under new incentive schemes (Rodwell and Patterson, 1994) might be able to support more of the potential insect populations than would be predicted by a species–area relationship, such as that described for the epigeal beetle species of woods in the Vale of York, England (Usher *et al.*, 1993). More diverse plant structure and species in the low vegetation of the woodland provided draught-free conditions that entrapped leaves and ensured that there were viable populations of phytophagous arthropods. The habitat diversity also promoted general arthropod diversity associated with woodland vegetation by providing microhabitat or microclimatic conditions for the species to feed, locate a mate or oviposit (section 18.4.2a; Dennis *et al.*, 1994).

The history of management of British woodlands has restricted the potential arthropod species available to benefit from an increase in the area of native woodlands (sections 18.2.1, 18.2.2 and 18.3.1) and it is necessary to retain undisturbed ancient or semi-natural woodland for the species and age diversity of trees and deadwood which provide for the saproxylic species least likely to occur in new woods and as a source of colonists for new native woodlands (sections 18.2, 18.3 and 18.4.2a).

Draught-free structurally diverse woodlands could be established with the necessary diversity of microhabitats and microclimate for the widest potential diversity of woodland arthropods. The results of the current study agree with guidelines on woodland management that woodlands should be managed to provide scrubby edges, open spaces and low vegetation (Insley, 1988). Further, the consideration of the larger-scale patterns of phytophages in woodlands suggested that woodlands should be established in locations where mutual shelter can be maximized – for instance, within established hedgerow or shelterbelt networks or close to existing woodlands. These rules could be incorporated by landscape designers into large-scale plans for forestry in order to promote the conservation of insect diversity (Anon., 1995).

ACKNOWLEDGEMENTS

This work was supported by a research grant awarded to Dr Allan D. Watt by the Forestry Research Co-ordination Committee as part of the

Natural Environment Research Council Special Topic Programme in Farm Forestry. Thanks to Allan Watt for his support during the project and paper writing. Thanks also to Drs Gary L.A. Fry and Richard J. Hobbs for interesting discussions about edge effects and habitat fragmentation.

REFERENCES

Abele, L. and Connor, E. (1979) Application of island biogeography theory to refuge design: making the right decision for the wrong reasons. In *Proceedings of the 1st Conference on Scientific Research in the National Parks* (ed. R. Linn), pp. 89–94, US Department of the Interior, Washington DC.

Anon. (1995) *Forest Landscape Design*, The Forestry Authority, HMSO, London.

Baguette, M. (1993) Habitat selection of carabid beetles in deciduous woodlands of southern Belgium. *Pedobiologia* **37**, 365–378.

Bale, J. (1984) Bud burst and success of the beech weevil *Rhynchaenus fagi* L.: feeding and oviposition. *Ecological Entomology* **9**, 139–148.

Bevan, D. (1987) *Forest Insects: a Guide to Insects Feeding on Trees in Britain*, HMSO, London.

Birks, H.J.B. (1980) British trees and insects: a test of the time hypothesis over the last 13 000 years. *American Naturalist* **115**, 600–605.

Brown, K. (1991) Conservation of neotropical environments: insects as indicators. In *The Conservation of Insects and their Habitats* (eds N. Collins and J. Thomas), pp. 350–404, Academic Press, London.

Burkey, T.V. (1989) Extinction in nature reserves: the effect of fragmentation and the importance of migration between reserve fragments. *Oikos* **55**, 75–81.

Buse, A. and Good, J.E.G. (1993) The effects of conifer forest design and management on abundance and diversity of rove beetles (Coleoptera: Staphylinidae): implications for conservation. *Biological Conservation* **64**, 67–76.

Cappuccino, N. and Root, R.B. (1992) The significance of host patch edges to the colonisation and development of *Corythucha marmorata* (Hemiptera: Tingidae). *Ecological Entomology* **17**, 109–113.

Cates, R. and Orians, G. (1975) Successional status and palatability of plants to generalised herbivores. *Ecology* **56**, 410–418.

Claridge, M.F. and Evans, H.F. (1990) Species–area relationships: relevance to pest problems of British trees? In *Population Dynamics of Forest Pests* (eds A.D. Watt, S.R. Leather, M.D. Hunter and N.A.C. Kidd), pp. 59-69, Intercept, Andover.

Cole, B. (1981) Colonizing abilities, island size and the number of species on archipelagos. *American Naturalist* **117**, 629–638.

Connor, E.F. and McCoy, E.D. (1979) The statistics and biology of the species–area relationship. *American Naturalist* **113**, 791–833.

Connor, E.F., Adams-Manson, R.H., Carr, T.G. and Beck, M.W. (1994) The effects of host plant phenology on the demography and population dynamics of the leaf-mining moth, *Cameraria hamadryadella* (Lepidoptera: Gracillariidae). *Ecological Entomology* **19**, 111–120.

Crist, T.O. and Wiens, J.A. (1994) Scale effects of vegetation on forager movement and seed harvesting by ants. *Oikos* **69**, 37–46.

Crist, T., Guertin, D., Wiens, J.A. and Milne, B.T. (1992) Animal movement in heterogeneous landscapes: an experiment with *Eleodes* beetles in shortgrass prairie. *Functional Ecology* **6**, 536–544.

Darlington, A. (1975) *Pocket Encyclopaedia of Plant Galls*, Blandford Press, Dorset.

Dempster, J. (1991) Fragmentation, isolation and mobility of insect populations. In *The Conservation of Insects and their Habitats* (eds N. Collins and J. Thomas), pp. 143–153, Academic Press, London.

Dempster, J.P., Atkinson, D.A. and Cheeseman, O.D. (1995) The spatial population dynamics of insects exploiting a patchy-food resource. 1. Population extinction and regulation. *Oecologia* **104**, 340–353.

Dennis, P. (1991) Fragmentation of farm woodland: processes affecting the distribution and fitness of arboreal insect species. First report to the FRCC Special Topic Programme in Farm Forestry. Institute of Terrestrial Ecology, Edinburgh Research Station, Bush Estate, Penicuik, Midlothian, Scotland EH26 0QB. 51 pp.

Dennis, P. (1993) Fragmentation of farm woodland: factors which limit the distribution of arboreal insect populations. In *Landscape Ecology in Britain* (ed. R.H. Haines-Young), pp. 125–126, Working Paper 21, Department of Geography, The University of Nottingham.

Dennis, P., Robinson, R. and Ferron, S.E. (1994) Functional consequences of farm woodland fragmentation on insects, small mammal and bird populations. In *Fragmentation in Agricultural Landscapes* (ed. J. Dover), pp. 145–152, International Association for Landscape Ecology (UK), Myerscough College, Preston.

Dennis, P., Usher, G.B. and Watt, A.D. (1995) Lowland woodland structure and pattern and the distribution of arboreal, phytophagous arthropods. *Biodiversity and Conservation* **4**, 728–744.

Diamond, J. (1975) The island dilemma: lessons of modern biogeographic studies for the design of natural reserves. *Biological Conservation* **7**, 129–146.

Drake, J. (1988) Biological invasions into nature reserves. *Trends in Ecology and Evolution* **3**, 186–187.

Duelli, P., Studer, M., Marchand, I. and Jakob, S. (1990) Population movements of arthropods between natural and cultivated areas. *Biological Conservation* **54**, 193–207.

Dunning, J.B., Danielson, B.J. and Pulliam, H.R. (1992) Ecological processes that affect populations in complex habitats. *Oikos* **65**, 169–175.

Emmet, A.M. (1976) Nepticulidae. In *The Moths and Butterflies of Great Britain and Ireland. Vol. 1. Micropterigidae–Heliozelidae* (ed. J. Heath), pp. 171–267, Blackwell Scientific Publications, Oxford, and Curwen Press Ltd, London.

Emmet, A.M., Watkinson, I.A. and Wilson, M.R. (1985) Gracillariidae. In *The Moths and Butterflies of Great Britain and Ireland. Vol. 2. Cossidae–Heliodinidae* (eds J. Heath and A.M. Emmet), pp. 244–362, Harley Books, Colchester, Essex.

Fahrig, L. and Merriam, G. (1985) Habitat patch connectivity and population survival. *Ecology* **66**, 1762–1768.

Forman, R. and Godron, M. (1986) *Landscape Ecology*, Wiley, New York.

Frampton, G.K., Cilgi, T. and Fry, G.L.A. (1995) Effects of grassy banks on the dispersal of some carabid beetles (Col.: Carabidae) on farmland. *Biological Conservation* **71**, 347–355.

Fuller, R.J. and Warren, M.S. (1995) Management for biodiversity in British woodlands – striking a balance. *British Wildlife* **7**, 26–37.

Gilpin, M. and Diamond, J. (1980) Subdivision of reserves and the maintenance of species diversity. *Nature* **285**, 567–568.

Gourov, A.V. (1994) Territorial mosaic and the problem of boundaries (in case of secondary succession). In *Landscape Ecology: Ecologia del paesaggio* (eds D. Cattaneo and P. Semenzato), pp. 97–117 Corso di Cultura in Ecologia, University of Padova.

Grace, J. (1988) Plant response to wind. *Agriculture, Ecosystems and Environment* **22/23**, 71–88.

Greatorex-Davies, J.N., Sparks, T.H., Hall, M.L. and Marrs, R.H. (1993) The influence of shade on butterflies in rides of coniferised lowland woods in southern England and implications for conservation management. *Biological Conservation* **63**, 31–41.

Greatorex-Davies, J.N., Sparks, T.H. and Hall, M.L. (1994) The response of Heteroptera and Coleoptera species to shade and aspect in rides of coniferised lowland woods in southern England. *Biological Conservation* **67**, 255–273.

Green, S.R., Hutchings, N.J., Grace, J. and Greated, C. (1991) Shelter effects in agroforestry. *Agroforestry in the UK* **2**, 14–17.

Haila, Y. (1986) On the semiotic dimension of ecological theory: the case of island biogeography. *Biology and Philosophy* **1**, 377–387.

Haila, Y. (1990) Toward an ecological definition of an island: a northwest European perspective. *Journal of Biogeography* **17**, 561–568.

Hambler, C. and Speight, M.R. (1995) Biodiversity conservation in Britain: science replacing tradition. *British Wildlife* **6**, 137–147.

Hanski, I. (1989) Metapopulation dynamics: does it help to have more of the same? *Trends in Ecology and Evolution* **4**, 113–114.

Hanski, I. (1991) Single-species metapopulation dynamics: concepts, models and observations. *Biological Journal of the Linnean Society* **42**, 17–38.

Hanski, I. (1994a) Patch-occupancy dynamics in fragmented landscapes. *Trends in Ecology and Evolution* **9**, 131–135.

Hanski, I. (1994b) A practical model of metapopulation dynamics. *Journal of Applied Ecology* **63**, 151–162.

Hanski, I. and Gilpin, M. (1991) Metapopulation dynamics: brief history and conceptual domain. *Biological Journal of the Linnean Society* **42**, 3–16.

Hanski, I. and Hammond, P. (1995) Biodiversity in boreal forests. *Trends in Ecology and Evolution* **10**, 5–6.

Harding, P.T. and Rose, F. (1986) *Pasture Woodlands in Lowland Britain. A Review of their Importance for Wildlife Conservation*, Institute of Terrestrial Ecology, Monks Wood, Huntingdon.

Harris, E. and Harris, J. (1991) *Wildlife Conservation in Managed Woodlands and Forests*, Blackwell Scientific Publications, Oxford.

Harris, L. (1984) *The Fragmented Forest. Island Biogeography Theory and the Preservation of Biotic Diversity*, University of Chicago Press, London.

Harrison, S., Murphy, D.D. and Ehrlich, P.R. (1988) Distribution of the Bay checkerspot butterfly, *Euphydryas editha bayensis*: evidence for a metapopulation model. *American Naturalist* **132**, 360–382.

Heatwole, H. and Levin, R. (1972) Trophic structure stability and faunal change during recolonisation. *Ecology* **53**, 531–534.

Heisler, G. and Dewalle, D. (1988) Effects of windbreak structure on wind flow. *Agriculture, Ecosystems and Environment* **22/23**, 41–69.

Higgs, A. (1981) Island biogeography theory and nature reserve design. *Journal of Biogeography* **8**, 117–124.

Higgs, A. and Usher, M. (1980) Should nature reserves be large or small? *Nature* **285**, 568–569.

Hobbs, R. (1992) The role of corridors in conservation: solution or bandwagon? *Trends in Ecology and Evolution* **7**, 389–392.

Insley, H. (1988) *Farm Woodland Planning*, HMSO, London.

Jarvinen, O. (1982) Conservation of endangered plant populations: single large or several small reserves. *Oikos* **38**, 301–307.

Kirby, P. (1992) *Habitat Management for Invertebrates: a Practical Handbook*, Royal Society for the Protection of Birds/Joint Nature Conservancy Committee, Peterborough.

Laurance, W. and Yensen, F. (1991) Predicting the impacts of edge effects in fragmented habitats. *Biological Conservation* **55**, 77–92.

Lawton, J. and Price, P. (1979) Species richness of parasites on hosts: agromyzid flies on the British Umbelliferae. *Journal of Animal Ecology* **48**, 619–638.

Lawton, J. and Schroder, D. (1977) Effects of plant type, size of geographical range and taxonomic isolation on number of insect species associated with British plants. *Nature* **265**, 137–140.

Levin, S.A. (1976) Population dynamic models in heterogeneous environments. *Ann. Rev. Ecol. Syst.* **7**, 287–310.

Liebhold, A.M., Rossi, R.E. and Kemp, W.P. (1993) Geostatistics and Geographic Information Systems in applied insect ecology. *Annual Review of Entomology* **38**, 303–327.

Lord, J.M. and Norton, D.A. (1990) Scale and the spatial concept of fragmentation. *Conservation Biology* **4**, 197–202.

MacArthur, R.H. and Wilson, E.O. (1967) *The Theory of Island Biogeography*, Princeton University Press, New Jersey.

Maiorana, V. (1978) What kind of plants do herbivores really prefer? *American Naturalist* **112**, 631–635.

Margules, C., Higgs, A.J. and Rafe, R.W. (1982) Modern biogeographic theory: are there any lessons for nature reserve design? *Biological Conservation* **24**, 115–128.

McDonnell, M. and Picket, S. (1988) Connectivity and the theory of landscape ecology. In *Connectivity in Landscape Ecology. Proceedings of the 2nd International Seminar of the International Association for Landscape Ecology* (ed. K. Schreiber), pp. 17–19, Munstersche Geographische Arbeiten, Munster.

Moran, V. and Southwood, T. (1982) The guild composition of arthropod communities in trees. *Journal of Animal Ecology* **51**, 289–306.

Murcia, C. (1995) Edge effects in fragmented forests: implications for conservation. *Trends in Ecology and Evolution* **10**, 58–62.

Naveh, Z. and Lieberman, A. (1984) *Landscape Ecology: Theory and Application*, Springer-Verlag, New York.

Nee, S. and May, R. (1992) Dynamics of metapopulations: habitat destruction and competitive coexistence. *Journal of Animal Ecology* **61**, 37–40.

Nielsen, B.O. and Ejlersen, X. (1977) The distribution pattern of herbivory in a beech canopy. *Ecological Entomology* **2**, 293–299.

Nisbet, R.M. and Gurney, W.S.C. (1982) *Modelling Fluctuation Populations*, John Wiley and Sons, London.

Olivieri, I., Couvet, D. and Gouyan, P.-H. (1990) The genetics of transient populations: research at the metapopulation level. *Trends in Evolution and Ecology* **5**, 207–210.

Pasek, J. (1988) Influence of wind and windbreaks on local dispersal of insects. *Agriculture, Ecosystems and Environment* **22/23**, 539–554.

Peterken, G. (1993) *Woodland Conservation and Management*, 2nd edn, Chapman & Hall, London.

Price, P. (1977) General concepts on the evolutionary biology of parasites. *Evolution* **31**, 405–420.

Pulliam, R. (1988) Sources, sinks and population regulation. *American Naturalist* **132**, 652–661.

Pullin, A.S. (1985) A simple life table study based on development and mortality in the beech leaf mining weevil, *Rhynchaenus fagi* L. *Journal of Biological Education* **19**, 152–156.

Quinn, J. and Hastings, A. (1987) Extinction in subdivided habitats. *Conservation Biology* **1**, 198–208.

Rodwell, J. and Patterson, G. (1994) *Creating New Native Woodlands*, HMSO, London.

Roland, J. (1993) Large-scale fragmentation increases the duration of tent caterpillar outbreak. *Oecologia* **93**, 25–30.

Saunders, D.A., Hobbs, R.J. and Margules, C.R. (1991) Biological consequences of ecosystem fragmentation: a review. *Conservation Biology* **5**, 18–32.

Scriber, J. (1978) The effects of larval feeding specialisation and plant growth form upon the consumption and utilisation of plant biomass and nitrogen: an ecological consideration. *Entomologia experimentalis et applicata* **24**, 694–710.

Shaffer, M. (1981) Minimum population sizes for species conservation. *Bioscience* **31**, 131–134.

Shaffer, M. and Samson, F. (1985) Population size and extinction: a note on determining initial population sizes. *American Naturalist* **125**, 144–152.

Shure, D.J. and Philips, D.L. (1991) Patch size of forest openings and arthropod populations. *Oecologia* **86**, 325–334.

Simberloff, D. (1976) Species turnover and equilibrium island biogeography. *Science* **194**, 572–578.

Simberloff, D. (1988) The contribution of population and community biology to conservation science. *Annual Review of Ecology and Systematics* **19**, 473–511.

Simberloff, D. and Abele, L. (1976) Island biogeography theory and conservation practice. *Science* **191**, 285–286.

Simberloff, D. and Abele, L. (1982) Refuge design and island biogeographic theory: effects of fragmentation. *American Naturalist* **120**, 41–50.

Simberloff, D. and Wilson, E. (1978) Experimental zoogeography of islands: the colonisation of empty islands. *Ecology* **50**, 278–296.

Southwood, T.R.E. (1961) The number of species of insect associated with various trees. *Journal of Animal Ecology* **30**, 1–8.

Sparks, T.H., Porter, K., Greatorex-Davies, J.N. et al. (1994) The choice of oviposition sites in woodland by the Duke of Burgundy butterfly *Hamearis lucina* in England. *Biological Conservation* **70**, 257–264.

Stamps, J.A., Buechner, M. and Krishnan, V.V. (1987) The effects of edge permeability and habitat geometry on emigration from patches of habitat. *American Naturalist* **129**, 533–552.

Strong, D. (1974) The insects of British trees: community equilibrium in ecological time. *Annals of the Missouri Botanical Garden* **61**, 692–701.

Strong, D.R., Lawton, J.H. and Southwood, T.R.E. (1984) *Insects on Plants. Community Patterns and Mechanisms*, Blackwell Scientific Publications, Oxford.

Taylor, P.D., Fahrig, L., Henein, K. and Merriam, G. (1993) Connectivity is a vital element of landscape structure. *Oikos* **68**, 571–573.

Terborgh, J. (1976) Island biogeography and conservation: strategy and limitations. *Science* **193**, 1029–1030.

Thomas, C.D. (1991) Habitat use and geographic ranges of butterflies from the wet lowlands of Costa Rica. *Biological Conservation* **55**, 269–282.

Thomas, C.D. and Harrison, S. (1992) Spatial dynamics of a patchily distributed butterfly species. *Journal of Animal Ecology* **61**, 437–446.

Thomas, C.D. and Jones, T.M. (1993) Partial recovery of a skipper butterfly (*Hesperia comma*) from population refuges: lessons for conservation in fragmented landscape. *Journal of Animal Ecology* **62**, 472–481.

Thomas, J.A. (1989) Ecological lessons from the reintroduction of Lepidoptera. *The Entomologist* **108**, 56–68.

Thomas, J.A. and Morris, M.G. (1994) Patterns, mechanisms and rates of extinction among invertebrates in the UK. *Philosophical Transactions of the Royal Society of London, Series B* **344**, 47–54.

Usher, M., Brown, A. and Bedford, S. (1992) Plant species richness in farm woodlands. *Forestry* **65**, 1–13.

Usher, M., Field, J. and Bedford, S. (1993) Biogeography and diversity of ground dwelling arthropods in farm woodlands. *Biodiversity Letters* **1**, 54–62.

Van Apeldoorn, R.C., Oostenbrink, W.T., van Winden, A. and van der Zee, F.F. (1992) Effects of habitat fragmentation on the bank vole, *Clethrionomys glareolus*, in an agricultural landscape. *Oikos* **65**, 265–274.

Van der Zee, F.F., Wiertz, J., Ter Braak, C.J.F. et al. (1992) Landscape change as a possible cause of the badger *Meles meles* L. decline in The Netherlands. *Biological Conservation* **61**, 17–22.

Van Horne, B. (1983) Density as a misleading indicator of habitat quality. *Journal of Wildlife Management* **47**, 893–901.

Verboom, B. and Van Apeldoorn, R. (1990) Effects of habitat fragmentation on the red squirrel, *Sciures vulgaris* L. *Landscape Ecology* **4**, 171–176.

Verboom, J., Schotman, A., Opdam, P. and Metz, J.A.J. (1991) European nuthatch metapopulations in a fragmented agricultural landscape. *Oikos* **61**, 149–156.

Waring, P. (1989) *General Requirements of Moths in Woodland*, Nature Conservancy Council Moth Conservation Project News, Bulletin no. 15, Nature Conservancy Council, Peterborough.

Warren, M.S. (1987) The ecology and conservation of the heath fritillary butterfly, *Mellicta athalia*. II. Adult population structure and mobility. *Journal of Animal Ecology* **24**, 483–498.

Warren, M.S. and Key, R.S. (1991) Woodlands: past, present and potential for insects. In *The Conservation of Insects and their Habitats* (eds N.M. Collins and J.A. Thomas), pp. 155–211, Academic Press, London.

Warren, M.S. and Thomas, J.A. (1992) Butterfly responses to coppicing. In *Ecology and Management of Coppice Woodlands* (ed. G.P. Buckley), pp. 249–270, Chapman & Hall, London.

Watkinson, A.R. and Sutherland, W.J. (1992) Sources, sinks and pseudo-sinks. *Journal of Animal Ecology* **64**, 126–130.

Wiens, J.A. and Milne, B.T. (1989a) Spatial scaling in ecology. *Functional Ecology* **3**, 385–397.

Wiens, J.A. and Milne, B.T. (1989b) Scaling of landscapes in landscape ecology, or, landscape ecology from a beetle's perspective. *Landscape Ecology* **3**, 87–96.

Wiens, J.A., Milne, B.T. and Crist, T.O. (1992) Animal movements and population dynamics in heterogeneous landscapes. *Landscape Ecology* **7**, 56–63.

Wiens, J.A., Crist, T.O. and Milne, B.T. (1993) On quantifying insect movements. *Environmental Entomology* **22**, 709–715.

Wilcox, B.A. and Murphy, D.D. (1985) Conservation strategy: the effects of fragmentation on extinction. *American Naturalist* **125**, 879–887.

Wilson, E. and Willis, E. (1975) Applied biogeography. In *Evolution of Communities* (eds M. Cody and J. Diamond), pp. 522–534, Belknap, Cambridge, Massachusetts.

With, K. and Crist, T.O. (1995) Critical thresholds in species responses to landscape structure. *Ecology* **76**, 2446–2459.

Woiwod, I. and Stewart, A. (1990) Butterflies and moths – migration in the agricultural environment. In *Species Dispersal in Agricultural Habitats* (eds R.H.B. Bunce and D.C. Howard), pp. 189–202, Belhaven Press, London.

Wood, P.A. and Samways, M.J. (1991) Landscape element pattern and continuity of butterfly flight paths in an ecologically landscaped botanic garden, Natal, South Africa. *Biological Conservation* **58**, 149-166.

FICUS: A RESOURCE FOR ARTHROPODS IN THE TROPICS, WITH PARTICULAR REFERENCE TO NEW GUINEA

Yves Basset, Vojtech Novotny and George Weiblen

19.1 INTRODUCTION

Ficus (Moraceae) represents an important component of tropical floras, in terms of both species richness and diversity of growth strategies. Estimates of species richness for *Ficus* range from 740 to 900 species (Janzen, 1979; Berg, 1989), but our survey of the primary taxonomic literature yielded a more conservative estimate of 712 (Corner, 1958, 1965; Berg and Wiebes, 1992). *Ficus* is pantropical in distribution although the Indo-Australian region is the main centre of diversity with over 500 species. A few of the approximately 105 African species are found in Asia. About 130 species are native to the Neotropics. The genus is divided into four subgenera, including the monoecious strangling-figs (subg. *Urostigma*), the monoecious free-standing figs (subg. *Pharmacosycea*), the (gyno)dioecious figs (subg. *Ficus*) and their monoecious relatives (subg. *Sycomorus*; Corner, 1958, 1962). All Neotropical *Ficus* species are monoecious, whereas in Africa a minority of species are (gyno)dioecious. The Indo-Australian region contains the broadest taxonomic distribution of *Ficus*. Each of the four subgenera is represented, numerous sections are endemic and (gyno)dioecious figs are prevalent. New Guinea and Borneo are major centres of diversity. New Guinea is particularly rich, with 135 described species (nearly 20% of the world species count) and a high degree of endemism (53% of species; Corner, 1967).

The high species richness of the genus is mirrored by the variety of growth forms and life histories (trees, shrubs, stranglers, hemiepiphytes, epiphytes, vines, rheophytic species), as well as the diversity of habitats that it occupies (coastlines, swamps, savannas, riparian, lowland and montane forests; Berg, 1989). In tropical forests, *Ficus* is a conspicuous element of both pioneer and climax vegetation. For example, in Indo-Australian secondary forests, (gyno)dioecious free-standing figs are among the most abundant tree species, whereas (gyno)dioecious climbing figs and monoecious stranglers commonly occur in primary forests. Figs may be important in reforesting areas of degraded pasture in the Neotropics (Williams-Linera and Lawton, 1995). Some species, such as *F. microcarpa*, have become invasive pest trees, particularly in Hawaii and Florida (McKey, 1989).

The key taxonomic feature of the genus is the syconium, a receptacle containing many minute flowers which is enclosed by a bract-lined opening (ostiole). This highly specialized inflorescence is pollinated by host-specific agaonid wasps (Hymenoptera: Agaonidae). Fig pollination has attracted considerable scientific interest world-

---

Forests and Insects. Edited by A.D. Watt, N.E. Stork and M.D. Hunter. Published in 1997 by Chapman & Hall, London. ISBN 0 412 79110 2.

wide and generated substantial literature, with particular reference to mutualism, antagonism, host specificity and coevolution (reviews in Janzen, 1979; Wiebes, 1979; Frank, 1989; Herre, 1989; McKey, 1989; Bronstein, 1992), but it is not our aim to treat this subject.

The interest in *Ficus* for conservation studies is equally considerable. The genus is known to attract a wide range of frugivorous animals, including many species of birds, bats, marsupials and primates. Observations of vertebrate feeding habits, the local abundance of figs and their distinctive fruiting schedules led Terborgh (1986) to hypothesize that figs are keystone species. Terborgh supposed that figs could provide a continual supply of food to frugivorous vertebrates and could be particularly important during periods of resource scarcity, due to pollination-level flowering asynchrony which results from the unique system of pollination (Janzen, 1979). However, Mills *et al.* (1993) indicated that the concept of 'keystone species' (Paine, 1969) is not accepted widely and is defined inadequately. Recently, an international workshop was organized to seek consensus on a definition of keystone species (Power and Mills, 1995). The participants re-defined a keystone species as one 'whose impact on its community or ecosystem are large, and much larger, than would be expected from its abundance' (Power and Mills, 1995).

In addition to fig crops being eaten by frugivorous vertebrates, arthropods are also conspicuous primary consumers of the syconia (= 'figs'), foliage and wood of *Ficus* trees. Arthropod consumers may be classified into four main categories, according to the type of resource and how it is used:

1. arthropods feeding on syconia, either internally or externally;
2. arthropods feeding on sap, tapping either the mesophyll, phloem or xylem;
3. insects chewing leaf tissues, either free-living, gall-making or leaf-mining;
4. insects boring into stems (living tissues) and wood (living or early decaying).

If we assume that certain *Ficus* species may be keystone for frugivorous vertebrates, it may be of particular interest to examine whether these same tree species could also be considered as keystone for the community of *Ficus*-feeding arthropods. Such fig species would be invaluable in maintaining local animal diversity.

In this chapter, information about arthropod consumers of *Ficus* is provided. Root-feeding arthropods are not treated because specific information on this guild could not be located. The standard definition of a keystone species set by the international workshop is followed and the impact of a *Ficus* species on the community of *Ficus*-feeding insects is examined in terms of the number of insect species feeding on it. More specifically, three questions are addressed:

1. Which arthropod families and genera are most likely to feed on *Ficus* and are these taxa relatively species-rich and/or host-specific?
2. Is the local composition of arthropod faunas feeding on *Ficus* in New Guinea different from elsewhere, when considered at a higher taxonomic level?
3. Can the keystone species concept be substantiated for certain *Ficus* species, with reference to the *Ficus*-feeding insect community in the Madang area, Papua New Guinea?

To answer these questions, the literature is reviewed and some preliminary data from an ongoing study of selected taxa feeding on 15 species of *Ficus* in New Guinea are presented.

19.2 ARTHROPOD PRIMARY CONSUMERS OF *FICUS*: A LITERATURE REVIEW

19.2.1 METHODS

General surveys and community studies of *Ficus*-feeding arthropods, particularly foliage feeders, are rare (e.g. Picard, 1919; Swailem and Awadallah, 1973; Ozar *et al.*, 1986; Basset *et al.*, 1996). Instead, most of the compiled literature was extracted from the CABAbstracts (1984–1994) and CABPestCD (1973–1995) databases (Centre for Agriculture and Biosciences International, Wallingford, UK). This information was supplemented by taxonomic monographs (indicated later in the text) and with various lists or catalogues providing host records (e.g. Houard, 1922; Lima, 1936; Beeson and Bhatia, 1937, 1939; Condit and

Enderud, 1956; Duffy, 1957, 1960, 1963, 1968; Guagliumi, 1966; Silva et al., 1968; Avidov and Harpaz, 1969; Bruner et al., 1975; Martorell, 1976; Mound and Halsey, 1978; Annecke and Moran, 1982; Bigger, 1988; Williams and Watson, 1988a,b, 1990; Zhang, 1995). Several colleagues supplied additional records and, in particular, G. Robinson and I. Kitching allowed us to extract lepidopteran records from their database 'Hosts' (The Natural History Museum, records as from August 1995). Fig wasps are included in our estimates of arthropod species richness feeding on *Ficus* but are not discussed since the literature is comprehensive (e.g. Wiebes, 1966, 1979, 1994; Hill, 1967; Ramírez, 1970; Boucek, 1988; Berg and Wiebes, 1992; Compton and Hawkins, 1992; Compton and van Noort, 1992). Arthropods feeding on fallen, or dried and stored, syconia are not treated. Arthropods feeding on syconia are treated by G.W., sap-sucking arthropods by V.N. and leaf-chewing and wood-boring insects by Y.B. The database, in which each 'record' represents a different arthropod species, is available from the authors on request.

19.2.2 LITERATURE REVIEW

(a) General patterns and limitations

In total, we found published records of 1875 species feeding on *Ficus*, including 742 species feeding on syconia, 481 sap-sucking species, 369 leaf-chewing species and 283 stem/wood-boring species. It is probable that the actual number of species feeding on *Ficus* is much higher than these figures. Arthropod records were obtained from only 286 species of *Ficus*, but these are likely to be biased towards *Ficus* species of economic importance – for example, edible and ornamental species: *F. carica* (151 arthropod records), *F. elastica* (51 records), *F. retusa* (42 records), *F. sycomorus* (35 records), *F. benghalensis* (35 records), *F. benjamina* (33 records), *F. religiosa* (29 records) – and arthropod pests.

The 20 most speciose arthropod families (Table 19.1), accounting for about 80% of arthropod records, include:

- some polyphagous pests prominent in the literature, such as certain Cerambycidae, Aleyrodidae, Coccidae, Pseudococcidae, Diaspididae, Bostrichidae, Lymantriidae and Noctuidae;
- some specialized and often speciose taxa that exploit particular resources, such as in certain Agaonidae, Drosophilidae, Curculionidae and Tephritidae, feeding internally on syconia; Homotomidae feeding on phloem; certain Phlaeothripidae feeding on mesophyll; and certain Nymphalidae, Crambidae, Bombycidae and Noctuidae, feeding on leaf tissues.

Although these patterns are more conspicuous when examined at generic level (Table 19.2), it should be noted that the number of congeneric species may depend on the availability of recent taxonomic revisions.

Considering broad biogeographic areas, the database included 892 species records for the Indo-Australian region, 331 Afrotropical, 310 Neotropical, 91 Palaearctic, 33 Nearctic and 76 cosmopolitan records. It is probable that Palaearctic records are inflated by arthropods feeding on cultivated figs (*F. carica* and *F. sycomorus*) and that the numbers of Neotropical and cosmopolitan species are underestimated. In particular, for the former, a compilation of Neotropical Agaoninae was not available. Approximately 5% of arthropod records came from New Guinea, in comparison with New Guinean species of *Ficus* contributing nearly 20% of the total. This suggests that arthropods feeding on *Ficus* in New Guinea may have been undersampled.

(b) Arthropods feeding on syconia

Internal feeders

The fig wasps (Agaonidae) are the largest group of internal feeders (Tables 19.1 and 19.2), consisting of pollinating wasps (335 records; all in Agaoninae), and 'parasitoids' (186 records). Many of the so-called parasitoid fig wasps are phytophagous, feeding on the seed contents of galled fig flowers, but the life history of many species is not well understood (e.g. Boucek, 1988; Compton and van Noort, 1992; Compton et al., 1994). For example, some species are capable of galling fig flowers in the absence of wasp larvae (i.e.

Table 19.1 The 20 most speciose arthropod familes feeding on *Ficus*, recorded from review of literature

Family	No. species	Family	No. species
1. Agaonidae (Hym.)	529	11. Diaspididae (Hem.)	31
2. Cerambycidae (Col.)	225	12. Nymphalidae (Lep.)	31
3. Aleyrodidae (Hem.)	82	13. Phlaeothripidae (Thys.)	30
4. Drosophilidae (Dipt.)	73	14. Triozidae (Hem.)	28
5. Curculionidae (Col.)	61	15. Aphididae (Hem.)	28
6. Noctuidae (Lep.)	52	16. Eriophyidae (Ac.)	27
7. Homotomidae (Hem.)	49	17. Chrysomelidae (Col.)	25
8. Coccidae (Hem.)	42	18. Tephritidae (Dipt.)	24
9. Crambidae (Lep.)	41	19. Bostrichidae (Col.)	22
10. Pseudococcidae (Hem.)	31	20. Lymantriidae (Lep.)	21

Apocryptophagus: Godfray, 1988; G.W., personal observation). It is accepted that each *Ficus* species has a host-specific pollinator (Wiebes, 1987; but exceptions were noted in Ramírez, 1970; Compton, 1990; Wiebes, 1994) but, as yet, wasps are known from only one-third of all *Ficus* species (Wiebes, 1994). 'Parasitoid' fig wasps appear to be less host specific than the pollinators (Compton and van Noort, 1992).

Many other arthropods feed internally on syconia, mostly in the Coleoptera, Diptera and Lepidoptera. Perrin (1992) reported that 35 African species of *Curculio* (Curculionidae) are specialists feeding on syconia, whilst others feed on Fagales. In Australia, some species also feed on syconia, but *C. bicruciatus* breed in the fruits of *Syzygium* (A. Howden, personal communication). Other weevil species also feed within syconia, such as African species of *Omophorus* and Neotropical species of *Ceratopus*, and, perhaps, of *Geraeus* (A. Howden, personal communication). One species of the former is reported to be a pest of cultivated figs in

Table 19.2 The 36 most speciose arthropod genera feeding on *Ficus*, recorded from review of literature

Genus	No. species	Genus	No. species
1. *Ceratosolen* (Agaonidae) *	62	19. *Euploea* (Nymphalidae)	14
2. *Pegoscaphus* (Agaonidae) *	45	20. *Asota* (Noctuidae)	14
3. *Drosophila* (Drosophilidae)	40	21. *Elisabethiella* (Agaonidae) *	14
4. *Curculio* (Curculionidae)	35	22. *Pleistodontes* (Agaonidae) *	14
5. *Apocrypta* (Agaonidae)	35	23. *Lissocephala* (Drosophilidae)	14
6. *Apocryptophagus* (Agaonidae)	30	24. *Courtella* (Agaonidae) *	13
7. *Blastophaga* (Agaonidae) *	29	25. *Eupristina* (Agaonidae) *	12
8. *Krabidia* (Agaonidae) *	23	26. *Macrohomotoma* (Homotomidae)	11
9. *Pauropsylla* (Triozidae)	22	27. *Aleuroplatus* (Aleyrodidae)	11
10. *Homotoma* (Homotomidae)	20	28. *Zaprionus* (Drosophilidae)	10
11. *Dialeurodes* (Aleyrodidae)	20	29. *Choreutis* (Choreutidae)	9
12. *Agaon* (Agaonidae) *	19	30. *Glyphodes* (Crambidae)	9
13. *Wiebesia* (Agaonidae) *	18	31. *Dolichoris* (Agaonidae) *	9
14. *Waterstoniella* (Agaonidae) *	17	32. *Ceroplastes* (Coccidae)	8
15. *Liporrhopalum* (Agaonidae) *	16	33. *Camarothorax* (Agaonidae)	8
16. *Platyscapa* (Agaonidae) *	15	34. *Aceria* (Eriophyidae)	7
17. *Gynaikothrips* (Phlaeothripidae)	14	35. *Stathmopoda* (Oecophoridae)	7
18. *Clusiosoma* (Tephritidae)	14	36. *Alfonsiella* (Agaonidae) *	7

* Pollinators

South Africa (Annecke and Moran, 1982). However, many fig species have yet to be examined for Curculionidae, particularly in Australasia.

In Diptera, records of Cecidomyiidae include several Indo-Australian genera and Nearctic species of *Ficiomyia*. Records of Lonchaeidae include several Palaearctic and Indo-Australian species, including some in *Lonchaea*. In Tephritidae, in addition to several rather polyphagous and cosmopolitan species, at least the genus *Clusiosoma* in New Guinea appears to be species-rich and specialized on *Ficus* (Hardy, 1986). Many Drosophilidae feed on syconia, but most records are Afrotropical. *Lissocephala* are obligate fig-breeders, with some species being restricted to a single species of *Ficus*. However, *Zapronius* and *Drosophila* appear to be feeding facultatively on yeast growing in the decaying syconia. Adults of *Lissocephala* lay eggs in and around the ostiole, through which first instar larvae enter the syconium cavity. Second and third instars feed inside the syconia and then emerge to pupate in the ground. Cohabiting *Lissocephala* species have slightly different oviposition sites, suggestive of niche partitioning. Temporal separation of niche was invoked to account for the succession of *Lissocephala*, *Zapronius* and *Drosophila* species feeding in the same syconia (Lachaise, 1977; Lachaise *et al.*, 1982; Couturier *et al.*, 1986). There are records of Phoridae reared from syconia, but larvae may feed mostly on dead fig wasps (Compton and Disney, 1991).

In Lepidoptera, families Crambidae, Oecophoridae and Pyralidae are also known to feed internally on syconia. Notably, many crambid records involve Indo-Australian species of *Cirrhochrista*, whereas oecophorid records include several Indo-Australian species of *Stathmopoda*.

External feeders

In comparison with internal feeders, there are relatively few records of arthropods that feed externally (only 35 out of 742 records) and a conspicuous part of this is species of Lygaeidae (Hemiptera). Many arboreal records are Afrotropical, including species of *Dinomachus* and *Appolonius*, the latter also known from Australia (Slater, 1972). Several polyphagous species of *Cotinis* (Coleoptera: Scarabaeidae: peach beetle) feed on cultivated figs in the United States (Ebeling, 1959). Several species of fruit-piercing moths in the Noctuidae also feed on syconia.

(c) Sap-sucking arthropods

Species feeding on *Ficus* are found within Acari, Hemiptera and Thysanoptera. All Acari and Thysanoptera are mesophyll feeders, whereas Hemiptera include phloem feeders (Stenorrhyncha: Coccoidea, Aleyrodoidea, Psylloidea and Aphidoidea; and most Auchenorrhyncha), mesophyll feeders (Heteroptera and Auchenorrhyncha: Typhlocybinae) and xylem feeders (Auchenorrhyncha: Cicadellinae, Cicadidae and Cercopidae).

Acari and Thysanoptera account for about 90% of all mesophyll-feeding records. The most species-rich families are gall-making thrips (Phlaeothripidae) and mites (Eriophyidae; Table 19.1). *Ficus* represents one of the most important host plant groups for the Phlaeothripidae (Ananthakrishnan, 1978). At least six genera contain *Ficus*-feeding species, including the species-rich, mainly Indo-Australian *Gynaikothrips*. One species, *G. ficorum*, has become a pest on ornamental *F. retusa* and *F. microcarpa* (Loche *et al.*, 1984; Paine, 1992). Gall-making thrips exhibit complex interspecific interactions, with only some species capable of gall induction, and others living in these galls as inquilines. In contrast with the Acari and Thysanoptera, only a few records of mesophyll-feeding hemipterans are available, namely Tingidae, Coreidae, Miridae and Cydnidae (Heteroptera), as well as several species of Typhlocybinae (Cicadellidae).

Phloem feeders are the most species-rich guild of sap-sucking herbivores on *Ficus*. They are dominated by three stenorrhynchous groups: Aleyrodoidea, Psylloidea and Coccoidea, while Aphidoidea and Auchenorrhyncha are poorly represented. In particular, three psyllid families – Psyllidae, Triozidae and Homotomidae – feed on *Ficus*, the latter family being limited almost exclusively to *Ficus*. Most of the homotomids are found in the Indo-Australian region (*Macrohomotoma*, *Mycopsylla*, *Homotoma* in part), though some are distributed also in the Afrotropical (e.g. *Pseudoeriopsylla*) and Neotropical (*Synoza*) regions. In Triozidae, the gall-making genus *Pauropsylla* is restricted to *Ficus* in the Old World tropics (Hollis and Broomfield, 1989). In general, psyllids exhibit narrow host specificity, and some genera are restricted to a single subgenus or section of *Ficus* (Hollis and Broomfield, 1989).

In contrast with generally host-specific psyllids, a large number of the scale insects and aleyrodids reported from *Ficus* include many polyphagous, often pest, species (e.g., *Ceroplastes rubens*, *Coccus viridis*, *Rastrococcus invadens*, *Dialeurodes citrifolii*; Mound and Halsey, 1978; Williams and Watson, 1988a,b, 1990). Polyphagous pests are numerous among the aphid species reported from *Ficus* (e.g. *Aphis craccivora*, *A. gossypii*, *A. fabae*, *Myzus persicae*; Blackman and Eastop, 1994). Although global species richness of phloem-feeding Auchenorrhyncha and Stenorrhyncha are similar, there are only 32 species of Auchenorrhyncha recorded from *Ficus*, in contrast with about 300 species of Stenorrhyncha. Eleven families of Auchenorrhyncha are represented, Cicadellidae being the most species-rich. Xylem-feeding groups are represented in the database by seven species on *Ficus*.

(d) Leaf-chewing insects

Free-living insects

Only one record of Orthoptera (Tettigoniidae) feeding on *Ficus* exists in the database. In Coleoptera, leaf-feeding records appear rather scattered, such as those of Melolonthinae (Scarabaeidae). Several species of Lamiinae (Cerambycidae) which bore the wood of *Ficus* perform maturation feeding by gnawing twigs and young leaves of *Ficus* (e.g. Basset *et al.*, 1996). To date, these records include mostly Indo-Australian species in the genera *Acololepta*, *Epepeotes*, *Olenecamptus* and *Rosenbergia*, but many other examples may exist. Records of Chrysomelidae are surprisingly few and include some rather host-specific Indo-Australian Galerucinae (within the genera *Atysa* and *Poneridia*, for example) and some Neotropical Alticinae (within the genus *Epitrix*). Several other chrysomelids (Eumolpinae: genera *Rhyparidella* and *Rhyparida*) feed on *Ficus*, particularly in New Guinea, but little is known about their host range. Records of Curculionidae feeding on foliage, particularly host-specific species, such as *Viticiina* (Viticiinae) from Papua New Guinea, are even rarer.

Most records of free-living insects chewing the leaves of *Ficus* involve Lepidoptera (at least 23 families and 127 genera recorded). Young leaves are tied together by Indo-Australian species of *Phycodes* (Brachodidae), *Choreutis* and *Tortyra*, and Neotropical species of *Hemerophila* and *Tortyra* (all Choreutidae). For example, *Choreutis nemorana* is considered to be a pest of cultivated figs in Crimea (Tkachuk, 1986). Most Crambidae (Pyraustinae) are leaf-rollers, such as Indo-Australian species of *Glyphodes* and *Talanga* and Afrotropical and Neotropical species of *Margaronia*. Skeletonizing occurs in Indo-Australian species of *Brenthia* (Choreutidae) which often exploit mature leaves by spinning a small web on the underside of leaves and chewing a hole to retreat rapidly to the adaxial surface in case of danger (Y.B., personal observation). Other skeletonizing species occur in Zygaenidae, particularly in Indo-Australian species of *Phauda* (Nageshchandra *et al.*, 1972). All Lycaenidae feeding on *Ficus* are within the subfamily Lycaeninae, which are not obligate myrmecophiles (Fiedler, 1991). Most records involve the genera *Myrina* (Africa), *Iraota* (Malaysia) and *Philiris* (New Guinea and Australia). In this last, the larvae feed on the tissue of the lower surface of leaves, leaving the upper epidermis intact (Parsons, 1984).

Most other lepidopterans eat the margin of leaves. In the Nymphalidae (Ackery, 1988), most records occur in the Limenitinae (Indo-Australian and Afrotropical species of *Cyrestis*, Neotropical species of *Marpesia*), Nymphalinae (*Hypolimnas*, Afrotropical) and, particularly, Danainae (*Euploea*, Indo-Australian, and *Lycorea*, Neotropical). Records of Bombycidae are mostly in the Indo-

Australian Bombycinae (*Gunda* and the '*Ocinara* group'). Several records of Saturniidae exist, but no genera appear to be particularly species-rich on *Ficus*. Records of *Ficus*-feeding Sphingidae are mostly Neotropical and Afrotropical (*Pseudoclanis*). Several polyphagous species in Lymantriidae, mostly in the genera *Euproctis*, *Lymantria* (Indo-Australian records) and *Dasychira* (Afrotropical and Indo-Australian records) feed on *Ficus* (e.g. Verma *et al.*, 1989). Arctiidae feeding on *Ficus* include mostly Neotropical species of *Ammalo* (Arctiinae) and New World species of *Lymire* (Ctenuchinae). Many Noctuidae are recorded from *Ficus* and the Indo-Australian and Afrotropical species of *Asota* appear particularly host-specific (Aganainae; placement of this subfamily follows Holloway, 1988). No other noctuid genera, particularly in the Catocalinae and Ophiderinae, appear particularly speciose and many records involve polyphagous species, like the pest *Achaea janata* in India (Prasad and Singh, 1984). Other lepidopteran records occur in families Psychidae, Oecophoridae, Blastobasidae, Gelechiidae, Metarbelidae, Megalopygidae, Limacodidae, Immidae, Pyralidae, Riodinidae, Lasiocampidae and Notodontidae. Surprisingly, there are very few records of Tortricidae and Geometridae (most Indo-Australian, some polyphagous species).

Several lepidopteran species deactivate the lactifers by chewing a narrow groove across the leaf lamina and mid-vein and then feeding on the distal portion of the leaves, thus avoiding contact with latex (Compton, 1989). This behaviour, known also as trenching (Dussourd and Denno, 1994), has been observed in African and Indonesian species of Sphingidae (*Pseudoclanis*) and Noctuidae (*Asota* and *Chrysodeixis*; Compton, 1989; S. Compton, personal communication). It may occur in some Danainae (Dussourd and Denno, 1994).

Gallers and leaf-miners

Records of leaf galls on *Ficus* are scarce and include mostly Cecidomyiidae, particularly some Indo-Australian species of *Horidiplosis* and *Pipaldiplosis*, as well as some Neotropical species of *Johnsonomya* (Barnes, 1948; Bruner *et al.*, 1975). One species of *Diplonearcha* (Tortricidae) has been recorded as a gall-maker on *Ficus* sp.

Leaf-mining insects are rather scarce on *Ficus*. One species of *Leiopleura* (Buprestidae) may be mining leaves of *Ficus* in Venezuela (Guagliumi, 1966). Other records include *Opostega* (Tineidae) on *F. carica* and *Opogona* (Lyonetiidae) on *F. elastica* in the Palaearctic region, and, most notably, several Indo-Australian species of *Acrocercops* (Gracillariidae).

(e) Stem- and wood-boring insects

Stem-borers

Several lepidopteran species bore into aerial roots, such as Indo-Australian species of *Scalmatica* and *Trachytyla* (Tineidae), as well as some species of *Meteoristis* (Gelechiidae). Records of stem- or twig-boring insects include several lepidopteran families. One species of *Paropta* (Cossidae) bores into the branches of *F. carica* in Israel (Avidoz and Harpaz, 1969). Indo-Australian species of *Indarbela* (Metarbelidae), *Copromorpha* and *Phycomorpha* (Copromorphidae), as well as Neotropical species of *Azochis* (Crambidae), are recorded in the database. Several records are from Sesiidae including some Indo-Australian species of *Carmenta* and *Tinthia*, as well as one Neotropical species of *Ficivora* (Gallego, 1971).

Wood-borers

Wood-borers are often less host specific than foliage-feeding arthropods and many records from the soft, easily rotted wood of *Ficus* (cf. Corner, 1967) include polyphagous species. All the available records are of Coleoptera, such as Buprestidae (some Neotropical species of *Colobogaster*). There are more records of Bostrichidae, particularly in the genera *Sinoxylon* (cosmopolitan) and *Dinoderus* (Indo-Australian). However, most wood-boring records include Cerambycidae. There are few records of Prioninae (Neotropical species of *Parandra*), whereas those of Cerambycinae are more common and include Neotropical species of

Cyllene, Indo-Australian species of *Xylotrechus*, and Palaearctic species of *Hesperophanes* and *Trichoferus*.

Lamiinae are the dominant wood-boring group on *Ficus*. Many species of *Acololepta, Batocera, Cotops, Dihammus, Epepeotes* and *Rosenbergia* are recorded from the Indo-Australian region. Species of *Anisopodus, Oncideres, Taeniotes* and *Trachyderes* are recorded from the Neotropical region, whereas records in Africa include the genera *Phryneta, Phrynetopsis* and *Sternotomis*. Notorious pest species also include some Lamiinae, such as *Psacothea hilaris* in Japan (Fukuda, 1992), *Batocera rufomaculata* in India (Mallikarjuna Rao and Mohan Rao, 1991) and *Phryneta spinator* in South Africa (Annecke and Moran, 1982).

Records of Scolytidae are rare and include several Indo-Australian and cosmopolitan species of *Xyleborus*. *Hypoborus ficus* is a relatively host-specific pest of cultivated figs in Israel and Turkey (Avidov and Harpaz, 1969). Other wood-boring species are scattered in the families Mordellidae, Curculionidae and Platypodidae.

19.3 HERBIVORES FEEDING ON SELECTED *FICUS* SPECIES IN THE MADANG AREA: PRELIMINARY DATA

19.3.1 METHODS

Our field work took place in the Madang area, particularly in primary and secondary forests near Baitabag (145°47' E, 5°8' S, *c.* 100 m) and Ohu (145°41' E, 5°14' S, *c.* 200 m) villages, as well as in coastal areas nearby and on islands close to the mainland. We chose 15 species of *Ficus* trees (Table 19.3), which were locally abundant and easy to recognize in the field, as in most cases insects were collected from infertile saplings or from trees devoid of mature figs. These species include trees of various architectures and which grow in different habitats.

Our collections targeted (a) fig wasps; (b) sap-sucking Auchenorrhyncha (hereafter 'leafhoppers'); and (c) leaf-chewing insects. For each *Ficus* species, we attempted to rear fig wasps from 10 individual trees. Syconia were collected when ripe, but prior to the emergence of the agaonids ('D' phase, according to the developmental series of Galil and Eisikowitch, 1968). Samples ranged from one to 20 syconia (depending on the size of the fig crop) and were stored in sealed plastic bags, where elevated CO_2 levels apparently hasten the emergence of the wasps. After 24–48 hours, wasps were removed, mounted and sorted to morphospecies.

Sap-sucking and leaf-chewing insects were collected by hand collection and beating. Since most tree species are small (< 10 m), trees were climbed or sampled from the ground. Larger trees were accessed with the single rope technique (Perry, 1978). Leafhoppers were collected during day-time from June to September 1995, whereas leaf-chewing insects were collected during both day and night, and from July 1994 to September 1995. Collecting effort was similar for each tree species and amounted, in average and for each *Ficus* species, to 1.5 hours spent in inspecting the foliage and 29 tree-inspections for sap-sucking insects (total 437 tree-inspections for all species), and to 10.9 hours and 165 tree-inspections for leaf-chewing insects (total 2472 tree-inspections for all species). Leaf-chewing species collected in the field were provided with fresh *Ficus* foliage in the laboratory to ensure that these species feed on the *Ficus* species from which they were collected. Leaf-chewing insects were raised to adults whenever possible. Collecting, rearing, mounting and sorting to morphospecies involved the authors, seven technical assistants and 12 collectors. The sampling programme, which surveyed individual trees in a variety of age classes and growing in various habitats, was optimized towards a better estimation of the total number of insect species feeding on the *Ficus* species studied. Assignment of morphospecies (hereafter 'species' for sake of brevity) was verified by various taxonomists.

The abundance of resource available to leaf-feeding insects was estimated as the standing volume of compact foliage per unit area (foliage volume in m^3 per hectare of forest). It was calculated for each of the 15 *Ficus* species as a product of (a) tree density, estimated from 167 surveys, each represented by a 20-minute walk covering an approx-

Table 19.3 Species of *Ficus* on which insect collections were made and their taxonomic placement, architecture, preferred habitats (Hab), leaf texture (Leaf), mode of reproduction (Rep), size of syconia (Syc) and abundance within the study area (Ab)

Ficus species	*Ficus* section	Habitus	Hab[a]	Leaf[b]	Rep[c]	Syc	Ab[d]
F. bernaysii King	Sycocarpus	Small evergreen	F	1	Di	Medium	62
F. botryocarpa Miq.	Sycocarpus	Medium evergreen	R	1	Di	Medium	90
F. conocephalifolia Ridley	Sycidium	Small evergreen	F	1	Di	Medium	140
F. copiosa Steud.	Sycidium	Medium evergreen	R	1	Di	Large	41
F. dammaropsis Diels	Sycocarpus	Small evergreen	R	2	Di	Large	81
F. hispidioides S. Moore	Sycocarpus	Medium evergreen	R	1	Di	Large	95
F. microcarpa L.	Conosycea	Large evergreen	C	2	Mo	Small	< 1
F. nodosa Teysm. & Binn.	Neomorphe	Large deciduous	R	2	Di	Large	67
F. phaeosyce Laut. & K. Schum.	Sycidium	Small evergreen	F	1	Di	Small	106
F. pungens Reinw. ex Bl.	Sycidium	Medium evergreen	R	1	Di	Small	185
F. septica Burm.	Sycocarpus	Small evergreen	R, C	2	Di	Medium	56
F. tinctoria Forst.	Sycidium	Medium evergreen	C	2	Di	Small	25
F. trachypison K. Schum.	Sycidium	Medium evergreen	R	1	Di	Small	66
F. variegata Bl.	Neomorphe	Large deciduous	F, R	2	Di	Large	835
F. wassa Roxb.	Sycidium	Small evergreen	F, R, C	1	Di	Small	290

[a] F = forest; R = regrowth; C = coastal. [b] 1 = scabrid; 2 = smooth. [c] Di = dioecious; Mo = monoecious. [d] Estimations of volume of foliage (m^3) per ha of forest (see text).

imate area of 380 × 4 m, during which all trees taller than 1 m were counted; and (b) tree size, measured during 25 such surveys. The area surveyed included all main sampling sites, amounting to about 25.4 ha, in which about 7200 trees were recorded.

19.3.2 RESULTS

(a) Fig wasps

During April to September 1995, 65% of the fig wasp sampling programme was completed (97 of 150 samples) and material from 14 *Ficus* species was obtained. Counts were based on the examination of a single collection representing each fig species and wasp species have not yet been compared among fig taxa. Surprisingly, few wasp species were found, ranging from two to four per fig species (Table 19.7 in section 19.5). It is probable that these data underestimate the richness of the local fig wasp fauna. However, Compton and Hawkins (1992), from species accumulation curves, showed that most (over 50%) wasp species could be obtained from a single collection of syconia. Table 19.4 lists the published records of the fig wasp associations for the 15 *Ficus* species, though these associations require confirmation in the Madang area. Anecdotal observations showed that, among other insects reared from syconia, Curculionidae, Cecidomyiidae, Phoridae and Crambidae are also present in the Madang area.

(b) Leafhoppers

In total, 5035 specimens of Auchenorrhyncha, representing 166 species from 18 families, were collected from the foliage of the 15 *Ficus* species. This figure does not include all the species feeding on these various *Ficus*, as new species were still being found at a high rate. In addition, it includes transient species, as no feeding experiments were performed. The Cicadellidae and Derbidae were the most species-rich families, while Aphrophoridae was the most abundant one (Table 19.5). Phloem feeders were by far the most species-rich guild (122 species from 15 families), followed by xylem feeders (23 species of Aphrophoridae, Cercopidae, Cicadidae and Cicadellinae) and mesophyll feeders (21 species of

Table 19.4 Fig wasps reported in the literature to be associated with the *Ficus* species studied in the Madang area (Ulenberg, 1985; Boucek, 1988; Wiebes, 1994)

Ficus species	Pollinator	'Parasitoids'
F. bernaysii	Ceratosolen hooglandi Wiebes	Apocrypata meromassa Ulenberg
F. botryocarpa	Ceratosolen corneri Wiebes	?
F. conocephalifolia	Kradibia jacobsi (Wiebes)	Sycoscapter conocephalus Wiebes
F. copiosa	Kradibia copiosae (Wiebes)	Grandiana armadillo Boucek
F. dammaropsis	Ceratosolen abnormis Wiebes	Tenka percaudata Boucek
F. hispidioides	Ceratosolen dentifer Wiebes	Apocrypta mega sp.
F. microcarpa	Eupristina verticillata Waterson	Acophila sp., Epichrysomalla sp., Walkerella 'kurandensis' Boucek, Odontofroggatia corneri Wiebes O. galili Wiebes, O. ishii Wiebes
F. nodosa	Ceratosolen nexilis Wiebes	?
F. phaeosyce	?	?
F. pungens	Ceratosolen nanus Wiebes	Sycoscapter spp.
F. septica	Ceratosolen bisulcatus (Mayr)	?
F. tinctoria	Liporrhopalum gibbosa Hill	Neosycophila sp.
F. trachypison	?	?
F. variegata	Ceratosolen appendiculatus (Mayr)	Apocrypta caudata (Girault)
F. wassa	Kradibia wassae (Wiebes)	Epichrysomalla 'atricorpus' Girault Grandiana wassae Wiebes

Typhlocybinae). Xylem feeders however dominated in terms of abundance.

Approximately half of the leafhopper species were collected from a single *Ficus* species, but this included many species found as singletons which could not be used in analysis of host preferences (Figure 19.1a). Even after the exclusion of singletons from the analysis, there was a strong correlation between the number of individuals collected (log transformed) and the number of fig hosts (Figure 19.1b; regression calculated by Bartlett's three group method, as the error of both the X and Y variables is unknown and the data are non-normal; Sokal and Rohlf, 1981). This regression predicted that a sample of only three individuals was needed in order to collect a species from two *Ficus* host species. In fact, all species collected in more than five individuals were found on at least two *Ficus* hosts (Figure 19.1b). On average, xylem feeders have a wider host plant range (7.7 *Ficus* spp. per species; singletons excluded from the analysis) than mesophyll and phloem feeders (3.8 and 4.6 *Ficus* spp. per species, respectively; host range differs significantly between feeding modes; Kruskal-Wallis test, $P < 0.05$; Figure 19.2).

Auchenorrhyncha were the only sap-sucking herbivores sampled systematically, but some attention was also paid to the aphids and psyllids. Despite a considerable sampling effort, no aphids and only seven psyllid species were found on the 15 *Ficus* species in the Madang area. However, we

Table 19.5 Species richness and abundance of the most important families and guilds of leafhoppers (Auchenorrhycha) collected in the Madang area

Family/guild	No. of species	No. of individuals
Cicadellidae	48	1267
Derbidae	33	170
Flatidae	18	270
Aphrophoridae	15	2474
Meenoplidae	10	97
Ricaniidae	10	410
Issidae	9	151
Phloem-feeders	122	1426
Xylem-feeders	23	3245
Mesophyll-feeders	21	364

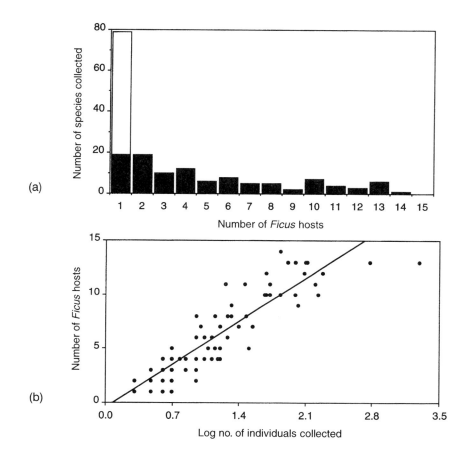

Figure 19.1 Host specificity of leafhoppers in the Madang area. (a) Number of species collected on particular number of *Ficus* species (for single *Ficus* records, empty bar shows number of singleton species). (b) Regression of log number of individuals collected against number of *Ficus* species on which they were collected (Bartlett's three group method: $y = 0.20 + 3.49x$; SE slope = 0.29).

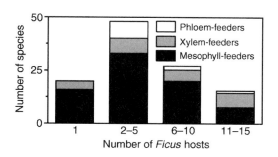

Figure 19.2 Host specificity of phloem-, xylem- and mesophyll-feeding Auchenorrhyncha on *Ficus* species in Madang area (singletons excluded).

recorded aphids feeding on these *Ficus* species in other locations in Papua New Guinea.

(c) Leaf-chewing insects

Overall, 6280 individuals representing 234 species from 21 families of leaf-chewing insects were collected from the 15 *Ficus* species in the Madang area. Despite the extensive sampling effort, the species accumulation curve showed that the number of new species collected grew steadily (Figure 19.3a; leaf-mining species excluded from these data). One explanation for this is that many species collected as singletons may feed only occasionally on *Ficus* (for example, most of the Cerambycidae involved in maturation feeding). In other terms, all the polyphagous species in the different locations visited may not have been sampled yet. However, when the species accumulation curve was plotted for species collected as five or more individuals only, the curve levelled out at about 80 species (Figure 19.3b).

The most important families, as well as the generic identity of some species, are summarized in Table 19.6. Chrysomelidae, Choreutidae, Crambidae (Pyraustinae) and, to a lesser extent, Cerambycidae dominated the samples. The species richness of different subguilds, with regard to the type of foliar damage, was distributed as follows: 66 species 'holing' leaves (chrysomelids, etc.), 58 spp. eating the leaf-margin (various lepidopteran families), 41 spp. gnawing leaves and twigs (cerambycids), 37 spp. tying leaves (choreutids and tortricids), 15 spp. rolling leaves (crambids), 6 spp. mining leaves (gracillariids, etc.) and 5 spp. skeletonizing leaves (choreutids and lycaenids). Many choreutids combined leaf-tying with skeletonizing. No galling insects were recorded.

Many leaf-chewing species were collected from a single species of *Ficus* (Figure 19.4a). However, many of these species were collected as singletons and nothing could be inferred from their host preferences. About half of the species were collected from and fed in the laboratory upon more than one *Ficus* species. On average, each species was feeding on 3.5 *Ficus* hosts (singletons excluded). For example, *Rhyparidella sobrina* (Bryant) and *Choreutis* sp. (an undescribed species: S.E. Miller, personal communication) were recorded on 14 and 13 out of the 15 *Ficus* species studied, respectively.

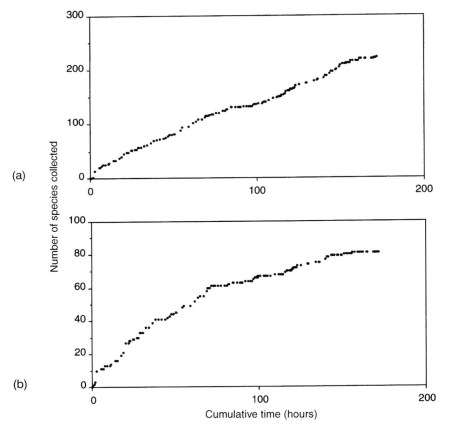

Figure 19.3 Species accumulation curves for leaf-chewing insects with regard to (a) all species collected and (b) species collected as five or more individuals.

Table 19.6 The most important families and guilds of leaf-chewing and stem-boring insects collected in the Madang area, their species richness, abundance and some representative genera identified to date

Family/guild	No. of species	No. of individuals	Genera
Chrysomelidae	50	2864	*Rhyparida, Rhyparidella, Atysa, Sastra*
Cerambycidae	40	149	*Rosenbergia, Epepeotes, Prosoplus*
Choreutidae	23	1793	*Choreutis, Brenthia, Saptha*
Tettigoniidae	16	57	*Sasima, Phyllophora*
Crambidae	15	853	*Talanga, Glyphodes, Parotis*
Tortricidae	14	38	*Adoxophyes*
Curculionidae	13	193	*Apirocalus, Oribius, Rhinoscapha*
Noctuidae	11	82	*Asota, Caryatis*
Phasmatidae	6	19	?
Lymantriidae	6	14	*Euproctis*
Acrididae	5	8	?
Nymphalidae	4	89	*Euploea, Cyrestis*
Pyrgomorphidae	3	21	?
Lycaenidae	2	43	*Philiris*
Eumastacidae	2	5	*?Mnesicles*
Stem-borers	7	15	?
Leaf-miners	6	15	?

Bartlett's three group regression between the number of individuals collected (log transformed, singletons excluded) and the number of *Ficus* hosts predicts that a minimum sample size of four individuals would be needed to collect a particular species from two *Ficus* hosts (Figure 19.4b). Thus, of the 126 species collected from a single *Ficus* species, only 17 species (collected as four or more individuals) were more likely to be monophagous. A further four species were unlikely to be very host specific, since they belong to distinctly polyphagous groups such as Acrididae and Limacodidae. Thus, 13 species (11% of the total number of species, excluding singletons) may be expected to be monophagous. It is probable that this figure is inflated since insects were collected from only 15 *Ficus* species, out of a conservative 40 *Ficus* species found in the Madang area.

19.4 FAUNAL COMPOSITION AND USE OF RESOURCES ON *FICUS*

New field data presented here show unequivocally that many families feeding on the foliage have been overlooked in the literature, namely Derbidae, Flatidae, Aphrophoridae, Chrysomelidae, Tettigoniidae and Tortricidae. This observation is supported by a field study of the insect fauna of *F. nodosa* at a different location (Wau, Papua New Guinea, 1200 m altitude) with different sampling methods (Basset *et al.*, 1996). This disparity between field data and published records is common in the tropics, unlike in countries where the insect fauna is well known (e.g. Southwood *et al.*, 1982, in the UK).

In Auchenorrhyncha, the low number of species reported as feeding on *Ficus*, in comparison with Stenorrhyncha, may be attributed to general lack of information on host plants for the former (see the recent and authoritative review by Wilson *et al.*, 1994). This is supported by the present field data, since all of the *Ficus* species were colonized by many leafhopper species, often at high population densities, whereas Stenorrhyncha were few in terms of both species and individuals. A related problem is the recording or extraction of the information on highly polyphagous taxa. For example, few or no records of Orthoptera, Phasmoptera or Aphrophoridae feeding on *Ficus* were found in the literature, but they are relatively common feeders on *Ficus* in the Madang area.

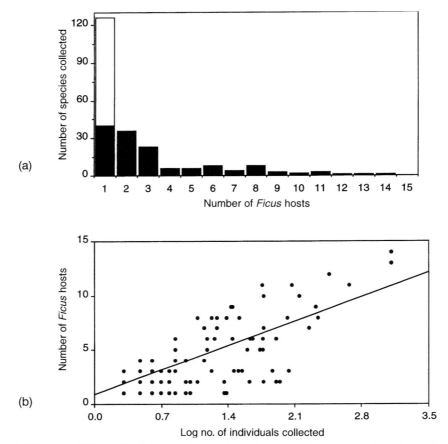

Figure 19.4 Host specificity of leaf-chewing insects in Madang area. (a) Number of species collected on particular number of *Ficus* species (for single *Ficus* records, empty bar shows number of singleton species). (b) Regression of log number of individuals collected against number of *Ficus* species on which they were collected (Bartlett's three group method: $y = -0.75 + 5.94x$; SE slope = 0.26).

Conversely, some groups, such as Psylloidea and Bombycidae, appeared under-represented or absent in the present samples, but relatively species-rich in the database, particularly in the Indo-Australian region. We do not have an explanation for this discrepancy, but Bombycidae appear to be rare in New Guinea (J.D. Holloway, personal communication). Some notably speciose phytophagous groups, such as Aphididae, leaf-feeding Curculionidae and Geometridae, appear to be rare on *Ficus*, as indicated by both the present samples and the database. Unlike psyllids, aphids do not appear to be diverse on *Ficus*. This may be a consequence of the tropical distribution of this genus and of the low species diversity of aphids in the tropics. Aphids appear to be relatively inefficient at finding their host plants in diverse tropical vegetation (Dixon *et al.*, 1987). As for the relative scarcity of the other groups, we cannot offer an explanation.

All resources provided by *Ficus* trees appear to be exploited by arthropods. We could not find specific information about root-feeding arthropods (other than some Tineidae and Gelechiidae feeding on aerial roots), but they exist (see discussion below about Eumolpinae). Perhaps the most distinctive feature of *Ficus* for its primary consumers, in comparison with other tropical trees, is the agglomeration of flowers, fruits and seeds into a single structure, the syconium, which is available throughout the year within local *Ficus* populations. This may account for the diversity of certain taxa consuming syconia, such as Agaonidae, Curculionidae, Drosophilidae and Tephritidae. Most leaf-gallers on *Ficus* are sap-sucking insects and, notably, galling thrips of the family Phlaeothripidae are exceptionally species-rich (Ananthakrishnan, 1978). Leaf-mining species appeared to be scarce, this being confirmed by both literature and field data. This may be related to the presence of lactifers and the probable inefficiency of the trenching strategy for a leaf-miner.

Although segregation of resources appears to prevail among arthropods feeding on *Ficus*, a few species are able to exploit several resources. For example, some sap-suckers, both as adults and larvae, feed on both leaves and syconia (*Acerias*, Eriophyidae, and *Oxycarenus*, Lygaeidae, notably); many Cerambycidae feed on wood as larvae and perform maturation feeding on leaves as adults; some crambid larvae (*Azochis*) feed both in syconia and inside stems; and some choreutid larvae (*Tortyra*) feed both on leaves and inside twigs. In addition, it is probable that many chrysomelids – particularly in the Eumolpinae, the dominant subfamily in the Madang samples – feed on roots as larvae and on the foliage as adults. Free-living chrysomelid larvae on *Ficus* leaves are rare and are found in *Atysa* and *Sastra* (Galerucinae).

19.5 COMPARISON OF ARTHROPOD COMMUNITIES FEEDING ON *FICUS*

19.5.1 BETWEEN ARTHROPOD GUILDS

The literature compilation indicated that arthropods feeding on syconia were much more speciose than sap-sucking and leaf-eating arthropods and, in particular, that records of fig wasps were more abundant than those of leaf-chewing insects. The preliminary field data do not support this: 40 putative fig wasp species (Table 19.7) were found as opposed to 234 leaf-chewing species (Figure 19.3a). A more probable figure may be twice as many leaf-chewing species as fig wasps (Figure 19.3b; the curve levelled out at about 80 species). These discrepancies between literature and field data may merely reflect higher scientific interest in fig wasp communities than foliage-feeding communities, or less complicated sampling procedures for the former. Whilst it is easy to define communities of arthropods feeding on syconia, it is harder to do so for communities of foliage feeders. Specifically, the minimum occurrence of a polyphagous species needed to justify its inclusion in a community of leaf-eating insects feeding on a particular species of *Ficus* is open to discussion. This is beyond the scope of this chapter (but see Basset, 1997).

In contrast with the present field data, mesophyll- and phloem-feeding leafhoppers, particularly those feeding on dicotyledoneous trees, are reported as being host-specific (Claridge and Wilson, 1981; Loye, 1992; Wilson *et al.*, 1994). However, this pattern usually emerges from a comparison of leafhopper assemblages on taxonomically distant host plants, in situations where large complexes of congeneric potential hosts either do not exist (Claridge and Wilson, 1981) or have not been studied (Loye, 1992). The broad host plant range of xylem-feeding leafhoppers on *Ficus* corroborates other data on the wide polyphagy of many xylem feeders. Xylem sap is low in nutrient and secondary metabolites (Raven, 1983) and xylem feeders appear to respond sensitively to nutrient quality of individual plants, but less so to their species identity (Novotny and Wilson, in press).

Although data about the host range of leaf-chewing insects outside *Ficus* are lacking, a study of *F. nodosa* in Papua New Guinea and comparison with nine other tree species belonging to different plant families (Basset *et al.*, 1996) suggests that *Ficus*-chewers are relatively specialized. Apparently more restricted host ranges of leaf-chewing than sap-sucking insects on *Ficus* (compare Figures 19.1 and 19.4, particularly the slopes of the regressions) may be explained at least in part by the different nature of each data set (that for leafhoppers includes transient species). However, mesophyll feeders and leaf-chewing insects appear to have a similar host range (average 3.8 and 3.5 *Ficus* spp., respectively) and this may reflect simi-

Table 19.7 Species of *Ficus* studied and the number of species of fig wasps, leafhoppers, leaf-chewing insects, birds and mammals that they support in the Madang area (in each of the four last taxa, the figures for the five most species-rich fig species are indicated in bold)

Ficus species	Wasps	Leafhoppers	Chewing insects	Birds	Mammals
F. bernaysii	3	40	30	**32**	14
F. botryocarpa	3	36	32	0	3
F. conocephalifolia	2	**49**	52	0	1
F. copiosa	2	**58**	62	2	14
F. dammaropsis	2	37	38	0	14
F. hispidioides	3	38	31	0	2
F. microcarpa	4	15	20	**39**	14
F. nodosa	2	45	44	1	**17**
F. phaeosyce	2	**49**	34	**27**	14
F. pungens	2	**48**	37	**40**	**15**
F. septica	4	32	22	0	11
F. tinctoria	4	10	23	15	**16**
F. trachypison	?	46	**45**	22	14
F. variegata	3	43	**59**	1	**16**
F. wassa	4	**67**	57	**32**	**15**
All *Ficus* spp	>40	166	234	59	17

lar ecological constraints in sucking cell contents and eating leaf tissues.

19.5.2 BETWEEN GEOGRAPHICAL LOCATIONS

If groups for which literature data are certainly under-represented and those which were not targeted in field collections are excluded, the composition of the Madang samples at a higher taxonomic level appear to be similar to that of the fauna that feeds on *Ficus* elsewhere, with two major differences.

First, in Costa Rica, Janzen (1979) observed that *Ficus* does not support a rich fauna of foliage-feeding insects. In the database, Neotropical records of leaf-chewing insects represent less than a quarter of similar Indo-Australian records. In Papua New Guinea, Basset *et al.* (1996) showed that *Ficus nodosa* supports a rather rich and specialized fauna of leaf-chewing insects, in comparison with other tree species. Since *F. nodosa* is not exceptional in this respect, in comparison with other *Ficus* species (our field data), this suggests that the fauna of foliage-feeding insects on *Ficus* in New Guinea may be relatively rich and diverse compared with elsewhere. This may be a result of the considerable diversity and endemism of *Ficus* in New Guinea. One possible example of a durable association may be between *Ficus* and the Choreutidae, the latter particularly well represented in our samples. However, whether this merely indicates uneven sampling between different locations is not known. Few data on choreutids are available, presumably because they are day-flying moths and caught rarely by light trapping. An alternative or additional explanation for a rather depauperate leaf-chewing fauna on *Ficus* in the Neotropics, in comparison with the Indo-Australian region, may be related to the rather coriaceous leaves of monoecious figs, prevalent in the former region (Berg, 1989). Unfortunately, comparative data are lacking to explore this possibility.

Second, the overall richness of the fig wasp fauna for the 15 *Ficus* species in the Madang area was considerably lower than that reported for southern African species (an average of three wasp species per fig species in Madang, 11 wasp species in southern Africa: Compton and Hawkins, 1992). One explanation for this difference may involve the different mating systems within *Ficus*. All but one species of *Ficus* in the African study were monoecious, whereas all but one species in Madang were (gyno)dioecious. In contrast with monoecious figs, only male trees within (gyno)dioecious figs represent a suitable resource for wasp pollinators (Weiblen et al., 1995). Overall, wasp assemblages of (gyno)dioecious figs may be rather depauperate, because many non-pollinating wasps depend on the presence of males of the pollinating species to chew exit holes (Bronstein, 1992). The only (gyno)dioecious species in the African study (F. capreifolia) supported only three wasp species, which is similar to our data for (gyno)dioecious figs in Madang. Furthermore, at least the published records for F. microcarpa (Table 19.4) suggest that this monoecious species supports a richer assemblage of fig wasps than its local (gyno)dioecious counterparts. However, it is unlikely that these considerations apply to other arthropods feeding on syconia (e.g. curculionids, drosophilids, pyralids, etc.).

19.6 KEYSTONE RESOURCES FOR *FICUS*-FEEDING INSECTS AND VERTEBRATES IN THE MADANG AREA

In Table 19.7, we presented a summary of the number of insect species supported by the 15 species of *Ficus* studied in the Madang area, as well as the number of bird and mammal species that regularly feed on the syconia of these species. The latter information was derived from discussion with villagers in Baitabag and Ohu, using the identification guides of Beehler et al. (1986) and Flannery (1990). These local records may be conservative and biased towards game species but, nevertheless, they suggest that the most important species for frugivorous birds are *F. pungens*, *F. microcarpa*, *F. bernaysii*, *F. wassa* and *F. phaeosyce*, whilst the most important species for frugivorous mammals are *F. nodosa*, *F. variegata*, *F. pungens*, *F. tinctoria* and *F. wassa*. No significant correlation existed between the number of bird and mammals species supported by each *Ficus* species ($r = 0.44$, $P = 0.10$). Conversely, there was strong correlation between the number of sap-sucking and leaf-chewing species supported by each *Ficus* species ($r = 0.80$, $P < 0.001$). The five most important species for foliage-feeding insects appear to be *F. wassa*, *F. copiosa*, *F. conocephalifolia*, *F. variegata* and *F. trachypison*.

Two of the important species for vertebrates (*F. microcarpa* for birds and *F. tinctoria* for mammals) appear particularly unattractive for foliage-feeding insects and two of the most important species for insects (*F. copiosa* and *F. conocephalifolia*) were not notably so for vertebrates. Further, the syconia of *F. variegata* and *F. nodosa*, while eaten by many mammals, are too big for many birds to swallow or seize. Thus, poor correlation between sap-sucking, leaf-chewing insects, birds and mammals suggests that the keystone species concept does not universally apply to all of these groups, whatever the abundance of the fig species is. In addition, the fig species supporting the highest number of consumer species of different guilds, *F. wassa*, sustains only 175 out of a total of 516 animal consumers in the system (34%, Table 19.7). *F. wassa* is a very common species in the study area (Table 19.3: it ranks second in abundance of all *Ficus* species) and its local impact on animal communities is predictable from its abundance. The present field data lead to the conclusion that the keystone concept cannot be applied across different guilds feeding on *Ficus* in the Madang area.

There are at least two further problems with the keystone concept when applied to *Ficus*. First, community-level studies in Borneo and peninsular Malaysia supported the vertebrate keystone idea (Leighton and Leighton, 1983; Lambert and Marshall, 1991), whereas other studies in Africa and India did not, as the dietary importance and seasonal availability of syconia were too low (Gautier-Hion and Michaloud, 1989; Borges, 1993). Second, the data for both herbivorous insects and vertebrate frugivores suggest substantial overlap between the faunas supported by the *Ficus* species studied (Table 19.7; compare the sum of the number of species supported by each *Ficus* with the number of species cross-checked in the last entry). Therefore, it is unlikely that any single species of *Ficus* could act as a reservoir for a specific and diverse fauna. In conclusion, we cannot confirm that any single species of *Ficus* can be considered as a keystone species for the *Ficus*-feeding insect community in the Madang area. However, whether the various species of *Ficus* could be viewed collectively as 'keystone', providing resources to a large and diverse array of consumers, remains to be considered. For example, we do not have sufficient data to discuss whether the genus *Ficus* harbours a rich and diverse insect fauna, distinct from that feeding on other speciose genera of trees in the tropics.

ACKNOWLEDGMENTS

We are pleased to acknowledge the help of Chris Dal, Martin Kasbal, Andrew Kinibel, Frankie Duadak, Barry Andreas, Roy Nanguai and Keneth Sehel-Molem (technical assistants); Manek Pius Balide, Tatau Wesly Molem, Nataniel Mataton, Kau Beck, Wamei Nataniel, Ulai Koil, Mataton Soru, Kellie Dewilog, Chris Kau, Brus Isua, Riki Devo and Loman Kau (collectors). The landowners Kiatik Batet and Hais Wasel kindly allowed us to collect in Baitabag and Ohu, respectively. Various colleagues assisted in the preparation of the database and commented on the manuscript: Robert Anderson, Frank Bonaccorso, Daniel Burckhardt, Konrad Fiedler, Keyt Fischer, Jeremy Holloway, Jaroslav Holman, Anne Howden, Tom Huddleston, Ian Kitching, Pavel Lauterer, James Menzies, Scott Miller, Gordon Nishida, Gaden Robinson, Neil Springate, Nigel Stork, Pavel Stys, Petr Svacha and Mike Wilson. Taxonomic help was provided by C. Berg, Jeremy Holloway, Paul Katik, Scott Miller, Olivier Missa, Al Samuelson, Kevin Tuck and Richard Vane-Wright. Neil Springate provided much appreciated linguistic and culinary assistance. Various aspects of the project were funded by the National Science Foundation (DEB-94-07297), the Christensen Research Institute, the National Geographic Society (grant no. 5398-94), the Czech Academy of Sciences (C6007501), the Arnold Arboretum of Harvard University, and the Otto Kinne Foundation. This is contribution no. 158 of the Christensen Research Institute.

REFERENCES

Ackery, P.R. (1988) Hostplants and classification: a review of nymphalid butterflies. *Biol. J. Linn. Soc.* **33**, 95–203.

Ananthakrishnan, T.N. (1978) Thrips galls and gall thrips. *Technical Monograph, Zoological Survey of India* **1**, 1–69.

Annecke, D.P. and Moran, V.C. (1982) *Insects and Mites of Cultivated Plants in South Africa*, Butterwoths, Durban/Pretoria.

Avidov, Z. and Harpaz, I. (1969) *Plant Pests of Israel*, Israel University Press, Jerusalem.

Barnes, H.F. (1948) *Gall Midges of Economic Importance. Vol. III. Gall Midges of Fruit*, Crosby Lockwood and Son Ltd, London.

Basset, Y. (1997) Species abundance and body size relationships in insect herbivores associated with New Guinea forest trees, with particular reference to insect host-specificity. In *Canopy Arthropods* (eds N.E. Stork, J.A. Adis and R.K. Didham), pp. 23–264, Chapman & Hall, London.

Basset, Y., Samuelson, G.A. and Miller, S.E. (1996) Similarities and contrasts in the local insect faunas associated with ten forest tree species of New Guinea. *Pacific Science* **50**, 157–183.

Beehler, B.M., Pratt, T.K. and Zimmerman, D.A. (1986) *Birds of New Guinea*, Princeton University Press, Princeton.

Beeson, C.F.C. and Bhatia, B.M. (1937) On the biology of the Bostrychidae (Coleopt.). *Indian Forest Records (New Series) Entomology* **2**, 1–323.

Beeson, C.F.C. and Bhatia, B.M. (1939) On the biology of the Cerambycidae (Coleopt.). *Indian Forest Records (New Series) Entomology* **5**, 1–235.

Berg, C.C. (1989) Classification and distribution of *Ficus*. *Experientia* **45**, 605–611.

Berg, C.C. and Wiebes, J.T. (1992) *African Fig Trees and Fig Wasps*, North-Holland, Amsterdam.

Bigger, M. (1988) *The Insect Pests of Forest Plantation Trees in the Solomon Islands*, Overseas Development Natural Resource Institute, Chatham, Kent.

Blackman, R.L. and Eastop, V.F. (1994) *Aphids on the World Trees. An identification and Information Guide*, C.A.B. International and The Natural History Museum, Wallingford/London.

Borges, R.M. (1993) Figs, malabar giant squirrels, and fruit shortages within two tropical Indian forests. *Biotropica* **25**, 183–190.

Boucek, Z. (1988) *Australasian Chalcidoidea (Hymenoptera). A Biosystematic Revision of Genera of Fourteen Families, with a Reclassification of Species*, C.A.B. International, Wallingford.

Bronstein, J.L. (1992) Seed predators as mutualists: ecology and evolution of the fig/pollinator interaction. In *Insect–Plant Interactions* (ed. E.A. Bernays), pp. 1–47, CRC Press, Boca Raton, Florida.

Bruner, S.C., Scaramuzza, L.C. and Otero, A.R. (1975) *Catalogo de los Insectos que Atacan a las Plantas Economicas de Cuba*, Academia de Ciencias de Cuba, Instituto de Zoologia, La Habana.

Claridge, M.F. and Wilson, M.R. (1981) Host plant associations, diversity and species–area relationships of mesophyll-feeding leafhoppers on trees and shrubs in Britain. *Ecol. Entomol.* **6**, 217–238.

Compton, S.G. (1989) Sabotage of latex defences by caterpillars feeding on fig trees. *S. Afr. J. Sci.* **85**, 605–606.

Compton, S.G. (1990) A collapse of host specificity in some African fig wasps. *S. Afr. J. Sci.* **86**, 39–40.

Compton, S.G. and Disney, R.H.L. (1991) New species of *Megaselia* (Diptera: Phoridae) whose larvae live in fig syconia (Urticales: Moraceae), and adults prey on fig wasps (Hymenoptera: Agaonidae). *J. Nat. Hist.* **25**, 203–219.

Compton, S.G. and Hawkins, B.A. (1992) Determinants of species richness in southern African fig wasp assemblages. *Oecologia* **91**, 68–74.

Compton, S.G. and van Noort, S. (1992) Southern African fig wasps (Hymenoptera: Chalcidoidea): resource utilization and host relationships. *Proc. Kon. Ned. Akad. v. Wetensch.* **95**, 423–435.

Compton, S.G., Rasplus, J.-Y. and Ware, A.B. (1994) African fig wasp parasitoid communities. In *Parasitoid Community Ecology* (eds B. Hawkins and W. Sheehan), pp. 393–368, Oxford University Press, Oxford.

Condit, I.J. and Enderud, J. (1956) A bibliography of the fig. *Hilgardia* **25**, 1–663.

Corner, E.J.H. (1958) An introduction to the distribution of *Ficus*. *Reinwardtia* **4**, 325–355.

Corner, E.J.H. (1962) The classification of Moraceae. *The Garden's Bulletin Singapore* **19**, 187–252.

Corner, E.J.H. (1965) Check-list of *Ficus* in Asia and Australasia with keys to identification. *The Garden's Bulletin Singapore* **21**, 1–186.

Corner, E.J.H. (1967) *Ficus* in the Solomon Islands and its bearing on the post-Jurassic history of Melanesia. *Proc. R. Soc. Lond. (B)* **253**, 23–159.

Couturier, G., Lachaise, D. and Tsacas, L. (1986) Les Drosophilidae et leur gite larvaires dans la forêt dense humide de Taï en Côte d'Ivoire. *Rev. Fr. Entomol.* **7**, 291–307.

Dixon, A.F.G., Kindlmann, P., Leps, J. and Holman, J. (1987) Why are there so few species of aphids, especially in the tropics? *Am. Nat.* **129**, 580–592.

Duffy, E.A.J. (1957) *A Monograph of the Immature Stages of African Timber Beetles (Cerambycidae)*, British Museum (Natural History), London.

Duffy, E.A.J. (1960) *A Monograph of the Immature Stages of Neotropical Timber Beetles (Cerambycidae)*, British Museum (Natural History), London.

Duffy, E.A.J. (1963) *A Monograph of the Immature Stages of Australasian Timber Beetles (Cerambycidae)*, British Museum (Natural History), London.

Duffy, E.A.J. (1968) *A Monograph of the Immature Stages of Oriental Timber Beetles (Cerambycidae)*, British Museum (Natural History), London.

Dussourd, D.E. and Denno, D.F. (1994) Host range of generalist caterpillars: trenching permit feeding on plants with secretory canals. *Ecology* **75**, 69–78.

Ebeling, W. (1959) *Subtropical Fruit Pests*, University of California, Division of Agricultural Sciences.

Fiedler, K. (1991) Systematic, evolutionary, and ecological implications of myrmecophily within the Lycaenidae (Insecta: Lepidoptera: Papilionoidea). *Bonner Zoologische Monographien* **31**, 1–210.

Flannery, T. (1990) *Mammals of New Guinea*, Robert Brown and Associates, Carina, Australia.

Frank, S.A. (1989) Ecological and evolutionary dynamics of fig communities. *Experientia* **45**, 674–680.

Fukuda, H. (1992) Control of the yellowspotted longicorn beetle *Psacothea hilaris* Pascoe on fig trees. *Proceedings of the Kanto Tosan Plant Protection Society* **39**, 253–255.

Galil, J. and Eisikowitch, D. (1968) On the pollination ecology of *Ficus sycomorus* in East Africa. *Evolution* **49**, 259–269.

Gallego M.F.L. (1971) *Ficivora leucoteles*, a new genus and species of fig tree pest. *Revista, Facultad Nacional de Agronomia, Medellin* **26**, 15–22.

Gautier-Hion, A. and Michaloud, G. (1989) Are figs always keystone resources for tropical frugivorous vertebrates – a test in Gabon. *Ecology* **70**, 1826–1833.

Godfray, H.C.J. (1988) Virginity in haplodiploid populations: a study on fig wasps. *Ecol. Entomol.* **13**, 283–291.

Guagliumi, P. (1966) *Insetti e Aracnidi delle Piante communi del Venezuela segnalati nel periodo 1938–1963*, Istituto Agronomico per l'Oltremare, Firenze.

Hardy, D.E. (1986) Fruit flies of the tribe Acanthonevrina of Indonesia, New Guinea, and the Bismarck and Solomon Islands (Diptera: Tephritidae: Trypetinae: Acanthonevrini). *Pac. Ins. Monogr.* **42**, 1–191.

Herre, E.A. (1989) Coevolution of reproductive characteristics in 12 species of New World figs and their pollinator wasps. *Experientia* **45**, 637–647.

Hill, D.S. (1967) Figs (*Ficus* spp.) and fig-wasps (Chalcidoidea). *J. Nat. Hist.* **1**, 413–434.

Hollis, D. and Broomfield, P.S. (1989) *Ficus*-feeding psyllids (Homoptera), with special reference to the Homotomidae. *Bull. Br. Mus. Nat. Hist. (Ent.)* **58**, 131–183.

Holloway, J.D. (1988) *The Moths of Borneo: Family Arctiidae, Subfamilies Syntominae, Euchromiinae, Arctiinae; Noctuidae (Camtoloma, Aganainae)*. Southdene, Kuala Lumpur.

Houard, C. (1922) *Les Zoocécidies des Plantes d'Afrique, d'Asie et d'Océanie.* T.1, Jules Hermann, Paris.

Janzen, D.H. (1979) How to be a fig. *Ann. Rev. Ecol. Syst.* **10**, 13–51.

Lachaise, D. (1977) Niche separation of African *Lissocephala* within the *Ficus* Drosophilid community. *Oecologia* **31**, 201–214.

Lachaise, D., Tsacas, L. and Couturier, G. (1982) The Drosophilidae associated with tropical African figs. *Evolution* **36**, 141–151.

Lambert, F.R. and Marshall, A.G. (1991) Keystone characteristics of bird-dispersed *Ficus* in a Malaysian lowland rain forest. *J. Ecol.* **79**, 793–809.

Leighton, M. and Leighton, D.R. (1983) Vertebrate responses to fruiting seasonality within a Bornean rain forest. In *Tropical Rain Forests: Ecology and Management* (eds S.L. Sutton, T.C. Whitmore and A.C. Chadwick), pp. 181–196, Blackwell, Oxford.

Lima, A.M. da Costa (1936) *Terceiro Catalogo dos Insectos que vivem nas Plantas do Brasil.* Ministerio da Agricultura, Directoria de Estatistica da Producção, Rio de Janeiro.

Loche, P., Piras, S., Piseddu, V. and Podda, P. (1984) Contribution to the knowledge of *Gynaikothrips ficorum* March. (Thysanoptera, Tubulifera), a thrips dangerous to *Ficus retusa* Desf. in Sardinia: trends in control. *Difesa delle Piante* **7**, 153–160.

Loye, J.E. (1992) Ecological diversity and host plant relationships of treehoppers in a lowland tropical rainforest (Homoptera: Membracidae and Nicomiidae). In *Insects of Panama and Mesoamerica* (eds D. Quintera and A. Aiello), pp. 280–289, Oxford University Press, Oxford.

Mallikarjuna Rao, D. and Mohan Rao, G.V.M. (1991) Fight the fig stem-borer menace. *Indian Horticulture* **36**, 18–19.

Martorell, L.F. (1976) *Annotated Food Plant Catalog of the Insects of Puerto Rico*, Department of Entomology, University of Puerto Rico, Agricultural Experiment Station.

McKey, D. (1989) Population biology of figs: applications for conservation. *Experientia* **45**, 661–673.

Mills, L.S., Soulé, M.E. and Doak, D.F. (1993) The keystone-species concept in ecology and conservation. *Bioscience* **43**, 219–224.

Mound, L.A. and Halsey, S.H. (1978) *Whitefly of the World. A systematic catalogue of the Aleyrodidae (Homoptera) with host plant and natural enemy data*, British Museum (Natural History) and John Wiley and Sons, Chichester.

Nageshchandra, B.K., Rajagopal, B.K. and Balasubramanian, R. (1972) Occurrence of slug caterpillar *Phauda flammans* Wlk. (Lepidoptera: Zygaenidae) on *Ficus racemosa* L. in South India. *Mysore Journal of Agricultural Sciences* **6**, 186–189.

Novotny, V. and Wilson, M.R. (in press) Why there are no small species among xylem-feeding arthropods. *Evol. Ecol.*

Ozar, A.I., Onder, P., Saribay, A. *et al.* (1986) Investigations on improving the measures for controlling pests and diseases occurring on figs in the Aegean Region. *Doga* **10**, 263–277.

Paine, R.T. (1969) A note on trophic complexity and community stability. *Am. Nat.* **103**, 91–93.

Paine, T.D. (1992) Cuban laurel thrips (Thysanoptera: Phlaeothripidae) biology in southern California: seasonal abundance, temperature dependent development, leaf suitability, and predation. *Annls Ent. Soc. Am.* **85**, 164–172.

Parsons, M. (1984) Life histories of four species of *Philiris* Röber (Lepidoptera: Lycaenidae) from Papua New Guinea. *J. Lep. Soc.* **38**, 15–22.

Perrin, H. (1992) Double radiation sur Fagales et sur *Ficus* (Moraceae) du genre *Curculio* (Coleoptera: Curculionidae). *C.R. Acad. Sci. Paris* **314**, 127–132.

Perry, D.R. (1978) A method of access into the crowns of emergent and canopy trees. *Biotropica* **10**, 155–157.

Picard, F. (1919) La faune entomologique du figuier. *Ann. Epi.* **6**, 34–174.

Power, M.E. and Mills, L.S. (1995) The Keystone cops meet in Hilo. *Trends Ecol. Evol.* **10**, 182–184.

Prasad, S. and Singh, D.R. (1984) *Achoea janata* Linn. (Lepidoptera: Noctuidae) as a pest of *Ficus* species. *Indian Journal of Agricultural Sciences* **54**, 216.

Ramírez, B.W. (1970) Host specificity of fig wasps (Agaonidae). *Evolution* **24**, 680–691.

Raven, J.A. (1983) Phytophages of xylem and phloem: a comparison of animal and plant sap-feeders. *Adv. Ecol. Res.* **13**, 136–234.

Silva, A.G. d'A, Goncalves, C.R., Galvao, D.M. *et al.* (1968) *Quarto Catalogo dos Insectos que vivem nas Plantas do Brasil seus parasitos e predatores*, 4 tomos, Ministerio da Agricultura, Rio de Janeiro.

Slater, J.A. (1972) Lygaeid bugs (Hemiptera: Lygaeidae) as seed predators of figs. *Biotropica* **4**, 145–151.

Sokal, R.R. and Rohlf, J.F. (1981) *Biometry*, 2nd edn, Freeman and Co., New York.

Southwood, T.R.E., Moran, V.C. and Kennedy, C.E.J. (1982) The assessment of arboreal insect fauna: comparisons of knockdown sampling and faunal lists. *Ecol. Entomol.* **7**, 331–340.

Swailem, S.M. and Awadallah, K.T. (1973) On the seasonal abundance of the insect and mite fauna on the leaves of sycamore fig trees. *Bulletin de la Société Entomologique d'Egypte* **57**, 1–8.

Terborgh, J. (1986) Keystone plant resources in the tropical forest. In *Conservation Biology. The Science of*

Scarcity and Diversity (ed. M.E. Soulé), pp. 330–344, Sinauer, Sunderland, Mass.

Tkachuk, V.K. (1986) The fig leafroller *Choreutis nemorana* Hbn. (Lepidoptera: Choreutidae) in the Crimea. *Trudy Gosudarstvennogo Nikitskogo Botanicheskogo-Sada* **99**, 101–110.

Ulenberg, S.A. (1985) *The Systematics of the Fig Wasp Parasites of the Genus Apocrypta Coquerel*, North-Holland, Amsterdam.

Verma, S.K., Yamdagni, R. and Daulta, B.S.A. (1989) Note on new hosts of *Euproctis fraterna* Moore (Lymantriidae: Lepidoptera). *Haryana Journal of Horticultural Sciences* **18**, 237–238.

Weiblen, G., Spencer, H. and Flick, B. (1995) Seed set and wasp predation in dioecious *Ficus variegata* from an Australian wet tropical rainforest. *Biotropica* **27**, 391–394.

Wiebes, J.T. (1966) Provisional host catalogue of fig wasps (Hymenoptera: Chalcidoidea). *Zool. Verh.* **83**, 1–44.

Wiebes, J.T. (1979) Co-evolution of figs and their insect pollinators. *Annu. Rev. Ecol. Syst.* **10**, 1–12.

Wiebes, J.T. (1987) Coevolution as a test of the phylogenetic tree. In *Systematics and Evolution: a Matter of Diversity* (ed. P. Hovenkamp), pp. 309–314, Utrecht University, Utrecht.

Wiebes, J.T. (1994) *The Indo-Australian Agaoninae (pollinators of figs)*, North-Holland, Amsterdam.

Williams, D.J. and Watson, G.W. (1988a) *The Scale Insects of the Tropical South Pacific Region. Part 1. The Armoured Scales (Diaspididae)*, C.A.B. International Institute of Entomology, Wallingford.

Williams, D.J. and Watson, G.W. (1988b) *The Scale Insects of the Tropical South Pacific Region. Part 2. The Mealybugs (Pseudococcidae)*, C.A.B. International Institute of Entomology, Wallingford.

Williams, D.J. and Watson, G.W. (1990) *The Scale Insects of the Tropical South Pacific Region. Part 3. The Soft Scales (Coccidae) and other families*, C.A.B. International Institute of Entomology, Wallingford.

Williams-Linera, G. and Lawton, R.O. (1995) The ecology of hemiepiphytes in forest canopies. In *Forest Canopies* (eds M.D. Lowman and N.M. Nadkarni), pp. 255–283, Academic Press, San Diego.

Wilson, S.W., Mitter, C., Denno, R.F. and Wilson, M.R. (1994) Evolutionary patterns of host plant use by delphacid planthoppers and their relatives. In *Planthoppers: their Ecology and Management* (eds R.F. Denno and T.J. Perfect), pp. 7–113, Chapman & Hall, New York.

Zhang, B.-C. (1995) *Index of Economically Important Lepidoptera* (electronic version), C.A.B. International, Wallingford.

Worldwide bibliography of arthropods feeding on the leaves, fruits and wood of *Ficus* spp. can be found at http://www.bishop.hawaii.org/bishop/natsailing/ngecol.html

PART SIX
INSECT CONSERVATION

ARTHROPODS OF COASTAL OLD-GROWTH SITKA SPRUCE FORESTS: CONSERVATION OF BIODIVERSITY WITH SPECIAL REFERENCE TO THE STAPHYLINIDAE

Neville N. Winchester

20.1 INTRODUCTION

The global biodiversity crisis continues to be fuelled by habitat loss (Wilson 1988, 1989, 1992; Soulé, 1991) and consequent extinctions of floral and faunal species assemblages that cannot adjust to rapid, and often large-scale, habitat alterations. Habitat loss is most pronounced in forests, which throughout the world are being compromised by human-induced perturbations. In northern temperate zones some of the last remaining tracts of intact old-growth coniferous forests occur in the Pacific Northwest of North America (Franklin, 1988; Kellogg, 1992). The fragmentation of these landscapes has heightened the awareness for a need to understand the endemic fauna and flora (Scudder, 1994) and apply system-based conservation approaches across a wide range of forest types (Murray *et al.*, 1993; Harding and McCullum, 1994).

Historically, little research concerning the conservation of biodiversity has been done in the primeval old-growth forests of the Pacific Northwest (Winchester, 1993; Winchester and Ring, 1996a), and research has generally failed to link results to ecosystem processes. In British Columbia these forests are thought to contain much of the biodiversity of the province (Fenger and Harcombe, 1989; Bunnell, 1990; Pojar *et al.*, 1990; Winchester and Ring, 1996a,b). They often have diffuse boundaries with other ecosystems, and this temporal and spatial mosaic creates a dynamic and complex set of habitats that are utilized by a variety of species. The faunal elements associated with these old-growth forests form a heterogeneous group, and nowhere is this more evident than in the Arthropoda. Arthropods, primarily insects, are an integral part of these old-growth systems and may comprise 80–90% of the total species (Asquith *et al.*, 1990). They play a primary role in the function of natural ecosystems (Lattin and Moldenke, 1990; Moldenke and Lattin, 1990; Wiggins *et al.*, 1991), may regulate nutrient cycling (Mattson and Addy, 1975; O'Neill, 1976; Moldenke and Lattin, 1990; Naeem *et al.*, 1994, 1995) and are now frequently mentioned as important components of diversity that need to be identified and applied to conservation issues (May, 1986; Wilson, 1988; di Castri *et al.*, 1992; Samways, 1992, 1994).

Within the forests of the Pacific Northwest only a few studies on old-growth forest invertebrates have been completed to date (Denison *et al.*, 1972; Pike *et al.*, 1972, 1975; Voegtlin, 1982; Schowalter, 1986, 1989; Winchester, 1996). These studies were carried out in the context of old-growth Douglas

---

Forests and Insects. Edited by A.D. Watt, N.E. Stork and M.D. Hunter. Published in 1997 by Chapman & Hall, London. ISBN 0 412 79110 2.

fir–western hemlock surveys and at present I know of no studies being conducted in old-growth Sitka spruce forests. Given the importance of arthropods in these old-growth forests coupled with the lack of taxonomic knowledge of this biotope, the objective of this chapter is to present results from the first year of a continuing study that aims to document the complete arthropod fauna from an old-growth Sitka spruce forest.

Staphylinidae is the largest family of Coleoptera in Canada and the least known of any large family of beetles in North America (Campbell and Peck, 1978). Many are associated with habitats found in forest interiors (Hatch, 1957; Campbell and Peck, 1978), and old-growth forests may act as source areas for several species (Campbell and Winchester, 1993). This study is concerned with Staphylinidae (excluding Aleocharinae) which will be used to illustrate source habitats and to address species turnover along a gradient from an old-growth forest to a harvested area. Based largely on Malaise trap results, within-site variability and temporal sequencing are discussed.

For each of the sampling methods used, data on abundance and proportional representation of insect orders are provided (*sensu* Hammond, 1990). Samples taken in the canopy are discussed in relation to habitat specificity of canopy microarthropods. Estimates of undescribed and endemic arthropods are viewed in terms of conservation of biodiversity.

20.2 STUDY AREA

The study area is located in the Upper Carmanah Valley drainage (48°44' N; 124°37' W) on the southwest coast of Vancouver Island, British Columbia, Canada (Figure 20.1). This typical U-shaped coastal valley, approximately 6731 ha in extent, is situated between the villages of Port Renfrew and Bamfield. The entire valley lies within the Coastal Western Hemlock Biogeoclimatic Zone, with the exception of two high-elevation areas (Meidinger and Pojar, 1991). The dominant conifers in the Carmanah drainage are Sitka spruce (*Picea sitchensis* (Bong) Carr.), western hemlock (*Tsuga heterophylla* (Rafn.) Sarg.), western red cedar (*Thuja plicata* D.Don) and Pacific silver fir (*Abies amabilis* (Dougl.) Forb.).

The sample area includes four study sites: old-growth canopy, old-growth forest floor (both old-growth sites are approximately 700 years old), transition zone (edge between old-growth and clearcut), and clearcut. All study sites are adjacent to each other along a 4 km transect. This watershed represents an intact, ancient forest that has evolved since the Wisconsin glaciation (*c.* 10 000 years before present). In 1985 the clearcut site (*c.* 4 ha in extent) was harvested and is the only area in the entire Carmanah watershed to be logged.

20.2.1 CANOPY ACCESS

A 2000 m linear transect was placed along Carmanah Creek and all Sitka spruce trees taller than 30 m were identified. From these trees, five were randomly chosen to be incorporated into a canopy access system. Access to the canopy is by means of a 2:1 mechanical advantage pulley system. Four wooden platforms strapped to the branches and trunk of the main tree provide consistent heights (31–67 m) from which to sample. A series of burma bridges provides access to four other Sitka spruce trees, complete with platforms (Ring and Winchester, 1995). At the inception of this study this station was the only permanent access system of this type available for long-term canopy work in northern temperate rainforests.

20.2.2 SURVEY DESIGN

Owing to the diverse nature of arthropods and their varied habits, no single survey method or sampling technique can be used for a complete study. The variety of techniques used in this study is summarized in Winchester and Scudder (1993) and in a report of the Biological Survey of Canada (1994).

20.3 SAMPLING PROTOCOL

20.3.1 BRANCH CLIPPINGS

The branch clipping programme was conducted in the five Sitka spruce trees within the fixed-access

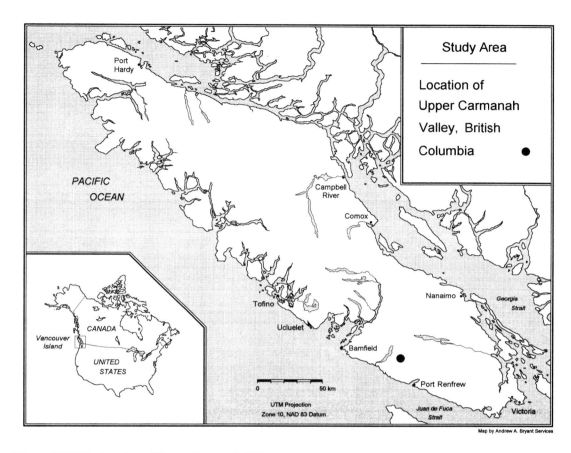

Figure 20.1 Map location of Upper Carmanah Valley research sites on Vancouver Island, British Columbia, Canada.

canopy system and the procedure was modified after Schowalter (1989). In each tree three samples were taken at each of three heights: 33, 45 and 54 m. A total of 45 branches was collected for each of six sample periods. Samples were collected at one-month intervals from May to October, 1991. All insects were removed from each sample and prepared for identification. Immatures of most species were placed in a rearing facility in an attempt to rear individuals to maturity. Using Tullgren funnels, Collembola and Acarina were extracted from single branches taken from each tree at each height. After sorting, all branches were dried and then total branch weight and needle weight were recorded.

20.3.2 MALAISE TRAPS

Five Malaise traps (Townes, 1962), randomly positioned along a linear transect, were placed at each of the three ground-level sites. In addition, five traps – one per Sitka spruce tree – were run in the canopy. Each canopy trap was anchored at around 35 m above ground level. Malaise traps were operated continuously between June 22 and October 17, a period of 18 weeks, and cleared at two-week intervals. Arthropods were collected into 70% ethyl alcohol to which six drops of ethylene glycol had been added. A total of 245 trap samples were collected and the arthropods from each sample were sorted and enumerated (Table 20.1).

Table 20.1 Samples of Arthropoda collected in 1991 from the Upper Carmanah Valley.

Sampling programme	Sample period	Canopy		Forest floor		Transition zone		Clearcut		Total
		1	2	1	2	1	2	1	2	
Malaise traps	3 July/27 Oct. 1991	45	13,928	45	79,318	45	275,144	45	189,144	557,534
Pan traps	3 July/27 Oct. 1991	–	–	45	11,509	38	16,311	44	18,781	46,601
Moss corings	3 July/27 Oct. 1991	58	3,864	27	776	29	1,324	29	1,334	7,298
Branch clippings	4 June/27 Oct. 1991	270	1,268	–	–	–	–	–	–	1,268
Total		373	19,060	117	91,603	112	292,779	118	209,259	612,701

The two columns of numbers under each sample site refer to:
(1) the number of samples available for study;
(2) the number of arthropods collected.

20.3.3 PAN TRAPS

In each of the three ground study sites, five white pan traps were placed at random along a linear transect and operated continuously for 18 weeks between June 22 and October 17. The pans (23 × 15 × 5 cm) were buried with their rims flush at ground level. A saturated salt solution was used as a preservative with a few drops of detergent added as a wetting agent. Traps were cleared at two-week intervals and a total of 120 trap samples were collected. All arthropods were washed in water, sorted and stored in 75% ethyl alcohol.

20.3.4 MOSS CORES

A hand-held moss/soil corer (3 cm × 5 cm) was used to collect five moss/soil cores at random from each sample site at monthly intervals from May to October. A total of 120 cores was collected. In the laboratory the arthropods were extracted from each core using Tullgren funnels for 48 hours. Samples were preserved in 75% ethyl alcohol. Volume displacement and dry weight were recorded for each core sample.

20.4 SAMPLE SORTING AND DATA ANALYSES

Taxonomic and habitat knowledge for terrestrial arthropods from these forests is at best sketchy and the information that is available is usually difficult to interpret because of the number of independent sources and the rather diffuse nature of the pertinent data. Exploring the species richness component in this study is restricted by the inability to identify specimens to species level. Lack of taxonomic resolution is a severe impediment to the documentation of patterns of natural processes and ecological roles that are mediated by arthropods (Danks, 1991; Danks and Ball, 1993; Ball and Danks, 1993). Progress to date in extracting arthropods from the sampling program is summarized in Table 20.1. Samples were initially processed to order and then, where possible, to family. Abundance and proportional representation of arthropods from each sampling protocol are outlined in Table 20.2. Material at the family level was then sorted to morphospecies and from this material over 150 000 individuals from a variety of taxonomic groups have been sent to 80 systematists for identification. The only complete set of species identifications, to date, is for Staphylinidae excluding Aleocharinae. As identifications become available, rare and endemic species are tabulated (Table 20.3).

Malaise trap catches of Staphylinidae are reported as means and 95% confidence limits for each replicate (five), at each study site (four), pooled over all collection dates (nine). A one-way analysis of variance was used to test for mean trap species differences within sites (SPSS, 1994). Numerical relationships between staphylinid species and study sites are graphed.

Abundance data for each species are compiled in a site by time matrix and analysed descriptively

Table 20.2 Abundance of arthropods in each of five Malaise traps from four study sites: (a) canopy, (b) forest floor, (c) transition zone and (d) clearcut

Taxa	Trap number									
	1		2		3		4		5	
	A	B	A	B	A	B	A	B	A	B
(a) Canopy										
Aquatics	1	1	1	3	1	1	1	2	<1	<1
Acarina	39	22	34	29	23	18	35	20	25	12
Aranaea	6	4	5	4	2	2	9	5	4	3
Coleoptera	2	3	6	4	7	8	5	6	1	1
Collembola	129	106	29	20	27	14	34	25	34	48
Diptera	235	185	292	190	134	157	224	168	35	38
Hymenoptera	18	16	42	25	12	8	18	16	1	1
Heteroptera	5	3	6	5	2	1	2	2	<1	<1
Lepidoptera	4	6	8	8	12	18	6	5	1	2
Phalangida	<1	<1	<1	0	0	0	<1	1	0	0
Psocoptera	6	8	12	10	4	3	3	4	2	3
Thysanoptera	<1	<1	1	1	2	4	<1	<1	0	0
Others	1	1	1	1	<1	1	0	0	<1	<1
(b) Forest floor										
Aquatics	1	1	1	2	1	1	2	3	2	3
Acarina	48	31	47	44	52	45	67	55	71	79
Aranaea	14	14	11	7	7	5	10	7	27	38
Coleoptera	39	25	42	22	39	19	42	26	56	32
Collembola	149	174	181	159	94	77	164	128	1010	1541
Diptera	1177	653	1461	984	945	422	1320	691	1218	746
Hymenoptera	108	73	82	57	54	26	125	98	119	97
Heteroptera	5	5	6	6	4	3	8	10	5	4
Lepidoptera	4	4	5	5	5	5	7	8	3	4
Phalangida	1	2	2	1	2	2	1	1	1	1
Psocoptera	23	23	20	25	12	15	17	18	13	16
Thysanoptera	1	1	<1	<1	<1	<1	1	1	<1	<1
Others	2	1	10	18	1	2	5	7	2	2
(c) Transition zone										
Aquatics	26	23	117	63	79	54	116	70	20	11
Acarina	105	106	43	19	79	56	105	105	79	127
Aranaea	13	6	9	5	8	5	13	7	10	2
Coleoptera	117	77	119	69	134	70	148	103	207	254
Collembola	178	124	63	64	103	74	100	79	260	200
Diptera	6018	3394	3863	1972	6503	2921	5746	3818	3219	1637
Hymenoptera	534	251	498	260	737	351	559	355	284	112
Heteroptera	57	30	30	5	33	8	79	56	14	12
Lepidoptera	16	11	14	11	20	14	16	15	8	9
Phalangida	1	1	1	1	2	2	111	1	2	2
Psocoptera	16	11	28	36	44	52	41	45	22	21
Thysanoptera	39	66	12	17	15	19	34	44	10	13
Others	2	3	3	3	3	2	2	3	4	3

Table 20.2 Continued

Taxa	Trap number									
	1		2		3		4		5	
	A	B	A	B	A	B	A	B	A	B
(d) Clearcut										
Aquatics	2	2	5	3	8	6	5	2	5	2
Acarina	63	69	77	94	85	79	121	70	46	48
Aranaea	5	4	4	4	5	4	6	4	5	5
Coleoptera	117	77	119	69	134	70	148	103	207	254
Collembola	328	235	105	81	132	135	138	104	139	114
Diptera	2303	1491	3708	2744	3334	2819	3901	2350	2447	2002
Hymenoptera	253	135	356	175	258	143	360	178	199	111
Heteroptera	79	52	85	41	75	39	131	65	55	43
Lepidoptera	10	9	28	24	20	19	17	15	18	17
Phalangida	<1	1	<1	<1	<1	<1	<1	1	0	0
Psocoptera	1	1	2	3	4	6	1	2	2	2
Thysanoptera	538	842	269	367	433	592	320	447	391	409
Others	2	3	4	2	6	5	7	5	2	3

A = mean number of individuals collected during nine two-week intervals.
B = standard error of the mean number of individuals collected during nine two-week intervals.

using indirect ordination techniques. A Q-analysis (a site-by-site comparison using both species composition and numerical contribution) was employed in all analyses. Principal components analysis (PCA) was applied to the data matrix (Cornell Ecological computer program, Ordiflex, CEP-25B). When assessing similarities in community structure between study sites, the following *a priori* computer options were chosen:

- Data matrix values (abundance) were transformed by log (X + 1).
- Non-centred adjustment for PCA with ordination scores scaled from 0 to 100.
- All samples were treated independently.
- Graphical displays for PCA used the first two axes.

Staphylinid species composition and abundance data for the forest floor were subjected to a non-centred PCA to look at temporal sequencing of staphylinids at this site. Results are discussed on the basis of information derived from the first two principal components. Sample similarities were subjectively decided upon, based on their proximity of sample coordinates within the ordination space.

20.5 DISCUSSION

20.5.1 TAXONOMIC COMPOSITION

Approximately 612 700 arthropods representing an undetermined number of species were collected during the 1991 season (Table 20.1). The complex variety of taxonomic groups in our study represents different patterns of association across habitat gradients where the numerically dominant arthropod groups are Diptera, Collembola, Hymenoptera, Thysanoptera and Acarina. Mean number of individuals caught for most taxa decrease in the following order: transition zone, forest floor, clearcut, canopy (Table 20.2). When total individuals are associated with sites, the transition zone (ecotone between old-growth forest and clearcut) contains the greatest number of groups that are highly vagile (e.g. Diptera, Hymenoptera) as well as the greatest numbers of aquatic groups, represented by Ephemeroptera, Plecoptera and Trichoptera. The forest floor site has the next highest level of abundance for many groups – in particular, Collembola, Coleoptera and Psocoptera. The canopy site is somewhat depauperate in most taxonomic groups, though Acarina are speciose and numerous. The most radically altered habitat, the clearcut, contains a high number of vagile taxa such as Diptera, Hymenoptera and Lepidoptera. This site also contains arthropods that exhibit rapid population increases (Winchester, unpublished data), a result that supports previous observations for arthropod increases in herb-dominated pioneer seral stages (Shure, 1973; Brown and Southwood, 1983; Schowalter et al., 1981; Godfray, 1985). For example, the number of thysanopterans present are several orders of magnitude greater than in the other sample sites.

The number of species identified was 1311, with a total of 80 (6.2% of identified species) being confirmed as new to science. Species identifications are lacking for many groups because of a lack of suitable taxonomists (e.g. Ichneumonidae, Proctotrupidae, etc.) which have been sorted only to genera and morphospecies. Given the fact that only a small percentage of the total number of specimens in this study have been identified to species level, it is likely that the number of described and undescribed species recorded presents a conservative estimate.

20.5.2 MALAISE TRAP DATA INTERPRETATION

Catches obtained from Malaise traps may be influenced by trap position and general site characteristics (Hammond, 1990; Noyes, 1989; Muirhead-Thompson, 1991; Winchester and Scudder, 1993). This, in turn, makes site comparisons relatively uninformative when a large proportion of the variation is a direct result of incident variables affecting trap performance. At each site, variability of mean numbers of staphylinid species per trap pooled over time is not significant (Figure 20.3). Therefore, comparisons between sites can be made without incorporating a significant trap bias, and consistency within site replicates based on species presence in terms of activity abundance can be related to habitat characteristics or to proximity to a source population.

20.5.3 STAPHYLINIDAE

Examination of the numerical relationships between staphylinid species and study sites (Figure 20.2) reveals that the number of species is greatest in the transition site (49), followed by the clearcut (29), the forest floor (23) and the canopy (six). The transition zone represents an ecotone and is subject to a variety of factors that act to create heterogeneous habitats that, in turn, affect species distributions (Wilcove et al., 1986, MacGarvin et al., 1986; Laurance, 1987; Laurance and Yensen, 1991; Wood and Samways, 1991; Samways, 1994). This is supported by the wide degree of variation illustrated in Figure 20.3 and the scatter of points depicted in the principal component plot (Figure 20.4). In this study the greater number of staphylinid species present in the transition zone arises from two areas: from the core habitat area within the forest interior (20 species in common) and the disturbed area – the clearcut (15 species in common) (Figure 20.2). Habitat changes that occur by logging of old-growth forests have been well documented, but the effect that this type of habitat alteration has on arthropods has been addressed only to a limited extent (McLeod, 1980; Chandler, 1987, 1991; Schowalter, 1989; Niemelä et al., 1988, 1993; Chandler and Peck, 1992).

Figure 20.4 summarizes the non-centred PCA ordination for the entire staphylinid species-site abundance matrix. The proximity of one point to another is an indication of how similar sites are with respect to their staphylinid populations. The variation captured by the first two principal components (64.22%) is sufficient to separate the old-growth forest sites from the transition zone and clearcut site. The tightest clusterings of points within the ellipsoids are contained within the old-growth forest floor and the clearcut.

The forest floor staphylinid fauna in this study are characterized by species that are associated with coniferous needle litter, fungi and different ages of coarse woody debris. Prominent among this fauna are several rare omaliine staphylinids that are recorded only from the old-growth forest interior, including *Pseudohaida rothi* Hatch, *Coryphium arizonense* (Bernhauer), *Subhaida ingrata* Hatch, *Tanyrhinus singularis* Mannerheim, *Trigonodemus fasciatus* Leetch, and eight new species of Omaliinae. Habitat specificity associated with characteristics of old-growth forests has also been

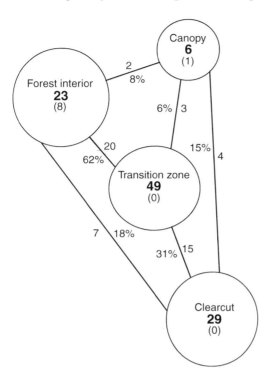

Figure 20.2 Numerical relationship between staphylinid species (excluding Aleocharinae) in study sites located in Upper Carmanah Valley. Data from 1991 Malaise trap programme. Numbers along lines represent those species shared in common between the sites; numbers in circles represent species occuring at that site.

shown for a variety of other beetles, such as Pselaphidae (Chandler, 1987, 1990), fungus feeding beetles (Chandler, 1991), Leiodidae (Chandler and Peck, 1992), Staphylinidae (Campbell and Winchester, 1993) and Carabidae (Niemelä et al., 1993).

Temporal sequencing for the forest floor site indicates that there is a gradual shift in species composition throughout the sampling period (Figure 20.5). The most obvious shift in species and abundances occurs with Omaliinae species, and this relates to the appearance of fungal fruiting bodies at the later sampling dates. Percentage similarity is lowest between the canopy and all ground sites (range 6–15%) (Figure 20.3) The canopy habitat does not appear to provide the conditions necessary for this group of beetles, though the most abundant species recorded in the canopy is a new species of Omaliinae that also occurs on the forest floor. This site overlap is depicted in Figure 20.4. This result is in direct contrast to the survey on *Pseudotsuga menziesii* (Mirb.) Franco conducted by Voegtlin (1982), where several staphylinid species (> 14) were recorded.

The majority of the species that occur in the clearcut are represented by singletons that are not recorded as being persistent over time. Little evidence has been found to support long-term habitat use in the clearcut. The depression in species occurrence in altered forest habitats such as the clearcut site has also been documented for fungus-feeding beetles (Chandler, 1991) and I hypothesize that the staphylinid species associated with fungi are absent in the clearcut.

20.5.4 ENDEMICS AND CONSERVATION

New emphasis has been placed on the conservation of arthropods (Wilson, 1987; Erwin, 1991; Collins and Thomas, 1991; Samways, 1994). The key components of biological diversity that are often identified for conservation are endemic and threatened species (Noss, 1990; Platnick, 1991; Erwin, 1991). In Canada, approximately half of the estimated 66 000 insect species have been described (Danks, 1993), but to date there has been no attempt to create an annotated systematic list of the potentially rare and endemic species. Knowledge of the regional diversity of insects in Canada is incomplete, though the western regions probably contain the majority of species (Danks, 1994). In British Columbia there may be as many as 40 000 arthropod species (Cannings, 1992), and many of these are undescribed (Winchester and Ring, 1996a). Scudder (1994) lists 168 endemic species and 203 potentially rare and endangered arthropods.

The 80 new species so far recorded in this study represent a contribution towards categorizing the endemic arthropod fauna of this old-growth forest (Table 20.3). It is expected that, with continued taxonomic resolution, this list will be significantly increased. Of particular importance to the maintenance of arthropod biodiversity are those species that are new to science or species that are specific to habitats that are only found in the old-growth

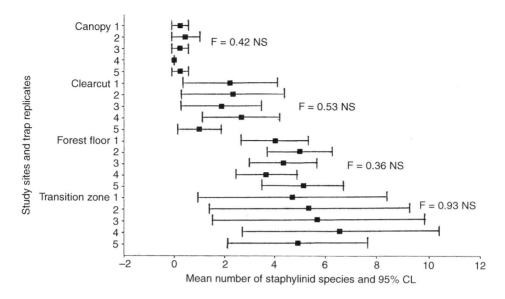

Figure 20.3 Mean number and 95% confidence intervals of Staphylinidae species for each Malaise trap in each study site over nine two-week collection periods. Results of a one-way ANOVA comparing means within each study site are shown (NS = $P > 0.05$).

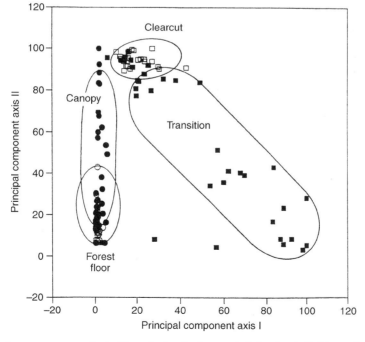

Figure 20.4 Principal components analysis (Q-analysis, site by species) for Staphylinidae collected in 1991 from Upper Carmanah Valley. Data matrix: 180 samples (4 areas, 9 sampling periods, 5 replicates); 80 taxa; data transform log $(x + 1)$; all samples treated independently. Analysis: sum % Eigenvalues = 64.22 (PC axes I and II); non-centred ordination (dispersion matrix); PCA scores relativized to 100.

374 *Arthropods of coastal old-growth Sitka spruce forests*

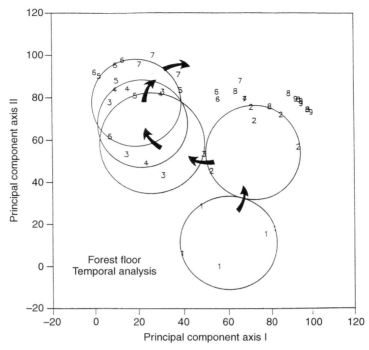

Figure 20.5 Principal components analysis (temporal analysis, trap by time) for Staphylinidae collected in 1991 from Upper Carmanah Valley. Data matrix: 45 samples (1 area, 9 sampling periods, 5 replicates); 80 taxa; data transform log ($x + 1$); all samples treated independently. Analysis: sum % Eigenvalues = 66.80 (PC axes I and II); non-centred ordination (dispersion matrix); PCA scores relativized to 100.

forest. Examples include *Hypogastrura arborea* Fjellberg, a collembolan that is known only from the Carmanah Valley, where it is confined to the thick rich moss mats that occur on the forest floor and in the high canopy (Fjellberg, 1992). Also in this study mycetophilids (Table 20.3) are represented by 15 new species (e.g. *Anacliliea vallis* Coher and *A. winchesteri* Coher) that are associated with habitats found only in the old-growth forest interior (Coher, 1995). Similar results have been found in Norway by Økland (1994) where semi-natural (i.e. the oldest) forests sustain more mycetophilids. This type of habitat specificity is well documented for the staphylinid beetles; examples include *Pseudohaida rothi*, which represents the first record for Canada (Campbell and Winchester, 1993), and *Trigonodemus fasciatus*, which is endemic to British Columbia (Scudder, 1994). Eight new species of Omaliinae staphylinid beetles appear to rely on old-growth forests as a source area to maintain reproductively viable subpopulations (Campbell and Winchester, 1993).

Perhaps the most interesting and least explored habitat in old-growth forests is the canopy (Winchester, in press). In the Sitka spruce canopy a key habitat feature is provided by the moss mats, 4–28 cm deep, which are supported by a well developed soil layer. These mats are primarily composed of three moss species: *Isothecium myosuroides* Brid., *Antitrichia curtipendula* (Hedw.) Brid. and *Dicranum fuscescens* Sm., which are also abundant on the forest floor. Soil microarthropods dominate this canopy soil/litter habitat, a fact that has not been well documented in these forests but has been noted in other canopy studies (Nadkarni and Longino, 1990; Paoletti *et al.*, 1990). From the oribatid mites identified to date there is strong evidence that we are dealing with a distinct arboreal fauna (Winchester and Ring, 1996b; Winchester, 1996). A high number of

Table 20.3 Samples of Arthropoda sent for identification

Taxonomic group	Total number of specimens sent for identification	Total number of confirmed species	Total number of confirmed new species
Acarina			
Non-oribatids	63	18	4
Oribatida	1112	71	26
Araneae	2657	75	2
Coleoptera			
Staphylinidae	3732	80	8
Pselaphidae	35	14	1
Helodidae	3	2	1
Collembola	1286	42	1
Diptera			
Mycetophilidae	9581	217	15
Tipulidae	2862	53	13
Tachnidae	89	24	1
Hymenoptera			
Braconidae	1008	145	?
Symphyta	120	27	1
Heteroptera			
Aphididae	30	8	1
Cicadellidae	2776	41	2
Psocoptera	513	13	1
Totals	25 867	820	80

species with low percentage similarity to ground sites is consistent with the findings of Moldenke and Lattin (1990); Behan-Pelletier et al.(1993); Walter et al. (1994).

The discovery of several new oribatid species (Table 20.3) is not surprising, given the scope of the study. Fifteen undescribed species appear to be confined to habitats found only in the old-growth forest canopy. For example, *Dendrozetes* has been recorded for the first time in North America and this new species has modifications for an arboreal existence (V. Behan-Pelletier, personal communication). *Parapirnodus*, *Paraleius* and *Anachiptera* are genera that are known to be arboreal (Behan-Pelletier, personal communication) and in this study each is represented by an undescribed, strictly arboreal species. Similarly, new species with unique habitat associations have been recorded in northern Venezuela (Behan-Pelletier et al., 1993), in Peru (Wunderle, 1992) and in Australia (Walter et al., 1994).

The microhabitats associated with the canopy of the ancient Sitka spruce trees in British Columbia are not found in any second-growth forest canopies that I have surveyed, and it is unlikely that such habitat features will develop in second-growth forests that are in an 80–120-year rotation. I suggest that there are enough differences in canopy microhabitat conditions to promote the development of taxonomically discrete species assemblages that will be lost if these canopy habitats are not retained or allowed to develop in second-growth forests.

20.6 CONCLUSION

Forest harvesting affects arthropod diversity by altering key patterns of natural processes that are inseparably linked to habitat diversity. In this study, evidence from several arthropod groups indicates that this old-growth forest acts as a source area for several species, many which are currently undescribed. Of the 80 staphylinid species recorded in this study, many show habitat specificity that is related to characteristics associated with the old-growth forest. Many species of Coleoptera are restricted to old-growth forests where two important conditions for survival are met: first, a supply of over-mature, fallen logs which are allowed to decay under natural conditions in the shade of the forest canopy; and second, the maintenance of deep layers of undisturbed forest floor litter that have not been eradicated by the extreme conditions of clearcutting and subsequent exposure to desiccation and erosion (Campbell and Winchester, 1993). Forest litter and decaying logs abound with a large variety of fungi, many of which serve as hosts for beetles.

Habitat specificity is most pronounced in the canopy where soil micro-arthropods such as the oribatid mites exhibit arboreal specificity. The microhabitats associated with the canopy of the ancient Sitka spruce trees are not found in any second-growth canopies surveyed to date and it is unlikely that these habitat features will develop in second-growth forests that are in an 80–120-year rotation (Winchester and Ring, 1996b).

ACKNOWLEDGEMENTS

I acknowledge with thanks the Western Canada Wilderness Committee who made available the research facility and the fixed canopy access system in the Upper Carmanah Valley. Field work was supported by FRDA research grants from the B.C. Ministry of Forests, Research Branch. Acknowledgement is made to A. Mackinnon and B. Nyberg of that Branch for their continued support.

I am indebted to numerous taxonomic experts for identifications and advice they have provided for this project. In particular, taxonomic identifications for this chapter were provided by: V. Behan-Pelletier (Oribatids), J.M. Campbell and A. Davies (Staphylinidae), H. Goulet, K.G.A. Hamilton, M.H. Sharkey, A. Smetana and D.M. Wood (Agriculture Canada, Ottawa); F. Brodo (Canadian Museum of Nature, Ottawa); E.I. Coher, Mycetophilidae (Long Island University, New York); D. Buckle (Saskatoon, Canada); E.L. Mockford, (Illinois State University, Normal, Il).

I also thank the many research assistants and laboratory technicians who have generously contributed large amounts of time to this project. To my climbers, Kevin Jordan, Stephanie Hughes and Nancy Prockiw, I extend a special thanks for providing technical expertise in the arboreal aspects of this project. Finally, to R.A. Ring, a special thanks for your continued support and involvement in all aspects of this project.

REFERENCES

Asquith, A., Lattin, J.D. and Moldenke, A.R. (1990) Arthropods, the invisible diversity. *Northwest Environ. J.* **6**, 404–405.

Ball, G.E., and Danks, H.V. (1993) Systematics and entomology: Introduction. In *Systematics and Entomology: Diversity, Distribution, Adaptation, and Application* (eds G.E. Ball and H.V. Danks), pp. 3–10, Mem. Ent. Soc. of Can. 165.

Behan-Pelletier, V.M. (1993) Diversity of soil arthropods in Canada: systematic and ecological problems. In *Systematics and Entomology: Diversity, Distribution, Adaptation, and Application* (eds G.E. Ball and H.V. Danks), pp. 11–50, Mem. Ent. Soc. of Can. 165.

Behan-Pelletier, V.M., Paoletti, M.G., Bissett, B. and Stinner, B.R. (1993) Oribatid mites of forest habitats in northern Venzuela. *Tropical Zoology*, Special Issue **1**, 39–54.

Biological Survey of Canada (1994) Terrestrial arthropod biodiversity: planning a study and recommended sampling techniques. *Supplement to the Bull. Ent. Soc. Canada* **26**, 1–33.

Brown, V.K. and Southwood, T.R.E. (1983) Trophic diversity, niche breadth and generation times of exopterygote insects in a secondary succession. *Oecologia* **56**, 220–225.

Bunnell, F.L. (1990) Forestry wildlife: w(h)ither the future? In *Proc. Symp. on Forests – Wild and Managed: Differences and Consequences* (eds A.F. Pearson and D.A. Challenger), pp. 163–176, Students for Forestry Awareness, U.B.C., Vancouver, British Columbia.

Campbell, J.M. and Peck, S.B. (1978) Staphylinoidea. In *Canada and its Insect Fauna* (ed. H.V. Danks), pp. 369–371, Mem. Ent. Soc. Can. 108.

Campbell, J.M., and Winchester, N.N. (1993) First record of *Pseudohaida rothi* Hatch (Coleoptera: Staphylinidae: Omaliinae) from Canada. *J. Ent. Soc. Brit. Columbia* **90**, 83.

Cannings, S. (1992) Arthropods – where diversity is really at. In *Methodology for Monitoring Wildlife Diversity in B.C. Forests* (ed. L.R. Ramsey), pp. 43–48, B.C. Environment, Wildlife Branch, Victoria, British Columbia.

Chandler, D.S. (1987) Species richness and abundance of Pselaphidae (Coleoptera) in old-growth and 40-year-old forests in New Hampshire. *Can.J. Zool.* **65**, 608–615.

Chandler, D.S. (1990) Insecta: Coleoptera. Pselaphidae. In *Soil Biology Guide* (ed. D.L. Dindal), J. Wiley, San Diego.

Chandler, D.S. (1991) Comparison of some slime-mold and fungus feeding beetles (Coleoptera) in old-growth and 40-year-old forest in New Hampshire. *Coleopt. Bull.* **45**, 239–256.

Chandler, D.S. and Peck, S.B. (1992) Diversity and seasonality of Leiodid beetles (Coleoptera: Leiodidae) in an old-growth and a 40-year-old forest in New Hampshire. *Env. Entomol.* **21**, 1283–1293.

Coher, E.I. (1995) A contribution to the study of the genus Anacliliea (Diptera: Mycetophiliea). *Ent. News* **106**(5), 257–262.

Collins, N.M. and Thomas, J.A. (eds) (1991) *The Conservation of Insects and their Habitats, 15th Symposium of the Royal Entomological Society of London*, Academic Press, London.

Danks, H.V. (1991) Museum collections: fundamental values and modern problems. *Collection Forum* **7**, 95–111.

Danks, H.V. (1993) Patterns of diversity in the Canadian insect fauna. In *Systematics and Entomology: Diversity, Distribution, Adaptation, and Application* (eds G.E. Ball and H.V. Danks), pp. 51–74, Mem. Ent. Soc. of Can. 165.

Danks, H.V. (1994) Regional diversity of insects in North America. *Am. Ent.* **40**, 50–55.

Danks, H.V. and Ball, G.E. (1993) Systematics and entomology: Some major themes. In *Systematics and Entomology: Diversity, Distribution, Adaptation, and Application* (eds G.E. Ball and H.V. Danks), pp. 257–272, Mem. Ent. Soc. of Can. 165.

Denison, W.C., Tracy, D.M., Rhoades, F.M. and Sherwood, M. (1972) Direct non-destructive measurement of biomass and structure in living, old-growth Douglas-fir. In *Proceedings, Research on Coniferous Forest Ecosystems, a Symposium* (eds J. Franklin, L.J. Dempster and R.H. Warings), pp. 147–158, USDA Forest Service, Pacific Northwest Forest and Range Experiment Station, Portland, Oregon.

di Castri, F., Vernhes, J.R and Younes, T. (1992) Inventorying and monitoring of biodiversity. A proposal for an international network. *Biol. Intl.* Special Issue Number **27**, 1–27.

Erwin, T.L. (1991) An evolutionary basis for conservation strategies. *Science* **253**, 750–752.

Fenger, M. and Harcombe, A. (1989) A discussion paper on old-growth forests, biodiversity, and wildlife in British Columbia. Prepared by B.C. Min. Environ., for the invitational workshop 'Towards an Old-Growth Strategy' Nov. 3–5, Parksville, B.C.

Fjellberg, A. (1992) *Hypogastrura (Mucrella) arborea* sp. nov., a tree-climbing species of Collembola (Hypogastruridae) from Vancouver Island, British Columbia. *Can. Ent.* **124**, 405–407.

Franklin, J.F. (1988) Structural and functional diversity in temperate forests. In *Biodiversity* (ed. E.O. Wilson), pp. 166–175, National Academy Press, Washington DC.

Godfrey, H.C.J. (1985) The absolute abundance of leaf miners on plants of different successional stages. *Oikos* **45**, 17–25.

Hammond, P.M. (1990) Insect abundance and diversity in the Dumoga-Bone National Park, N. Sulawesi. In *Insects and the Rain Forest of South East Asia (Wallacea)* (eds W.J. Knight and J.D. Holloway), pp. 197–255, Royal Entomol. Soc. Lond.

Harding, L.E. and McCullum, E. (eds) (1994) *Biodiversity in British Columbia: Our Changing Environment*, Canadian Wildlife Service, Environment Canada, Ottawa.

Hatch, M.H. (1957) *The Beetles of the Pacific Northwest. Part II. Staphyliniformia*, Univ. Wash. Publ. Biol. 16.

Kellogg, E. (ed.) (1992) *Coastal Temperate Rain Forests: Ecological Characteristics, Status and Distribution Worldwide*, Ecotrust and Conservation International Occassional Paper No. 1, 64 pp.

Lattin, J.D. and Moldenke, A.R. (1990) Moss lacebugs in northwest conifer forests: adaptation to long-term stability. *Northwest Env. J.* **6**, 406–407.

Laurance, W.F. (1987) The rainforest fragmentation project. *Liane* **25**, 9–12.

Laurance, W.F. and Yensen, E. (1991) Predicting the impacts of edge effects in fragmented habitats. *Biol. Cons.* **55**, 77–92.

MacGarvin, M., Lawton, J.H. and Heads, P.A. (1986) The herbivorous insect communities of open and woodland bracken: observations, experiments and habitat manipulations. *Oikos* **47**, 135–148.

Mattson, W.J. and Addy, N.D. (1975) Phytophagous insects as regulators of forest primary production. *Science* **190**, 515–522.

May, R.M. (1986) The search for patterns in the balance of nature: advances and retreats. *Ecology* **67**, 1115–1126.

McLeod, J.M. (1980). Forests, disturbances, and insects. *Can. Ent.* **112**, 1185–1192.

Meidinger, D. and Pojar, J. (1991) *Ecosystems of British Columbia*, B.C. Ministry of Forests, Victoria, B.C.

Moldenke, A.R. and Lattin, J.D. (1990) Density and diversity of soil arthropods as 'biological probes' of complex soil phenomena. *Northwest Env. J.* **6**, 409–410.

Muirhead-Thompson, R.C. (1991) *Trap Responses of Flying Insects: the Influence of Trap Design on Capture Efficiency*, Academic Press,

Murray, C.L., Berry, T.M. and Gardner, J. (1993) Biodiversity Conservation Strategy for British Columbia – Discussion Paper on Strategy Development. Prepared by ESSA Ltd, Victoria and Vancouver, and Dovetail Consulting Inc., Vancouver, for the Inter-Ministry Biodiversity Group, Province of British Columbia, Victoria.

Naeem, S., Thompson, L.J., Lawler, S.P. *et al.* (1994) Declining biodiversity can alter the performance of ecosystems. *Nature* **368**, 734–736.

Naeem, S., Thompson, L.J., Lawler, S.P. *et al.* (1995) Empirical evidence that declining species diversity may alter the performance of terrestrial ecosystems. *Phil Trans. R. Soc. Lond.* **347**, 249–262.

Nadkarni, N.M. and Longino, J.T. (1990) Invertebrates in canopy and ground organic matter in a Neotropical montane forest, Costa Rica. *Biotropica* **22**(3), 286–289.

Niemelä, J., Haila, Y., Halme, E. *et al.* (1988) The distribution of carabid beetles in fragments of old coniferous taiga and adjacent managed forest. *Ann. Zool. Fenn.* **25**, 107–119.

Niemelä, J., Langor, D. and Spence, J.R. (1993) Effects of clear-cut harvesting on boreal ground-beetle assemblages (Coleoptera: Carabidae) in western Canada. *Cons. Biol.* **7**, 551–561.

Noss, R.F. (1990) Indicators for monitoring biodiversity: a hierarchical approach. *Cons. Biol.* **4**, 355–364.

Noyes, J.S. (1989) A study of five methods of sampling Hymenoptera (Insecta) in a tropical forest with special reference to the Parasitica. *J. Nat. Hist.* **23**, 285–298.

Økland, B. (1994) Mycetophilidae (Diptera), an insect group vulnerable to forestry practices? A comparison of clearcut, managed and semi-natural spruce forests in southern Norway. *Biodiversity and Conservation* **3**, 68–85.

O'Neill, R.V. (1976) Ecosystem perturbation and heterotrophic regulation. *Ecology* **57**, 1244–1253.

Paoletti, M.G., Stinner, B.R., Stinner, D. *et al.* (1990) Diversity of soil fauna in the canopy of Neotropical rain forest. *J. Tropical Ecology* **7**, 135–145.

Pike, L.H., Tracey, D.M., Sherwood, M. and Nielsen, D. (1972) Estimates of biomass and fixed nitrogen of epiphytes from old-growth Douglas-fir. In *Research on Coniferous Ecosystems. Proceedings, Symposium Northwest Science Association* (eds J.F. Franklin, L.J. Dempster and R.H. Waring), pp. 177–187.

Pike, L.H., Dennison, W.C., Tracy, D.M. *et al.* (1975) Floristic survey of epiphytic lichens and bryophytes growing on old growth conifers in western Oregon. *Brylogist* **78**, 389–402.

Platnick, N.L. (1991) Patterns of biodiversity: tropical vs. temperate. *J. Nat. Hist.* **25**, 1083–1088.

Pojar, J., Hamilton, E., Meidinger, D. and Nicholson, A. (1990) Old-growth forests and biological diversity in British Columbia. In *Symposium of Landscape Approaches to Wildlife and Ecosystem Management*, Vancouver, B.C.

Ring, R.A. and Winchester, N.N. (1995) Coastal temperate rainforest canopy access systems in British Columbia, Canada. *Selbyana* **17**, 22–26.

Samways, M.J. (1992) Some comparative insect conservation issues of north temperate, tropical, and south temperate landscapes. *Agriculture, Ecosystems and Environment* **40**, 137–154.

Samways, M.J. (1994) *Insect Conservation Biology*, Chapman & Hall, London.

Schowalter, T.D. (1986) Ecological strategies of forest insects: the need for a community-level approach to reforestation. *New Forests* **1**, 57–66.

Schowalter, T.D. (1989) Canopy arthropod community structure and herbivory in old-growth and regenerating forests in western Oregon. *Can. J. For. Res.* **19**, 318–322.

Schowalter, T.D., Webb, J.W. and Crossley, D.A. Jr (1981) Community structure and nutrient content of canopy arthropods in clearcut and uncut forest ecosystems. *Ecology* **62**, 1010–1019.

Scudder, G.G.E. (1994) *An annotated systematic list of the potentially rare and endangered freshwater and terrestrial invertebrates in British Columbia*, Occasional paper 2, Entomol. Soc. Brit. Columbia.

Shure, D.J. (1973) Radionuclide tracer analysis of trophic relationships in an old field ecosystem. *Ecol. Monogr.* **43**, 1–19.

Soulé, M.E. (1991) Conservation: tactics for a constant crisis. *Science* **253**, 744–750.

SPSS Inc. (1994) *SPSS Advanced Statistics 6.1*, SPSS Inc., Chicago, Il.

Townes, H. (1962) Design for a Malaise trap. *Proc. Entomol. Soc. Wash.* **64**, 253–262.

Voegtlin, D.J. (1982) *Invertebrates of the H.J. Andrews Experimental Forest, western Cascade Mountains, Oregon: a survey of arthropods associated with the canopy of old-growth* Pseudotsuga menziesii, Forest Res. Lab., Oregon State University, Corvallis, 29 pp.

Walter, D.E., O'Dowd, D.J. and Barnes, V. (1994) The forgotten arthropods: foliar mites in the forest canopy. *Mem. Queensland Museum* **36**. 221–226.

Wiggins, G.B., Marshall, S. and Downes, S.A. (1991) The importance of research collections of terrestrial arthropods. *Bulletin Ent. Soc. Can.* **23**, 1–16.

Wilcove, D.C., McLellan, C.H. and Dobson, A.P. (1986) Habitat fragmentation in the temperate zone. In *Conservation Biology: the Science of Scarcity and Diversity* (ed. M.E. Soulé), pp. 237–256, Sinauer, Sunderland, Massachusetts.

Wilson, E.O. (1987) The little things that run the world (the importance and conservation of invertebrates). *Cons. Biol.* **1**, 344–346.

Wilson, E.O. (1988) The current state of biological diversity. In *Biodiversity* (ed. E.O. Wilson), pp. 3–17, National Academy Press, Washington, DC.

Wilson, E.O. (1989) Threats to biodiversity. *Scientific American* **261**, 109–116.

Wilson, E.O. (1992) *The Diversity of Life*, Harvard University Press, Cambridge, Mass.

Winchester, N.N. (1993) Coastal Sitka spruce canopies: conservation of biodiversity. *Bioline* **11**(2), 9–14.

Winchester, N.N. (1996) Canopy arthropods of coastal Sitka spruce trees on Vancouver Island, British Columbia, Canada. In *Canopy Arthropods* (eds N.E. Stork, J.A. Adis and R.K. Didham), pp. 151–168, Chapman & Hall, London.

Winchester, N.N. and Ring, R.A. (1996a) Northern temperate coastal Sitka spruce forests with special emphasis on the canopies: studying arthropods in an unexplored frontier. *Northwest Science* Special Issue **70**, 94–103.

Winchester, N.N. and Ring, R.A. (1996b) Centinelan extinctions: extirpation of Northern Temperate old-growth rainforest arthropod communities. *Selbyana* **17**(1), 50–57.

Winchester, N.N. and Scudder, G.G.E. (1993) *Methodology for Sampling Terrestrial Arthropods in British Columbia*, Resource Inventory Committee, B.C. Ministry of Environment, Lands and Parks.

Wood, P.A. and Samways, M.J. (1991) Landscape element pattern and continuity of butterfly flight paths in an ecologically landscaped botanic garden, Natal, South Africa. *Biol. Con.* **58**, 149–166.

Wunderle, I. (1992) Die baum – und bodenbewohnenden Oribatiden (Acari) im Tieflandregenwald von Panguana, Peru. *Amazoniana* **XII**(1), 119–142.

CONSERVATION CORRIDORS AND RAINFOREST INSECTS

21

Christopher J. Hill

21.1 INTRODUCTION

The idea that habitat fragments connected by corridors of natural vegetation might be subject to lower extinction rates was first proposed by Diamond (1975) and Wilson and Willis (1975). They hypothesized that the presence of corridors would encourage the movement of individuals between habitat fragments, thus increasing rates of immigration. Increased rates of immigration may reduce the chances of extinction due to genetic deterioration and chance demographic events (Brown and Kodric-Brown, 1977; Soulé and Simberloff, 1986) and increase the probability of re-establishment after local extinctions have occurred (Simberloff and Cox, 1987). Subsequently this idea, largely untested, has been adopted in conservation texts (e.g. Harris, 1984; Bennett, 1990). The need for empirical information with which to assess the effectiveness of conservation corridors is generally recognized (Simberloff and Cox, 1987; Hobbs, 1992; Simberloff *et al.*, 1992). However, collecting such information requires experimental designs which, whilst theoretically possible (Nicholls and Margules, 1991; Inglis and Underwood, 1992), either do not occur with adequate replication in natural landscapes or are ethically undesirable (Crome, 1994). Consequently, research effort has focused at one ecological level on the movement of individuals through linear 'corridor-like' vegetation strips (Burel, 1991; Saunders and de Rebeira, 1991; Ruefenacht and Knight, 1995; Bennett *et al.*, 1994; Vermeulen, 1994) and at another level on describing the assemblages of species found in linear vegetation strips (Keals and Majer, 1991; Keller *et al.*, 1993; Claridge and Lindenmayer, 1994; Crome *et al.*, 1994; Hill, 1995).

Common to both approaches is the aim to identify those attributes of linear strips of vegetation that enhance either the movement of individuals or the species richness and abundance of animal assemblages. Factors that have been considered important in increasing either the corridor function or the conservation value of linear strips include width (Baur and Baur, 1992; Harrison, 1992; Keller *et al.*, 1993; Ruefenacht and Knight, 1995), continuity (Laurance, 1990; Bennett *et al.*, 1994; Ruefenacht and Knight, 1995), topographic diversity (Lindenmayer and Nix, 1993; Claridge and Lindenmayer, 1994), floristic composition (Bennett *et al.*, 1994), vegetation structure (Bennett *et al.*, 1994; Crome *et al.*, 1994) and distance from the presumed source population (Laurance, 1990; Dunning *et al.*, 1995). The majority of the work to date has focused on vertebrates (mostly birds and mammals) and has been conducted in temperate systems. Invertebrate contributions to this research area have mostly concerned the use of roadside vegetation as potential dispersal corridors (e.g. Keals and Majer, 1991; Munguira and Thomas, 1992; DeMers, 1993; Vermeulen, 1994) and have also been carried out

Forests and Insects. Edited by A.D. Watt, N.E. Stork and M.D. Hunter. Published in 1997 by Chapman & Hall, London. ISBN 0 412 79110 2.

in temperate systems. Clearly the applicability of conservation corridors needs to be assessed in a wider variety of habitats across a range of taxa. This study, on dung beetles (Coleoptera: Scarabaeinae) in tropical rainforest of northern Australia, involves both a vegetation type and a taxon generally under-represented in research on this topic. Dung beetles were selected as the study organism because they are taxonomically well known (in Australia), standard techniques exist for quantitative sampling, and they have often been proposed as an indicator taxon because of their habitat specificity and close links to the vertebrate fauna on which they rely for food (Halffter and Favila, 1993; Kremen et al., 1993).

There is also an increasing realization that research emphasis should be placed on species that are most abundant in the interior of their habitat (or edge-aversive species), because it is these species that are most likely to be negatively affected by the process of habitat fragmentation (Lovejoy et al., 1986; Chapter 17, this volume) and therefore most likely to benefit from conservation corridors (Hobbs, 1992; Hill, 1995). If habitat interior species are shown to occur in linear (corridor-like) vegetation strips outside their preferred habitat but not in the surrounding matrix, then such a result supports the concept of conservation corridors since increased dispersal between fragments is more likely to occur.

Using this approach, Hill (1995) identified four species of rainforest interior insect (two dung beetle species and two butterfly species) and found that two of these species (one dung beetle and one butterfly) did occur in the linear strips but not in the surrounding arable land, whereas the remaining two species were confined to the rainforest habitat. It was not possible to determine whether the linear strips were being used for dispersal or as residential habitat by the rainforest interior species and this chapter describes the subsequent work carried out as part of this study. The aims of the work are to determine whether linear strips of rainforest vegetation are used for dispersal or as habitat by dung beetles, and to investigate the effect of width of linear strip vegetation on dung beetle assemblages.

21.2 STUDY SITES AND METHODS

21.2.1 STUDY SITES

The study took place in North Queensland, Australia, 145°53'E, 17°11'S. This region has a tropical climate with the wet season falling between November and April. Average monthly temperatures range from 20.7 to 28.9°C, and there is an average annual rainfall of 2500 mm. The study made use of existing linear strips of rainforest connected to large tracts of rainforest surrounded by arable land. The rainforest has been classified as complex mesophyll vine forest (type 1a) by Tracey (1982) and the arable land is mostly cultivated for sugar cane. Figure 21.1 shows the distribution of the vegetation at the five sites used in this study. Each site comprised a continuous linear strip of rainforest vegetation at least 750 m long connected to a large area of rainforest at one end, with the other end terminating in the arable land. The most distant sites were separated by a distance of 90 km.

21.2.2 SAMPLING METHODS

Several studies have investigated the movement of animals in linear strips or corridors in order to determine whether the presence of conservation corridors can increase dispersal between habitat fragments. Because such work is labour intensive, these studies have largely been confined to a single site or single species (e.g. Burel, 1991; Vermuelen, 1994; Bennett et al., 1994; Ruefenacht and Knight, 1995). A different approach is taken in this study which allows a greater number of species and sites to be investigated at the expense of data on individual movements.

The approach predicts that if the presence of linear strips or corridors encourages dispersal by habitat interior species then, in general, individuals from the source population in the primary habitat should be moving through the linear strip away from the primary habitat. In other words, large areas of primary habitat act as source habitats whereas linear strips or corridors act as sink habitats (sensu Pulliam, 1988). This prediction is based on a number of assumptions that conform to those

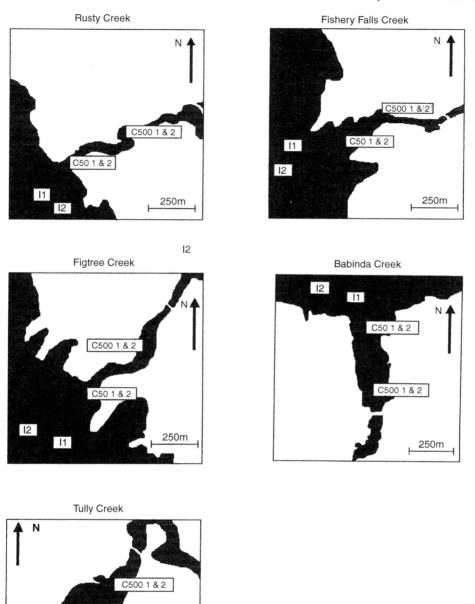

Figure 21.1 Positions of sampling locations at five study sites in North Queensland, Australia. Shaded areas indicate rainforest vegetation; unshaded areas are arable land – sugar cane or pasture.

made in other studies on this topic (Vermeulen and Opdam, 1995). The first assumption is that individuals move at random within their habitat. The second is that the boundary of the habitat is not absolute and so a proportion of individuals cross habitat boundaries and are 'lost' to the system. Finally, having entered the corridor, dispersing individuals are assumed to continue moving in the same manner as they would in their primary habitat. This scenario would result in a net movement of individuals in a particular direction through the linear strip. With equally large source populations on each end of a corridor one would expect the net movement of individuals through the corridor to be the same from both directions. But in linear strips attached to a large area of primary habitat at one end only (as in this study), there should be more individuals moving away from the source population than towards it. If this is true, then a trap (positioned in a linear vegetation strip) that faces the source population should capture significantly more individuals compared with a trap facing in the opposite direction. Alternatively, if linear strips are used as habitat, or act as source habitats (Pulliam, 1988) – i.e. for feeding, mating, reproduction etc. rather than solely for dispersal – by the species in question, then the position of the catching surface of a trap should have no consistent effect on the numbers of individuals caught.

The effect of the width of the vegetation of the linear strip on dung beetle species richness and abundance was determined by trapping at set distances from the presumed source population in linear strips of variable widths.

Flight intercept traps (Peck and Davies, 1980) were used since this method has been shown to be effective in sampling dung beetles in rainforest vegetation (Howden *et al.*, 1991; Hill, 1993) and can be installed so as to trap individuals flying in a particular direction. The traps comprised a vertical, transparent plastic sheet (1300 × 900 mm) underneath and on each side of which were positioned three rectangular containers so that each side of the trap was sampled independently. The containers were filled with diluted formalin which acted as a preservative. Two traps were positioned in the interior of the rainforest (> 250 m from the forest edge and > 20 m between traps) at each site. The traps were positioned so that the vertical catching surface was at 90° to the direction of the linear strip. Two traps were similarly positioned at 50 and 500 m along the linear strips of rainforest vegetation – that is, with catching surfaces facing towards and away from the rainforest interior. In summary, there were five study sites each comprising a continuous linear strip of rainforest vegetation connected at one end to a large area of relatively undisturbed rainforest. At each study site there were three sampling locations: rainforest interior, 50 m away from the rainforest edge in the linear strip, and 500 m along the linear strip. At each sampling location were two flight intercept traps positioned so that the catching surfaces were at 90° to the direction of the linear strip and therefore capable of distinguishing between individuals flying towards or away from the rainforest interior at the time of capture.

Sampling times were selected to coincide with the wet season, which is the period of greatest adult dung beetle activity in the North Queensland wet tropics (Hill, 1993, and personal observation). The traps were installed on 24/25 January 1994 at Rusty Creek, Fishery Falls, Figtree Creek and Babinda Creek. They were cleared on 7–9 February, 24/25 February, 15/16 March, 6/7 April and 26/27th April. At Tully Creek, the equivalent dates were 16 March, 6 April, 26 April, 17 May and 8 June 1994. Catches from each side of a trap were kept separate and all dung beetles captured were counted and identified to species using keys provided by Matthews (1972, 1974, 1976) and Storey and Weir (1990); the species nomenclature follows these sources. Data analyses were conducted on the average value across the sampling times for each side of a particular trap.

The width of the vegetation at each sampling point in the linear strip was calculated from aerial photographs. Visual estimates were made of the percentage canopy cover (> 5 m), the percentage shrub cover (< 5 m), the percentage grass cover, the percentage ground cover (including grasses), and the percentage rock cover for a 10 × 10 m quadrat at each trap site. Canopy cover and shrub cover were combined to give an index of total shading.

21.2.3 ANALYSES

To test for a movement effect the data for each site were analysed separately. Data were first log $(x + 1)$ transformed (Zar, 1984) and a nested ANOVA was conducted with location of trap (interior, 50 m or 500 m) and direction (side of trap) as independent variables, with direction nested within location.

To examine the effect of width of vegetation in the linear strips, the proportional abundance and species richness for each trap (proportion of the total number of individuals or species richness at that study site) was calculated in order to standardize comparisons between sites. Data were square root arcsine transformed for subsequent analyses (Zar, 1984). A single value for each sampling location was obtained by calculating the mean proportional abundance and species richness at each sampling location (mean of each side of two traps at each location, $n = 4$). This value was plotted against linear strip width at that location. For this experimental design the interior locations represent the expected maximum values for beetle abundance and species richness and therefore should be included in the analyses. To achieve this, the interior locations were allocated a nominal width of 500 m. This width was based on the work of Laurance (1991) in which he showed that vegetational edge effects in North Queensland rainforests extend some 200 m into the forest interior. Therefore non-edge rainforest vegetation is only likely to exist in linear strips with a width of at least 400 m.

Non-linear regression was then used to examine the relationship between linear strip width and beetle species richness and abundance. The equation used was $y = $ constant $ + \log (x)*$beta. This equation makes more biological sense than a linear regression because it reaches an asymptote (finite number of species/abundance) and is the same equation used to investigate species–area relationships. Analysis of variance was used to determine whether the regression provided a significant fit to the data.

The relationships between vegetation structure, linear strip width and dung beetle species richness and abundance were examined using principal components analysis. All analyses were carried out on the software package SYSTAT (Wilkinson et al., 1992).

21.3 RESULTS

21.3.1 MOVEMENT

A total of 2303 individuals belonging to 15 species of dung beetle were recorded during the study. Species richness, total abundance and abundance of the four commonest dung beetle species were analysed at each site. There were significant differences between sampling locations for most of the variables tested, and these will be described further in the analysis of linear strip width. There were only two instances in which there were significant differences due to trap direction. At Rusty Creek there was a significant difference in beetle species richness between different sides of the traps (Table 21.1). At the interior and 500 m locations, species richness was greatest on the side of the traps facing away from the intact rainforest but the opposite was true at the 50 m location (Figure 21.2a). At Babinda Creek there was a significant difference in trap direction for *Onthophagus furcaticeps* (Table 21.1). At this site, the side of the traps facing the rainforest interior had the highest values at the interior and 500 m locations, with the opposite result at the 50 m location (Figure 21.2b). No directional effects were observed at the Fishery Falls, Figtree Creek, and Tully Creek study sites (Table 21.1).

21.3.2 WIDTH

For the investigation of the effect of linear strip width, the analyses were confined to species richness, total abundance and the five most abundant species across all five study sites. The results of the non-linear regressions are shown in Table 21.2. There was a significant positive relationship between linear strip width and proportional species richness (Figure 21.3a), total abundance (Figure 21.3b), *Coptodactyla depressa* (Figure 21.4a) and *Temnoplectron politulum* (Figure 21.4b). In contrast the proportional abundance of *Lepanus globulus*, *L. nitidus* and *O. furcaticeps* did not show a significant relationship with linear strip width. For those variables that do show a significant relationship with linear strip width, the fitted regression line tends to increase sharply as width increases from 0 to approximately 200 m and then more

Table 21.1 The results of nested analyses of variance of species richness, total abundance and the four most abundant dung beetle species at Rusty Creek, Fishery Falls, Figtree Creek, Babinda Creek and Tully Creek, with location of sampling site (LOC) and side of intercept trap (DIR) as independent variables

Study site	Variable	Source	SS	DF	MS	F	P
Rusty Creek	Species richness	LOC	10.663	2	5.331	36.303	0.001
		DIR	3.436	3	1.145	7.799	0.017
	Total abundance	LOC	13.060	2	6.530	5.845	0.039
		DIR	1.620	3	0.540	0.483	0.706
	C. depressa	LOC	10.252	2	5.126	3.319	0.107
		DIR	1.180	3	0.393	0.255	0.855
	L. nitidus	LOC	0.538	2	0.269	7.872	0.021
		DIR	0.348	3	0.116	3.399	0.094
	O. yiroyont	LOC	0.017	2	0.008	0.549	0.604
		DIR	0.165	3	0.055	3.638	0.084
	L. globulus	LOC	6.057	2	3.029	8.869	0.016
		DIR	0.112	3	0.037	0.109	0.951
Fishery Falls	Species richness	LOC	6.081	2	3.040	8.860	0.016
		DIR	1.045	3	0.348	1.015	0.449
	Total abundance	LOC	1.376	2	0.688	5.920	0.038
		DIR	0.366	3	0.122	1.051	0.436
	L. globulus	LOC	2.627	2	1.313	9.956	0.012
		DIR	0.330	3	0.110	0.834	0.522
	C. depressa	LOC	0.838	2	0.419	4.383	0.067
		DIR	0.509	3	0.170	1.772	0.252
	L. nitidus	LOC	0.222	2	0.111	2.544	0.158
		DIR	0.081	3	0.027	0.621	0.627
	O. furcaticeps	LOC	0.028	2	0.014	0.326	0.734
		DIR	0.202	3	0.067	1.555	0.295
Figtree Creek	Species richness	LOC	4.842	2	2.421	14.973	0.005
		DIR	0.791	3	0.264	1.631	0.279
	Total abundance	LOC	2.657	2	1.328	9.430	0.014
		DIR	1.681	3	0.560	3.977	0.071
	O. furcaticeps	LOC	1.142	2	0.571	4.434	0.066
		DIR	1.034	3	0.345	2.676	0.141
	C. depressa	LOC	1.658	2	0.829	12.500	0.007
		DIR	0.363	3	0.121	1.827	0.243
	T. politulum	LOC	1.332	2	0.666	7.742	0.022
		DIR	0.954	3	0.318	3.697	0.081
	L. globulus	LOC	4.873	2	2.436	30.839	0.001
		DIR	0.296	3	0.099	1.249	0.372
Babinda Creek	Species richness	LOC	2.401	2	1.200	0.862	0.469
		DIR	7.031	3	2.344	1.683	0.269
	Total abundance	LOC	0.700	2	0.530	1.601	0.277
		DIR	2.944	3	0.981	4.487	0.056
	O. furcaticeps	LOC	0.847	2	0.424	5.185	0.049
		DIR	1.699	3	0.566	6.935	0.022
	C. depressa	LOC	0.317	2	0.159	1.080	0.398
		DIR	1.445	3	0.482	3.281	0.100
	O. palumensis	LOC	2.272	2	1.136	6.504	0.031
		DIR	0.428	3	0.143	0.817	0.530
	T. politulum	LOC	1.601	2	0.801	3.840	0.084
		DIR	1.201	3	0.400	1.920	0.228

Table 21.1 (*continued*)

Study site	Variable	Source	SS	DF	MS	F	P
Tully Creek	Species richness	LOC	8.660	2	4.330	8.040	0.020
		DIR	4.132	3	1.377	2.557	0.151
	Total abundance	LOC	0.750	2	0.375	18.369	0.003
		DIR	0.062	3	0.021	1.020	0.447
	T. politulum	LOC	1.269	2	0.634	25.562	0.001
		DIR	0.057	3	0.019	0.771	0.551
	L. nitidus	LOC	0.377	2	0.189	11.000	0.010
		DIR	0.008	3	0.003	0.149	0.926
	C. depressa	LOC	0.796	2	0.398	47.564	0.001
		DIR	0.090	3	0.030	3.579	0.086
	O. capelliformis	LOC	0.062	2	0.031	3.573	0.095
		DIR	0.005	3	0.002	0.184	0.903

gradually to the maximum values found at the rainforest interior locations.

Principal components analysis of the vegetation structure at the sampling locations (Table 21.3 and Figure 21.5) suggested that linear strips that are narrow (see Figure 21.1) tended to have more grass and ground cover, be less shaded and have fewer rocks (Factor 1). Some sites tended to have a greater proportion of shrub cover compared with canopy cover (Factor 2). When species richness, total abundance and abundance of *C. depressa* and *T. politulum* are plotted against the first factor of the PCA analysis (Figure 21.6) there are significant positive relationships between these variables.

However, only 37–42% of the variance is explained by these regressions (Figure 21.6).

21.4 DISCUSSION

21.4.1 MOVEMENT

If the presence of linear strips of vegetation attached to large areas of rainforest increases dispersal rate of rainforest dung beetles, then greater numbers of beetles should be caught on the sides of flight intercept traps facing the rainforest habitat (the presumed source population of dung beetles) compared with traps facing in the opposite direc-

Table 21.2 Values for the equation y = constant + log (x)*beta fitted to the proportional values of species richness, total abundance and abundance of dung beetles species at the study sites (significant fits of the regressions are determined by analysis of variance)

Variable	Constant	Beta	RSQ	F(df1,13)	P
Species richness	−13.076	6.468	0.498	12.922	0.003
Total Abundance	−12.971	4.021	0.398	8.607	0.012
C. depressa	−20.904	5.519	0.492	12.598	0.004
L. globulus	−18.253	5.019	0.252	4.390	0.056
L. nitidus	−2.907	2.122	0.083	1.178	0.297
O. furcaticeps	2.877	1.030	0.017	0.225	0.640
T. politulum	−21.526	5.636	0.371	7.674	0.016

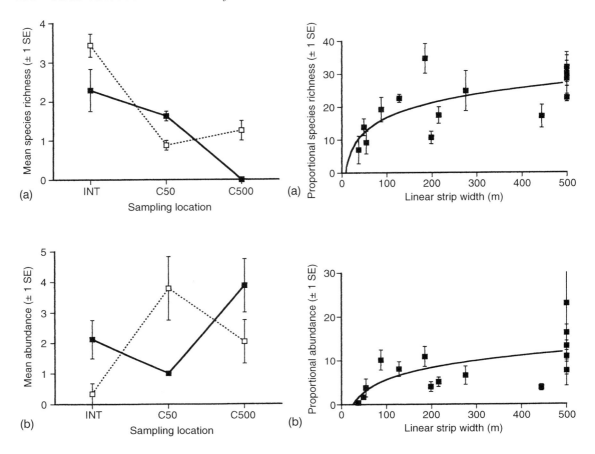

Figure 21.2 Mean values for (a) species richness at Rusty Creek and (b) abundance of *O. furcaticeps* at Babinda Creek. ■, side of trap facing rainforest interior; □, traps facing in opposite direction.

Figure 21.3 Proportional values and fitted regression (solid line) for (a) species richness and (b) total abundance of dung beetles at the study sites. Interior sampling locations have been allocated a nominal width of 500 m.

tion. The same methodology has been used to examine dung beetle assemblages across a natural rainforest–woodland ecotone (Hill, 1996): it was found that the sides of flight intercept traps facing the rainforest interior caught significantly more individuals of some species of rainforest dung beetle (compared with the sides of traps facing the edge of the rainforest), indicating that the methodology is capable of detecting differential beetle movements. In this study some directional effects were found but they did not support the dispersal hypothesis, and in the majority of cases there was no evidence of a net movement of dung beetles through the linear strips in a particular direction.

Therefore either the methodology used in this study was not capable of discerning net movement of individuals, or the linear strips are being used by the rainforest interior species as habitat rather than as dispersal corridors. Dung beetles, and other invertebrates (Margules *et al.*, 1994), may be responding to their environment at a finer scale than that examined in this study, and within-habitat heterogeneity or patchiness may play an important role in determining their movements and local abundance. Clearly a definitive answer to this question requires detailed knowledge of the movements of particular individuals. Vermuelen (1994),

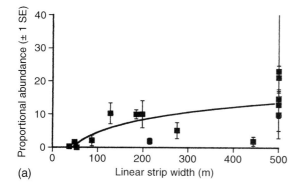

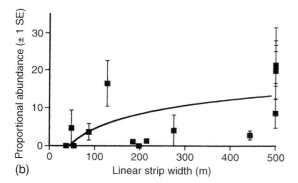

Figure 21.4 Proportional abundance values and fitted regression (solid line) for (a) *C. depressa* and (b) *T. politulum* at study sites. Interior sampling locations have been allocated a nominal width of 500 m.

working on carabid beetles, has found that heathland species will enter linear habitats, such as roadside verges, but that the distances moved were less than 100 m. He concluded that if such habitats were to function as corridors and increase dispersal over greater distances, then reproduction must be possible within the linear strips, i.e. the corridors must function as habitat for the beetles.

21.4.2 WIDTH

It is generally recognized that the width of conservation corridors is likely to be an important factor if they are to function effectively (Soulé and

Table 21.3 The component loadings from a principal components analysis of the vegetation structure at the three sampling locations (forest interior, linear strip 50 and 500 m) at the five study sites

Variable	Factor 1	Factor 2
Shade	0.81	0.07
Grass	−0.71	0.08
Ground cover	−0.69	−0.21
Rocks	0.53	0.60
Shrubs	0.28	−0.92
Canopy	0.31	0.91
% Variance explained	34.79	34.72

Gilpin, 1991; Hobbs, 1992), but there is little empirical information on the effect of corridor width on the abundance of organisms. Corridor width has been shown to have a potential effect on dispersal for organisms such as snails (Baur and Baur, 1992) and carabid beetles (Vermuelen and

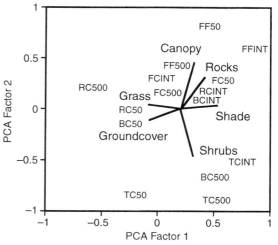

Figure 21.5 Results of principal component analysis of structure of vegetation at sampling locations of five study sites (see Figure 21.1). Solid lines indicate component loadings of categories of vegetation structure.

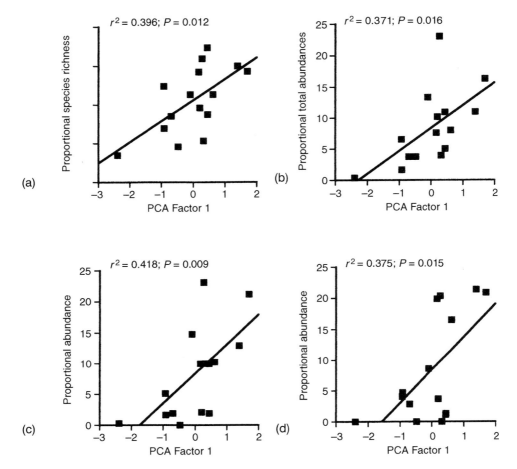

Figure 21.6 Results of linear regressions of principal component factor 1 with (a) proportional species richness of dung beetles, (b) proportional total abundance of dung beetles, (c) proportional abundance of *C. depressa*, and (d) proportional abundance of *T. politulum* at five study sites.

Opdam, 1995), but had no effect on dispersal in deermice (Ruefenacht and Knight, 1995). If the corridors act as habitat (albeit marginal or suboptimal) for the species in question rather than solely as conduits for dispersal, then corridor width is likely to be more important than other factors such as corridor length. In this study the species richness of dung beetles, the total abundance of dung beetles and the abundance of *C. depressa* and *T. politulum* were found to be related to the width of the linear strip habitat. Both *C. depressa* and *T. politulum* have been shown to be rainforest interior species (Hill, 1995). The results suggest that linear strip width must reach 200 m before species richness and abundance values approach those recorded from forest interiors. Interestingly, Keller *et al.* (1993), working on avian assemblages in riparian linear strips, found that the species richness of area-sensitive species increased rapidly with increasing vegetation width up to 200 m and then showed a more gradual increase.

The structure of the vegetation showed some relationship to corridor width, with narrower corridors tending to have more ground cover and less canopy cover, and positive relationships were established between elements of vegetation structure and dung beetle species richness, total dung beetle abundance and the abundance of *C. depressa* and *T. politulum*. However, these relationships only accounted for approximately 40% of the total variance. Therefore, whilst vegetation structure may play some role in determining the abundance of rainforest dung beetles in linear strips, it does not appear to be an overriding factor. Other factors not investigated in this study may include soil attributes, dung beetle behaviour and resource availability.

In summary, previous work has shown that some rainforest interior species do occur in linear vegetation strips, but are not present in the surrounding arable land, which suggests that conservation corridors could play a positive role in a fragmented rainforest landscape by increasing connectivity between isolated patches of habitat (Hill, 1995). The results of this study suggest that any increase in connectivity would be brought about by corridors being used by rainforest interior species as habitat (rather than solely for dispersal), thereby linking populations of species across the landscape. If corridors are to play this role, then the data gathered in this study suggest that such corridors should preferably be at least 200 m wide. Since dung beetle biology is known to be closely linked to the vertebrate fauna on which they rely for food (Hanski and Cambefort, 1991), a width of 200 m might also be appropriate for the rainforest vertebrates of this region.

21.4.3 FUTURE WORK

Clearly the usefulness of conservation corridors in fragmented forest landscapes remains in question. There is increasing evidence that linear (or corridor-like) vegetation strips of sufficient width can provide habitat for species that are more commonly found in habitat interiors (Keller *et al.*, 1993; Crome et al., 1994; Vermeulen, 1994; Hill, 1995 and this study). But such habitat might be more effectively provided by increasing the area of existing habitat fragments. Increased dispersal and/or recolonization via conservation corridors remains to be proved. Whether emphasis should be placed on increasing the area of existing habitat fragments, or on linking fragments with linear vegetation features, remains a point of contention. The answer, in part, will depend on the dispersal abilities of the species in question, and the perceived importance of immigration and emigration rates for the long-term persistence of populations. Accurately measuring dispersal using conventional mark and recapture techniques is notoriously difficult (Dempster, 1991). Furthermore, the opportunities for field studies which assess recolonization after extinctions, along the lines suggested by Nicholls and Margules (1991), are likely to be few because of the constraints of the experimental design and the destructive nature of the experiment. Nevertheless if, as Dempster (1991) suggests, increased habitat fragmentation and isolation leads to a reduced ability to disperse for some species, then a better understanding of the factors that govern dispersal may be critical to ensuring the long-term conservation of endangered species.

ACKNOWLEDGEMENTS

Thanks are due to Mike Cermak for invaluable field assistance and to Ross Storey and Tom Weir for helping to identify some of the dung beetles. Steve Compton, Bill Laurance and Chris Margules suggested improvements to the original manuscript. This study was funded by a Save the Bush Program grant administered by the Australian Nature Conservation Agency.

REFERENCES

Baur, A. and Baur, B. (1992) Effect of corridor width on animal dispersal: a simulation study. *Global Ecology and Biogeography Letters* **2**, 52–56.

Bennett, A.F. (1990) *Habitat Corridors: their Role in Wildlife Management and Conservation*, Department of Conservation and Environment, Melbourne.

Bennett, A.F., Henein, K. and Merriam, G. (1994) Corridor use and the elements of corridor quality: chipmunks and fencerows in a farmland mosaic. *Biological Conservation* **68**, 155–165.

Brown, J.H. and Kodric-Brown, A. (1977) Turnover rates in insular biogeography: effect of immigration and extinction. *Ecology* **58**, 445–449.

Burel, F. (1991) Ecological consequences of land abandonment on carabid beetles distribution in two contrasted grassland areas. *Optiones Méditerranéennes* **15**, 111–119

Claridge, A.W. and Lindenmayer, D.B. (1994) The need for a more sophisticated approach toward wildlife corridor design in the multiple-use forests of south-eastern Australia: the case for mammals. *Pacific Conservation Biology* **1**, 301–307.

Crome, F.H.J. (1994) Tropical forest fragmentation: some conceptual and methodological issues. In *Conservation Biology in Australia and Oceania* (eds C. Moritz and J. Kikkawa), pp. 61–76, Surrey Beatty and Sons, Chipping Norton, Australia.

Crome, F., Isaacs, J. and Moore, L. (1994) The utility to birds and mammals of remnant riparian vegetation and associated windbreaks in the tropical Queensland uplands. *Pacific Conservation Biology* **1**, 328–343.

DeMers, M.N. (1993) Roadside ditches as corridors for range expansion of the western harvester ant (*Pogonomyrmex occidentalis* Cresson) *Landscape Ecology* **8**, 93–102.

Dempster, J.P. (1991) Fragmentation, isolation and mobility of insect populations. In *The Conservation of Insects and their Habitats. 15th Symposium of the Royal Entomological Society of London* (eds N.M. Collins and J.A. Thomas), pp. 143–153, Academic Press, London.

Diamond, J.M. (1975) The island dilemma: lessons of modern biogeographic studies for the design of nature reserves. *Biological Conservation* **7**, 129–146.

Dunning, J.B. Jr, Borgella, R. Jr, Clements, K. and Meffe, G.K. (1995) Patch isolation, corridor effects, and colonisation by a resident sparrow in a managed pine woodland. *Conservation Biology* **9**, 542–550.

Halffter, G. and Favila, M.E. (1993) The Scarabaeinae (Insecta: Coleoptera), an animal group for analysing, inventorying and monitoring biodiversity in tropical rainforest and modified landscapes. *Biology International* **27**, 15–21.

Hanski, I. and Cambefort, Y. (1991) Species richness. In *Dung beetle ecology* (eds I. Hanski and Y. Cambefort), pp. 350–365, Princeton University Press, Princeton, New Jersey.

Harris, L.D. (1984) *The Fragmented Forest*, University of Chicago Press, Chicago.

Harrison, R.L. (1992) Toward a theory of inter-refuge corridor design. *Conservation Biology* **6**, 293–295.

Hill, C.J. (1993) The species composition and seasonality of an assemblage of tropical Australian dung beetles (Coleoptera: Scarabaeidae: Scarabaeinae) *Australian Entomologist* **20**, 121–126.

Hill, C.J. (1995) An assessment of the function of linear strips of rainforest vegetation as dispersal corridors for selected taxa of rainforest insects. *Conservation Biology* **9**, 1559–1566.

Hill, C.J. (1996) Habitat specificity and food preferences of an assemblage of tropical Australian dung beetles. *Journal of Tropical Ecology* **12**, 449–460.

Hobbs, R.J. (1992) The role of corridors in conservation: solution or bandwagon? *Trends in Ecology and Evolution* **7**, 389–392.

Howden, H.F., Howden, A.T. and Storey, R.J. (1991) Nocturnal perching of scarabaeine dung beetles (Coleoptera: Scarabaeidae) in an Australian tropical rainforest. *Biotropica* **23**, 51–57.

Inglis, G. and Underwood, A.J. (1992) Comments on some designs proposed for experiments on the biological importance of corridors. *Conservation Biology* **6**, 581–586.

Keals, N. and Majer, J.D. (1991) The conservation status of communities along the Wubin-Perenjori corridor. In *Nature Conservation 2: The role of corridors* (eds D.A. Saunders and R.J. Hobbs), pp. 387–393, Surrey Beatty and Sons, Chipping Norton, Australia.

Keller, C.M.E., Robbins, C.S. and Hatfield, J.S. (1993) Avian communities in riparian forests of different widths in Maryland and Delaware. *Wetlands* **13**, 137–144.

Kremen, C., Colwell, R.K., Erwin, T.L. et al. 1993) Terrestrial arthropod assemblages: their use in conservation planning. *Conservation Biology* **7**, 796–808.

Laurance, W.F. (1990) Comparative responses of five arboreal marsupials to tropical forest fragmentation. *Journal of Mammalogy* **71**, 641–653.

Laurance, W.F. (1991) Edge effects in tropical forest fragments: application of a model for the design of nature reserves. *Biological Conservation* **57**, 205–219.

Lindenmayer, D.B. and Nix, H.A. (1993) Ecological principles for the design of wildlife corridors. *Conservation Biology* **7**, 627–630.

Lovejoy, T.E., Bierregaard, R.O. Jr, Rylands, A.B. et al. (1986) Edge and other effects of isolation on Amazon forest fragments. In *Conservation Biology: The science of scarcity and diversity* (ed. M.E. Soulé), pp. 257–285, Sinauer Associates, Sunderland, Massachusetts.

Margules, C.R., Milkovits, G.A. and Smith, G.T. (1994) Contrasting effects of habitat fragmentation on the scorpion *Cercophonius squama* and an amphipod. *Ecology* **75**, 2033–2042.

Matthews, E.G. (1972) A revision of the scarabaeine dung beetles of Australia I. Tribe Onthophagini. *Australian Journal of Zoology*, Supplementary Series No. **9**, 3–330.

Matthews, E.G. (1974) A revision of the scarabaeine dung beetles of Australia II. Tribe Scarabaeini. *Australian Journal of Zoology*, Supplementary Series No. **24**, 1–211.

Matthews, E.G. (1976) A revision of the scarabaeine dung beetles of Australia III. Tribe Coprini. *Australian Journal of Zoology*, Supplementary Series No. **38**, 1–52.

Munguira, M.L. and Thomas, J.A. (1992) Use of road verges by butterfly and burnet populations, and the effects of roads on adult dispersal and mortality. *Journal of Applied Ecology* **29**, 316–329.

Nicholls, A.O. and Margules, C.R. (1991) The design of studies to demonstrate the biological importance of corridors. In *Nature Conservation 2: The role of corridors* (eds D.A. Saunders and R.J. Hobbs), pp. 49–61, Surrey Beatty and Sons, Chipping Norton, Australia.

Peck, S.B. and Davies, A.E. (1980) Collecting small beetles with large-area 'window' traps. *Coleopterist's Bulletin* **34**, 237–239.

Pulliam, H.R. (1988) Sources, sinks, and population regulation. *American Naturalist* **132**, 652–661.

Ruefenacht, B. and Knight, R.L. (1995) Influences of corridor continuity and width on survival and movement of deermice. *Biological Conservation* **71**, 269–274.

Saunders, D.A. and de Rebeira, C.P. (1991) Values of corridors to avian populations in a fragmented landscape. In *Nature Conservation 2: The role of corridors* (eds D.A. Saunders and R.J. Hobbs), pp. 221–245, Surrey Beatty and Sons, Chipping Norton, Australia.

Simberloff, D. and Cox, J. (1987) Consequences and costs of conservation corridors. *Conservation Biology* **1**, 63–71.

Simberloff, D., Farr, J.A., Cox, J. and Mehlman, D.W. (1992) Movement corridors: conservation bargains or poor investments? *Conservation Biology* **6**, 493–503.

Soulé, M.E. and Gilpin, M.E. (1991) The theory of wildlife corridor capability. In *Nature Conservation 2: The role of corridors* (eds D.A. Saunders and R.J. Hobbs), pp. 3–8, Surrey Beatty and Sons, Chipping Norton, Australia.

Soulé, M.E. and Simberloff, D. (1986) What do genetics and ecology tell us about the design of nature reserves? *Biological Conservation* **35**, 19–40.

Storey, R.I. and Weir, T.A. (1990) New species of *Onthophagus* Latreille (Coleoptera: Scarabaeidae) from Australia. *Invertebrate Taxonomy* **3**, 783–815.

Tracey, J.G. (1982) *The Vegetation of the Humid Tropical Region of North Queensland*, Commonwealth Scientific and Industrial Research Organisation, Melbourne, Australia.

Vermeulen, H.J.W. (1994) Corridor function of a road verge for dispersal of stenotopic heathland ground beetles Carabidae. *Biological Conservation* **69**, 339–349.

Vermeulen, H.J.W. and Opdam, P.F.M. (1995) Effectiveness of roadside verges as dispersal corridors for small ground-dwelling animals: A simulation study. *Landscape and Urban Planning* **31**, 233–248.

Wilkinson, L., Hill, M. and Vang, E. (1992) *SYSTAT: Statistics, version 5.2 edition*, Systat Inc., Evanston, Il.

Wilson, E.O. and Willis, E.O. (1975) Applied biogeography. In *Ecology and Evolution of Communities* (eds M.L. Cody and J.M. Diamond), pp. 522–534, Belknap, Cambridge, Mass.

Zar, J.H. (1984) *Biostatistical Analysis*, 2nd edn, Prentice-Hall International, New Jersey, pp. 236–244.

INSECT CONSERVATION 22

Thomas E. Lovejoy, Nathaniel Erwin and Sarah Boren

The symposium represented by this volume is not only filled with good intellectual fare, but also explicitly includes environmental change and conservation. Too many biologists seem quite content to browse the shelves of the library of life on earth and to bury themselves in individual volumes as if oblivious to the conflagration that is destroying it. The importance of forests and insects in the overall field of conservation is undeniable (Collins and Thomas, 1991; Forey *et al.*, 1994; Samways, 1994). Wherever one sits in the discussion of how many species there are (Erwin, 1991; May, 1992; Stork, 1988 and 1993) or whether God has an inordinate fondness for beetles, flies or soil organisms, forests and insects include a significant part of biological diversity. The topic of conservation of forests and insects breaks naturally into two parts. The first includes the present and near-term changes being wrought in the environment by human activity – the various forms of direct habitat alteration and the indirect ones through local and regional pollution. Increased UV-B radiation may also fit in this category (Chapter 13). The second part is the most massive of all environmental change, namely climate change.

Maureen Docherty presented interesting indications of diverse responses of trees and tree insect interactions to elevated CO_2 (Chapter 13) and it was correctly noted that CO_2 concentrations are not the only change to be anticipated. Temperature will increase and rainfall patterns will change. While we think of the changes as being greater in higher latitudes, there is every indication that there will be marked changes at lower latitudes (as we know was the case when natural change occurred in the other direction during glacial periods).

The various elements in plant communities are likely to respond differently. In the longest running field experiment, 10 years with elevated (double) CO_2, Bert Drake (1992) has been studying a very simple marsh community with two plant species: a sedge and a grass, each with a different way of responding to CO_2. As would be expected for a plant with a C3 photosynthetic pathway, the sedge has increased in biomass to the detriment of the C4 grass species. Less predictable will be the differential response within a group of C4 or C3 species. Obviously more complex experiments are needed in terms of plant communities as well as in terms of additional factors such as elevated temperature. In one elevated soil temperature experiment John Harte is transforming an alpine meadow in Colorado into chaparral (Harte and Shaw, 1995).

Preliminary branch chamber experiments by Catherine Lovelock and Klaus Winter (personal communication), using the Smithsonian Tropical Research Institute's canopy access system, show that two months of elevated CO_2 around branches of the canopy tree *Luchea seemannii* causes leaves to turn yellow and exhibit reduced photosynthetic competence (Lovelock and Winter, 1996). It appears that excessive accumulation of carbohydrates leads to feedback inhibition of photosynthesis which in turn increases susceptibility to high irradiation stress, photodamage and loss of chlorophyll. The rest of the tree does not seem to serve as a sink for the excess photosynthates. While this is intriguing, it most likely is not comparable to the

Forests and Insects. Edited by A.D. Watt, N.E. Stork and M.D. Hunter. Published in 1997 by Chapman & Hall, London. ISBN 0 412 79110 2.

physiology of a tree that has grown in a high CO_2 environment continuously. In any case, the above findings certainly indicate that internal dynamics of plant communities will be significantly affected by climate change, with implications in turn for insects, and perhaps even more complex interactions. Although there has been some consideration of this topic (Dennis and Shreve, 1991; Harrington and Stork, 1995; Harrington et al., 1995), paleoecological evidence indicates that forest communities in Europe and North America did not move as units in response to climate change during glacial/interglacial swings but, rather, that individual tree species moved at different rates and in different directions, so that communities disassembled and reassembled into different communities (Davis and Zabinski, 1992). There is evidence for similar disassembly and reassembly in lower latitudes as well. Clearly such dynamics carry major implications for insects and forest–insect interactions which probably can be no more than dimly perceived at this point. For example, a species like the aspen blotch leaf-miner (*Phyllonorycter salicifoliella*), which feeds on quaking aspen (*Populus tremuloides*) and winters under the bark of the red pine (*Pinus resinosa*), and therefore occurs only where the two tree species occur together (Martin, 1956; Auerbach, 1991; Auerbach and Alberts, 1992; Auerbach et al., 1995), could be in jeopardy.

The coming decades, of course, will affect how much of current forest and insect diversity will survive to be exposed to the challenge of climate change. Deforestation, still large in the tropics but also accelerating in Siberia and resurging in North America, is obviously a major factor. It is nothing short of a scandal that with remote sensing capabilities there is as yet no global vegetation map or precise estimate of status and trends of forests. In any case, the reduction of forest area should have some effect on the numbers and kinds of insect species that will survive in various diminished forests.

Deforestation has to have severe implications for insect conservation, though there is very limited information of the sort provided by Allan Watt and colleagues (Chapter 15). There are 511 described species out of a total of approximately 1000 of the magnificent picture-winged Drosophila of Hawaii – the so-called birds of paradise of the fly world because of their elaborate courtship displays and territorial defence studied so well by Hampton Carson (1992). None are recorded as extinct, though 18 are proposed for listing under the Endangered Species Act and some, collectable by the hundred 10 years ago, have been seen only once or not at all in recent years. In this case extinction is hard to prove without extensive surveys.

Nonetheless, when (if) the so-called Belém refugium in southeastern Amazonia is finally razed, it is reasonable to assume that the endemic butterflies will largely go too, as well as a myriad of undescribed insect species (Brown and Brown, 1992). Similarly, islands such as St. Helena, virtually shorn of their native vegetation, must have lost a major portion of an undescribed entomofauna.

The Brazil nut, *Bertholettia excelsa*, has become almost mythical and symbolic of forest tree–insect interactions with implications for conservation. Brazil nuts are pollinated by a number of robust bees, and without bees essentially no fruits are produced. The bees depend on the forest and it has been thought that Brazil nut trees out in the open will have poor fruit set as a consequence and plantations would be impractical. Recent research (Ortiz, 1991, 1995) has finally added some data to the speculation. He has demonstrated a threshold level of pollination below which a Brazil nut fruit aborts and that the number of pollinators is a key factor. Nonetheless he has recorded Brazil nut fruits on trees isolated 2–3 km by pasture. There is a different pollinator involved, and it probably occurs in such low numbers as to preclude Brazil nut plantations.

Habitat fragmentation has major implications for forest and insect conservation (Burkey, 1989; Klein, 1989; Brown and Hutchings, in press; Powell and Powell, 1987; Didham et al., 1996). It will not be dwelt on here, since the topic has been dealt with in Chapter 17, except to note the implications for important functional groups: for pollinators (Powell and Powell, 1987), decomposers (Klein, 1989), seed predators and parasitoids. In this connection Didham's (in press) recent analysis provides a useful new way to look at the forest–insect conservation problem.

The flip side of fragmentation is the elimination of a barrier between previously isolated groups of non-forest organisms. North American vertebrate examples include the spread of the cowbird beyond the great plains eastward and the genetic introgression of the red wolf and the coyote (the latter recently recorded on Cape Cod). The introgression of the moth *Hyalophora cecropiae*, which spread westward via riverine woodlands and shelterbelt plantings, with *H. gloveri* of the Rocky Mountains is probably such a case with tree planting 'fragmenting' prairie habitat (Cockerell, 1929a,b; Sweadner, 1937). This kind of interaction may be a more frequent consequence of habitat destruction, fragmentation and modification than is currently appreciated.

Forests, of course, are being seriously modified in a great variety of ways around the world with particular consequences for late succession/mature forest species and for saproxylic species that depend on rotting wood. There seems to be a serious need for real experiments (or fortuitous ones) to analyse the impact of selective logging, other forms of forest management, fragmentation, corridors and the like on forests and their insects. Insects themselves can produce interesting experiments: the near elimination or severe repression of hemlock in the northeastern United States by the woolly adelgid is mimicking the temporary disappearance of that tree species 8000 years ago. What, in fact, are the implications of defoliation by gypsy moths on other forest insects?

Forest aquatic insects probably present their own set of problems. If 40 kg of deadwood/m^2 of streams is said to be typical of North America (Chapter 16), and if this has been so reduced in parts of the northwest that it is a factor in the decline of salmon runs, surely there are implications for aquatic insects as well. This means deliberately including unstable elements (which will provide tree falls) to a larger ecosystem. Migratory insects present special problems. The realization that certain moth species in the dry forest of northwestern Costa Rica migrate into montane forest for parts of the year led to major redesign and expansion of Santa Rosa National Park into the Guanacaste National Park. The migration of the monarch butterflies east of the Rocky Mountains to wintering areas in fir forest in Michoacan, Mexico – surely one of the wonders of the world – depends on, among other things, a very special forest condition. The forest acts as a blanket keeping the forest microclimate intact and the butterflies from losing too much heat by radiation at night. It also acts as an umbrella protecting butterflies from dew formation which would also increase their likelihood of freezing. Even minor forest thinning would be likely to lead to major monarch mortality (Calvert *et al.*, 1982; Alonso-Meier *et al.*, 1992; Anderson and Brower, 1996). Would they survive climate change?

Keystone species are often thought of as species that are important in supporting others, in elevating biological diversity. They can also play key ecosystem roles, as the American oyster did until recently by filtering a volume equal to the entire Chesapeake Bay once a week (Newell, 1988). Moths of the subfamily Oecophorinae seem to play a similar functional role. An estimated 5500 species in Australia are overwhelmingly dependent on Myrtaceae, including a very high percentage on eucalyptus leaves; they probably radiated with eucalyptus. Many specialize on leaf litter. The leaves are tough, leathery, resistant to fungal breakdown when dry, poor in nitrogen and rich in phenolic compounds and tannins (Common, 1994). Largely responsible for the breakdown of eucalyptus leaves, the moths might be thought of as a keystone family. An interesting further elaboration occurs in three species that live entirely on the scat of the eucalyptus-feeding koala (Common and Horak, 1994). The keystone family notion probably should not be pushed very far but there are instances like this one where a functional group is composed of closely related organisms. David Nickle thinks that katydids in neotropical forests, which support vertebrate species and function as forest plankton, might be another case (Nickle, personal communication). Recognition of such groups is important for ecosystem function and management. We have always been intrigued by the index of Fisher *et al.* (1943), and rarefaction and the ability to estimate the overall pool of diversity from which a sample is taken and their potential utility for conservation. Such an index certainly could measure whether the overall diversity of a conservation unit was declining dramatically or holding steady. Its drawback is that it will not reflect whether the species are the desired set, another one or a mix of the two, but in the absence of sufficient financial resources these methods might provide a useful rough first evaluation of the status and trends of a conservation area. This could be something that entomology may have to contribute to conservation generally. This leads to the subject of biogeographical priorities indicator species or groups, some of the debate about which sometimes seems oversimplified (Prendergast *et al.*, 1993) and somewhat overwrought. Is it surprising that butterflies and dragonflies should show different hot spots, whether total species richness or concentrations of endemics is the particular goal? With such different life histories, that would seem to be predictable. For areas of the world where computerized databases are available the wisest approach would be to take as inclusive an approach as possible – and, failing that, to choose indicator groups with as wide a variety of life history types as is available. Where computer databases are still vestigial, a useful approach is that taken by Workshop 90 for conservation priorities in the Amazon. Basically it involved a week-long workshop with most of the authorities on the Amazonian flora and fauna sharing their knowledge, discussing the priorities that their own particular taxa indicated and refining them to produce overall priorities. This approach is not intended as the be-all or end-all (nor should computer-based analyses be) but it provides a reasonable basis for a start. It can work at a variety of scales: the Brazilian state of Amapa is about to convene its own more modest version of Workshop 90 to establish a set of priorities for a state network of conservation units. What such analyses do not necessarily take into account is the ecosystem context, but that can be taken up in the actual planning of units.

One approach in setting priorities is based on maximum genetic difference, an approach that would give priority, for example, to a family with but a single species. The difficulty with this approach is that communities in nature are not built to provide clear choices among genetic differences, but the key is to make sure such differences are not overlooked. It is important to go beyond hot spots of species richness or centres of endemism and include different communities and different assemblages of species so as to conserve relationships between species and processes.

The United States is currently experimenting with the ecosystem management approach. South Florida, southern California and the northwest forests are among the particular ones on which to focus. Dealing with large units of landscape, ecosystem management recognizes that many of our conservation problems are the consequence of multiple decisions made in isolation from one another, and that the way to avoid or solve the problems is with decisions made in consultation. The goal is maintenance of characteristic biodiversity and ecosystem processes. It is a tough challenge in the Pacific northwest because this approach has started so late relative to the decline of those old growth forests, but with a lot of field work and a good patch dynamics model it could work. Ecosystem management must also bear in mind the reality that, even if human-driven climate change is minimized, natural climate change will inevitably occur – so landscapes have to be managed to facilitate, not discourage, dispersal. In the end, sound ecosystem management is fundamental to sustainable development and central to conservation of both forests and insects. And that requires valuing species not only for what they are but also for what they do – much as a computer chip is valued for what it can do rather than its silicon content.

ACKNOWLEDGEMENTS

We are grateful to Lincoln Brower, Ebbe Neilsen, Charles Remington and Daniel Simberloff for their help.

REFERENCES

Alonso-Meier, A., Arellano-Guillermo, A. and Brower, L.P. (1992) Influence of temperature, surface body moisture and height above ground on survival of monarch butterflies overwintering in Mexico. *Biotropica* **24**, 415–419.

Anderson, J.B. and Brower, L.P. (1996) Freeze-protection of overwintering monarch butterflies in Mexico: critical role of the forest as a blanket and an umbrella. *Ecological Entomology* **21**, 107–116.

Auerbach, M. (1991) Relative impact of interactions within and between trophic levels during an insect outbreak. *Ecology* **72**, 1599–1608.

Auerbach, M. and Alberts, J. (1992) Occurrence and performance of the aspen blotch miner, *Phyllonorycter salicifoliella*, on three host tree species. *Oecologia* **89**, 1–9.

Auerbach, M.J., Connor, E.F. and Mopper, S. (1995) Minor miners and major miners: population dynamics of leaf-mining insects. In *Population Dynamics: New Approaches and Synthesis* (eds N. Cappuccino and P.W. Price), pp. 83–110, Academic Press, London and New York.

Brown, K.S. Jr and Brown, G.G. (1992) Habitat alteration and species loss in Brazilian forests. In *Tropical Deforestation and Species Extinction* (eds T.C. Whitmore and J. Sayer), pp. 119–142, Chapman & Hall, London.

Brown, K.S. Jr and Hutchings, R.W. (in press) Disturbance, fragmentation and the dynamics of diversity in Amazonian forest butterflies. In *Tropical Forest Remnants: Ecology, Management and Conservation of Fragmented Communities* (eds W.F. Laurance and R.O. Bierregaard Jr), University of Chicago Press, Chicago.

Burkey, T.V. (1989) Extinction in nature reserves: The effect of fragmentation and the importance of migration between reserve fragments. *Oikos* **55**, 75–81.

Calvert, W.H., Zuchowski, W. and Brower, L.P. (1982) The impact of forest thinning on microclimate in monarch butterfly (*Danaus plexippus* L.) overwintering areas of Mexico. *Bol. de la Sociedad. Botanica de Mexico* **42**, 11–18.

Carson, H.L. (1992) Inversions in Hawaiian Drosophila. In *Drosophila Inversion Polymorphism* (eds C.B. Krimbas and J.R. Powell) pp. 407–439, CRC Press, Bocca Raton, Florida.

Cockerell, T.D.A. (1929a) The westward spread of *Samia cecropia* (L). *J. Economic Entomology* **22**, 704.

Cockerell, T.D.A. (1929b) The spread of *Samia cecropia* (L). *J. Economic Entomology* **22**, 985.

Collins, N.M. and Thomas, J.A. (eds) (1991) *The Conservation of Insects and Their Habitats*, Academic Press, London and San Diego, 450 pp.

Common, I.F.B. (1994) Oecophorine genera of Australia I. The Wingia group (Lepidoptera, Oecophoridae). *Monographs on Australia Lepidoptera* **3**, 28–35.

Common, I.F.B. and Horak, M. (1994) Four new species of *Telanepsia turner* (Lepidoptera, Oecophoridae) with larvae feeding on Koala and possum scats. *Invertebrate Taxonomy* **8**, 809–828.

Davis, M.B. and Zabinski, C. (1992) Changes in geographical range resulting from greenhouse warming: effects on biodiversity in forests. In *Global Warming and Biological Diversity* (eds R.L. Peters and T.E. Lovejoy), pp. 297–308, Yale University Press, New Haven and London.

Dennis, R.L.H. and Shreeve, T.G. (1991) Climatic change and the British butterfly fauna: opportunities and constraints. *Biol. Conservation* **55**, 1–16.

Didham, R.K. (in press) The influence of edge effects and forest fragmentation on leaf-litter invertebrates in central Amazonia. In *Tropical Forest Remnants: Ecology, Management and Conservation of Fragmented Communities* (eds W.F. Laurance and R.O. Bierregaard Jr), University of Chicago Press, Chicago.

Didham, R.K., Ghazoul, J., Stork, N.E. and Davis, A.J. (1996) Insects in fragmented forests – a functional approach. *Trends in Ecology & Evolution* **11**, 255–260.

Drake, B.G. (1992) A field study of the effects of elevated CO_2 on ecosystem processes in a Chesapeake Bay Wetland. *The Australian Journal of Biology* **40**, 579–595.

Erwin, T.L. (1991) How many species are there? Revisited. *Conservation Biol.* **5**, 1–4.

Fisher, R.A., Corbet, A.S. and Williams, C.B. (1943) The relation between the number of species and the number of individuals in a random sample of an animal population. *J. Animal Ecology* **12**, 42–58.

Forey, P.L., Humphries, C.J. and Vane-Wright, R.I. (1994) *Systematics and Conservation Evaluation*, Oxford University Press, Oxford, 438 pp.

Harrington, R. and Stork, N.E. (1995) *Insects in a Changing Environment*, Academic Press, London and San Diego, 535 pp.

Harrington, R., Bale, J.S. and Tatchell, G.M. (1995) Aphids in a changing climate. In *Insects in a Changing Environment* (eds R. Harrington and N.E. Stork), pp. 125–155, Academic Press, London and San Diego.

Harte, J. and Shaw, R. (1995) Shifting dominance within a montane vegetation community: results of a climate-warming experiment. *Science* **267**, 876–880.

Klein, B.C. (1989) Effects of forest fragmentation on dung and carrion beetle communities in Central Amazonia. *Ecology* **70**, 1715–1725.

Lovelock, C.E. and Winter, K. (1996) Oxygen-dependent electron-transport and protection from photoinhibition in leaves of tropical tree species. *Planta* **198**, 580–587.

Martin, J.L. (1956) The bionomics of the Aspen Blotch Miner, *Lithocolletis salicifoliella* Cham.

(Lepidoptera, Gracillariidae). *Canadian Entomologist* **88**, 155–168.

May, R.M. (1992) How many species inhabit the earth? *Scientific American* **267**(4), 42–48.

Newell, R.I.E. (1988) *Ecological changes in Chesapeake Bay: are they the result of overharvesting the American oyster* Crassostrea virginica*? Proceedings of a conference*, Chesapeake Research Consortium Publication, Baltimore, Maryland.

Ortiz, E.G. (1991) Early recruitment of Brazil nut trees (*Bertholletia excelsa* Humb. & Bonp.): Preliminary results, discussion and experimental approach. A proposal-report to Wildlife Conservation International.

Ortiz, E.G. (1995) Survival in a nutshell. *Americas* **42**(5), 6–17.

Powell, A.H. and Powell, G.V.N. (1987) Population dynamics of male englossine bees in Amazonian forest fragments. *Biotropica* **19**, 176–179.

Prendergast, J.R. Quinn, R.M., Lawton, J.H. *et al.* (1993) Rare species, the coincidence of diversity hotspots and conservation strategies. *Nature* **365**, 335–337.

Pyle, R., Bontzien, M. and Opler, P. (1981) Insect conservation. *Ann. Rev. of Entomology* **26**, 233–258.

Samways, M.J. (1994) *Insect Conservation Biology*, Chapman & Hall, London, 358 pp.

Stork, N. (1988) Insect diversity: facts, fiction and speculation. *Biol. Journal Linneaean Society* **35**, 321–337.

Stork, N. (1993) How many species are there? *Biodiversity and Conservation* **2**, 215–232.

Swedner, W.R. (1937) Hybridization and the phylogeny of the genus *Platysamia*. *Annals of the Carnegie Museum* **XXV**, 174–229.

INDEX

Abies alba 188
Abies balsamea 287
Abies fraseri 237
Abundance, insect
  aphids 7
  forest structure and 321–36
  impact of forest loss on 273–84
  termites 109–10, 125
Acacia mangium 213, 215
Acanthotermes acanthothorax 121, 125
Acer negundo 154, 168
Acer saccharum 240
Acetogenesis 111
Acid precipitation 237
Acronicta spp. 168
Acucampestre spp. 7
Acyrthosiphon spp. 4, 233
Adaptation, evolutionary 322
Adelges cooleyi 189
Adelges picea 237
Adelgid, woolly 397
Agaonidae, see Fig wasps
Agrilus opulentus 214
Agrilus sexsignatus 217
Agroforestry 207
Air pollution
  effects on host plants 234–5
  effects on insects 229, 235–8, 241
Alder 69
Alsophila pometaria 16, 17, 29
Amitermes laurensis 122
Anacliliea spp. 374
Anoplognathes spp. 145–6
Anthropogenic disturbance 162
Ants
  canopy-dwelling 281
  forest fragmentation and 305
  leaf-litter 109, 280–1
  on oak 87
Apanteles spp. 163, 169, 308
Aphid
  Arctic 233
  green spruce, see *Elatobium abietinum*
  grey pine 231
  pea 235
  willow–carrot 9–10
Aphids 1–13
  abundance 7
  adaptation to forests 5–11
  birth size 4
  colour 4–5, 13

  equilibrium density 7
  fecundity 9
  fundatrix specialization 9
  gonad size 7
  host alternation 8–11, 12
  host location 7–8, 11
  host use, optimal 9–10
  hosts, secondary 9, 12
  migration 5–6
  polyphenic 5
  population growth rates 9
  proboscis 4
  stylets 3–4
  wings 5–8, 13
Aphis cytisorum 4
Aphis nerii 4
Aphis pomi 239
Apple budmoth 90–1
Aptery 12
Aquatic insects 397
Arachnidomyia aldrichi 99–100, 101–2
Argynnis paphia 17
Arthropods
  diversity in tropical forests 273–4
  roles within forests 283
Aspen, trembling, see *Populus tremuloides*
Assemblages, prehistoric/historic 321
Astratotermes spp. 121
Auchenorrhyncha 5, 353
  see also Leafhoppers
Autumnal moth, see *Epirrita autumnata*
Avetianella sp. 218
Azadirachta indica 219

Ballooning 17
Bark beetles 177, 180, 193
  spruce, see *Dendroctonus micans*; *Ips typographus*
Beauveria spp. 164, 221–2
Beech aphis 237
Beetle
  elm leaf, see *Xanthogalerucea luteola*
  rove 329
  scarab 145–6
  tuber flea, see *Epitrex tuberis*
Beetles
  abundance/diversity in boreal mixed-wood 287–98
  assemblages 290–6
  dung 382–91

  forest fragmentation 305–6, 311–12
Belém refugium 396
Bernoullia flammea 139
Bertholettia excelsor 396
Betula papyrifera 240, 287
Betula pubescens 70,71–2, 74–6, 77
BIOCAT database 62
Biodiversity, conservation of 365–76
Biogeographical priorities indicator species 398
Biological control 49–52, 62–4, 217
Biological Dynamics of Forest Fragments Project 306
Biotypes, insect 191–2
Birch
  leaf-out 74
  mountain, see *Betula pubescens*
  paper, see *Betula papyrifera*
  Alaskan 71
Birth rate
  and density 83
Blepharipa scutellata 163
Bolorea selene 323
Boreal mixed-wood forest 287–98
  biting flies in 288
  harvesting 289, 290, 297–80
Bottom-up effects 82
  see also Top-down/bottom-up
Box elder, see *Acer negundo*
Brachymeria intermedia 99
Branch clippings 366–7
Brazil nut, see *Bertholettia excelsor*
Browntail moth, see *Euproctis chrysorrhoea*
Budburst 16, 18, 20, 167, 230
Budmoth
  larch, see *Zeiraphera diniana*
  tufted apple, see *Platynota idaeusalis*
Budworm
  jack pine, see *Choristoneura pinus*
  spruce, see *Choristoneura orea*
  eastern, see *Choristoneura fumiferana*
  western, see *C. occidentalis*
Bupalus piniaria 38, 42–5, 238
Butterflies
  fritillary 323, 329
  in Mbalmayo Forest Reserve 281
  monarch 397

CAB Compendium 220
Calluna vulgaris 189

Calosoma sycophanta 163
Camera spp. 253
Cameroon 275–6
Campopleginae 257
Cankerworm, fall, *see Alsophila pometaria*
Canopy studies 140, 281–2
Carabidae 291, 292–3, 295
Carbon cycle
  global 109, 131
Carbon dioxide
  elevated 230–3, 240–1, 395–6
  enrichment, free air, *see* FACE
  production, measurement of 118–21, 121–2
  scaling up 130–1
Carbon flux 109, 111, 113, 116–18, 121
  scaling up to global 122–7
  termites as mediators 109–131
Carbon/nutrient
  availability hypothesis 77
  balance 88–9
Cardamine cordifolia 92
Casebearer, larch, *see Coleophora laricella*
Caterpillars
  forest tent, *see Malacosoma disstria*
  large moth 253–69
Cavariella aegopodii 9–10
Cedrela spp. 221–2
Cedrus atlantica 188
Cellulase, endogenous 110, 111
Cellulose, digestion of 110, 113
Cembran pine, *see Pinus cembra*
Cephalcia abietes 185
Cephalcia lariciphila 187
Cephalotermes rectangularis 118
Ceratopus spp. 344
Chemical defences, plants/trees 18, 82–93, 165, 211, 213, 231
Chiggers 309
Choristoneura fumiferana 38–42, 288
Choristoneura murinana 188
Choristoneura occidentalis 41, 93, 103–4
Choristoneura orea 238
Choristoneura pinus 42, 98–9
Chromatomyia suikazuirae 61
Cinara cupressi 217
Cinara pilicornis 181, 235, 236
Cinara pinea 61, 235, 237
Cinara pini 237
Cinarini 4–5
Citrus jamnhri 233, 234
Cleome serrulata 92
Climate change 229–41
Clusia spp. 139
Clusiosoma spp. 345
CO_2, *see* Carbon dioxide
Coccinella septempunctata 308

Coleophora laricella 52, 61, 190
Coleoptera
  boreal mixed-wood 287–98
  canopy-dwelling 281–2
Compsilura concinnata 99, 163, 168, 169
Computer databases 220
Cone pyralids/weevils 181
Conifers
  distribution 23–4, 29
  overlap 24, 26
Connectivity 322–4
Conservation, insect 395–8
Conservation corridors 381–91
Control, pest 162–3, 177
  see also Biological control
Coppicing 323
Coptodactyla depressa 385–7, 389–91
Coptotermes lacteus 122
Core-area model 303
Corridors, conservation 381–91
Corythucha marmorata 310
Costa Rica 251–69, 356
Cotton 69
Cryptophion spp. 253, 255, 256, 257, 258, 260, 261–7
Cubitermes spp. 120, 121, 122
Cupressus lusitanica 217
Cupressus semperivens 184
Curculio spp. 344

Dasyneura laricis 190
Decomposers
  forest fragmentation and 308
Decophorinae 397
Deductive modelling 52–4
Defences, *see* Chemical defences
Defoliation
  effects of 183
  and nutrient loss 71–2
Deforestation 396
  tropical 274–5
Dendrocnide excelsa 141
Dendroctonus spp. 236
Dendroctonus micans 177, 181, 182, 187–8
Dendrozetes spp. 375
Density
  equilibrium
  aphids 7
  herbivores 83, 92
Density-dependent mortality 54–6, 57, 60
Density-dependent predation 83, 88
Density-independent mortality 56, 57, 60
Diabrotica spp. 305
Dictyonotus spp. 257–8
Diet breadths 15–19, 20
Dioryctria spp. 181
Diprion pini 177
  and pollution 238

DIR 70, 71–2, 75–6, 77
DIS 70–1, 75–6, 77
Divergence, genetic 15–16
Diversity index 397–8
Doryphora sassafras 140
Douglas fir 93
Drepanosiphidae 5, 7
Dreyfusia nordmannianae 188
Drosophila
  edge effects 309
  and *Ficus* 345
  forest fragmentation 306
  picture-winged 396
Dryas octopetala 233
Duke of Burgundy butterfly 329
Dung beetles 382–91

Ecosystem management 398
Ectropus crepuscularia 238
Edge effects 304, 308–12, 330–2
Egg hatch 20, 21, 26, 29, 232
Elatobium abietinum 177, 180–1, 190–1, 235, 236
Elm leaf beetle 237
Encounter frequency 27, 29
Enemies, natural 49–64
  and forest fragmentation 307–8
  inefficiency of 215
  of introduced species 163–4
  and pollutants 238–9
  ravine 53, 54, 57, 62
Enicospilus lebophagus 253, 255
Epicerura pergrisea 218
Epinotia tedella 308
Epirrita autumnata 70, 71–2, 76
Epitrix tuberis 310
Equilibrium population density 83, 92
Ericaceae 21
Ericospilus genus-group 256, 260, 261–9
Eriocrania spp. 239
Eryinnis ello 253
Eucalyptus 142–6, 274, 397
  dieback 145
Eucalyptus blakelyi 139, 140, 141
Eucalyptus nova-anglica 139, 141, 145–6
Euceraphis betulae 237
Eulachnus agilis 237
Eupithecia sp. 16
Eupteromalus nidulans 163
Euproctis chrysorrhoea 154–8, 171
  coastal distribution 167–8
  host plant quality 165–9
  overwintering 167
  parasitoids 163
Eurema blanda 218
European forests
  tree pests 178–82
  tree species distribution 178

Euura spp. 57
Exposure frequency hypothesis 26, 28

FACE 229, 240
Fagus sylvatica 9
  and phytophagous insects 326–34
Fall cankerworm, *see Alsophila pometaria*
Fecundity 83, 98–9
Ficiomyia spp. 345
Ficus
  arthropod consumers, primary 342–8
  as keystone species 342, 357–8
  in New Guinea 341–58
Fig wasps 341, 343–4, 348, 349, 355, 356–7
Fire 208–9
Flea beetle, tuber 310
Flies
  biting 288
  tephritid 15
Flight period, timing of 16
Flightlessness 17
Fluoride 238
Foliage chemistry, spatial variation in 84–9, 89–91
Foliar astringency 85–8
Foliar nitrogen 230
Forest tent caterpillar, *see Malacosoma disstria*
Forests
  disturbance of 289
  dry 259–64
  European 177–98
  Finnish 323
  loss/regeneration of 273–84
  see also Deforestation
  structure and insect abundance 321–36
  tropical 207
    arthropod diversity 273–4
    deforestation/reforestation 273–4
    pest management 208, 209–16, 219–21
Forestry
  village/social 207
Forestry management/operations
  and pests 192–4
  tropical 211–13, 216–19, 207–20
Fragmentation
  forest 303–16, 396
  geographical 321, 322–3
  Midlothian study 325
  structural 321–2, 323
  woodland 322
Fritillary
  heath, *see Mellicta athalia*
  small pearl-bordered, *see Boloria selene*
Frosts 232
Fungus gardens 112
Furanocoumarins 17, 92–3, 233

Gall wasps 27
Gall-forming insects 85, 88
  on *Ficus* 347, 355
Gallotannins 239
  see also Tannins
Gastrophysa viridula 231
Genetic divergence 15–16
Genetic diversity
  arthropods in tropical forests 273–4
  trees 189
Genetic variance
  Lymantrids 159–62
Geographical range
  and abundance 21–6, 29
  and niche breadth 22
Geraeus spp. 344
Gilpinia hercyniae 181, 187
GLIM 24
Glycine max 234
Gmelina arborea 274–5
Great spruce bark beetle, *see Dendroctonus micans*
Green River study 38, 41
Growth/differentiation hypothesis 77
Growth rates, relative 235, 236
Guanacaste Conservation Area 251, 252, 397
Guilds 322
Gypsy moth, *see Lymantria dispar*

Habitat area 304
Habitat diversity hypothesis 27
Habitat islands 304–5
Habitat structure 98
Hairstreak, black 323
Hamaeris lucina 329
Hartigiola annulipes 327, 331–2
Harvesting, tree 192
  in boreal mixed-wood 288, 290, 297–8
Hazard rating 194–5
Health regulations, plant 214
Heathland invertebrates 309–10
Heavy metals 238
Hedgerows 332–4
Herbivore-induced responses 69–77
Herbivores 135–47
  density 83
  movement rates 99
Herbivory
  impact on trees 135–6
  measurement methods 138–9
  spatial variation 136, 139
Heteropsylla cubana 215, 217
Hormonal regulation, plant 73–4, 76
Hosts
  abundance/distribution 18
  alternation 8–11, 12
  -associated differentiation/speciation 15–16
  defences, *see* Chemical defences

  location 7–8, 11
  plant architecture 18
  plant quality 83–4, 92, 164–71
  range 15
    expansion of 19–28
  secondary 9, 12
  shifts 15, 17
  taxonomy affinity 17
  use
    lability in 16–17
    optimal 9–10
  vigour, loss of 209, 211
Hot spots 97, 398
Hyalophora cecropiae 397
Hyblaea puera 218
Hylobius abietis 177, 181, 182, 193–4, 195
Hymenoptera hyperparasitoids 40
Hypebaeus flavipes 323
Hypsipyla spp. 218, 221–3
Hypogastrura arborea 374
Hyposoter genus-group 256, 257

Ichneumonid parasitoids 253, 257–69
  koinobiont 253
Impact studies 208–9
Insect assemblages, historical 321
Insect biotypes 191–2
Insect checklists, country 220
Insect conservation 395–8
Insecticides 163, 218–19
  see also Control, pest
Insects
  forest aquatic 397
  migratory 397
  as pests 177–98
  phytophagous 3–13, 326–34
  pollinating 305, 307
  range extension 187–8
INSPIRE 220
Integrated pest management, *see* IPM
IPM 216–17, 221
Ips grandicollis 208–9
Ips typographus 177, 180, 182, 187, 188, 193, 194–5
IR 70
IS 70–1
Isaria sp. 164
Island biogeography theory 27, 303, 323–4

Jack pine budmoth, *see Choristoneura pinus*
Jugositermes tuberculatus 121

K-strategists 53, 57, 61, 62, 63
K-values
Katydids 397
Key factor analysis 54–5
Keystone species 342, 357–8, 397

Khaya spp. 221

Labidotermes 119, 125
Lachnidae 5
Landscape approach 98
Larch
    European, *see Larix decidua*
Larch budmoth, *see Zeiraphera diniana*
Larch casebearer, *see Coleophora laricella*
Larix decidua 70, 177, 184, 190
Laris kaempferi 190
Larva dispersal 29
Leaf-chewers 87, 88
    on *Ficus* 346–7, 348, 351–3, 355–6
Leafhoppers 27
    on *Ficus* 348, 349–51, 355
Leaf-miners 27, 86, 88
    on *Ficus* 347, 355
Leaf-out 74
Lepanus spp. 385–7
Lepidoptera 11, 12
    see also individual species
Leucaena leucocephala 214–15, 217
Leucoma salicis 154–8m 171
    host plant quality 169–70
    natural enemies 163–4
Life table studies 38
Light levels 140
Linear vegetation strips 381
    width effect 385–7, 389
Lissocephala spp. 345
Lobaria pulmonaria 289
Loblolly pine, *see Pinus taeda*
Lodgepole pine, *see Pinus contorta*
Logistic regression analysis 24–6
Log-normal distribution 16
Lymantria dispar 17, 29, 93, 97, 98, 99, 146, 147, 154–8, 161–2, 171, 180, 192
    and conservation 397
    devoliation by 165–7
    host plant quality 164–5
    natural enemies 163
    and pollution 236, 239–40
Lymantria monacha 177, 184, 185
    and fluoride 238
Lymantrids 153–70
    control methods 162–4
    genetic variance 159–62
    host plants 164–71
    invasion patterns/spread 153–8
    natural enemies 163–4
    outbreak species 154
    'weedy' species 162

Macroptery 5–8
Macrosiphum spp. 4
Macrotermes spp. 110, 118, 122, 125

Madang 348
Mahogany 221
Malacosoma disstria 99–104, 330
Malus spp. 89–91
Manaus project 306
Manolepidoptera 22–3
Mass spectrometry 121
Mating
    assortative 16
    plants as locations 16
Matsucoccus josephi 188
Mbalmayo Forest Reserve 109, 113–14, 275–82
Megacyllene robiniae 309
Megastigmus spp. 181
Melarnargia galathea 17
Mellicta athalia 323
Metapopulation dynamics 324–5
Metarrhizium sp. 221
Meteorus versicolor 164
Methane 110, 112, 114–18
    measurement of production 121–2
    scaling up 127–30
Microtermes spp. 119, 121, 122
Midlothian fragmentation study 325
Migration 5–6
Models
    reductive 52–4
    synoptic population 53
Monarch butterfly 397
Monoculture 189, 213–14
Moran effect 41–2
Mortality 83
    density-dependent 54–6, 57, 60
    density-independent 56, 57, 60
    spatial patterns 98, 99
Moth
    autumnal, *see Epirrita autumnata*
    browntail, *see Euproctis chrysorrhoea*
    gypsy, *see Lymantria dispar*
    nun, *see Lymantria monacha*
    pine beauty, *see Panolis flammea*
    pine processional, *see Thaumetopoea pityocampa*
    pine shoot, *see Rhyacionia buoliana*+
    rusty tussock, *see Orgyia antiqua*
    satin, *see Leucoma salicis*
    vapourer, *see Orgyia antiqua*
    winter, *see Operophtera brumata*
Myricaceae 21

Nasutitermes spp. 111, 116
Natural enemies, *see* Enemies, natural
Neem tree 219
Neodiprion sertifer 53, 57, 177, 185, 237, 239
Neotheronia tacubaya 253, 256
New Guinea 341–58
Niche breadth 22

Nitrogen 238
    foliar 230
Nitrogen oxides 234, 235
Noditermes spp. 121
Norway spruce, *see Picea abies*
NPV, *see* Viruses
Nun moth, *see Lymantria monacha*
Nursery management 213

Oak, *see Quercus* species
    black, *see Quercus velutina*
    chestnut, *see Quercus prinus*
    red, *see Quercus rubra*
Odontopera bidentata 20
Ooencyrtus kuwanai 99, 102
Old-growth stands 289–90, 294–7
Omophorus spp. 344
Onthophagus furcaticeps 385–8
Operophtera brumata 17, 18, 20, 28, 29, 181, 186–7, 189
    parasitoid impact 52, 56, 97
    population dynamics 38
Orgyia antiqua 154–8, 189
Outbreak species 154
Overwintering 20–1
Oviposition
    defoliation and 166
    site preference 17
Ozone 234, 236, 240–1

Pachypappa vesicalis 187, 235
Pachypappella lactea 187
Pacific Northwest 365
Panolis flammea 177, 183, 184, 186, 190, 238
Papilio machaon 17
Paraserianthes spp. 218
Parasitism 40, 42, 217–18
    forest fragmentation and 308
    ichneumonid 253–69
    impact on insect populations 49–64
    spatial ecology 97–104
    see also individual species
Parsnip webworm 93
Parthenogenesis 231, 233
Passer spp. 153
Patch quality/size 82, 92
Patellona pachypyga 99–101
Pathogens 218
Pemphigidae 5
Pericapritermes spp. 125
Periphyllus spp. 4
Pests, insect 177–98
    exotic tree species and 188–91
    forestry operations and 192–4
    origins of susceptibility 184–8
    prevention of 209–16, 219–21
    site factors 184–5
    tree damage by 182–4

tropical 207–20
 weather effects 185
Petrova resinella 239
Phenolics 93, 165
 accumulation of 77
 apple 89–91
 and carbon dioxide 231
 oak 85–6
 and UV-B protection 233
Pheromone traps 194
Philodendron spp. 139
Phloem sap 3–4
Phloretin 89–91
Phloridzin 89–91
Photocantha semipunctata 215, 218
Photoperiod 232
Phyllaphis fagi 237
Phyllonorycter maestingella 331–2
Phytophages, beech 326–334
Picea abies 187, 189, 190, 191
 and pollution 234
Picea glauca 287
Picea mariana 287
Picea orientalis 187–8
Picea sitchensis 20, 28, 180–1, 186–7, 188–9, 191, 195
 old-growth coastal 365–76
 and pollution 235, 236
Pieris spp. 17, 153
Pimplopterus dubius 308
Pine
 cembran, see *Pinus cembra*
 loblolly, see *Pinus taeda*
 lodgepole, see *Pinus contorta*
 Scots, see *Pinus sylvestris*
 white, see *Pinus strobus*
Pine bast scale 188
Pine beauty moth, see *Panolis flammea*
Pine looper, see *Bupalus piniaria*
Pine processional moth, see *Thaumetopoea pityocampa*
Pine resin gall, see *Petrova resinella*
Pine sawflies 177
 see also *Diprion pini*; *Neodiprion sertifer*
Pine shoot moth, see *Rhyacionia buoliana*
Pine weevil, see *Hylobius abietis*
Pineus similis 189
Pinus brutia 188
Pinus caribaea 213, 274–5
Pinus cembra 177, 191
Pinus contorta 186, 189, 190, 191
Pinus halepensis 188
Pinus negra 190
Pinus ponderosa 236
Pinus strobus 185, 240
Pinus sylvestris 77, 190, 234, 235
Pinus taeda 233, 234

Pissodes spp. 181, 189
Pityogenes chalcographus 180
Pityogenes phrygianus 237
Plagiodera versicolora 236
Plant health regulations 214
Plant-hoppers 11
Platynota idaeusalis 90–1
Poland, nun moth control in 177
Pollinators
 fig 341
 forest fragmentation and 305, 307
Pollution, air
 effects on host plants 234–5
 effects on insects 229, 235–8, 241
Polyphagy 265
Poplar
 balsam, see *Populus balsamifera*
 herbivore-induced responses in 69
Population bottlenecks 159
Population density, equilibrium 83, 92
Population dynamics 37–45, 97
Population growth rates
 aphids 9
Population model, synoptic 53
Population regulation
 bottom-up, top-down 56–61
Populus spp. 10, 167, 170
Populus balsamifera 100, 287
Populus tremula 287
Populus tremuloides 100, 170, 239, 287, 330
Predators
 density-dependent 83, 88
 and habitat structure 98
 impact on insect populations 49–64
 movement rates 99
 see also Enemies, natural; Biological control
Preference–performance correlation 17
Prehistoric determinants 321
Proanthocyanidin 85–8
Procubitermes arboricola 121
PROSPECT 220
Protermes prorepens 121
Provenance 189
Prunus maritima 168
Pseudohaida rothi 374
Pyruvate metabolism 111

q-values 52
Quarantine 214
Queensland 208–9, 381–91
Quercus spp. 9
 herbivores and plant chemistry 84–9
Quercus prinus 239
Quercus rubra 85–9
Quercus velutina 85–9

r-strategists 53, 57, 61, 63

Rainforest insects 381–91
Reforestation, tropical 274, 280
Resin acids 54
Resistance
 induced 70–2, 75–7
 and plant physiology 72–6
 tree, mechanisms 211–13
Resseliella spp. 181
Rhagoleitis spp. 15–16
Rhopalosiphum padi 238
Rhyacionia buoliana 184, 185, 190, 215
Rhynchaenus fagi 332
Rhynchophion spp. 256, 257, 258, 260, 261, 265
RIR 70
Risk/hazard rating systems 194–5
Rosa rugosa 168
Rove beetles 329
Rumex obtusifolius 231
Rusty tussock moth, see *Orgyia antiqua*

Saddleback worm, see *Ectropus crepuscularia*
Salix spp. 9, 10, 170
Salix nigra 154, 168
Sampling
 discrete 136
 long-term 137
Santa Rosa sector 252–64, 397
Sap-suckers 86–7, 88
 on *Ficus* 345–6, 348, 355
 see also Leafhoppers
Satin moth, see *Leucoma salicis*
Saturniidae 253, 257–69
Savanna 113, 114
Sawflies 177, 185, 187
 pine, see *Neodiprion sertifer*
Scandinavian forests 289
Scarab beetle 145–6
Scarification 194
Schizolachnus pineti 231
Scorpions 306
Scots pine, see *Pinus sylvestris*
Scots pine lacnhid 61
Seed stands 184
Shift species 19–20, 21
Sink/source hypothesis 74–5, 76, 77
 see also Source/sink models
Sitka spruce, see *Picea sitchensis*
Sitobion avenae 239
Sitobion pineti 235, 237
Source/sink models 82, 325, 382
Spatial ecology 81–93, 325
 parasitoid attack 97–104
Spatial variation
 foliage chemistry 84–9, 89–91
 herbivory 136, 139
 insect densities 82–3
Speciation, prehistoric 321

Species–area relationship 27
Species loss, British 323
Species richness
  on *Ficus* 341
  prehistoric factors 322
Sphingidae 253, 257–69
Spicaria spp. 164
Spiders, edge effects and 309
Spodoptera 231
Spruce
  black, see *Picea mariana*
  Norway, see *Picea abies*
  white, see *Picea glauca*
Spruce aphid, green, see *Elatobium abietinum*
Spruce bark beetles, see *Dendroctonus micans*; *Ips typographus*
Spruce budworm
  eastern, see *Choristoneura fumiferana*
  western, see *Choristoneura occidentalis*
Spruce sawfly, see *Gilpinia hercyniae*
Staphylinidae 291, 293–4, 295, 366, 368, 371–2, 374
Stem-borers, fig 347
Stigmella tityrella 331–2
Stomaphis quercus 4
Stress, tree 209, 211
Strobilomyia cone flies 181
Strymonidia pruni 323
Sulphur dioxide 234–7, 239, 240
Survivorship studies 38
Swietenia macrophyllap 221
Syconium 341, 355
  feeders on 343–5
Syndemis musculana 20
Syssphinx molina 265

Tannins 18, 87, 88, 89, 213
  and lymantrids 165
TCRS 20, 25
Technomyrmax sp. 281
Temnoplectron politulum 385–7, 389–91
Temperature, elevated 231–3, 240, 241

Tephritid flies 15
Termes hospes 118, 121, 122, 125
Terminalia spp. 218, 276, 277–8
Termites 109–131, 277, 279–80
  abundance 109–10, 125
  as decomposers 113
  digestion 110–12
  ecosystem functions affected by 110
  forest fragmentation and 305
  microbe relationship 110–11
  respiration 110–12
  root-feeding 213, 218
  trace gas production 114–16
  trophic specializations 112
Terpenes 93
Thaumetopoea pityocampa 180, 185
Thinning 193, 214
Thoracotermes macrothorax 121, 122
Thyreodon genus-group 256, 257–8, 260, 261–8
Thyreodon atriventris 253, 254
TIGER Programme 113, 275
Tobacco 69
Tomato 69
Tomicus piniperda 193
Toona ciliata 141, 221–2
Top-down/bottom-up effects 82, 83, 92, 93
  population regulation 56–61
Tortrix viridana 180
Townsendiellomyia nidicolor 168
Tropical pests 207–20
Trees
  definition of 3
  European species
  distribution of 178
  pests of 178–82
  exotic species and pests 188–91
  genetic diversity, narrowing of 189
  herbivore-induced responses in 69–77
  modularity of 73–5
  pest damage to 182–4
  resistance mechanisms of 211, 213
  resource availability for 77

  species selection 211, 213
  stress in 209, 211
  vigour, loss of 209, 211
Trichogramma minutum 169
Trigonodemus fasciatus 374
Triplochiton ni 234
Tsuga spp. 238

Upper Carmanah Valley 366
Uroleucon nigrotuberculatum 307
UV-B exposure 233–4, 241

Vapourer moth, see *Orgyia antiqua*
Village forestry 207
Viruses, polyhedrosis 218, 237, 239
Volatiles 234
Volcanic activity 238

Wasps, agaonid 341
Water availability 92
Weather
  effects on pests 185
  insect outbreak records 37
Web destruction 162
'Weedy' species 162
Willow, black, see *Salix nigra*
Windthrow 193
Winkler sampling method 306
Winter moth, see *Operophtera brumata*
Winter moth syndrome 17, 29
Wood-borers, fig 347–8
Woodland management, historical 322
Woolly adelgid 397
Workshop-90 398

Xanthogalerucea luteola 237

Yponomeuta spp. 16

Zale spp. 168
Zapronius spp. 345
Zeiraphera diniana 41, 70, 76, 177, 183, 190, 191–2
Zootermopsis angusticollis 114

Coláiste na hOllscoile Gaillimh